1ᶜ   89.95
                       65B

D1175359

Kin recognition

Kin recognition, the ability to identify and respond differentially to ones genetic relatives, is one of the fastest growing and most exciting areas of ethology. Dr. Hepper has brought together leading researchers in the field to create a thought-provoking and critical analysis of our current knowledge of the phenomenon, with particular emphasis on the underlying processes involved, and their significance for the evolution of social behaviour. Students of animal behaviour and evolutionary biology will find this book an invaluable source of information and ideas.

Kin recognition

Kin recognition

edited by
PETER G. HEPPER

School of Psychology
The Queen's University of Belfast

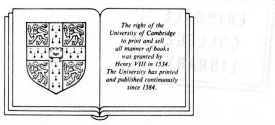

The right of the
University of Cambridge
to print and sell
all manner of books
was granted by
Henry VIII in 1534.
The University has printed
and published continuously
since 1584.

CAMBRIDGE UNIVERSITY PRESS

CAMBRIDGE

NEW YORK   PORT CHESTER

MELBOURNE   SYDNEY

Published by the Press Syndicate of the University of Cambridge
The Pitt Building, Trumpington Street, Cambridge CB2 1RP
40 West 20th Street, New York, NY 10011–4211, USA
10 Stamford Road, Oakleigh, Melbourne 3166, Australia

First published 1991

Printed in Great Britain at the University Press, Cambridge

British Library cataloguing in publication data

Kin recognition.
1. Kinship
I. Hepper, Peter, G.
306.83

Library of Congress cataloguing in publication data

Kin recognition / edited by Peter G. Hepper.
    p.   cm.
Includes index.
ISBN 0 521 37267 4 (hard)
1. Kin recognition in animals. 2. Kin recognition. I. Hepper,
Peter G.
QL761.5.K544   1991
591.56–dc20   90-21776   CIP

ISBN 0 521 37267 4 hardback

SE

Contents

viii   *Contents*

Contributors

P. Aldhous
Nature
4 Little Essex Street
London, WC2R 3LF, UK

J. Bard
Department of Microbiology and
    Immunology
University of Arizona
Tucson, AZ 85724, USA

C. J. Barnard
Department of Zoology
University of Nottingham
Nottingham, NG7 2RD, UK

G. K. Beauchamp
Monell Chemical Senses Center
3500 Market Streett
Philadelphia, PA 19104, USA

M. D. Beecher
Department of Psychology
University of Washington
Seattle, WA 98195, USA

I. S. Bernstein
Department of Psychology
University of Georgia
Athens, GA 30602, USA

E. A. Boyse
Department of Microbiology and
    Immunology
University of Arizona
Tucson, AZ 85724, USA

R. W. Elwood
Department of Biology
The Queen's University of Belfast
Belfast, BT7 1NN, Northern Ireland

W. M. Getz
Department of Entomology and
    Div. of Biological Control
University of California at Berkeley
Berkeley, CA 94720, USA

Z. T. Halpin
Department of Biology
University of Missouri-St Louis
St. Louis, MO 63121-4499, USA

P. G. Hepper
School of Psychology
The Queen's University of Belfast
Belfast, BT7 1NN, Northern Ireland

P. Jaisson
Laboratoire D'Ethologie et Sociobiologie
Universite Paris XIII
93430, Villetaneuse, France

M. H. Johnson
Department of Psychology
Carnegie Mellon University
Pittsburg, PA 15213-3890, USA

J. D. Ligon
Department of Biology
University of New Mexico
Albuquerque, NM 87131, USA

R. H. Porter

George Peabody College
Vanderbilt University
Nashville, TN 32203, USA

S. R. Robinson

Department of Zoology
Oregon State University
Corvallis, OR 97331, USA

W. P. Smotherman

Department of Psychology
SUNY-Binghamton
Binghamton, NY 13901, USA

B. Waldman

Biological Laboratories
Harvard University
16 Divinity Avenue
Cambridge, MA 02138, USA

K. Yamazaki

Monell Chemical Senses Center
3500 Market Street
Philadelphia, PA 19104, USA

Preface

Studies of kin recognition have progressed rapidly during the past decade. One of the most exciting aspects of this research is that the ability to recognize kin has been found, in some form or other, throughout the animal kingdom, from single-celled organisms to man. The increase in studies reporting kin recognition has led to a widespread acceptance of the role of kinship in behaviour. There is, of course, good theoretical reasons, provided by kin selection theory and mate choice theory, why this should be so, however, the importance of kinship for behaviour has often been unquestioned. One of the reasons for compiling this volume was to critically assess the role of kinship in behavioural interactions: is kin recognition responsible for the many observed differential interactions between kin and non-kin? Whilst many species have been demonstrated to recognize their kin, little attention has been given to determining how this is achieved and consequently the mechanisms underlying this ability are poorly understood. A second goal of this book was to present research which has investigated how individuals recognize their kin.

Rather than provide a taxonomic discussion of kin recognition, I have aimed to provide a book which deals with particular themes. Leading researchers in these areas were asked to discuss these issues with respect to their own expertise and species or group studied. The book may be broadly divided into two sections – that dealing with function and that dealing with mechanisms. With respect to function chapters deal with issues of nepotism, co-operation, fellowship, mate choice, parent–offspring recognition, genetic components of recognition and kinship in amphibians. As regards mechanisms, the cues of recognition, state mediated recognition, prenatal learning, imprinting, the development of recognition and the bee as a model recognition system, are issues discussed. Kin recognition in humans, for whom kin recognition has been established for the longest but was the most poorly understood, is also dealt with. Each chapter concentrates upon a particular theme and illustrates it with reference to the species or group of interest to the author. This has the benefit that particular issues are discussed in detail and because of the widespread occurrence of kin recognition the relevence of this discussion is not restricted to one species or group but has implications for kin recognition in all species. In a rapidly expanding field this volume has attempted to consider some of the most important issues and questions facing researchers of kin recognition.

The book is intended for all those interested in behaviour, the fact that kinship does influence behaviour means all students of behaviour must consider its influence. Authors were asked to provide chapters which dealt with the most up-to-date issues but also to provide some background material, and as such the book will be of interest to students, researchers and teachers.

I acknowledge the support of a number of people without whom this book could never have got past the planning stage. I thank all the authors for taking the time and putting in much effort in contributing to this volume. I thank Prof. Michael Morgan, who as my Ph.D. thesis supervisor, first stimulated my interest in the area of kin recognition and Prof. Ken Brown who has enabled me, with continued support, to pursue studies of kin recognition. I thank Ken Brown, Jennifer Cleland, Roddie Cowie, Fiona Hepper, and Ian Sneddon for the refereeing of the chapters. Finally I thank my wife Fiona for all her help.

Peter G. Hepper
1990

Introduction

Peter G. Hepper

The importance of kinship for human societies has long been recognized, indeed the interest in kinship is reflected in the large number of works, plays, books, operas, etc., which have kinship, often mistaken, as their central theme. Whilst there has been little doubt that humans recognize their kin and respond differentially to them, the ability of animals to recognize and respond differentially to kin has received little attention. In recent years, however, study of the influence of kinship on the social behaviour of animals has increased dramatically. Much of the impetus for this research can be attributed to the seminal works of Hamilton (1964a,b) and later Wilson (1975). Evidence that kinship, or genetic relatedness, influences an individual's behaviour has now been documented in all the major groups of animals – from single-celled organisms (e.g. Grosberg & Quinn, 1986) to man (e.g. see Porter, this volume). Reptiles remain an exception to this and only recently have studies appeared providing evidence of kin recognition in this group (e.g. Werner *et al.*, 1987), which probably reflects a lack of empirical investigation rather than a lack of ability to recognize kin. As diverse as the groups of animals which exhibit evidence of kin recognition are the behaviours in which individuals respond differentially on the basis of kinship; colonization patterns, mating, play, aggression, feeding, schooling, swarming, defence, etc. all are influenced by kinship. Investigations of the ability to recognize kin have revealed that individuals have well developed capabilities to recognize kin; siblings, half-siblings, cousins, parents, offspring, grandparents, aunts and uncles are all capable of being discriminated. Kinship thus appears to influence behaviour throughout the animal kingdom.

The theoretical importance of kinship for behaviour derives from the concept of inclusive fitness outlined by Hamilton (1964a,b; see Grafen, 1984 for full discussion). Essentially Hamilton argued that since related individuals have a number of genes in common that are identical by descent, an

1

individual may pass on its genes to the next generation via its relations. Thus, an individual's fitness, its *inclusive fitness*, should include its own reproductive success (not including that due to the actions of others) *and* the reproductive success of related conspecifics arising as a result of its actions. Thus, by responding differentially to kin and non-kin, an individual may achieve greater gains in fitness than individuals unable to do so. Mate choice and optimal outbreeding (e.g. Bateson, 1983) provide a further reason for expecting kinship to exert an effect on behaviour. The long established disadvantages of inbreeding and the more recently established problems of extreme outbreeding indicate that kinship may play an important role in the choice of a mate. It may, therefore, be expected that individuals would respond differentially to kin and non-kin, an expectation that is supported by empirical observations. To achieve this, individuals may be expected to have some means of recognizing their kin. The study of how individuals recognize their kin received little attention until the late 1970s. Whilst investigations assessed whether individuals responded differently to kin and non-kin, little attention was paid to the question of how individuals achieved this. One major exception, however, is the field of parent-offspring recognition where this ability had long been studied in a variety of species.

Kin recognition may be divided into two broad areas, the exhibition of differential behaviour to kin and non-kin, referred to as *kin discrimination*, and the mechanisms used to identify individuals as kin, non-kin, or a particular degree of relatedness, referred to as *kin recognition*. A distinction between the two is important because logically inferences drawn from results in one area may not provide information about the other. Thus, individuals who do not respond differentially to kin and non-kin, i.e. show no *kin discrimination*, may be unable to recognize kin, alternatively they may be perfectly well able to recognize kin but do not exhibit a discrimination in this situation. No *kin discrimination* need not imply no *kin recognition*. This places the emphasis squarely on the researcher and the methodology used to demonstrate that the animal in question can recognize its kin rather than simply relying on the animal to show that it can recognize its kin. More recently concern has been expressed as to whether evidence of kin discrimination is a result of kin recognition. Particular concern has been expressed over kin bias in mate choice where this may not necessarily be the result of kin recognition (see Barnard and Aldhous, this volume). Thus, it is important to distinguish between *kin discrimination* and *kin recognition*. A variety of other terms are used in the kinship literature and because of differences in their use by various authors no attempt has been made to provide a standard glossary of terminology in this book. Readers are referred to the definitions provided by each author.

Over the years, research has successfully demonstrated the ability of individual organisms to recognize their kin and to discriminate between kin and non-kin. The importance of kinship in the individual's behaviour and its prevalence throughout the animal kingdom is, therefore, not in question. More recently, a more critical appraisal of kinship and behaviour has been undertaken and both the evidence for discriminative responding and the ability of individuals to recognize their kin has been subject to detailed examination. It is the aim of this volume to consider these recent developments. Each chapter in the volume concentrates on a particular theme and although often illustrated with reference to a particular taxonomic group, the considerations discussed should apply to all animal groups. Very broadly the first part of the book deals with kin discrimination and the second with kin recognition.

Although it is often taken for granted that animals do exhibit kin discrimination, the first part of this volume examines the evidence for this and considers the situations in which kinship can be expected to exert an effect. The strongest evidence of differential responsiveness to kin and non-kin has often been thought to be provided by the non-human primates. Perhaps because of their close similarity with man, for whom kinship plays an important role, primates of all animal groups were expected to exhibit evidence of kin discrimination. Poor methodology and inappropriate analysis however have raised questions regarding the evidence of kin discrimination in primates. The chapter by Bernstein assesses the evidence for kinship influences on behaviour in the non-human primates. The following chapter by Ligon deals with co-operation and reciprocity in birds and mammals. Co-operation is common both to human and animal behaviour, it is often highly developed and is clearly frequently found in the animal kingdom. However, such behaviour may not require kinship for its maintenance and the role of kinship in co-operative behaviour is critically assessed by Ligon. As has often been pointed out, the complex societies of ants, wasps and bees pose special problems for Darwin's evolutionary theory. Jaisson discusses, with examples from the ants and wasps, how these can be overcome by reference to kinship and the 'superorganism' concept and illustrates the behavioural mechanisms used in colonies of ants and wasps to recognize their kin. The costs of recognizing kin are considered by Beecher using the example of parent–offspring recognition. It is often overlooked that recognizing kin and being recognized as kin involves costs, e.g. being identified as non-kin may result in rejection, increased aggression, etc. This chapter considers how the costs of kin recognition and discrimination will influence the evolution and development of kin recognition systems. Kinship has often been proposed as a factor in determining mate choice, however, Barnard and Aldhous discuss how kin bias in mating behaviour may have arisen for reasons that are unrelated to kinship. This chapter serves as a timely reminder that caution should be exerted in ascribing kin bias to kin discrimination. The genetic structure which has received most attention in studies of kin recognition is that of the major histocompatibility complex (MHC), which has been implicated in recognition in both invertebrates and vertebrates. The chapter by Boyse, Beauchamp, Yamazaki and Bard, reviews studies of the MHC concentrating on its role in influencing mating preferences and the production of the individual's odour signature. The kin recognition capabilities of amphibians are well understood but questions have been raised as to the functional significance of kin recognition for this group. Waldman examines the possible functions of this ability for amphibians.

Studies of how individuals recognize conspecifics have been criticized (e.g. Hepper, 1986) for providing little detail of how these processes take place. The second part of this volume examines research attempting to elucidate how individuals recognize their kin. In order to be recognized as kin, individuals must have some cue which signifies them as kin. The cues of kin recognition are the subject of the chapter by Halpin, where the information content, the origins and the sensory modality of cues used in the recognition of kin are discussed. Hepper discusses an approach which can be better used to determine how individuals recognize their kin. This approach enables the underlying basis of kin recognition, especially its development and how and where cues of kinship are classified, to be better understood. The chapter demonstrates that there are a wide variety of means available to the individual to enable it to recognize its kin. The motivational state of the individual has often been overlooked in the study of kin recognition yet this may play an important role in identifying kin. The fact that fathers enter a paternal state and refrain from killing their own young is discussed by Elwood detailing how the individual's state can be used to recognize kin. The possibility that the female may use this to deceive the male is also considered. Most studies of kin recognition have concluded that individuals have to learn about their kin in order to recognize them and, therefore, have to ensure that the information they learn is from related individuals. One means of ensuring that individuals learn about their kin is to start learning before birth. This, however, requires a certain level of sophistication in the sensory and learning capabilities as well as information regarding kinship being available to the fetus. This is the subject dealt with by Robinson and Smotherman. Underlying the ability to recognize is some neural process which converts the perceived cue to a measure of relatedness. For most species little is known about these processes. However, in one case, that of imprinting, the underlying neural processes are being elucidated. The chapter by Johnson examines what is known about the underlying physiological basis to imprinting and concentrates upon the role of the intermediate and medial parts of the hyperstriatum ventrale. Given the many factors involved in recognizing kin, the construction of a kin recognition system can be expected to be somewhat complex. Yet Getz, using the honey bee as his example, proposes a kin recognition model for this animal and reviews tl recognition in the bee. The final chapter returns to humans kinship has been acknowledged to play an important role in hu yet very little is known about how recognition is achieved. Pc trating on the mother–infant dyad, examines the development anu sensory mediation of kin recognition and its adaptive significance for this dyad.

Kin recognition encompasses a wide diversity to topics from the nepotistic aiding behaviour of certain primates to the neural processes involved in classifying individuals as kin. The aim of this book is to bring together a collection of papers that illustrate the breadth of the area as well as providing critical analysis of topics within the area. There is no doubt that the study of kin recognition is exciting. The fact that a wide range of species (from single-celled organisms to man) appear to exhibit the same complex patterns of behaviour is fascinating and prompts many questions. Also this book has aimed to provide an insight into the many questions currently being asked to understand the individual's ability to recognize its kin.

References

Bateson, P. P. G. (1983). Optimal outbreeding. In *Mate Choice*, ed. P. P. G. Bateson, pp. 257–277. Cambridge: Cambridge University Press.

Grafen, A. (1984). Natural selection, kin selection and group selection. In *Behavioural Ecology*, ed. J. R. Krebs & N. B. Davies, pp. 62–4. Oxford: Blackwell.

Grosberg, R. K. & Quinn, J. F. (1986). The genetic control and consequences of kin recognition by the larvae of a colonial marine invertebrate. *Nature*, **322**, 456–9.

Hamilton, W. D. (1964a). The genetical evolution of social behaviour I. *J. Theor. Biol.*, **7**, 1–16.

Hamilton, W. D. (1964b). The genetical evolution of social behaviour II. *J. Theor. Biol.*, **7**, 17–52.

Hepper, P. G. (1986a). Kin recognition: functions and mechanisms. A review. *Biol. Rev.*, **61**, 63–93.

Werner, D. I., Baker, E.M., Gonzalez, E. del C., Sosa, I.R. (1987). Kinship recognition and grouping in hatchling green iguanas. *Behav. Ecol. Sociobiol.*, **21**, 83–9.

Wilson, E. O. 1975. *Sociobiology*. Cambridge, MA: Belknap Press.

1

The correlation between kinship and behaviour in non-human primates

Irwin S. Bernstein

In 1986 Carolyn Ehardt and I published a paper, 'The influence of kinship and socialization on aggressive behaviour in rhesus monkeys (*Macaca mulatta*)'. The title of this paper is not unusual; it asserts that kinship is an independent variable that influences aggressive behaviour, as a dependent variable. This suggests that manipulation of the first will alter the second. Naturally we would be quick to deny a causal relationship and would repeat that science only demonstrates correlations, not causation. But why then did we not title the paper, 'The influence of aggressive behaviour on kinship'? If we had done so, many would have quickly pointed to this as an example of a backward causal argument. In fact, despite our denials, we do attempt to suggest causal relationships in our correlations and, in this case, as in many others, there is a clear correlation between two variables but there may be no direct causal relationship.

Perhaps we have acknowledged the contributions of Tinbergen (1951) only to fail to apply them to our own thinking. All too often we confuse function with proximal cause, evolutionary cause, and even structure (as in definitions like 'the process leading to'). Functional outcomes are argued to be the motivational cause of behaviour (the animal was 'trying to drive the predator away') and if something functions in an adaptive fashion we often assume that this is because it was selected for this function during evolution (birds developed wings in order to fly south in the winter). Teleological reasoning can be very seductive, whether we are looking for the cause of behaviour by examining the consequences, or we are looking for the answer to questions about evolutionary processes by searching for an adaptive explanation for observed behaviour.

Ever since Wilson (1975) published *Sociobiology: The New Synthesis* and called our attention to Hamilton's paper (1964) on inclusive fitness theory, we have been keenly aware of the selective pressures that would favour individuals that selectively benefited kin, at a cost to themselves less than the benefit received by their kin, devalued by the degree of relatedness. Degree of relatedness is often equated with kinship but, in fact, these terms are not synonymous. Even disregarding human kinship terms used for genetically unrelated people (your mother's sister's husband is your 'uncle') unrelated individuals do not necessarily have zero genes in common, nor do cousins necessarily share only half as many genes in common as do aunt and nephew. It is true, on average, for genes identical by descent from a common grandparent, but how many genes are shared in common by members of a population of the same species? When we measure degree of inbreeding we typically examine only those loci known to be heterogeneous in the population and regard as uninteresting loci that show no variation. Moreover, if there are three alleles for human ABO blood groups, then do we regard as 'kin', of some degree, all of those who share the same phenotype as our own and therefore share one or both alleles in common with ourself? If we could trace back far enough would we find that we were both carrying the same alleles derived from common ancestors, albeit very distant ancestors?

Kinship discussions are often marred by facile assumptions that degree of genetic relatedness and degree of kinship are synonymous, where kinship is defined as counting the number of links back to a common ancestor. When we use kinship in the second sense, then surely only humans determine kinship by counting generations to a common ancestor. All other animals must use some other criteria that correlates with this measure, or show no correlation between kinship and behaviour. In inbred strains we might expect considerable differences in the degree to which kinship influenced behaviour, if one strain literally could measure the proportion of genes in common whereas the other strain responded differentially to others based on association with the same mother, and the associates of that mother.

The actual pattern of correlations between behaviour and kinship, phenotypic factors and experiential factors may tell us much more about mechanism than any number of postulated possible mechanisms. The mere fact that all individuals, not genetically identical, are unique phenotypes does not mean that degrees of phenotypic variation are related one to one with genetic relatedness. We may recognize each individual as having unique dermatoglyphics, voice signature, visible anatomical properties, retinal patterns and chemical properties. We may succeed in recognizing similarities among some individuals and may be able to develop a classification scheme based on objective criteria. None of this is to say that our classification either has a direct genetic correlation, nor that individuals in a population actually respond to the variation that we recognize so easily. Even when it would be adaptive to do so, and obviously so easy to do from our perspective, the actual demonstration of the phenomena may be elusive.

As an analogy, I have often noted that rhesus monkeys treat members of the two sexes differently and that there is, what appears to be, extraordinary interest in the genitalia of infants on the part of adults. This is such an obvious sex identification procedure for humans that it seems almost foolish to ask if monkeys do indeed behave differently towards other monkeys based on the anatomical differences between males and females. Is it possible, however, that mothers respond towards their infants in accordance with the infants' behaviour and not their anatomy? Is it possible that monkeys respond primarily to the sexual differences in behaviour of males and females rather than to the differences in their anatomy? Is it possible that monkeys receive information on individual identity by examining all bodily orifices and chemical cues, and that data on sex and kinship are not obtained from these cues, even though it is theoretically possible to obtain such data from these cues? Is it possible that primates may fail to respond to cues that salmon, frogs and bees discriminate so keenly?

Non-human primate behaviour and kinship correlations

Our first task must be to establish the existence of a phenomenon before we try to understand mechanism (cause), function, ontogeny and evolution. In this case, it is relatively easy to demonstrate a strong correlation between certain classes of kinship and a wide variety of primate behavioural patterns. In multiple primate taxa, individuals in physical proximity, or huddled together in the cold, or resting together, or in physical contact or actively engaged in social grooming, have all been found to be more likely to be kin than would be expected by chance (Sade, 1965; Ransom & Rowell, 1972; Vessey, 1973; Loy & Loy, 1974; Kurland, 1977; Mori, 1977; Cheney, 1978; Baxter & Fedigan, 1979; Ehardt & Bramblett, 1980; Hanby, 1980; Walters, 1980; Silk et al., 1981; Berman, 1982; Silk, 1982, 1984; Takahata, 1982b; Furuichi, 1983; Giacoma, 1983; Baker & Estep, 1985; Colvin, 1985; Colvin & Tissier, 1985; Fairbanks & McGuire, 1985; Hornshaw, 1985; Koyama, 1985; Taylor & Sussman, 1985; Glick et al., 1986a,b; Quiatt, 1986). Although most of the data are from Old World monkeys, particularly macaques and baboons, wherever substantial data on kinship and behaviour are available positive correlations, with maternal kinship at least, have been reported. The data for play and mounting behaviour show weaker correlations with kinship, but age and sex also strongly correlate with play and mounting behaviour and, for example, the availability of male peers among kin as play partners may only be likely in large matrilines (Dittus, 1979). Mounting behaviour is also strongly influenced by age, sex and familiarity, and this may account for observations reporting that older males mount their kin proportionately less often than do younger males (Glick et al., 1986b; Ruehlmann et al., 1987) rather than a maturational kinship effect per se.

'Altruistic' behaviour is difficult to explain in evolutionary terms unless it is directed primarily towards kin (in which case it is not altruistic as it is in the genetic interests of the individual to do so). Altruistic behaviour like food sharing, co-operation and alarm calling all appear to be strongly correlated with kinship. Cheney & Seyfarth (1980, 1985a,b) report that vervets (*Cercopithecus aethiops*) not only alarm call and respond to the alarm calls of others in a manner demonstrating clear kinship biases, but that unrelated females also will look towards the mother in response to the playback of an infant's distress vocalization. Moreover, Marler (1985) reports that there are differences in the structure of certain vocalizations when directed towards kin.

The clearest evidence of primates engaging in behaviour which benefits another at some risk and/or cost to themselves exists in the area of third party interferences in agonistic encounters. Although one might argue that joining a victor in attacking a defeated victim entails little risk, and in no way benefits the original aggressor, it may help the interfering animal to acquire status relative to the victim, or to establish a coalition with the aggressor. Individuals, however, will also come to the aid of victims against aggressors that have repeatedly defeated the aiding animal in the past. This second type of defence surely entails some risk to the interfering animal, and this type of agonistic interference correlates strongly with sex (females do more of it) and matrilineal kinship. Although males do much less aiding of kin than do females, and seldom will aid an individual against an opponent who is dominant to them, the few times a male does so is almost always in defence of his kin (Koford, 1963; Sade, 1965, 1968; Kaplan, 1976, 1977, 1978; de Waal, 1977; Kurland, 1977; Massey, 1977; Lee & Oliver, 1979; Watanabe, 1979; Hardy, 1981; Bramblett, 1973; Datta, 1983a; Horrocks & Hunte, 1983a,b; Estrada, 1984; Seyfarth & Cheney, 1984; Bernstein & Ehardt, 1985; Colvin, 1985; Colvin & Tissier, 1985; Fairbanks & McGuire, 1985; Netto & Van Hooff, 1986; Reinhardt *et al.*, 1986). The decrease in aid of kin by males, as they become adult, is not matched by a decrease in the aid that natal adult males receive from kin. (Eaton, 1976).

The consequences of third party interferences in agonistic encounters goes beyond the immediate protection of the animal being aided. Kawai (1965) and Kawamura (1965) have proposed a theory of the social inheritance of dominance in primates and it has been demonstrated to predict social dominance relationships in several species (Koyama, 1967; Sade, 1969; Loy & Loy, 1974; Loy, 1975; Kawanaka, 1977; Dittus, 1979; Lee & Oliver, 1979; Cheney & Seyfarth, 1980; Walters, 1980; Cheney, *et al.*, 1981; Datta, 1983a,b; Horrocks & Hunte, 1983a,b). Predictions based on the theory (Chapais & Shulman, 1980) are sometimes contradicted by data (Missakian, 1972; Angst, 1975; Chikazawa *et al.*, 1979; Hrdy, 1981; Hausfater *et al.*, 1982; Chapais, 1985; Paul & Kuester, 1987), but it seems clear that dominance relationships among females, in at least some species, are strongly influenced by patterns of third party interference in agonistic encounters. Mothers are particularly likely to aid their offspring, but other matrilineal kin can also play important roles. In those cases where a young adult female comes from a matriline that was overthrown by one or more other matrilines after her birth, the female does not achieve the rank of her mother at the time of her birth (Ehardt & Bernstein, 1986). It seems essential for a young female to have at least one living matrilineal relative that has retained high rank, and will come to her aid, for a female to acquire the rank predicted at her birth (Chapais, 1985). If a female's family has not been overthrown, on the other hand, even an orphan female may rise to her mother's former rank with the aid of one or more others willing to support her (Altmann, 1980; Walters, 1980).

Aggression and kinship

The data reported so far indicate that animals behave towards matrilineal kin in ways that may reasonably be presumed to benefit the recipients of the behaviour. Such correlations are readily accommodated by inclusive fitness theory relating selective pressures to the genetic concomitants of the behaviour. In order to accommodate data on aggression and kinship to evolutionary theory so facilely, we would need to demonstrate a negative correlation between aggressive targets and genetic relatedness. Whereas Silk *et al.*, (1981) have claimed that monkeys show less aggression towards their kin, the vast majority of the data indicates that animals receive far *more* aggression from their matrilineal kin than would be expected by chance. Moreover, it is the more severe forms of aggression involving physical contact which are the more strongly biased towards kin (Kurland, 1977; Horrocks & Hunte, 1983a; Niemeyer & Chamove, 1983; Colvin, 1985; Eaton *et al.*, 1985, 1986; Bernstein & Ehardt, 1986). Furuichi (1983) concluded that due to insufficient data, the hypothesis of reduced aggression towards kin is not tenable. Attempts to explain the data away by weighting the frequency of aggression received by time in proximity in no way nullifies the fact that individuals, especially younger animals, are most likely to receive the more consequential forms of aggression from their own kin. Mothers are particularly prone to bite their own infants, although such biting may subjectively appear to be inhibited (Bernstein & Ehardt, 1986). Walters (1987), in his review, examined the proximity explanation hypothesis and noted that it would also have to be applied to grooming and agonistic aiding, as well as aggression. He concluded, however, that even with a compensation for time in proximity, data for grooming, agonistic aiding and aggression remain significantly correlated with kinship.

Aggressive interactions may result from competitive conflicts which will arise most often among animals in close proximity, but aggression does not necessarily result in individuals avoiding one another as a consequence of the aversive consequences of prior close proximity. Bolwig (1980) noted that aggressive interactions were often followed by intensive social interactions. Kummer (1971) has stressed that primates have few ecological specializations beyond sociality and Nagel (1979) has proposed a model which characterizes many social interactions as functioning primarily to establish and maintain social bonds. de Waal and his colleagues have proposed a theory of reconciliation (1977, 1979, 1983) which provides for mechanisms to overcome the aversive consequences of agonistic encounters engendered by competitive conflicts between individuals who spend a lot of time together. Their data indicate that kin and other close associates are significantly more likely to reconcile immediately after an agonistic interaction. Whereas there is no negative correlation between kinship and aggression, we can account for the continued association of individuals that interact agonistically, and Bernstein & Ehardt (1986) have proposed a positive socialization function that may be served by the socialization of kin through punishment of actions which might provoke more violent response in non-kin.

Kinship and inbreeding

Another body of data relating kinship to primate behaviour involves the reported decreased probability of inbreeding based on a behavioural mechanism resulting in avoidance of kin as sexual partners, i.e. a negative correlation between mounting, sexual invitations and other sexual behaviour and the degree of genetic relatedness between individuals. A calculation of the coefficient of inbreeding might be more relevant to evolutionary theories invoking selective pressures against inbreeding in otherwise outbred strains, but such a measure also reflects multiple mechanisms promoting outbreeding in the relatively small social units formed by non-human primates. Altogether, the various mechanisms do appear effective in reducing inbreeding.

Aoki & Nozawa (1984) demonstrated that the average degree of relatedness between troops of *Macaca fascicularis* was 0.09, whereas it was only 0.17 within troops (where mothers and dependent offspring are resident). Males are the dispersing sex in this genus, and the average degree of relatedness among the males was 0.11, whereas the figure for females was 0.19. In some insular populations, however, the degree of relatedness within a group was as high as 0.4.

Other studies also suggest that macaque troops are not generally inbred, and that male tenure is seldom greater than three years in a group (Melnick *et al.*, 1984). Langur females in a group may be more inbred than are macaques, according to Hrdy (1981), and this has been suggested to be one of the factors accounting for differences in 'nepotism' among macaques and langurs. Although some macaque species have been described to form inbred troops, the data are indirect and the conclusion is controversial (Mehlman, 1986). The costs of dispersal are high and only in outbred populations could inbreeding costs possibly exceed the costs of dispersal (Moore & Ali, 1984).

Primate troops in expanding populations do fission, and sons are less likely to join the same fission party as the rest of the matriline (Chepko-Sade & Sade, 1979). Matrilines are otherwise cohesive and, in expanding populations, matrilines form the subgroups which provide for the lines of fission when a troop divides. The fission process, if anything, would produce troops with much higher degrees of genetic relatedness than that which exists between troops, but the dispersal of virtually every member of one sex seems to effectively preclude extensive inbreeding. Troop fissions are far less frequent than individual transfers.

Male transfers, on the other hand, are not random with regard to kinship. Males are reported to be more likely to join a troop that a matrilineal brother is already resident in, and brothers may transfer together and ally together in a new troop (Boelkins & Wilson, 1972; Meikle & Vessey, 1981). Males are also more likely to join groups that contain males from their original natal group, or to transfer in the company of male peers, some of which may be patrilineal kin, simply because the number of breeding males in a group is usually much less than the number of breeding females (Drickamer & Vessey, 1973; Itoigawa, 1975; Cheney & Seyfarth, 1983; van Noordwijk & van Schaik, 1985; Paul & Kuester, 1985). Some have attempted to argue that a cohort of macaque males are likely patrilineal kin, by assuming that dominant males father the vast majority of offspring in a given year. Shively (1985), however, has demonstrated that assertions of preferential mating based on dominance are not supported by available data based on biochemical paternity testing.

Since the vast majority of males leave their natal group prior to reaching full body size (Koford, 1963; Itani, 1975; Kawanaka, 1977; Kurland, 1977; Packer, 1979; Cheney & Seyfarth, 1983; Fukuda, 1983; van Noordwijk & van Schaik, 1985), it might seem moot to ask if there was evidence of a behavioural mechanism involving avoidance of close kin as mating partners. Although male transfer is the rule in the groups described, a small number of females also transfer (Ransom & Rowell, 1972; van Noordwijk and van Schaik, 1985) and in some species it is primarily, but not exclusively, females that transfer (Glander, 1980; Pusey, 1980; Crockett, 1984). Furthermore, some males do return to their natal troops (Drickamer & Vessey, 1973; Itoigawa, 1975) and some males never leave their natal troops (Paul & Kuester, 1985). Moreover, many young males are sexually mature for several years prior to transferring to a new troop. The possibility of inbreeding between a male and his matrilineal kin does exist and certainly males with prolonged tenure in a group may have the opportunity to mate with their own daughters.

Measuring the frequency of patrilineal inbreeding is rarely accomplished because so few studies have data on patrilineal kinship, but many long term studies have an adequate data base to investigate the frequency of matrilineal inbreeding. The earliest reports simply stated that mother–son mating was rare or never observed, but such data are meaningless when the number of natal males, their reproductive behaviour, and the number of breeding females and total copulations are unreported. The frequency of sexual behaviour in natal males has been reported to be significantly less than that of non-natal males (Missakian, 1973; Hanby & Brown, 1974; Packer, 1979; Cheney & Seyfarth, 1983; Ruehlmann et al., 1987). A general decrease in sexual activity among natal males might reduce the probability of inbreeding, but would not shed any light on a possible behavioural mechanism suggesting the specific avoidance of kin as sexual partners.

The first systematic study of mother–son matings in rhesus monkeys reported low frequencies of such episodes in previous reports and concluded that mother–son mounting and copulations (rhesus monkeys are multi-mount ejaculators) were rare (Sade, 1968). The data, however, indicate that mother–son mounting accounted for 31 per cent of all natal male mounting, although mother–son dyads constituted only 9 per cent of natal male–female dyads. The difficulties in specifying the statistical null hypothesis, when expected frequencies are low, has deceived many into thinking that a statistical test was unnecessary. In this case the data actually lean towards the opposite conclusion.

Other authors have also reported that there is no evidence of avoidance of kin as sexual partners and even a *preference* for kin when natal males did engage in sexual behaviour (Giacoma, 1983; Hanby & Brown, 1974; Missakian, 1973; Smith, 1986; Ruehlmann et al., 1987). Other studies, however, confirmed little or no mating observed among kin (Baxter & Fedigan, 1979; Cheney, 1978; Eaton, 1976; Enomoto, 1978; Fedigan & Gouzoules, 1978; Glick et al., 1986a; Murray & Smith, 1983; Packer, 1979; Pusey, 1980; Takahata, 1982a; Taylor & Sussman, 1985) but only a few (most notably Paul & Keuster, 1985) present appropriate statistical tests. Paul & Kuester reported that in 22 observed copulations none were mother–son matings, although they calculated that the expected would have been seven. Several of the other reports do include mention of multiple mother–son or brother–sister matings and some of the data presented actually contradict the author's conclusions.

Explanations for less mating among kin than would be expected by chance include mechanisms that permit matings among kin involving young males but suggest that such behaviour decreases as a function of male age. Other mechanisms include theories of developmental dependence precluding mating. Others have demonstrated that animals with high scores for grooming and general social association are significantly less often mating partners, even in female same sex consortships (Baxter & Fedigan, 1979; Eaton et al.,

1985; Enomoto, 1978; Fedigan & Gouzoules, 1978; Itoigawa *et al.*, 1981; Packer, 1979; Sugiyama, 1984; Takahata, 1982a,b). Since matrilineal kin associate and groom with one another, this might indeed result in low mating scores for matrilineal kin but this would not necessarily mean that there was a specific avoidance of matrilineal kin as mating partners. Wilson & Gordon (1979) clearly demonstrated that adult female rhesus monkeys prefer new male immigrants as sexual and social partners during the breeding season. In this species, at least, females initiate and maintain consortships and may refuse matings. A female preference for less familiar adult male mating partners would decrease matings with matrilineal kin, as well as matings with all natal or long term resident males.

Paternal kinship and behaviour

With clear correlations between matrilineal kinship and many forms of social behaviour (even if the data on sexual behaviour do not show a clear influence of matrilineal kinship) we may now ask if patrilineal kinship also correlates with behaviour. Some primate species live in groups consisting of a single reproductive pair, or consisting of several reproductive females and a single reproductive male. Although primate social units are not necessarily breeding units (Cords *et al.*, 1986), these resident males are likely to be the fathers of several offspring, and age peers in the group are, therefore, likely to be related patrilineally. Arguments explaining the special relationship of a single adult male to immature animals in the group, based on kinship, cannot be evaluated by comparison with behavioural interactions between unrelated adult males and infants since there are no other adult males in the group. Intruder males, or extra group males, interacting with these infants may behave differently towards them because of their status as extra group or intruder males rather than because of their unrelated status. As such, although Walters (1987) notes that males are *not* infanticidal towards their offspring, this sheds little light on the relationship between male behaviour and paternity. After all, infanticide has not been observed in very many cases and the failure to note many cases of paternal infanticide may simply reflect sampling. It should be noted, however, that some infanticides have been reported by related males.

In multi-male groups it is easier to control for resident status when comparing the behaviour of related and unrelated males to offspring. Some authors have indeed suggested that fathers show preferential care of their offspring or that infants may seek out their own fathers (e.g. Altmann, 1984; Dunbar, 1984; Stein, 1984a; Taub, 1984) but in these studies paternity was also inferred by the behaviour between male and infant. Lacking independent measures of paternity it is not surprising that males, assumed to be fathers of infants because they behaved preferentially towards those infants, behaved preferentially towards infants that they had presumably fathered. When paternity was inferred due to the male's close association with the mother, the male was, of course, also closely associated with her infant. Busse (1985) suggests that males behave preferentially towards the infants of their female associates and that this association is inadequate evidence for paternity. Moreover, Stein (1984b) and Strum (1984) have both noted cases of preferential male treatment towards infants conceived prior to the male's entry into the troop. In my own observations of geladas (*Theropithecus gelada*, Bernstein, 1975), I reported that young adult bachelor males frequently carried and interacted with infants that they could not possibly have sired, whereas one-male-unit leaders seldom contacted infants at all.

Macaques generally live in multi-male groups and macaque males may show selective attention to particular infants. Japanese macaque (*Macaca fuscata*) males have been noted to protect matrilineal kin (H. Gouzoules, 1984) whereas Barbary macaque (*M. sylvanus*) males do not seem to prefer their own matrilineal kin (Kuester & Paul, 1986).

Attempts to demonstrate that patrilineal kinship correlates with behaviour are often based on measuring the independent variable, patrilineal kinship, by measuring male dominance and/or mating frequencies. Such measures have been repeatedly discredited (Fedigan, 1983; S. Gouzoules, 1984; Shively & Smith, 1985). Moreover, even when males show clear preferences for infants, this has not been correlated with a consistent mating pattern (H. Gouzoules, 1984; Hasegawa & Hiraiwa, 1980). To make matters worse, when female mating, rather than male mating patterns, are examined, it turns out that most females in these species copulate with multiple adult males at the height of their conception cycle (Fedigan, 1983; Shively, 1985; Taub, 1984). Male association with females that they had mated with might still increase the likelihood of their associating with their own offspring, but when male biochemical paternity data are available, it turns out that males show little consistency in their choice of mating partners from year to year (Smith, 1982). The final straw in discrediting studies measuring male association patterns with females to infer paternity, should have been the demonstration of *inverse* relationships between male association with a female lineage and his copulation rates with females in that lineage (Sugiyama, 1984). Male associates of a female were, nonetheless, most likely to adopt her orphaned infant (Takahata, 1982b).

In addition to adoption, protection and other preferential male behaviour towards infants, males interact with infants, in situations which do not benefit the infant, in ways other than direct infant killing. Deag & Crook (1971) first described agonistic buffering, where males use an infant in an agonistic interaction in a manner which appears to reduce their own chance of injury but which may place the infant at some risk. Hrdy (1976) reviewed male patterns of infant exploitation and suggested that males do not use related infants as agonistic buffers, but the data are far from conclusive. Strum (1984) suggests that males use infants, in agonistic buffering situations, with whom they have already established complex social relationships. These relationships, as a totality, are mutualistic. Measuring costs and benefits of a single interaction and attempting to correlate that with inclusive fitness theory is unlikely to prove fruitful. Busse (1984) and Hamilton (1984) have both indicated that long term resident males carrying infants in the presence of high ranking newcomer males may not be using the infants as agonistic buffers but may be actively protecting the infants from the newcomer males.

Patrilineal kinship has either been measured in some captive groups by biochemical paternity exclusion methods, or has been established by single male timed matings with females. In these cases the measure of the independent variable seems secure. The dependent variables used to see if patrilineal relatedness relates to behaviour have included a spatial approach and similar positive interactions. Wu *et al.*, (1980) published one of the most exciting studies of this sort – they found that 13 of 16 paternal half-sibling dyads, conceived separately and raised apart, showed significantly more positive approaches than expected by chance, in one experiment. A second experiment, in the same paper, failed to show a significant correlation with kinship. The discrepancy between the data in the two experiments undoubtably led to attempts at replication and one of the co-authors, Sackett, has repeatedly failed to replicate the original finding (Fredrickson & Sackett, 1984; Sackett & Fredrickson, 1987) and concluded that this original positive finding was due to a type one statistical error.

A less direct study reported by Small & Smith (1981) suggested that mothers were less likely to resist the attempts of an infant's paternal half sibling to remove and carry the infant. Walters, in his 1987 review, does not recognize an intrinsic paternity recognition mechanism but cites this paper as suggesting a bias towards paternal kin. An examination of the data, however, reveals that for the six focal infants, 18 of 105 (17%) paternal half-siblings attempted to 'grab' the infant whereas 24 of 107 (22%) non-relatives did so. Of the grabs attempted, mothers resisted 16 of 49 (33%) attempts by paternal half-siblings and 34 of 92 (37%) attempts by non-relatives. Although 34 is greater than 16, the proportion of resisted grabs is virtually identical for paternal half-siblings and non-relatives. The original analysis did a simple chi-square on the number of resists and failed to adjust for the number of attempts. With the appropriate adjustment the original data only provides evidence indicating that full siblings grabbed infants more often than paternal half-siblings or non-relatives, but even the ratio of maternal resistance of full sibling grabs (21 of 67, or 31%) did not differ from the ratio for paternal half-siblings or non-relatives. Moreover, what genetic mechanism would favour a female that permitted an individual unrelated to herself to gain access to and possibly abuse her infant? Paternal half-siblings are not related to the mother.

Despite clear demonstration of matrilineal kinship correlations with behaviour we have no clear evidence of patrilineal kinship correlating with behaviour. Berenstain *et al.*, (1981) in noting that adult males did associate with their own offspring somewhat more, also reported that paternity accounted for less than five per cent of the variance. Such association could also have been produced by males associating with infants of females that they associated with, when the females and their infants were not together. Such male association is not clearly related to kinship and some have even suggested that males may gain sexual access to females after first associating with their offspring.

Kinship recognition mechanisms

The theoretical importance of kinship recognition to theories of inclusive fitness and male strategies is such that authors dealing with agonistic buffering, infanticide, infant protection and similar topics, have settled for less than definitive data on paternity, and/or merely postulated possible kin identification mechanisms without experimental verification. Lancaster (1984a,b) simply assumes paternity recognition. Zeller (1985) proposes kin recognition based on similarities in behavioural patterns, whereas others explain cyclical swelling of a female's perineum during her reproductive cycle as a means of insuring 'paternity certainty' (Hamilton, 1984; Kurland & Gaulin, 1984). There are obviously several hidden premises in this logical argument. Are males supposed to: (1) monopolize females at peak swelling (often asserted but seldom demonstrated), (2) remember which females they successfully monopolized a gestation period later, and then (3) treat preferentially the infants of these females during the next months and years? Even with the usual denial of any cognitive recognition of paternity and reproductive processes on the part of the animals, there is no way to eliminate the cognitive feat involved in differentiating and remembering successful and unsuccessful monopolizations of copulations with females at maximum turgescence.

Of course, primates do have impressive cognitive capacities and Cheney & Seyfarth (1985a,b) argue that these capacities are especially well developed in the social domain. There is no doubt that primates establish long term social relationships and remember former social associates for long periods of time. None of this, however, suggests that primates remember specific events, finely discriminated from similar events (copulations not at peak swelling or copulations at peak swelling where the female was not monopolized) for prolonged periods of time (Old World monkeys and chimpanzees (*Pan troglodytes*), showing swellings, have a minimum of five to seven months for gestation). Seyfarth & Cheney (1984) have also reported that vervets respond to the vocalizations of a *recent* grooming partner. They also respond to the vocalizations of their matrilineal kin, even if somewhat less strongly. They make no mention of responses to regular but not recent grooming partners who are not also kin.

Theoretical comparisons suggesting that some primate taxa are more paternal than others, and that this may be based on the degree of paternal certainty of kinship, are far too preliminary to receive any credence (Taub, 1985). Other theoretical formulations regarding females that deceive males into thinking that they are the father of an infant (Lancaster, 1984b) go far beyond my understanding. Again, even denying the necessity for the impressive cognitive ability which would allow a male to literally believe that he was the father (he would have to know that males were parents, a fact that many human teenagers and some adults do not always know) the evolution of successful 'deception' would counter any selective pressure favouring a male that preferentially treated infants that he thought he had fathered. If a female could deceive a male by using behaviour that the male used to establish 'paternity', then the mechanism would not work for the male and males preferentially favouring infants based on such a mechanism would not be favouring their own offspring. This bewilderment seems unnecessary considering the fact that neither innate nor cognitive paternal kinship recognition mechanisms can be supported by the available evidence for non-human primates.

Hepper (1986) has described several other possible mechanisms that might result in individuals showing preferential treatment of their own close relatives. One of these, phenotype matching, has a certain appeal. Perhaps it is because we believe that we are quite good at recognizing family resemblances among humans, and even among monkeys and apes with whom we have become familiar. Of course most human beings search for family resemblances among related individuals only after obtaining independent information of the kinship relationship. We closely examine the features of our sons and daughters looking for family 'traits' and we look at the successful members of a matriline and recognize similar physical features. Even granting that we get quite good at this, such that it is *possible* to recognize phenotypic similarity, this is not evidence that non-human primates either have the ability to make the same abstractions, or that they do so and use this information in modifying their behaviour. Some contrary information exists in that most primates (an order in which vision appears to dominate the sensory modalities) fail to recognize their own mirror image (Gallup 1970; 1979; Gallup *et al.*, 1980). Although monkeys can recognize a cage mate's mirror image (Ledbetter & Basen, 1982) agonistic responses to their own image persist. This would seem to argue against any predisposition to act affinitively towards any monkey matching their own phenotype even less accurately than does a mirror image. Matching against mother's phenotype remains a possibility, but this seems unnecessary to postulate given the existence of much better mechanisms based on: (1) the long period of primate biological dependence, (2) the long term social relationships maintained by primates, and (3) the consequent, essentially matrilineal, subgroupings seen in many primate social units, especially in expanding populations.

Explanations for preferential treatment of relatives by non-human primates based on association can be supported using several types of data. Adoptions have been reported in wild troops and, although allomothering and adoption may occur primarily among matrilineal kin (Johnson *et al.*, 1980), it is not restricted to kin. Deets (1974) has reported successful cross-fostering experiments and I have seen several cases of females kidnapping unrelated infants and successfully rearing them, sometimes alongside their own live infants, born within a few days of the adopted infants. Despite all of our efforts to distinguish the adopted from the natural infants we could never detect anything on the part of mother or infants that would aid us in differentiating the two infants. In those cases where the infants had not been marked prior to the adoption, we never succeeded in identifying which was the adopted infant.

The fact that matrilineal kin reared together, as well as unrelated 'adopted' animals, all treat each other preferentially, whereas patrilineal kin do not, suggests an association mechanism. Walters (1987) and MacKenzie *et al.*, (1985) have pointed to the period of early infant association as sufficient to account for most social preferences among primates. The infant is continually associated with the mother in many primate species for the first part of its life. (It would be interesting to study kinship and behaviour correlates in taxa where this is less true, e.g. some colobines.) The bond between mother and infant persists after the birth of her next infant. As a consequence, the youngest infant's first prolonged contacts are likely to be with mother and mother's other still dependent offspring. The bond between mother and child (especially female offspring) may be prolonged and enduring so that infants are likely to find older siblings also in proximity. Moreover, mother's mother may also still be in association, and mother's female siblings may be in proximity as well. The infant's earliest social contacts are, therefore, almost all with matrilineal kin. Social relationships involving mutualism and reciprocal altruism will develop with these kin. Such relationships will also be selected for in accordance with inclusive fitness theory, but the proximal mechanism can be based on the early association pattern, and not in any way related to direct genetic recognition mechanisms, or less direct but more elaborate mechanisms involving phenotype matching and abstract abilities. The pattern of errors will, once again, be most instructive. The treatment of a close associate of mother, that is not biological kin, as if it were kin, argues powerfully for an association mechanism. In fact, in many studies of wild populations, field workers use close association as a measure of kinship. There is the assumption that whatever errors that this method might allow does not negate the recognition of a functional matrilineal network (Paul & Kuester, 1987; Smuts, 1985; Walters, 1981).

The association hypothesis also can be invoked to account for special associations among infants and adult males. When an adult male begins an association with a female he, of necessity, begins an association with her dependent young and, to a lesser extent, her entire matriline. Maintaining this association with the female will be dependent upon not antagonizing the female by inviting her to come to the defence of members of her matriline, and the male may be expected to show tolerance towards all of her usual associates. These females may, or may not, be the same females that the male breeds with, but if he continues to associate with the same female for a prolonged period, he will associate with her next infants as well. If a tendency can be demonstrated for females to prefer less familiar males as mating partners, then this most certainly will mean the male has a *decreased* probability of being the father of her infant and it virtually precludes long term association between a male and the offspring of those females with whom he breeds. Obviously there may be many benefits to individual associations other than reproduction and protection of a specific genetic investment. Macaque and baboon males, notably protect troop infants by threatening any extra-group source of disturbance to any infant.

Individuals in close physical proximity may not only choose each other as grooming and resting partners, but they may also come into competition for the same resources at the same time. These competitions may increase the chances of confrontation and agonistic conflict. Individuals in frequent close physical proximity will have increased opportunities to provoke aggression from one another in a number of ways. Social individuals, however, have mechanisms to cope with conflicts with their close associates without destroying the social relationship. de Waal (1977) indicates that kin are often preferred social associates and, that following an agonistic episode, close social associates reconcile with one another preferentially, and also restrain the vigour of those aggressive responses directed towards each other. Bernstein & Ehardt (1986), also suggested that although mothers may be particularly likely to bite their own offspring that such biting seemed, subjectively, to be strongly inhibited. Despite the potential damage that a female could inflict in biting an infant's arm, there was rarely any indication of physical injury to the infant, despite its vigorous screaming and convulsive body jerking in response to the bite.

Kinship correlations and evidence for kin recognition

Having noted a strong correlation between many types of primate behaviour and genetic relatedness we may now proceed with attempts to account for this relationship. A direct causal relationship would imply that behavioural biases would be correlated with the degree of genetic relatedness regardless of maternal or paternal origin and independent of ontogenetic factors depending on environmental inputs. The available data do not support a direct causal relationship. There is a clear correlation between matrilineal kinship and preferential behaviour, but patrilineal kinship has not been demonstrated to correlate with preferential behaviour in the same manner. Likewise, no causal mechanism based on intrinsic individual factors correlated with kinship can be supported.

The 'errors' in kin recognition indicating that adopted infants and close associates of mother (like an adult male) may also receive the same preferential treatment as observed towards biological matrilineal relatives support a mechanism based on early association as the most likely explanation for the correlation between kinship and preferential association and behaviour in non-human primates. The same conclusion was also reached by Walters (1987) in his review.

Given that primates (1) have a long period of biological dependence, (2) are noted for the establishment and maintenance of long term social bonds, and (3) are capable of modification of behaviour as a function of experience and long term retention of this learning, it should not be any surprise that they form intense social bonds with their earliest associates and that these bonds are maintained for many years. Such bonds will be formed with matrilineal kin as a function of long term infant dependence on the mother. Other frequent associates of the mother may also be included in these social bonds as most primates are capable of forming polyadic complex social networks. As Kummer puts it 'Nonsocial ecological techniques are poorly developed in primates. Their specializations must be sought in the way they act as groups' (p. 38, 1971). The social group begins with the mother in mammals.

References

Altmann, J. (1980). *Baboon Mothers and Infants*. Cambridge, MA: Harvard University Press.

Altmann, J. (1984). Sociobiological perspectives on parenthood. In *A Psychodynamic Perspective*, ed. R. S. Cohen, B. J. Cohler, & S. H. Weissman, pp. 9–23. New York: Guilford Press.

Angst, W. (1975). Basic data and concepts on the social organization of *Macaca fascicularis*. *Primate Behav.*, **4**, 325–88.

Aoki, K. & Nozawa, K. (1984). Average coefficient of relationship within troops of the Japanese monkey and other primate species with reference to the possibility of group selection. *Primates*, **25**, 171–84.

Baker, S. C. & Estep, D. Q. (1985). Kinship and affiliative behavior patterns in a captive group of Celebes black apes (*Macaca nigra*). *J. Comp. Psychol.*, **99**, 356–60.

Baxter, M. J. & Fedigan, L. M. (1979). Grooming and consort partner selection in a troop of Japanese monkeys (*Macaca fuscata*). *Arch. Sex. Behav.*, **8**, 445–58.

Berenstain, L., Rodman, P. S. & Smith, D. G. (1981). Social relations between fathers and offspring in a captive group of rhesus monkeys (*Macaca mulatta*). *Anim. Behav.*, **29**, 1057–63.

Berman, C. M. (1982). The ontogeny of social relationships with group companions among free-ranging infant rhesus monkeys. I. Social networks and differentiation. *Anim. Behav.*, **30**, 149–62.

Bernstein, I. S. (1975). Activity patterns in a gelada monkey group. *Folia Primatol.*, **23**, 50–71.

Bernstein, I. S. & Ehardt, C. L. (1985). Agonistic aiding: kinship, rank, age and sex influences. *Am. J. Primatol.*, **8**, 37–52.

Bernstein, I. S. (1986). The influence of kinship and socialization on aggressive behaviour in rhesus monkeys (*Macaca mulatta*). *Anim. Behav.*, **34**, 739–47.

Boelkins, R. C. & Wilson, A. P. (1972). Intergroup social dynamics of the Cayo Santiago rhesus (*Macaca mulatta*) with special reference to changes in group membership by males. *Primates*, **13**, 125–40.

Bolwig, N. (1980). Early social development and emancipation of *Macaca nemestrina* and species of *Papio*. *Primates*, **21**, 357–75.

Bramblett, C. A. (1973). Social organization as an expression of role behavior among Old World monkeys. *Primates*, **14**, 101–12.

Busse, C. (1984). Triadic interactions among male and infant chacma baboons. In *Primate Paternalism*, ed. D. M. Taub, pp. 186–212. New York: Van Nostrand Reinhold Company.

Busse, C. D. (1985). Paternity recognition in multi-male primate groups. *Am. Zool.*, **25**, 873–81.

Chapais, B. (1985). An experimental analysis of a mother–daughter rank reversal in Japanese macaques (*Macaca fuscata*). *Primates*, **26**, 407–23.

Chapais, B. & Schulman, S. R. (1980). An evolutionary model of female dominance relations in primates. *J. Theor. Biol.*, **82**, 47–89.

Cheney, D. L. (1978). Interactions of immature male and female baboons with adult females. *Anim. Behav.*, **26**, 389–408.

Cheney, D. L., Lee, P. C. & Seyfarth, R. M. (1981). Behavioral correlates of non-random mortality among free-ranging female vervet monkeys. *Behav. Ecol. Sociobiol.*, **9**, 153–61.

Cheney, D. L. & Seyfarth, R. M. (1980). Vocal recognition in free-ranging vervet monkeys. *Anim. Behav.*, **28**, 362–67.

Cheney, D. L. & Seyfarth, R. M. (1983). Nonrandom dispersal in free-ranging vervet monkeys: social and genetic consequences. *Am. Nat.*, **122**, 393–412.

Cheney, D. L. & Seyfarth, R. M. (1985a). Vervet monkey alarm calls: manipulation through shared information? *Behaviour*, **94**, 150–66.

Cheney, D. L. & Seyfarth, R. M. (1985b). Social and non-social knowledge in vervet monkeys. *Phil. Trans. R. Soc. London,* B308, 187–201.

Chepko-Sade, B. D. & Sade, D. S. (1979). Patterns of group splitting within matrilineal kinship groups: Study of social group structure in *Macaca mulatta* (Cercopithecidae, Primates). *Behav. Ecol. Sociobiol.,* 5, 67–86.

Chikazawa, D., Gordon, T. P., Bean, C. A. & Bernstein, I. S. (1979). Mother–daughter dominance reversals in rhesus monkeys (*Macaca mulatta*). *Primates,* 20, 301–5.

Colvin, J. D. (1985). Breeding-season relationships of immature male rhesus monkeys with females: I. Individual differences and constraints on partner choice. *Int. J. Primatol.,* 6, 261–87.

Colvin, J. & Tissier, G. (1985). Affiliation and reciprocity in sibling and peer relationships among free-ranging immature male rhesus monkeys. *Anim. Behav.,* 33, 959–77.

Cords, M., Mitchell, B. J., Tsingalia, H. M. & Rowell, T. E. (1986). Promiscuous mating among blue monkeys in the Kakamega Forest, Kenya. *Ethology,* 72, 214–26.

Crockett, C. M. (1984). Emigration by female red howler monkeys and the case for female competition. In *Female Primates: Studies by Women Primatologists,* ed. M. F. Small, pp. 159–73. New York: Alan R. Liss, Inc.

Datta, S. B. (1983a). Patterns of agonistic interference. In *Primate Social Relationships: An Integrated Approach.* ed. R. A. Hinde, pp. 289–97. Sunderland, MA: Sinauer Associates.

Datta, S. B. (1983b). Relative power and maintenance of dominance. In *Primate Social Relationships: An Integrated Approach.* ed. R. A. Hinde, pp. 103–12. Sunderland, MA: Sinauer Associates.

Deag, J. M. & Crook, J. H. (1971). Social behaviour and 'agonistic buffering' in the wild barbary macaque (*Macaca sylvania*). *Folia Primatol.,* 15, 183–200.

Deets, A. C. (1974). Age mate or twin sibling: effects on interactions between monkey mothers and infants. *Develop. Psychol.,* 10, 748–63.

de Waal, F. B. M. (1977). Organization of agonistic relations within two captive groups of Java monkeys (*Macaca fascicularis*). *Z. Tierpsychol.,* 44, 225–82.

de Waal, F. B. M. & van Roosmalen, A. (1979). Reconciliation and consolation among chimpanzees. *Behav. Ecol. Sociobiol.,* 5, 55–66.

de Waal, F. B. M. & Yoshihara, D. (1983). Reconciliation and redirected affection in rhesus monkeys. *Behaviour,* 85, 224–41.

Dittus, W. P. J. (1979). The evolution of behaviors regulating density and age-specific sex ratios in a primate population. *Behaviour,* 69, 265–302.

Drickamer, L. C. & Vessey, S. H. (1973). Group changing in free-ranging male rhesus monkeys. *Primates,* 14, 359–68.

Dunbar, R. I. M. (1984). Infant-use by male gelada in agonistic contexts: Agonistic buffering, progeny protection or soliciting support? *Primates,* 25, 28–35.

Eaton, G. G. (1976). The social order of Japanese macaques. *Sci. Am.,* 235, 97–106.

Eaton, G. G., Johnson, D. F., Glick, B. B. & Worlein, J. M. (1985). Development in Japanese macaques (*Macaca fuscata*): sexually dimorphic behaviour during the first year of life. *Primates,* 26, 238–47.

Eaton, G. G., Johnson, D. F., Glick, B. B. & Worlein, J. M. (1986). Japanese macaques (*Macaca fuscata*) social development: sex differences in juvenile behavior. *Primates*, **27**, 141–50.

Ehardt, C. L. & Bernstein, I. S. (1986). Matrilineal overthrows in rhesus monkey groups. *Int. J. Primatol.*, **7**, 157–81.

Ehardt, C. L. & Bramblett, C. A. (1980). The structure of social space among a captive group of vervet monkeys. *Folia Primatol.*, **34**, 214–38.

Enomoto, T. (1978). On social preference in sexual behavior of Japanese monkeys (*Macaca fuscata*). *J. Hum. Evol.*, **7**, 283–93.

Estrada, A. (1984). Male-infant interactions among free-ranging stumptail macaques. In *Primate Paternalism*, ed. D. M. Taub, pp. 56–87. New York: Van Nostrand Reinhold.

Fairbanks, L. A. & McGuire, M. T. (1985). Relationships of vervet mothers with sons and daughters from one through three years of age. *Anim. Behav.*, **33**, 40–50.

Fedigan, L. M. (1983). Dominance and reproductive success in primates. *Yb. Phys. Anthrop.*, **26**, 91–129.

Fedigan, L. M. & Gouzoules, H. (1978). The consort relationships in a troop of Japanese monkeys. In *Recent Advances in Primatology, vol. 1*, ed. D. J. Chivers, J. Herbert, pp. 493–5. London: Academic Press.

Fredrickson, W. T. & Sackett, G. P. (1984). Kin preferences in primates (*Macaca nemestrina*): relatedness or familiarity? *J. Comp. Psychol.*, **98**, 29–34.

Fukuda, F. (1983). Relationship between age and troop shifting in male Japanese monkeys. *Jap. J. Ecol.*, **32**, 491–8.

Furuichi, T. (1983). Interindividual distance and influence of dominance on feeding in a natural Japanese macaque troop. *Primates*, **24**, 445–55.

Furuichi, T. (1983). Symmetrical patterns in non-agonistic social interactions found in unprovisioned Japanese macaques. *J. Ethol.*, **2**, 109–19.

Gallup, G. D., Jr. (1970). Chimpanzees self-recognition. *Science*, **167**, 86–7.

Gallup, G. G. (1979). Self-awareness in primates. *Am. Sci.*, **67**, 417–21.

Gallup, G. G., Jr., Wallnau, L. B. & Suarez, S. D. (1980). Failure to find self-recognition in mother–infant and infant–infant rhesus monkey pairs. *Folia Primatol.*, **33**, 210–19.

Giacoma, C. (1983). Early social interactions of juvenile pigtail macaques, *Macaca nemestrina*. *Boll. Zoo.*, **50**, 41–5.

Glander, K. E. (1980). Reproduction and population growth in free-ranging mantled howling monkeys. *Am. J. Phys. Anthrop.*, **53**, 25–36.

Glick, B. B., Eaton, G. G., Johnson, D. F. & Worlein, J. M. (1986a). Social behavior of infant and mother Japanese macaques (*Macaca fuscata*): effects of kinship, partner sex, and infant sex. *Int. J. Primatol.*, **7**, 139–55.

Glick, B. B., Eaton, G. G., Johnson, D. F. & Worlein, J. M. (1986b). Development of partner preferences in Japanese macaques (*Macaca fuscata*): effects of gender and kinship during the second year of life. *Int. J. Primatol.*, **7**, 467–79.

Gouzoules, H. (1984). Social relations of males and infants in a troop of Japanese monkeys: a consideration of causal mechanisms. In *Primate Paternalism*, ed. D. M. Taub, pp. 127–45. New York: Van Nostrand Reinhold.

Gouzoules, S. (1984). Primate mating systems, kin associations, and cooperative behavior: Evidence for kin recognition? *Yb. Phys. Anthropol.*, **27**, 99–134.

Hamilton, W. D. (1964). The genetical theory of social behaviour. *J. Theor. Biol.*, **7**,

1–52 (1–16, 17–52).

Hamilton, W. J., III (1984). Significance of paternal investment by primates to the evolution of adult male–female associations. In *Primate Paternalism*, ed. D. M. Taub, pp. 309–35. New York: Van Nostrand Reinhold.

Hanby, J. P. (1980). Relationships in six groups of rhesus monkeys. I. Networks. *Am. J. Phys. Anthrop.*, **52**, 549–64.

Hanby, J. P. & Brown, C. E. (1974). The development of sociosexual behaviours in Japanese macaques, *Macaca fuscata*. *Behaviour*, **49**, 152–96.

Hasegawa, T. & Hiraiwa, M. (1980). Social interactions of orphans observed in a free-ranging troop of Japanese monkeys. *Folia Primatol.*, **33**, 129–58.

Hausfater, G., Altmann, J. & Altmann, S. (1982). Long-term consistency of dominance relations among female baboons (*Papio cynocephalus*). *Science*, **217**, 752–5.

Hepper, P. G. (1986). Kin recognition: functions and mechanisms. A review. *Biol. Rev.*, **61**, 63–93.

Hornshaw, S. G. (1985). Proximity behavior in a captive group of lion-tailed macaques (*Macaca silenus*). In *The Lion-Tailed Macaque: Status and Conservation*. ed. P. G. Heltne, pp. 269–92. New York: Liss.

Horrocks, J. A. & Hunte, W. (1983a). Maternal rank and offspring rank in vervet monkeys: an appraisal of the mechanisms of rank acquisition. *Anim. Behav.*, **31**, 772–82.

Horrocks, J. A. & Hunte, W. (1983b). Rank relations in vervet sisters: a critique of the role of reproductive value. *Am. Nat.*, **122**, 417–21.

Hrdy, S. B. (1976). Care and exploitation of nonhuman primate infants by conspecifics other than the mother. *Adv. Study Behav.*, **6**, 101–58.

Hrdy, S. B. (1981). 'Nepotists' and 'altruists': the behavior of old females among macaques and langur monkeys. In *Other Ways of Growing Old: Anthropological Perspectives*, ed. P. T. Amoss & S. Harrell, pp. 59–76. Stanford: Stanford University Press.

Itani, J. (1975). Twenty years with Mount Takasaki monkeys. In *Primate Utilization and Conservation*, ed. G. Bermant & D. G. Lindburg, pp. 101–125. New York: John Wiley & Sons.

Itoigawa, N. (1975). Variables in male leaving a group of Japanese macaques. In *Proc. Symp. Fifth Cong. Int. Primatol. Soc.*, ed. S. Kondo, M. Kawai, A. Ehara & S. Kawamura, pp. 233–245. Tokyo: Japan Science Press.

Itoigawa, N., Negayama, K. & Kondo, K. (1981). Experimental study on sexual behavior between mother and son in Japanese monkeys (*Mucaca fuscata*). *Primates*, **22**, 494–502.

Johnson, C., Koerner, C., Estrim, M. & Duoos, D. (1980). Alloparental care and kinship in captive social groups of vervet monkeys (*Cercopithecus aethiops sabaeus*). *Primates*, **21**, 406–15.

Kaplan, J. R. (1976). Interference in fights and control of aggression in a group of free-ranging rhesus monkeys. *Am. J. Phys. Anthrop.*, **44**, 189.

Kaplan, J. R. (1977). Patterns of fight interference in free-ranging rhesus monkeys. *Am. J. Phys. Anthrop.*, **47**, 279–88.

Kaplan, J. R. (1978). Fight interference and altruism in rhesus monkeys. *Am. J. Phys. Anthrop.*, **49**, 241–9.

Kawai, M. (1965). On the system of social ranks in a natural troop of Japanese monkeys. I. Basic rank and dependent rank. In *Japanese Monkeys*, ed. S. A. Altmann, pp. 66–86. Chicago: University of Chicago Press.

Kawamura, S. (1965). Matriarchal social ranks in the Minoo B troop: a study of the rank system of Japanese monkeys. In *Japanese Monkeys*, ed. S. A. Altmann, pp. 105–12. Chicago: University of Chicago Press.

Kawanaka, K. (1977). Division of males in a Japanese monkey troop on the basis of numerical data. *Bull. Hiruzen Res. Inst.*, **3**, 11–44.

Koford, C. B. (1963). Rank of mothers and sons in bands of rhesus monkeys. *Science*, **141**, 356–7.

Koyama, N. (1967). On dominance rank and kinship of a wild Japanese monkey troop in Arashiyama. *Primates*, **8**, 189–216.

Koyama, N. (1985). Playmate relationships among individuals of the Japanese monkey troop in Arashiyama. *Primates*, **26**, 390–406.

Kuester, J. & Paul, A. (1986). Male–infant relationships in semi-free-ranging Barbary macaques (*Macaca sylvanus*) of Affenberg Salem/FRG: testing the 'male care' hypothesis. *Am. J. Primatol.*, **10**, 315–27.

Kummer, H. (1971). *Primate Societies. Group Techniques of Ecological Adaptation.* Chicago: Aldine-Atherton.

Kurland, J. A. (1977). Kin selection in the Japanese monkey. *Contr. Primatol.*, **12**, i–x, 1–145.

Kurland, J. A. & Gaulin, S. J. C. (1984). The evolution of male parental investment: Effects of genetic relatedness and feeding ecology on the allocation of reproductive effort. In *Primate Paternalism*, ed. D. M. Taub, pp. 259–308. New York: Van Nostrand Reinhold.

Lancaster, J. B. (1984a). Introduction. In *Female Primates: Studies by Women Primatologists*, ed. M. F. Small, pp. 1–10. New York: Alan R. Liss.

Lancaster, J. B. (1984b). Evolutionary perspectives on sex differences in the higher primates. In *Gender and Life Course*, ed. A. S. Rossi, pp. 3–27. New York: Aldine.

Ledbetter, D. H. & Basen, J. A. (1982). Failure to demonstrate self-recognition in gorillas. *Am. J. Primatol.*, **2**, 307–10.

Lee, P. C. & Oliver, J. I. (1979). Competition, dominance and the acquisition of rank in juvenile yellow baboons (*Papio cynocephalus*). *Anim. Behav.*, **27**, 576–85.

Loy, J. (1975). The descent of dominance in *Macaca*: Insights into the structure of human societies. In *Socioecology and Psychology of Primates*, ed. R. H. Tuttle, pp. 153–80. The Hague: Mouton Publishers.

Loy, J. & Loy, K. (1974). Behavior of an all juvenile group of rhesus monkeys. *Am. J. Phys. Anthrop.*, **40**, 84–95.

MacKenzie, M. M., McGrew, W. C. & Chamove, A. S. (1985). Social preferences in stump-tailed macaques (*Macaca arctoides*): effects of companionship, kinship, and rearing. *Develop. Psychobiol.*, **18**, 115–23.

Marler, P. (1985). Representational vocal signals of primates. *Fortschr. Zoo.*, **31**, 211–21.

Massey, A. (1977). Agonistic aids and kinship in a group of pigtail macaques. *Behav. Ecol. Sociobiol.*, **2**, 31–40.

Mehlman, P. (1986). Male intergroup mobility in a wild population of the Barbary macaque (*Macaca sylvanus*), Ghomaran Rif Mountains, Morocco. *Am. J.*

Primatol., **10**, 67–81.

Meikle, D. B. & Vessey, S. H. (1981). Nepotism among rhesus monkey brothers. *Nature*, **294**, 160–1.

Melnick, D. J., Pearl, M. C. & Richard, A. F. (1984). Male migration and inbreeding avoidance in wild rhesus monkeys. *Am. J. Primatol.*, **7**, 229–43.

Missakian, E. A. (1972). Genealogical and cross-genealogical dominance relations in a group of free-ranging rhesus monkeys (*Macaca mulatta*) on Cayo Santiago. *Primates*, **13**, 169–80.

Missakian, E. A. (1973). Genealogical mating activity in free-ranging groups of rhesus monkeys (*Macaca mulatta*) on Cayo Santiago. *Behaviour*, **45**, 225–41.

Moore, J. & Ali, R. (1984). Are dispersal and inbreeding avoidance related? *Anim. Behav.*, **32**, 94–112.

Mori, A. (1977). Intra-troop spacing mechanism of the wild Japanese monkeys of the Koshima troop. *Primates*, **18**, 331–57.

Murray, R. D. & Smith, E. O. (1983). The role of dominance and intrafamilial bonding in the avoidance of close inbreeding. *J. Hum. Evol.*, **12**, 481–6.

Nagel, V. (1979). On describing primate groups as systems: The concept of Ecosocial Behaviour. In *Primate Ecology and Human Origins*, ed. I. S. Bernstein & E. O. Smith, pp. 313–39. New York: Garland STPM Press.

Netto, W. J. & van Hooff, J. A. R. A. M. (1986). Conflict interference and the development of dominance relationships in immature *Macaca fascicularis*. In *Primate Ontogeny, Cognition and Social Behaviour*, ed. J. G. Else & P. C. Lee, pp. 291–300. Cambridge: Cambridge University Press.

Niemeyer, C. L. & Chamove, A. S. (1983). Motivation of harassment of matings in stumptailed macaques. *Behaviour*, **87**, 298–323. (French summary).

Packer, C. (1979). Intertroop transfer and inbreeding avoidance in *Papio anubis*. *Anim. Behav.*, **37**, 1–36.

Paul, A. & Kuester, J. (1985). Intergroup transfer and incest avoidance in semifree-ranging Barbary macaques (*Macaca sylvanus*) at Salem (FRG). *Am. J. Primatol.*, **8**, 317–22.

Paul, A. & Kuester, J. (1987). Dominance, kinship and reproductive value in female Barbary macaques (*Macaca sylvanus*) at Affenberg Salem. *Behav. Ecol. Sociobiol.*, **21**, 323–31.

Pusey, A. E. (1980). Inbreeding avoidance in chimpanzees. *Anim. Behav.*, **28**, 543–52.

Quiatt, C. (1986). Juvenile/adolescent role functions in a rhesus monkey troop: an application of household analysis to nonhuman primate social organization. In *Primate Ontogeny, Cognition and Social Behaviour*, ed. J. G. Else & P. C. Lee, pp. 281–9. Cambridge: Cambridge University Press.

Ransom, T. W. & Rowell, T. E. (1972). Early social development of feral baboons. In *Primate Socialization*, ed. F. E. Poirier, pp. 105–44. New York: Random House.

Reinhardt, V., Dodsworth, R. & Scanlan, J. (1986). Altruistic interference shown by the alpha-female of a captive troop of rhesus monkeys. *Folia Primatol.*, **46**, 44–50.

Ruehlmann, T. E., Bernstein, I. S. & Judge, P. G. (1987). The influence of adult male sexual behavior and inbreeding in rhesus macaques (*Macaca mulatta*). Paper presented at Animal Behavior Society in Williams College, MA.

Sackett, G. P. & Fredrickson, W. T. (1987). Social preferences by pigtailed macaques:

Familiarity versus degree and type of kinship. *Anim. Behav.*, **35**, 603–6.

Sade, D. S. (1965). Some aspects of parent–offspring and sibling relations in a group of rhesus monkeys, with a discussion of grooming. *Am. J. Phys. Anthrop.*, **23**, 1–18.

Sade, D. S. (1968). Inhibition of son–mother mating among free-ranging rhesus monkeys. *Sci. Psychoanal.*, **12**, 18–37.

Sade, D. S. (1969). An algorithm for dominance relations among rhesus monkeys: Rules for adult females and sisters. *Am. J. Phys. Anthrop.*, **31**, 261.

Seyfarth, R. M. & Cheney, D. L. (1984). Grooming, alliances and reciprocal altruism in vervet monkeys. *Nature*, **308**, 541–3.

Shively, C. (1985). The evolution of dominance hierarchies in nonhuman primate society. In *Power, Dominance, and Nonverbal Behaviour*, ed. S. L. Ellyson & J. F. Dovidio, pp. 67–87. New York: Springer-Verlag.

Shively, C. & Smith, D. G. (1985). Social status and reproductive success of male *Macaca fascicularis*. *Am. J. Primatol.*, **9**, 129–35.

Silk, J. B. (1982). Altruism among female *Macaca radiata*: explanations and analysis of patterns of grooming and coalition formation. *Behaviour*, **79**, 162–88.

Silk, J. B. (1984). Measurement of the relative importance of individual selection and kin selection among females of the genus *Macaca*. *Evolution*, **38**, 553–9.

Silk, J. B., Samuels, A. & Rodman, P. S. (1981). The influence of kinship, rank, and sex on affiliation and aggression between adult female and immature bonnet macaques (*Macaca radiata*). *Behaviour*, **78**, 111–37.

Small, M. F. & Smith, D. G. (1981). Brief report: interactions with infants by full siblings, paternal half-siblings, and nonrelatives in a captive group of rhesus macaques (*Macaca mulatta*). *Am. J. Primatol.*, **1**, 91–4.

Smith, D. G. (1982). A test of the randomness of paternity of members of maternal sibships in six captive groups of rhesus monkeys (*Macaca mulatta*). *Int. J. Primatol.*, **3**, 461–8.

Smith, D. G. (1986). Inbreeding in the maternal and paternal lines of four captive groups of rhesus monkeys (*Macaca mulatta*). In *Current Perspectives in Primate Biology*, ed. D. M. Taub, F. A. King, pp. 214–25. New York: Van Nostrand Reinhold.

Smuts, B. (1985). *Sex and Friendship in Baboons*. New York: Aldine.

Stein, D. M. (1984a). Ontogeny in infant–adult male relationships during the first year of life for yellow baboons (*Papio cynocephalus*). In *Primate Paternalism*, ed. D. M. Taub, pp. 213–43. New York: Van Nostrand Reinhold.

Stein, D. M. (1984b). *The Sociobiology of Infant and Adult Male Baboons*. Norwood: Ablex Publishing Corporation.

Strum, S. C. (1984). Why males use infants. In *Primate Paternalism*, ed. D. M. Taub, pp. 146–85. New York: Van Nostrand Reinhold.

Sugiyama, Y. (1984). Recent advances in the field study of male and female ranking order in the Japanese monkey (*Macaca fuscata*). In *Current Primate Researches*, ed. M. L. Roonwal, S. M. Mohnot & N. S. Rathore, pp. 417–22. Jodhpur: University of Jodhpur.

Takahata, Y. (1982a). The socio-sexual behavior of Japanese monkeys. *Z. Tierpsychol.*, **59**, 89–108.

Takahata, Y. (1982b). Social relations between adult males and females of Japanese monkeys in the Arashiyama B. Troop. *Primates*, **23**, 1–23.

Taub, D. M. (1984). Male caretaking behavior among wild barbary macaques (*Macaca sylvanus*). In *Primate Paternalism*, ed. D. M. Taub, pp. 20–55. New York: Van Nostrand Reinhold.

Taub, D. M. (1985). Male–infant interactions in baboons and macaques: A critique and reevaluation. *Am. Zool.*, **25**, 861–71.

Taylor, L. & Sussman, R. W. (1985). A preliminary study of kinship and social organization in a semi-free ranging group of Lemur catta. *Int. J. Primatol.*, **6**, 601–14.

Tinbergen, N. (1951). *The Study of Instinct*. Oxford: Clarendon Press of Oxford University Press.

van Noordwijk, M. A. & van Schaik, C. P. (1985). Male migration and rank acquisition in wild long-tailed macaques (*Macaca fascicularis*). *Anim. Behav.*, **33**, 849–61.

Vessey, S. H. (1973). Night observations of free-ranging rhesus monkeys. *Am. J. Phys. Anthrop.*, **38**, 613–19.

Walters, J. R. (1980). Interventions and the development of dominance relationships in female baboons. *Folia Primatol.*, **34**, 61–89.

Walters, J. R. (1981). Inferring kinship from behaviour: maternity determinations in yellow baboons. *Anim. Behav.*, **29**, 126–136.

Walters, J. R. (1987). Kin recognition in non-human primates. In *Kin Recognition in Animals*, ed. D. J. C. Fletcher & C. D. Michener, pp. 359–93. Chichester: John Wiley & Sons.

Watanabe, K. (1979). Alliance formation in a free-ranging troop of Japanese macaques. *Primates*, **20**, 459–74.

Wilson, E. O. (1975). *Sociobiology: The New Synthesis*. Cambridge, MA: Belknap Press of Harvard University Press.

Wilson, M. E., Gordon, T. P. (1979). Sexual activity of male rhesus monkeys introduced into heterosexual group. *Am. J. Phys. Anthrop.*, **50**, 515–23.

Wu, H. M. H., Holmes, W. G., Medina, S. R. & Sackett, G. P. (1980). Kin preference in infant *Macaca nemestrina*. *Nature*, **285**, 225–7.

Zeller, A. (1985). Component patterns in gesture formation in *Macaca sylvanus* of Gibraltar. *Can. J. Anthropol.*, **4**, 35–42.

2

Co-operation and reciprocity in birds and mammals

J. David Ligon

Co-operation can be defined as any mutually beneficial interaction between two or more individuals. By this standard definition, many, probably most, birds and mammals exhibit co-operation in one or more contexts. Co-operation can range from a simple apparently incidental group effect, such as the simultaneous mobbing of an owl by individuals of two species of songbirds, to complex mutual dependence, such as the rotating sentinel systems of some group-living animals (e.g. McGowan, 1987). Virtually all researchers recognize co-operation when they see it and its widespread occurrence is not a matter of controversy (Axelrod & Hamilton, 1981).

The phenomenon of co-operative behaviour is a fascinating one, for several reasons. First, co-operation is a universal human trait and the vast majority of interactions among individual humans are co-operative to a greater or lesser degree. Thus, it is easy for people to empathize with the co-operative behaviours they observe in animals.

Second, highly developed intra-group co-operation is often seen in species also characterized by high levels of inter-group competition. Our own species is the prime example of this relationship. Killing of conspecifics may be viewed as indicative of extreme intraspecific competition, and human warfare, the searching out and killing of male chimpanzees (*Pan troglodytes*) by groups of males (Goodall, 1986), the expulsion or killing of male pride-holding lions, (*Panthera leo*) by invading coalitions of males (e.g. Packer & Pusey, 1982) and the running down and killing of a lone wolf (*Canis lupis*) by a pack (Mech, 1970), illustrate this relationship between co-operation and competition. Although these mammalian species have the means to kill effectively, levels of aggressive intent probably are equally high in many group-living birds, as indicated by the intensity of their attacks on experimentally introduced 'intruders' into their territories (pers. obs.).

Third, although co-operation clearly is common in the natural world, its origins and maintenance provide theoretical problems that the prisoner's dilemma game has been used to illustrate (Axelrod & Hamilton, 1981; Brown, 1983). In brief, the assumptions of the game between two individuals are that (1) each can co-operate or defect, with the pay-off measured in terms of individual fitness, (2) the selfish choice of defection yields a higher pay-off than co-operation, and (3) if both defect, both do worse than if both had co-operated. If two individuals are to play only once, the best strategy is always to defect (Axelrod & Hamilton, 1981). Axelrod & Hamilton (1981) discuss two factors that can alter the costs and benefits of co-operation versus defection: the probability of repeated future interactions (clustering) and kinship. For social birds and mammals, both repeated interactions among pairs of individuals and genetic ties among group members are typical. Axelrod & Hamilton (1981) thus provide a theoretical solution to the problem of co-operation, a solution that is paralleled reasonably closely by some animal societies (Ligon, 1983; Wiley & Rabenold, 1984).

In nature, additional factors also are present that further alter the costs and benefits of defection. For example, many group-living birds are highly territorial, suggesting that either space or some specific resource is in limited supply. In such species, vigorous displays and brief skirmishes between groups occur frequently at territorial boundaries (e.g. green woodhoopoe, *Phoeniculus purpureus*, Ligon & Ligon 1978b, 1982). In green woodhoopoes, if group sizes are about equal, and especially if two or more adult males are present in each group, the displays typically do not escalate into serious, damaging fights and boundaries may remain stable for long periods of time. However, if one male, instead of participating in the group's boundary defence, flies away ('defects' from its group), or otherwise is absent from the display, its group may then be attacked, driven back and vigorously pursued. Once the 'balance' between the two groups is destroyed, the result may be permanent loss of a significant portion of the losing group's territory. Under these circumstances, the defector clearly is not rewarded for its defection, whether or not it is genetically related to other flock members and whether or not it will interact with them in the future. Rather, it suffers a loss of resources along with the rest of its social unit. If this pattern is repeated over a period of time, the end result may be the dissolution of the flock and the disappearance of the territory (Ligon & Ligon 1983, 1988).

Takeover of territorial space by larger groups at the expense of smaller ones, when group sizes are highly unequal also occurs. In these latter cases, defection also can be observed: one bird first 'breaks ranks' and retreats in the face of the numerically superior opponents. This often triggers attack by the larger group, with the smaller one then being routed.

Observations like these indicate that ecological factors such as restricted space or resources, also can play a role in elevating the costs of defection. Thus, for a woodhoopoe, kinship ties and future interactions with fellow group members, plus the costs associated with possible loss of resources, all promote a high level of intra-group co-operation.

My goals in this chapter are threefold. First, I address the ecological bases of group-living, the essential precursor of on-going co-operation between individuals. Second, I review and consider the most controversial kind of co-operation, reciprocal altruism or reciprocity. And finally, I briefly describe selected case studies of birds and mammals to illustrate that the roles of reciprocity, broadly defined, and kinship vary in importance from species to species in the development and maintenance of complex co-operative behaviour.

Ecological bases of group-living

Co-operation can only develop among individuals that are in close proximity. It also is most likely to occur when individuals have the opportunity to interact repeatedly, but for a number of times unknown to them (Axelrod & Hamilton, 1981). Since dying of 'old age' is rare in most wild animals, and is unpredictable in any case, both of these conditions are typical of nearly all stable societies.

The critical precursors to the development of co-operative bonds are those factors that lead to group-living in the first place (e.g. Waser, 1988). Such factors are ecological in nature. Alexander (1974) recognized only three that can override the inherent disadvantages of group-living. Group-living is favoured when it critically increases (1) protection from predators, (2) access to food, or (3) access to an important and limited resource. These categories are not mutually exclusive and for many species more than one may be important. A few examples of each type of benefit will illustrate the comprehensiveness of Alexander's short list.

Protection from predators

Avoidance of predators is one the most widespread factors leading to group-living in vertebrates (Alexander, 1974), although it may usually be the least obvious to the researcher. Co-operative behaviours directly related to this selective pressure include alarm calling, mobbing behaviour, sentinel behaviour, and group defence.

The role of predation as a selective agent favouring group-living is widely recognized for mammals, especially large species, where predation is conspicuous and dramatic to the human observer (e.g. Jarman, 1974). It can also be an equally important force behind the sociality of small mammalian species (e.g. Rood, 1983, 1986). However, predation as a principle factor molding complex, co-operative avian social systems may be less widely appreciated. Nevertheless, again it may be the single most important factor favouring group-living in small birds, including co-operatively breeding species (e.g. Ford et al., 1988).

Certain lineages of birds, such as the New World jays, some of the African laniid shrikes, and the babblers in Africa, Asia, and Australia, are characterized by group-living and co-operative breeding (e.g. Zack & Ligon, 1985a). What ecological factor favouring sociality do all of these species share – all forage on the ground in open woodland. This mode of life in this kind of habitat apparently makes these birds extremely vulnerable to predators, particularly avian ones. Supporting this suggestion is the fact that many of these species possess highly sophisticated sentinel systems (e.g. Gaston, 1977; McGowan, 1987). Individual flock members take turns scanning the environment from elevated perches, while other group members forage below. Not only do adults take turns as look-outs, they also recognize the alarm calls of individual group members, and ignore those of youngsters, which are likely to sound the alarm in response to inappropriate stimuli (McGowan, 1987).

Another, similar line of evidence suggesting that ground-foraging in open woodland is usually related to group-living can be seen in certain Australian birds. Both co-operative and non-cooperative species occur in treecreepers (Climactoridae) and thornbills (Acanthizidae) and some of these occur sympatrically and syntopically. In both, the group-living species inhabit open woodland and forage on or near the ground, whereas their non-social congeners either forage high in the foliage of trees or in brushy thickets (Ford *et al.*, 1988).

To summarize, a diverse array of birds that live in structurally similar habitats and that forage on or near the ground exhibit two, apparently convergent, behavioural traits related to risks of predation: group-living and sentinel behaviour. The small, diurnal dwarf mongoose has developed essentially identical responses to avian predators (Rasa, 1983; Rood, 1983, 1986).

Increased access to food

Here, too, group-living can benefit each individual in a variety of ways that range from simple and apparently incidental, to complex 'teamwork' where, to the human observer, it appears that each individual carries out an assignment (e.g. flush and ambush hunting tactics of lions, pers. obs.). A list of behavioural patterns that appear to fall under this category would include the following:

1 Group foraging, where individuals move in a co-ordinated manner to individually capture and/or consume prey, e.g. some species of pelicans (See Alcock, 1984, p. 288).
2 Information exchange concerning the location of food, e.g. swallows (Brown, 1986), vultures (Rabenold, 1983, 1986), and ravens, *Corvus corax* (Heinrich, 1988).
3 Co-operative defence of food, e.g. bell miners *Manorina melanophrys* of Australia (Loyn, 1987).

4 Co-operative group hunting, as seen in wolves (Mech, 1970), spotted hyenas *Crocuta crocuta* (Kruuk, 1972), lions (Schaller, 1972), African wild dogs *Lycaon pictus* (Frame *et al.*, 1979), and in at least one bird, the Harris' hawk, *Parabuteo unicinctus* (Bednarz, 1988).

When considering the relationship between food acquisition and group-living, most biologists probably think first of the large mammalian carnivores listed above. Recently, however, Packer (1986) has suggested that for lions group-living might not be related primarily to increased food capture. Rather, lion sociality may result instead from a combination of three factors: preference for large prey (although members of larger prides do not obtain more food on a per capita basis), openness of habitat, and high population density.

Whether or not Packer's suggestion will prove to hold true for the other large, social mammalian carnivores, increased food intake per individual as group size increases has been shown for the co-operatively hunting Harris' hawk (Bednarz, 1988). In the southwestern United States, these hawks live in groups of up to six. During the breeding season social units apparently contain only a single breeding pair, plus from zero to three non-breeding helpers. Helpers exert no detectable effect on production of young hawks (Bednarz, 1987). Harris' hawks do not display territorial behaviour and vacant, previously occupied territories are regularly present. Similarly, none of several other possible ecological constraints appear to play a role in the unusual sociality (for a bird of prey) seen in this species (Bednarz & Ligon, 1988). Rather, all group members benefit through co-operative hunting (Bednarz, 1988). The hawks feed principally on desert cottontails, (*Sylvilagus auduboni*), and black-tailed jackrabbits (*Lepus californicus*), both of which are swift, elusive, and potentially dangerous prey. (An adult jackrabbit is about three times as heavy as a male hawk.) Capture rate goes up with increased group size and it does so to a degree that leads to more food per bird per unit time. Thus for Harris' hawks, group-living is clearly related to procurement of food. It is probable that this benefit is most critical during the autumn and winter, when callow young-of-the-year are present in many groups. Young hawks require time and experience to perfect techniques appropriate for the safe capture of hard-kicking prey larger than themselves. In addition, they are allowed first access to the carcasses, which, as in wild dogs (Malcolm & Marten, 1982) and wolves (Mech, 1970), strongly suggests that extended parental care is an important aspect of group-living and group-hunting (Bednarz & Ligon, 1988).

Access to a critical and limited resource
This general benefit, like the preceding two, can be simple at one extreme and complex at the other. The resource can be either unmodified by

the users (e.g. sleeping cliffs of Hamadryas baboons, *Papio hamadryas*, (Kummer, 1968)) or constructed by them. Also included here are 'resources' such as wasp nests that protect the nests of yellow-rumped casiques, *Cacicus cela* (Robinson, 1985). The most common resources are physical structures that are critical to the species that use them, and that are inheritable, such as secure dwelling sites. For some species the time and energy devoted to constructing a burrow system or cavity dramatically increases the value of an area. Some rodents use the same burrow systems for many years (e.g. Waser, 1988). Armitage (1988) has recorded burrows of yellow-bellied marmots (*Marmota flaviventris*) being used for at least 24 consecutive years. He also cites evidence that rodent burrows may vary greatly in quality, and thus in value to the animals. Similarly, red-cockaded woodpeckers (*Picoides borealis*) excavate cavities in living pines. Such cavities may require a year or so to complete, thus the effort expended makes them very valuable (Walters *et al.*, 1988). Again, the same cavity may be passed from generation to generation for many years. The same pattern probably also holds true for the lodges of beavers (*Castor fiber* and *C. canadensis*).

For some other social species the critical resource is not self-constructed, but is just as important to the welfare of the individuals that depend on it. The critical limited resource is roost cavities for green woodhoopoes (Ligon & Ligon, 1988). These birds apparently require cavities to protect them from low night-time temperatures (Ligon *et al.*, 1988). However, cavities are in short supply, partly as a result of the presence of numerous other species that also require them. Because of the requirements for cavity roosts by the woodhoopoes, and because secure, vacant cavities are scarce, the young birds remain indefinitely in their parents' territory, where known roost sites are available. This dependence on cavities probably is also the basis for the extremely conservative dispersal of green woodhoopoes, when they do emigrate from their natal territories: all 33 male woodhoopoes that had attained breeding status did so either in their own natal territory or in a territory that was one or two territories away from their natal territories. Similarly, 33 of 38 females bred either in their natal territory or only one to two territories away. Dispersal decisions probably are based in part on knowledge of the location, number and quality of cavities in neighbouring territories (Ligon & Ligon, 1988). Thus a physiological trait which requires that the birds pass the night in cavities has had a major effect on the social behaviour and population structure of green woodhoopoes.

Natal philopatry

Students of complex social organization sometimes focus on benefits of group-living without considering the ecological factors that initially promote sociality. However, as Waser & Jones (1983) point out, an important part of understanding how social groups have arisen is recognizing the conditions that lead to natal philopatry. For this reason I discuss here in some detail the selective backgrounds that favour natal philopatry. The ideas presented may apply equally well to birds and mammals.

Natal philopatry, where a young animal remains in the territory of its birth until physiological maturity, or even for its lifetime, is characteristic of co-operative breeders (e.g. Rood, 1983; Emlen, 1984), although it is not unique to them. Typically only one sex or the other exhibits long-term natal philopatry (usually males in birds, females in mammals, Greenwood, 1980). In the avian literature, in particular, the most frequently invoked explanation for natal philopatry is the 'Habitat saturation' hypothesis, first proposed over 20 years ago (Selander, 1964). The idea is that young birds are 'forced' to remain in their natal territories as a result of an absence of suitable, unoccupied habitat. Stacey & Ligon (1987) have considered the ecologies and dispersal patterns of a number of co-operatively breeding birds, and have concluded that the habitat saturation model is not well supported (e.g. K. N. Rabenold, 1984; Zack & Ligon, 1985b). They have developed an alternative hypothesis to account for natal philopatry in co-operatively breeding birds (Stacey & Ligon, 1990). This hypothesis is termed the 'benefits of philopatry' model, to emphasize the point that an important benefit or benefits, reflected in lifetime gains in fitness, may be obtained by remaining in the natal territory, and to contrast it with the 'habitat saturation' model which predicts that young birds would be better off, in terms of lifetime fitness, if they could disperse and breed at one year of age (e.g. Koenig, 1981; Koenig & Mumme, 1987; Emlen, 1982; Emlen & Vehrencamp, 1983; Woolfenden & Fitzpatrick, 1984). The benefit gained can either be access (1) to a critical physical resource, which can be natural or constructed by the animals, or (2) to other individuals, when group size effects are important to the well-being of each group member (see preceding section). The second part of the Stacey–Ligon model emphasizes high variance among territories, as reflected by various fitness parameters of the occupants. This perspective de-emphasizes overall population averages for various measures of fitness and focuses instead on the differences among individuals in lifetime fitness based on the territories they occupy.

For many social species, natal philopatry seems to be based on access to a critically important physical resource (Table 2.1). Why a particular species developed a strong dependence on a particular resource, while others have not, may in some cases be the result of an evolutionary idiosyncrasy. For example, the co-operatively breeding anteater chat (*Myrmecocichla aethiops*), a small open-country African passerine bird, depends on holes made by aardvarks, (*Orycteropus afer*). Other sympatric small birds are not co-operative and do not require such resources. Once a dependence on a restricted resource is established, it can create a number of effects. For a young anteater chat, continuing access to aardvark holes is essential, and the

Table 2.1. *Selected examples of natal philopatry and sociality based on a critical and inheritable resource.*

Constructed Physical Resource	
Acorn woodpecker	granary tree
Red-cockaded woodpecker	roost cavities in living pines
Banner-tailed kangaroo rat	mound system
Yellow-bellied marmot, and many other terrestrial rodents	burrow system
Beaver	lodge and dam
Non-self Constructed Physical Resource	
Green woodhoopoe	roost cavities in trees
Anteater chat	aardvark holes
Hamadryas baboon	sleeping cliffs

surest means of maintaining such access and having holes as an adult is not to disperse, but rather to remain on its natal territory. Second, because of the variation in numbers and/or quality of the essential resource, large variation in quality among territories also exists.

This argument is strengthened by comparing co-operative and non-co-operative species. Comparisons of variation among territories in several measures of fitness (total young birds produced over time, young produced/breeder, mean young produced/year, mean lifespan of breeders) showed that coefficients of variation were significantly greater in two co-operative species than in a non-co-operative one (Stacey & Ligon, 1990). What this suggests, in brief, is that the essential precursor of co-operative breeding – natal philopatry – occurs in species with an unusually high reliance on some resource (sometimes this 'resource' is other group members, e.g. Rabenold, 1984). Territories vary in the amount or quality of the critical resource which then leads to great variation in fitness among breeders on different territories. An expected outcome of this is that natal philopatry too will vary from territory to territory, with many young animals remaining on the high quality territories and most or all dispersing from the low quality ones.

Demography

The preceding discussion addresses those factors behind the first and possibly most critical precursor of co-operative living – close physical proximity of individuals for extended periods of time. In addition, in social as in other species, the demographic environment can also be viewed as an important agent of selection (e.g. Ligon, 1981, 1983; Rowley, 1983; Woolfenden & Fitzpatrick, 1984). The mortality rate characteristic of a species or population also can be important to the development of co-operative behaviours (Trivers, 1971; Alexander, 1974; West-Eberhard, 1975; Axelrod & Hamilton, 1981). Trivers (1971) first pointed out that a long lifetime was one condition that favoured the evolution of reciprocal altruism, and Axelrod & Hamilton (1981) emphasized the importance of a high probability that two interacting individuals will meet again, but for an unknown number of times. Many co-operatively breeding birds and mammals meet the two requirements for reciprocity mentioned here – low annual rates of mortality relative to similar species (e.g. Zack & Ligon, 1985a) and a high probability of interacting repeatedly over an unknown period of time, because they live in stable social groups.

Alternatively, under certain specific conditions, a relatively high mortality rate might favour an initial act of co-operation which could set in motion the *tit for tat* scenario developed by Axelrod & Hamilton (1981). This seems especially relevant for certain co-operatively breeding species where unrelated, same sex individuals live in the same social units and exhibit co-operative behaviours (e.g. Ligon & Ligon, 1983; Rood, 1983; Reyer, 1980, 1984). Earlier (Ligon, 1983), I suggested that in certain social environments, where benefits of group-living are very high, individuals without genetically related allies might be more willing to form co-operative alliances with non-relatives than in populations where mortality is low and related allies are nearly always available. In green woodhoopoes in Kenya, for example, annual mortality is high, about 40 per cent for males and 30 per cent for females (Ligon & Ligon, 1988). This means that individuals often lose their same-sex flockmates. In addition, females in particular sometimes migrate singly from their natal territories. Each of these two factors leads to the regular occurrence of lone non-territory-owning female woodhoopoes. Two females commonly merge to colonize territories without females, and at least 50 per cent of these coalitions are composed of non-relatives (Ligon & Ligon, 1988: Table VII).

Whether or not the causal factors suggested here are correct, unrelated individuals of the same sex do form social bonds and exhibit co-operative behaviours on a regular basis in several species that also exhibit high rates of annual mortality (e.g. acorn woodpecker, *Melanerpes formicivorus*, Stacey, 1979; pied kingfisher, *Ceryle rudis*, Reyer, 1980, 1984; lions, Packer & Pusey, 1982; green woodhoopoe, Ligon & Ligon, 1983, 1988; dwarf mongoose, *Helogale parvula*, Rood, 1983, 1986). In the mongooses, woodhoopoes and woodpeckers the older individual is always dominant and becomes the breeder in the occupied territory if both survive and remain in the territory until breeding occurs.

The question of reciprocity

Possibly the most interesting, and certainly the most controversial, aspect of the study of co-operation is the phenomenon labelled reciprocal altruism or reciprocity (Trivers, 1971; Alexander, 1974; West-Eberhard, 1975; Axelrod & Hamilton, 1981; Taylor & McGuire, 1988), hereafter abbreviated as RA. From its inception to the present, RA has generated disagreements among writers considering this subject (see Rothstein & Pierotti, 1988). Similarly, the relationship between complex forms of co-operation such as RA and kinship has remained controversial for well over a decade. It, therefore, seems appropriate to provide a brief historical review of the factors behind these disagreements and to consider current thinking on the subject.

In large part, arguments in the literature have been based on terminology and on different shades of meaning that different authors give to certain words. The word/concept responsible for much of the contention is 'altruism'. (This word must be responsible for more arguments and misunderstandings in sociobiology than all others combined.) Trivers (1971) opened his classic paper on RA with this definition: 'Altruistic behaviour can be defined as behaviour that benefits another organism, not closely related, while being apparently detrimental to the organism performing the behaviour, benefit and detriment being defined in terms of contribution to inclusive fitness.' Trivers went on to say: '. . . that under certain conditions natural selection favours these altruistic behaviours because in the long run they benefit the organism performing them' (Trivers, 1971, p. 35). Note that the 'apparently detrimental' behaviour, in terms of inclusive fitness, actually leads to a long-term benefit to inclusive fitness. These quotes may illustrate the origin of the most persistent problem: some writers have emphasized the short-term, (apparently) costly beneficent behaviour, while others have concentrated only on the eventual, long-term outcome of the behaviour. This difference, together with the small costs of most helpful behaviours (there usually is no reason whatsoever to assume they actually involve a loss of direct fitness) recorded by empiricists, is the basis for much of the variation in perspective concerning RA, even though other early influential papers on the subject were clear concerning this point. Alexander (1974) equated the term reciprocity with Trivers' RA, and he then stated: 'Reciprocity does not actually involve altruism, except in some temporary sense. In systems of reciprocity each individual is in effect gambling that his investments will increase his inclusive fitness, perhaps usually through benefits returned to his own phenotype, but feasibly through benefits reciprocated to his offspring or other relatives as well' (Alexander, 1974, p. 337). West-Eberhard (1975) recognized four classes of intra-specific mutualisms, including mutualism maintained by reciprocal-altruistic selection. Each donor of a beneficent behaviour expects future reciprocation '. . . so as to result in net gain in classical [personal or direct] fitness of both participants' (West-Eberhard, 1975, p. 19). In all of these foundation papers it is clear that RA was viewed primarily as a personal gain strategy. As Trivers (1971, p. 35) stated: 'Models that attempt to explain altruistic behaviour in terms of natural selection are models designed to take the altruism out of altruism.' At this stage in the evolution of the concept of RA, three reports appeared that seemed to provide empirical support for it. Packer (1977), Rood (1978), and Ligon & Ligon (1978a) described co-operative behaviours in olive baboons (*Papio anubis*), dwarf mongooses and green woodhoopoes, respectively, where donation of aid by unrelated individuals later was, or was likely to be, paid back by the original recipients.

The early 1980s saw the appearance of several theoretical papers treating the concepts of co-operation and reciprocity. Rothstein (1980) suggested that RA and kin selection were not separable phenomena. Axelrod & Hamilton (1981) demonstrated via computer simulations that under certain, specific conditions co-operation, including reciprocity, was a robust, stable, evolutionary strategy that could evolve independently of kinship. Waltz (1981, p. 588) offered a set of criteria for what he termed narrowly defined reciprocity: '(1) One individual aids another, (2) in anticipation that the recipient will return the favour, (3) benefitting the actor at some time in the future.' Wasser (1982), like Trivers (1971), emphasized that co-operation assumes no time-lag between donating and receiving aid, whereas reciprocity does refer to a time-lag between donation and receipt of aid. Several papers describing behaviours that apparently fulfilled these requirements appeared at about this time (e.g. Pierotti 1980; Emlen 1981; Ligon 1981; Ligon & Ligon, 1983; Conner & Norris, 1982; Rood, 1983; Russell, 1983).

Ligon (1983) and Ligon & Ligon (1983) recognized that the behaviour described for baboons by Packer (1977) differed from that of green wood-hoopoes and used the phrase 'opportunistic reciprocity' to indicate those cases where individuals made decisions to co-operate or not on a case-by-case basis. 'Obligate reciprocity', on the other hand, was used to refer to situations where older animals – helpers – aided younger ones, with the ecological, social and demographic environments virtually assuring that at least some of the younger animals eventually would provide aid to the original helper.

The term 'pseudo-reciprocity' was introduced by Conner (1986) to refer to situations where individual A aids B at a cost, and later A gains an incidental benefit, when B performs a behaviour primarily useful to itself. Conner's point is that because B is acting in its own interest, there will be no selection for cheating by B, thus B is not truly reciprocating the cost incurred by A. Conner's suggestion recently has been endorsed by Koenig (1988) and Rothstein & Pierotti (1988).

Recently, an entire issue of the journal *Ethology and Sociobiology* was devoted to an assessment of RA (Taylor & McGuire, 1988). A reading of those papers evaluating empirical studies describing putative cases of RA leads to an obvious conclusion: RA is in grave danger of being defined out of existence by some writers. For example, according to the review and definitions of Rothstein & Pierotti (1988), Wilkinson's (1984) study of food sharing in the vampire bat (*Desmodus rotundus*) is the only clear non-human example of RA under natural conditions. What factors have led to the conceptual decline of RA over the past 15 years?

First, as stated above, the term/concept altruism continues to muddy the waters. For some, RA requires that 'both the original act and the reciprocal act must entail a cost in direct fitness to the individuals involved' (Koenig, 1988, p. 74). Taken literally, this means that the acts must decrease the lifetime reproductive success of both parties. Presumably, Koenig actually means either that the original beneficent behaviour must involve some degree of risk to the donor that potentially decreases its lifetime reproduction, or that some sort of temporary loss of fitness occurs. (However, a net loss of lifetime fitness cannot be temporary and it is net fitness that is important. See below.) As there is no reason to assume that avian helpers lose direct fitness by helping (one possible exception, Reyer, 1984, see below), and a good deal of evidence to suggest that in most cases both the original donor and recipient gain in terms of *lifetime* reproductive output, it is not surprising that by his criteria Koenig (1988) found no cases of RA in birds. Koenig's emphasis on a short-term cost, rather than on the net effect, differs from the way earlier writers used the concept (e.g. Trivers 1971; Alexander, 1974; West-Eberhard, 1975; Ligon & Ligon, 1978a, 1983), and illustrates the nature of the 'threat' to the concept of RA. The different and sometimes apparently arbitrary assignment of costs and benefits by different authors, together with the different time-frames emphasized, form the basis for the rejection of nearly all putative empirical examples of RA. For example, Ligon & Ligon (1978a, 1983) and Rood (1978, 1983) described long-term effects of exchanges of aid among green woodhoopoes and dwarf mongooses, respectively, and, based on the definitions of Trivers (1971), and especially Alexander (1974), West-Eberhard (1975) and Waltz (1981), suggested that certain behaviours of these animals might be examples of reciprocity. However, Conner (1986), Koenig (1988) and Rothstein & Pierotti (1988) argue that woodhoopoes actually exhibit pseudo-reciprocity (Conner 1986). This leads to my next point concerning the current state of the RA concept.

Second, 'Ever since Trivers (1971) initiated it, the literature on RA has followed a tortured path. RA has been narrowly or broadly defined and nearly all proposed examples have been criticized and reinterpreted by subsequent authors as representing something other than RA' (Rothstein & Pierotti, 1988, pp. 193–4). As one whose examples have been criticized and reinterpreted, (e.g. see Rothstein & Pierotti, 1988), I believe this to be an accurate summary of the history of RA. This situation is due basically to changing perspectives over time (see above). Theorists have their views of the ways co-operation and reciprocity should work, and empiricists attempt to obtain data bearing on the theorists' original concepts. However, by the time the empiricists have published their findings and interpretations, other theorists may have changed the rules of the game and the empiricist may then be reproved for misusing the word or concept in question. This kind of situation probably is inevitable in any area of science where ideas so far outpace (as measured by number of published papers) the gathering of critical empirical data, which in this case, often require years to obtain (e.g. genealogical data).

In their overview of the current status of RA, Taylor & McGuire (1988) address the still-worsening problems of definition, and make some important suggestions, three of which I quote here: (1) 'Possibly, one could sidestep the issue of definition by recasting reciprocal altruism in other terms . . .' (2) 'For evolution it is largely the payoffs on the interactions that matter. Time delays and expectations appear irrelevant to evolutionary dynamics . . .' (3) '. . . little is to be gained by forcing a wide variety of social interactions into any specific definitional mold. At this time it seems best not to over-worry about definition, but instead to view social interactions as phenomena in their own right – keeping in mind the sense originally intended, a time-lag, the need for positive action by the co-participants, and role reversal, although not necessarily strictly in kind' (Taylor & McGuire, 1988, p. 71). Because virtually every species has a unique wrinkle in its behavioural repertoire, it is unlikely that a single set of strict criteria will match the empirical data for most species. For this reason it appears to me that the suggestions of Taylor & McGuire are appropriate and realistic. Thus, in this chapter I use RA in the broad sense, to refer to beneficent exchanges of aid over time.

An integration, rather than separation, of RA and kin selection

After the appearance of papers by Trivers (1971), Alexander (1974) and West-Eberhard (1975), students of certain vertebrate social systems became interested in the possibility of assessing the relative importance of RA and kin selection. Following Trivers (1971), they approached this problem by focusing on beneficent behaviours that occurred between unrelated animals (e.g. Packer, 1977; Rood, 1978, 1983; Ligon & Ligon, 1978a, 1983; Pierotti, 1980; Emlen, 1981; Russell, 1983). Because aid-giving behaviour occurred between unrelated as well as related animals and because significant 'repayments' subsequently were received by the donors as a result of their beneficence, it appeared that kin selection probably was not adequate to account for such exchanges of aid. However, partly because aid-giving behaviour (and all complex social behaviour, for that matter) occurs so much more frequently among relatives than non-relatives, and partly because of the changing perspectives and definitions over time, these arguments have not convincingly separated RA and kin selection. Rather, as Rothstein (1980) first argued, it appears that they usually may not be separable phenomena. Both the empirical data and the theoretical considerations presented in Taylor & McGuire (1988) suggest that this is the case. Nevertheless, focusing on helpful behaviours between unrelated individuals was valuable, for at least two reasons. (1) It slowed the uncritical rush in the 1970s to embrace kin selection as a sufficient explanation for co-operation, by showing that helpful individuals could reap critically important fitness rewards as a result of their behaviour, independent of shared genes (see Axelrod & Hamilton, 1981, refs. 13–15). (2) The fact that unrelated helpers in a variety of species benefit personally, as a result of their helping, adds support to Wilkinson's (1988, p. 98) recent reassertion of the argument by Trivers (1971) that reciprocal aid giving 'can generate a substantial selective force independent of kin selection even when performed among related animals.'

As measures of the kinship component of inclusive fitness, in particular, become ever more realistic (e.g. Grafen, 1982), attempts to separate RA and kin selection by emphasizing helpful behaviours between unrelated individuals (e.g. Ligon & Ligon, 1978a, 1983; Rood, 1978, 1983; Conner & Norris, 1982; Seyfarth & Cheney, 1984) may be replaced by the other approaches. Whether or not Rothstein's (1980) genetical arguments are correct, his suggestion that RA and kin selection are not clearly separable phenomena currently is supported by most empirical studies (but see Wilkinson, 1988).

To illustrate that pure RA and pure kin selection may be viewed as opposite and rarely realized ends of a continuum, some of the case histories discussed below are characterized by complex co-operation among unrelated animals, while others involve both unrelated and related individuals. This perspective can be integrated with Grafen's (1982) suggestion concerning comparative measurements of lifetime reproductive success plus kinship effects: '. . . simply count the number of the animal's offspring' (Grafen, 1982, p. 425). The idea here is that although animal A helps relatives to produce more of their offspring, the same or other relatives also help A to produce more of A's offspring (i.e. A's aid is reciprocated). Thus, in systems where young animals are virtually obligate helpers (Ligon, 1983), that is, where cheating is unlikely, the costs and benefits of giving and receiving help in effect cancel each other out, leaving only personal reproduction. Grafen suggests that his number-of-offspring procedure is likely to prove more suitable for field studies than for calculation of inclusive fitness proper. The number-of-offspring rule suggests that in societies where helping behaviour is the norm, the net costs and benefits of helping to a helper unrelated to the recipient may be equivalent to those of a related helper, and thus that the rewards of helping are similar in both cases. Of course, unrelated helpers may be 'exploiting' a system that originally developed among kin (see below).

Memory and individual recognition

Another separate problem with invoking RA, is the assumption that only a few mammals, such as man and some other higher primates, and possibly cetaceans, have the mental powers necessary to recall the relevant past behaviour of fellow group members (Trivers, 1971, 1985). However, evidence is accumulating that even some small birds have both the ability to recognize many other individuals and to remember past events. For example, Clark's nutcrackers (*Nucifraga columbiana*), can remember the location of caching sites for at least 90 days after they have been removed from the caching arena (R. P. Balda & A. C. Kamil, unpub. data). Under the challenging experimental conditions used, this would be an impressive feat even for humans.

Pinyon jays (*Gymnorhinus cyanocephalus*) are highly monogamous and exhibit strong, permanent pair bonds (Marzluff & Balda, 1988a,b). Apparently related to this trait, these birds may possess long-term memory with regard to individual recognition. I separated a mated pair of pinyon jays in November, such that they could neither see nor hear each other. The male was isolated with a new female, while the original female of the pair remained in a large outdoor aviary containing jays of both sexes. By April of the following year the original female had a new mate and the male had been isolated with another female for almost six months. At that time I replaced the male in the outdoor aviary, and by later on the same day he and his former mate had reunited and were exhibiting the behaviour characteristic of a firmly bonded pair. Marzluff (1988 and pers. comm.) recorded a similar case. These anecdotal observations suggest that, with regard to social relationships, we usually do not know the extent to which free-living animals can recall prior behavioural exchanges with other individuals. Among birds and mammals, this ability probably will vary from species to species, but not necessarily along taxonomic lines.

Co-operation and relatedness: empirical studies

Long-term data on individually recognizable animals are now available for a considerable number of birds and mammals. These data permit preliminary, empirically-based generalizations concerning the bases and costs and benefits of group-living. In this section I consider the relationship between kinship and complex co-operation, in particular RA, broadly defined (by more restrictive definitions most of these cases involve either pseudo-reciprocity (Conner, 1986) or return benefit altruism (Trivers, 1985)). This approach produces some strange ecological-behavioural bed-fellows, such as unisexual teams or coalitions of invading lions and green woodhoopoes. My hope is that across-species comparisons will illuminate some fundamental patterns of long-term or recurring co-operation among birds and mammals.

The social units featured in this section range from (1) different species, to (2) unrelated and related conspecifics, where direct personal gain appears to be the primary force behind the behaviour of donors of aid, to (3) close kin, where indirect fitness benefits may be the prime advantages to aid-givers. This approach is meant in part to underscore Wilkinson's (1988, p. 98) suggestion that repeated exchanges among animals in relatively large groups – such as lions, elephants, dwarf mongooses, brown hyenas, and others – are likely to have an RA component of inclusive fitness that exceeds the component due to kin selection. It also empirically illustrates that the relative importance of these two features varies among group-living animals.

Co-operation without kinship

Hornbills (*Tockus deckeni* and *T. flavirostris*) and dwarf mongooses

These animals have developed a mutualistic anti-predation relationship (Rasa, 1983). This is an especially interesting case because, unlike grooming or food-sharing, major, life-and-death costs are involved. The hornbills accompany groups of mongooses as they forage and warn the mongooses of the approach of predators. The birds gain primarily by the mongooses' ability to locate and flush prey, and the mongooses too gain increased foraging efficiency, largely as a result of the reduced requirement of guarding or sentinel behaviour. Of special significance here are the facts that (1) the hornbills warn of predators dangerous to the mongooses, but not to the birds; i.e. the benefit to the mongooses is not merely an incidental side-effect of a bird–bird system, and (2) the birds do not warn of the approach of raptor species that are not dangerous to the mongooses; i.e. the hornbills are not simply giving a generalized predator-alarm call. In addition to the specific nature of the alarm system, the hornbills wait for the mongooses to appear before beginning to forage, and vice versa.

The degree of interdependence and mutual communication between the hornbills and mongooses is striking. If the actors in this interplay were conspecifics, the co-operation (including RA, by most writers on the subject) shown probably would be attributed to kin selection. The hornbill–mongoose relationship illustrates why Trivers (1971) chose to illustrate his perception of RA by use of interspecific examples.

Coalitions or teams of manakins of the genus *Chiroxiphia*

Male manakins of several species perform multi-male courtship displays at traditional or permanent lekking sites (Foster, 1977, 1981; Snow, 1977). The level of co-operation among a displaying group of males is both high and important, in that females are attracted to the total display performance, produced by all male team members. Perhaps the most interesting point is that only one of the males, the dominant, normally mates with the female, thus the subordinates receive no immediate benefits to compensate them for the time and energy they expend in the displays. Moreover, there is no evidence that the males are genetically related, and some good reasons, based on manakin breeding biology, to assume that normally they are not. If not, what may the basis of this system be?

1 At this stage in the evolutionary history of manakins, groups of males are probably required to attract any females (i.e. lone males have no chance to mate).
2 In a female-attracting group, there is always the chance that a subordinate male will have the opportunity to 'steal' a copulation.
3 Lekking sites are traditional and individual females return to them.
4 Dominant males generally are older, thus, other things being equal, they normally will die before the subordinates.

Moreover, the dominant acts as a sentinel and as a result may be more conspicuous to predators. When a high-ranking bird disappears, all those subordinate to him move up a step in the hierarchy. 'Thus, if a male can live long enough, he eventually will acquire dominance on a lek of his own' (Foster, 1981, p. 176. Also see Wiley & Rabenold, 1984). These manakins provide perhaps the best known example of a long-term, co-operative waiting strategy where the eventual fitness pay-offs apparently compensate for an unpredictably long wait. Once an individual has attained alpha status, it mates with virtually all the females attracted to its lek-site for the rest of its life.

Pied kingfisher, secondary helpers

Pied kingfishers have two types of male nest helpers: primary ones, that feed the offspring of one or both parents, and secondary helpers, that feed unrelated offspring (Reyer, 1980, 1984). For this reason I consider the two types separately. Secondary helpers, which make up 50 per cent of the helper population where food is scarce, are allowed to join the breeding pair only after the young have hatched. At another, richer site, unrelated, unmated males also attempted to feed nestlings, but never were allowed to do so. The behaviour of breeding males at both sites suggest that secondary helpers potentially are costly to male breeders in some way, and that only in situations where their aid is critical are they accepted. The contributions of one helper *doubles* the number of young fledged/nest, a conspicuously large benefit to the breeding pair.

Since indirect inclusive fitness benefits are not a possible factor, what does the secondary helper gain as a result of its efforts? Evidence relating to this question is meager. However, several potential gains have been identified. First, females are fewer in number than males, thus male–male competition for females is intense. Second, mortality of both sexes is unusually high (annual turnover 65% for males, and 75% for females) as compared with all other small, co-operatively breeding birds for which comparable data are available. Third, both sexes appear to be highly philopatric, returning annually to the same colony to breed for as long as they live. In an unpredictable social and demographic environment these three factors favour the employment of a variety of alternative tactics to obtain and maximize future breeding success. Apparently, attachment to a site in a breeding colony leads to a high likelihood of reoccupancy of that site in the following year. This can lead to earlier breeding and a greater probability of acquiring a secondary helper, which as noted above, doubles reproductive success. Despite the high disappearance rate of both sexes, secondary males sometimes acquire the female helped in year A as a mate in year B. In addition, some of the young that the (former) helper aids may become helpers for it in year B, providing strong fitness benefits to the former helper, as described above (Reyer, 1980, 1984).

Co-operation occurring among both kin and non-kin

Dwarf mongoose and green woodhoopoe

The co-operative breeding system of the dwarf mongoose (Rood, 1978, 1983, 1986) is strikingly similar to that of certain co-operatively breeding birds, such as the green woodhoopoe (Ligon & Ligon, 1982, 1983, 1988). Although the primary ecological factors promoting group living probably differ in these two African species (predator avoidance in the mongoose and limited roost sites in the woodhoopoes; see above), their social biology is probably about as similar as it could be, given that one is a mammal and the other a bird (Table 2.2).

Two major categories of co-operative behaviour shown by the mongooses and green woodhoopoes are helping behaviour and unisexual team migration (Table 2.2). Helping behaviour – guarding and feeding small young, and guiding and protecting them when they begin to move with the group – is highly developed in both species, and in both, unrelated helpers uncommonly, but regularly, provide high-quality care to juveniles. In each species, both related and unrelated helpers can eventually reap the same personal benefits, provided that they live long enough to attain breeding status. If this occurs, the young animals that they helped to rear 'repay' them in one or more ways: (1) by assisting in the retention of breeding status by the older, former helpers in the original territory (e.g. helping to defend the territory from invasion by a would-be breeder and its allies), or (2) by moving to a new

Table 2.2. *Shared sociobiological traits of group-living dwarf mongooses and green woodhoopoes.*

1 Diurnal, group-foraging insectivores.

2 Groups contain a single, dominant breeding pair, breeding suppressed in other group members.

3 Groups also contain subordinate adults, yearlings and juveniles.

4 Most genetic lineages that persist for extended periods of time are matrilineal.

5 High reproductive rates: up to three litters or broods per year.

6 Non-breeding helpers of both sexes provide food, protection and guidance to young animals

7 Unrelated as well as related helpers of regular occurrence.

8 Long-term survival key to attaining breeding status.

9 Natal philopatry by some non-breeders of both sexes, which will eventually inherit breeding status if they live long enough.

10 'Team' migrations by two or more same-sex group members to neighbouring groups or unoccupied areas.

area with the former helper and assisting it in acquiring and holding breeding status there, and by (3) helping to feed and guard the offspring of the former helper. In the future, some of those young animals will in turn provide the current helper with the same kinds of assistance listed above.

Similarly, when two or more same-sex individuals, again either related or unrelated, merge to occupy a territorial vacancy, only one, the older and dominant, breeds right away, while the subordinate must wait to breed (Ligon & Ligon, 1983). There is no evidence, however, that the subordinate, individual is behaving in a fitness-reducing manner, given its circumstances. Rather, to use the overworked phrase, it is 'making the best of a bad job'. If the beta individual lives long enough, it eventually will attain breeding status in the territory and subordinate helpers (youngsters that it helped to rear while serving as a helper itself) for its own breeding tenure.

Although both kin and non-kin are involved in the social systems of dwarf mongooses and green woodhoopoes, the systems we see today probably developed as extensions of family units. The following scenario describes the probable sequence of events leading to the current social system of each species, including the regular occurrence of unrelated helpers.

In both the mongooses and the woodhoopoes a demographic feature – high mortality – appears to have set the stage for the first and most critical step in the hypothetical evolutionary development of (1) co-operative breeding, and later, (2) reciprocity among non-kin. In the woodhoopoes most mortality occurs at roost cavities. Because safe roost sites are scarce, mated pairs of woodhoopoes allowed their grown offspring to remain in the parents' territory and to use roost sites therein (Ligon & Ligon, 1983, 1988). In mongooses predation probably is an even more serious problem, with group-living being the response to it (Rood, 1986). Retention of matured offspring, beneficial to both the breeders and to grown young, had at least two critical consequences for the breeders. (1) With more than two adults per group, opportunities for territorial expansion were increased relative to neighbouring pairs without allies. This meant an increase in territory quality. In addition, more eyes and ears were available to detect predators. (2) Additional adults in the group represented potential help to the breeders: potential defenders of the territory and potential feeders and protectors of the youngsters. The probable reduction of the costs of breeding to the alpha male and female, as a result of the presence of additional individuals, meant greater potential annual reproduction. In addition, when opportunities arose to occupy a new territory, older adult non-breeders in a social unit now had allies composed of same-sex, younger, subordinate group members, whose assistance increased the migrants' chances both of obtaining and retaining ownership of a new territory. This clearly benefits the original breeders because one or more (in sequence) of their offspring gain breeding status as a result of unisexual sibling-group migration. Depending on unpredictable circumstances, each individual migrant also stands to gain by this group movement (Ligon & Ligon, 1983).

Once interdependence became established among related individuals for territorial enlargement and defence, reproduction, and later, acquisition of new territorial space, the unaided pair, or single individual, was placed at an overwhelming competitive disadvantage. However, lone individuals without supportive kin do occur in both species because of high mortality during dispersal or displacement of a singleton from a territory by a group. Thus, because of the group-based population structure, the best option open to most lone mongooses and woodhoopoes, if they are to gain territory ownership and eventually to breed, is to procure subordinate allies of their sex, as well as a mate. This they can sometimes do by joining social units containing no relatives (mongoose) or by allowing younger floaters of their sex to join them (woodhoopoe). These unrelated helpers provide critical aid in feeding and guarding the offspring of the dominant breeders. By such co-operation they in turn will eventually obtain the benefits described above, provided that they live long enough.

Lion

As in the preceding two species, unisexual, co-operative coalitions occur in lions of both sexes. In male–male coalitions 42 per cent contain non-relatives. These male teams co-operate to gain access to groups of females and after having gained females, to defend them against other groups of males (Packer & Pusey, 1982; also see subsequent discussion and elaboration by Bertram, 1983, and Packer & Pusey, 1983).

I was struck by certain similarities between coalitions of male lions and those of male and female green woodhoopoes (Ligon & Ligon, 1983, 1988). In both species mortality is high, thus many individuals lose the option of co-operating with a relative. In a social environment where aggressive team competition for territorial space and breeding status is the norm, the best remaining option to a singleton without same-sex relatives is to form a co-operative coalition with a non-relative, as discussed above.

Since the classic study by Schaller (1972), most studies of lions have taken place in the Serengeti, Tanzania, and all have characterized prides of female lions as invariably containing only closely related individuals. Thus it was surprising to learn that in another area, females form prides consisting of unrelated animals (M. J. Owens & D. D. Owens 1984). This is an important as well as interesting point, in that it clearly suggests that group-living in female lions, as in males, has advantages sufficient to promote co-operation among non-kin. (See Packer, 1986, for a recent discussion of the advantages of group-living in lions.)

Coalitions of unrelated females are rare in lions and common in green woodhoopoes, while teams of unrelated males are common in lions and rare in the woodhoopoes. These apparently contrasting patterns appear to be based on a common causal factor, namely, the risks associated with dispersal. Among lions the males are the ones that typically migrate from their natal prides, whereas among woodhoopoes some females disperse greater distances than any males are known to do (see Greenwood, 1980, and Ligon & Ligon, 1988). This re-emphasizes the point made above. When mortality is high and when intra-sexual competition typically occurs among coalitions or teams, selection should favour lone individuals that are willing to co-operate with non-relatives.

Brown hyenas, *Hyaena brunnea*

This species exhibits a unique social system where female clan members give birth to litters in isolated dens (D. D. Owens & M. J. Owens, 1979, 1984). Later, after two and a half to three months, mothers carry the cubs to a communal den where all young are provisioned by clan females, plus a few males. This system, where distant (second cousins, $r = 0.03$) as well as close female kin provide food for the cubs, is similar to that of the dwarf mongoose and green woodhoopoe in the following way: females who received help as cubs later provisioned their helpers' offspring. This cross-generation aid by females appears to be driven in part by the natal philopatry of females. In addition, adult females of similar ages provisioned one another's cubs. In contrast, subadult males did not always provision their helpers' offspring, although they were as closely related to the cubs as their sisters were. This probably is related to the fact that males almost always leave their natal clan, and thus have no 'expectation' of later repayment.

Because mothers of small cubs frequently are killed by other predators, this system also can be viewed as a form of 'life insurance'. Rather than dying when their mothers are killed, orphaned cubs are reared to independence by other clan members (M. Owens & D. Owens 1984).

Blood sharing in vampire bats

Female vampire bats live in groups that occupy hollow trees. Female offspring tend to remain in their maternal social groups, whereas males do not do so, thus female groups contain some close relatives, as well as some unrelated individuals. (Average coefficients of relatedness within female groups are low, averaging 0.11, or less.) Vampires regurgitate blood to fellow group members that did not obtain food on their previous foraging bout. Such donations are important to the future survival of recipients and are distributed preferentially both to close relatives and to individuals having high past roost associations. Captive, unrelated roostmates also exchange food (Wilkinson, 1984).

By use of both computer simulations and empirical field data, Wilkinson (1988) investigated the effects of vampire bat food sharing on their probability of survival. An intriguing outcome of these analyses was that reciprocity contributed more to an individual's inclusive fitness than kinship, regardless of relatedness. Wilkinson concluded that RA is important independent of kin selection even when it occurs among related animals.

The evidence presented by Wilkinson (1984, 1988) suggests that, in an evolutionary sense, the blood sharing of vampires may be based on an extension of maternal care; e.g. 33 per cent of bats less than two years old failed to feed, while only 7 per cent of older bats did not obtain meals. Of 110 bouts of regurgitation, 77 involved a mother and her dependent offspring, while the other 33 occurred between either two adults or an adult and juvenile (Wilkinson, 1988). If this is correct, it supports the suggestion made above for dwarf mongooses and green woodhoopoes, that complex co-operation among kin, perhaps especially parents and offspring, can be viewed as 'preadaptations' that set the stage for co-operation between unrelated individuals.

Strong kinship effects: kin selection, a major evolutionary force

For kin selection properly to be invoked as a major factor in the evolution of complex co-operation, such as aid provided by helpers, it must be shown that the beneficent behaviours of individuals towards kin make a difference; i.e. more related individuals must be produced, or aided relatives must have increased survival and reproductive success of their own (Grafen, 1982). Even when interactions among kin yield these results, however, other factors also may be involved. The studies of Woolfenden & Fitzpatrick (1984) and their co-workers on Florida scrub jays (*Aphelocoma c. coerulescens*) clearly illustrate this point. When non-breeding helpers are present, survival of jays in *all* social categories (breeders, other helpers, immatures) increases. However, the food provided to nestlings by helpers (helping behaviour *per se*) does not affect the number of young jays fledged. Rather, the effect of helpers on productivity is due to reduced predation on young birds both before and after they have become integrated flock members. Thus, in a strict, narrow sense, it may be erroneous to conclude that helpers help. Rather, a general relationship exists between group size and survival, and this general benefit is shared by all flock members. (On the other hand, a broader interpretation would be that since flock members normally are relatives and since helpers positively affect the number of young jays surviving to maturity, that a kinship effect (Grafen, 1982) is important.) For this kind of reason, in this section I consider two examples of co-operation where helping behaviour *per se* does make a great difference (case 1), and where explanations of helpers based on indirect fitness benefits are particularly compelling (case 1 and 2).

Pied kingfisher, primary helpers

Primary helpers typically are one-year-old offspring of one or both of the breeders, thus they are closely related to the nestlings they provision with fish. These helpers provide significantly more and higher quality fish to nestlings than secondary helpers do. The effort expended in capturing and delivering fish is costly; Reyer (1984) found an inverse relationship between energy expended in helping and survival to the next year: 73 per cent of the secondary helpers returned, versus only 47 per cent of primary helpers. (For comparison, 70 per cent of non-helping males returned the next year, suggesting that secondary helpers do not work hard enough to affect negatively their well-being.) Based on the metabolic costs of feeding nestlings, Reyer (1984) makes the only convincing case to date for a co-operatively breeding bird that helpers are actually altruistic, in that they expend so much effort helping that their probability of future survival is measurably lowered; i.e. it is 20 per cent more likely that the lifetime direct fitness of a primary helper will be 0, directly as a result of its helping effort (Reyer, 1984, Table III). This fitness cost, plus the magnitude of a helper's effect (doubling the number of fledglings/nest), make this study a thus far unique example among vertebrates of altruistic, kin-selected helping behaviour as it was often envisioned in the early 1970s. (Unfortunately, however, age differences between primary and secondary helpers leave even this case open with regard to the causes of the differential mortality and thus to the question of altruism.)

Silver-backed jackal, *Canis mesomelas*, helpers

In this species, all helpers were full siblings of the pups and almost all helpers remained in their parents' territories and helped for only one year before migrating (Moehlman, 1979). The importance of helpers is reflected in pup survival. On average, a helper added 1.5 surviving pups to the litter. Unlike most other helper systems, the *per capita* effect of helpers also was pronounced: 1 pup/litter, versus 0.5 pup/litter in unhelped pairs. The consistently close genetic relationship between helpers and pups, the strong positive effect of helpers, and the evidence that differences in food availability among territories probably did not differ to an important degree, together provide a convincing case in these jackals for the importance of indirect fitness benefits to the helper.

Discussion and summary

Sociality is firstly a behavioural response to particular ecological conditions and the specific benefits associated with group-living vary from species to species. Nevertheless, some patterns do exist. Animals live in groups for one or more of three basic reasons: response to predators, increased access to food, and increased access to some critical and limited resource (Alexander, 1974). Once group-living develops, co-operation of many sorts becomes possible, and in some birds and mammals complex co-operation, including delayed repayment of aid rendered, reciprocal altruism or reciprocity (RA), occurs (by the definitions of Trivers, 1971; Alexander, 1974 and West-Eberhard, 1975; Axelrod & Hamilton, 1981, and by Taylor & McGuire 1988).

The subject of RA in non-humans has been a contentious one, for several reasons. First and foremost, to date no consensus has been reached concerning how it is be defined and recognized (Taylor & McGuire, 1988). Many empirical observations fit those definitions of RA that do not involve 'altruism' (loss of direct fitness), while, when altruism is required, almost no cases of reciprocity are recognized (e.g. Conner, 1986; Koenig, 1988; Rothstein & Pierotti, 1988). According to Rothstein & Pierotti (1988), for example, the only example of RA in free-living animals is the food-sharing of vampire bats (Wilkinson, 1984). However, even this case is open to question, if altruistic loss of direct fitness is required to invoke RA, as Koenig (1988) and Rothstein & Pierotti (1988) suggest. In his simulation studies, Wilkinson (1988) assumed that food-sharing had no direct cost, and he supported this assumption by noting that captive bats invariably do not share blood if they do not have sufficient reserves to survive at least 24 hours.

In line with the strict-narrow and flexible-broad perspectives of RA discussed above, de Waal & Huttrell (1988, p. 103) suggest two categories of RA, 'symmetry-based' and 'calculated'. The former involves exchanges between closely bonded individuals who help each other without stipulating equivalent returns. Here the exchange of aid may be one-sided for a period, but will often balance out over a lifetime. The second type is regulated by feedback: the continuation of helpful behaviour is contingent upon the partner's reciprocation. These categories correspond fairly closely to Ligon's (1983) 'obligate' and 'opportunistic', and to Sahlins' (1965, in de Waal & Luttrell, 1988) 'generalized' and 'balanced' reciprocity in human gift-exchange relationships.

As discussed above, whether or not reciprocity even exists in animals depends on the definition followed. It is obvious that in a variety of birds and mammals important exchanges of aid over time do occur between pairs of individuals. Because aid is rendered and later repaid in one form or another by the original recipient, a broad, rather than narrow, definition of reciprocity may be the most productive approach to the study of sociality and co-operation (Taylor & McGuire, 1988). The giving and receiving of aid over time is a fascinating aspect of social behaviour, whatever label is applied to it.

I review certain co-operative behaviours in a variety of birds and mammals in an attempt to make three points. First, complex exchanges of aid between different species (e.g. Rasa, 1983) unambiguously illustrate that complex co-operation, including RA, can develop and be maintained in the absence of kinship.

Second, co-operation occurs most commonly among kin, and in these situations it usually may not be possible to separate and assess the relative importance of each selective agent (Rothstein, 1980), although Wilkinson's (1988) computer simulations provide a promising technique for doing so. For most species, studying the inter-relationship between co-operation and kinship may usually be a more productive framework than attempting to tease them apart, as has been attempted in the past by emphasizing the co-operative interactions of unrelated individuals in societies where kin groups are the norm. However, the two discrete categories of helpers in the pied kingfisher – unrelated (secondary) and closely related (primary) – suggest that RA and kin selection can be separate phenomena, even though in most cases, they currently are not separable.

Third, in a few cases among vertebrates, the kinship or indirect benefits of providing aid to relatives apparently can account, to a major extent, for the aid given by one individual to another. All of the case studies presented in this chapter strongly suggest that the pathway followed by an individual species in the evolution of its social system is unique to a greater or lesser extent, and that the relative importance of RA and kinship will vary from case to case.

Acknowledgements

I thank S. H. Ligon, D. A. McCallum, P. B. Stacey, and R. Thornhill for providing constructive criticisms of this chapter. Grants from the National Science Foundation and the National Geographic Society have provided support for my studies of avian social systems.

References

Alcock, J. (1984). *Animal Behavior. An Evolutionary Approach.* Sunderland MA: Sinauer.

Alexander, R. D. (1974). The evolution of social behavior. *Ann. Rev. Ecol. Syst.*, **5**, 325–83.

Armitage, K. B. (1988). Resources and social organization of ground-dwelling squirrels. In *The Ecology of Social Behaviour*, ed. C. N. Slobodchikoff, pp. 131–55. New York: Academic Press.

Axelrod, R. & Hamilton, W. D. (1981). The evolution of co-operation. *Science*, **211**, 1390–6.

Bednarz, J. C. (1987). Pair and group reproductive success, polyandry, and co-operative breeding in Harris' hawks. *Auk*, **104**, 393–404.

Bednarz, J. C. (1988). Co-operative hunting in Harris' hawks. *Science*, **239**, 1525–7.

Bednarz, J. C. & Ligon, J. D. (1988). A study of the ecological basis of cooperative breeding in the Harris' hawk. *Ecology*, **69**, 1176–87.

Bertram, B. C. R. (1983). Cooperation and competition in lions. *Nature*, **302**, 356.

Brown, C. R. (1986). Cliff swallow colonies as information centers. *Science*, **234**, 83–5.

Brown, J. L. (1983). Cooperation – a biologist's dilemma. *Adv. Stud. Behav.*, **13**, 1–37.

Conner, R. C. (1986). Pseudo-reciprocity: investing in mutualism. *Anim. Behav.*, **34**, 1562–6.

Conner, R. C. & Norris, K. S. (1982). Are dolphins reciprocal altruists? *Am. Nat.*, **119**, 396–409.

de Waal, F. B. & Luttrell L. M. (1988). Mechanisms of social reciprocity in three primate species: symmetrical relationship characteristics or cognition? *Ethol. Sociobiol.*, **9**, 101–18.

Emlen, S. T. (1981). Altruism, kinship, and reciprocity in the white-fronted bee-eater. In *Natural Selection and Social Behavior: Recent Research and New Theory*, ed. R. D. Alexander & D. W. Tinkle, pp. 217–30. New York: Chiron Press.

Emlen, S. T. (1982). The evolution of helping. I. An ecological constraints model. *Am. Nat.*, **119**, 29–39.

Emlen, S. T. (1984). Cooperative breeding in birds and mammals. In *Behavioural Ecology: An Evolutionary Approach*, ed. J. R. Krebs & N. B. Davies, pp. 305–39. Sunderland, MA: Sinauer.

Emlen, S. T. & Vehrencamp, S. L. (1983). Cooperative breeding strategies among birds. In *Perspectives in Ornithology*, ed. A. H. Brush & G. A. Clark, Jr., pp. 93–133. Cambridge: Cambridge University Press.

Ford, H. A., Bell, H., Nias, R. & Noske, R. (1988). The relationship between ecology and the incidence of cooperative breeding in Australian birds. *Behav. Ecol. Sociobiol.*, **22**, 239–49.

Foster, M. S. (1977). Odd couples in manakins: a study of social organization and cooperative breeding in *Chiroxipha linearis*. *Am. Nat.*, **111**, 845–53.

Foster, M. S. (1981). Cooperative behaviour and social organization of the swallow-tailed manakin (*Chiroxiphia caudata*). *Behav. Ecol. Sociobiol.*, **9**, 167–77.

Frame, L. H., Malcolm, J. R., Frame, G. W. & van Lawick, H. (1979). Social organization of African wild dogs (*Lycaon pictus*) on the Serengeti Plains, Tanzania 1967–1978. *Z. Tierpsychol.*, **50**, 225–49.

Gaston, A. J. (1977). Social behavior within groups of jungle babblers (*Turdoides striatus*). *Anim. Behav.*, **25**, 828–48.

Goodall, J. (1986). *The Chimpanzees of Gombe: Patterns of Behavior.* Cambridge, MA: Harvard University Press.

Grafen, A. (1982). How not to measure inclusive fitness. *Nature*, **298**, 425–6.

Greenwood, P. J. (1980). Mating systems, philopatry and dispersal in birds and mammals. *Anim. Behav.*, **28**, 1140–62.

Hamilton, W. D. (1964). The genetical evolution of social behaviour. I, II. *J. Theor. Biol.*, **7**, 1–52.

Heinrich, B. (1988). Food sharing in the raven, *Corvus corax*. In *The Ecology of Social Behavior*, ed. C. N. Slobodchikoff, pp. 285–311. New York: Academic Press.

Jarman, P. J. (1974). The social organization of antelope in relations to their ecology. *Behaviour*, **48**, 215–67.

Koenig, W. D. (1981). Reproductive success, group size, and the evolution of cooperative breeding in the acorn woodpecker. *Am. Nat.*, **117**, 421–43.

Koenig, W. D. (1988). Reciprocal altruism in birds: a critical review. *Ethol. Sociobiol.*, **9**, 73–84.

Koenig, W. D. & Mumme, R. L. (1987). *Population Ecology of the Cooperatively Breeding Acorn Woodpecker.* Princeton: Princeton University Press.

Kruuk, H. (1972). *The Spotted Hyaena: A Study of Predation and Social Behavior.* Chicago: University of Chicago Press.

Kummer, H. (1968). *Social Organization of Hamadryas Baboons.* Chicago: University of Chicago Press.

Ligon, J. D. (1981). Demographic patterns and communal breeding in the green woodhoopoe (*Phoeniculus purpureus*). In *Natural Selection and Social Behavior: Recent Research and New Theory*, ed. R. D. Alexander & D. W. Tinkle, pp. 231–43, New York: Chiron Press.

Ligon, J. D. (1983). Cooperation and reciprocity in avian social systems. *Am. Nat.*, **121**, 366–84.

Ligon, J. D., Carey, C. & Ligon, S. H. (1988). Cavity roosting, philopatry and cooperative breeding in the green woodhoopoe may reflect a physiological trait. *Auk.* **105**, 123–7.

Ligon, J. D. & Ligon, S. H. (1978a). Communal breeding in the green woodhoopoe as a case for reciprocity. *Nature*, **176**, 496–8.

Ligon, J. D. & Ligon, S. H. (1978b). The communal social system of the green woodhoopoe in Kenya. *Living Bird.* **17**, 159–97.

Ligon, J. D. & Ligon, S. H. (1982). The cooperative breeding behavior of the green woodhoopoe. *Sci. Am.*, **247**, 126–34.

Ligon, J. D. & Ligon, S. H. (1983). Reciprocity in the green woodhoopoe (*Phoeniculus purpureus*). *Anim. Behav.*, **31**, 480–9.

Ligon, J. D. & Ligon, S. H. (1988). Territory quality: key determinant of fitness in the group-living green woodhoopoe. In *The Ecology of Social Behavior*, ed. C. N. Slobodchikoff, pp. 229–54. New York: Academic Press.

Loyn, R. H. (1987). The bird that farms the dell. *Nat. Hist.*, **96**, 54–60.

Malcolm, J. R. & Marten, K. (1982). Natural selection and the communal rearing of pups in African wild dogs (*Lycaon pictus*). *Behav. Ecol. Sociobiol.*, **10**, 1–13.

Marzluff, J. M. (1988). Vocal recognition of mates by breeding pinyon jays, *Gymnorhinus cyanocephalus*. *Anim. Behav.*, **36**, 296–8.

Marzluff, J. M. & Balda, R. P. (1988a). Pairing patterns and fitness in a free-ranging population of pinyon jays: what do they reveal about mate choice? *Condor*, **90**, 201–13.

Marzluff, J. M. & Balda, R. P. (1988b). The advantages of, and constraints forcing, mate fidelity in pinyon jays. *Auk*, **105**, 286–95.

Maynard Smith, J. (1964). Group selection and kin selection. *Nature*, **20**, 1145–7.

McGowan, K. J. (1987). Social development in young Florida scrub jays (*Aphelocoma coerulescens*). Ph.D. dissertation, University South Florida.

Mech, L. D. (1970). *The Wolf: The Ecology and Behaviour of an Endangered Species.* Garden City: The Natural History Press.

Moehlman, P. D. (1979). Jackel helpers and pup survival. *Nature*, **277**, 382–3.

Owens, D. D. & Owens M. J. (1979). Communal denning and clan associations in brown hyenas of the central Kalahari Desert. *Afr. J. Ecol.*, **17**, 35–44.

Owens, D. D. & Owens M. J. (1984). Helping behavior in brown hyenas. *Nature*, **308**, 843–5.

Owens, M. J. & Owens, D. D. (1984). *Cry of the Kalahari.* Boston: Houghton Mifflin.

Owens, M. J. & Owens, D. D. (1984). Kalahari lions break the rules. *Int. Wildl.*, **14**, 4–13.

Packer, C. (1977). Reciprocal altruism in *Papio anubis*. *Nature*, **265**, 441–3.

Packer, C. (1986). The ecology of sociality in felids. In *Ecological Aspects of Social Evolution. Birds & Mammals*, ed. D. I. Rubenstein & R. W. Wrangham, pp. 429–51. Princeton: Princeton University Press.

Packer, C. & Pusey, A. E. (1982). Cooperation and competition within coalitions of male lions: kin selection or game theory. *Nature*, **296**, 740–2.

Packer, C. & Pusey, A. E. (1983). Cooperation and competition in lions. *Nature*, **302**, 356.

Pierotti, R. (1980). Spite and altruism in gulls. *Am. Nat.*, **115**, 290–300.

Rabenold, K. N. (1984). Cooperative enhancement of reproductive success in tropical wren societies. *Ecology*, **65**, 871–85.

Rabenold, P. P. (1983). The communal roost in black and turkey vultures – an information center? In *Vulture Biology and Management*, ed. S. R. Wilbur & J. A. Jackson, pp. 303–21. Berkeley: University of California Press.

Rabenold, P. P. (1986). Family associations in communally roosting black vultures. *Auk*, **103**, 32–41.

Rasa, O. A. E. (1983). Dwarf mongoose and hornbill mutualism in the Taru Desert, Kenya. *Behav. Ecol. Sociobiol.*, **12**, 181–90.

Reyer, H. V. (1980). Flexible helper structure as an ecological adaption in the pied kingfisher (*Ceryle rudis rudis* L.). *Behav. Ecol. Sociobiol.*, **6**, 219–27.

Reyer, H. V. (1984). Investment and relatedness: a cost/benefit analysis of breeding and helping in the pied kingfisher (*Ceryle rudis*). *Anim. Behav.*, **32**, 1163–78.

Robinson, S. K. (1985). Coloniality in the yellow-rumped cacique as a defense against nest predators. *Auk*, **102**, 506–19.

Rood, J. P. (1978). Dwarf mongoose helpers at the den. *Z. Tierpsychol.*, **48**, 277–87.

Rood, J. P. (1983). The social system of the dwarf mongoose. In *Recent Advances in the Study of Mammalian Behaviour*, ed. J. F. Eisenberg & D. G Kleiman, pp. 454–88. American Society of Mammalogists: Special publication No. 7.

Rood, J. P. (1986). Ecology and social evolution in the mongooses. In *Ecological Aspects of Social Evolution*, ed. D. I. Rubenstein & R. W. Wrangham, pp. 131–52. Princeton: Princeton University Press.

Rothstein, S. I. (1980). Reciprocal altruism and kin selection are not clearly separable phenomena. *J. Theor. Biol.*, **87**, 255–61.

Rothstein, S. I. & Pierotti, R. (1988). Distinctions among reciprocal altruism, kin selection, and cooperation and a model for the initial evolution of beneficent behavior. *Ethol. Sociobiol.*, **9**, 189–209.

Rowley, I. (1983). Commentary on cooperative breeding strategies among birds. In *Perspectives in Ornithology*, ed. A. H. Brush & G. A. Clark, pp. 127–33. Cambridge: Cambridge University Press.

Russell, J. K. (1983). Altruism in coati bands: nepotism or reciprocity? In *Social Behavior of Female Vertebrates*, ed. S. K. Wasser, pp. 263–90. New York: Academic Press.

Schaller, G. B. (1972). *The Serengeti Lion*. Chicago: University of Chicago Press.

Selander, R. K. (1964). Speciation in wrens of the genus *Campylorhynchus*. *Univ. Cal. Publ. Zool.*, **74**, 1–305.

Seyfarth, R. M. & Cheney, D. L. (1984). Grooming, alliances and reciprocal altruism in vervet monkeys. *Nature*, **308**, 541–3.

Seyfarth, R. M. & Cheney, D. L. (1988). Empirical tests of reciprocity theory: problems in assessment. *Ethol. Sociobiol.*, **9**, 181–7.

Snow, D. W. (1977). Duetting and other synchronized displays of the blue-back manakins, *Chiroxiphia* spp. In *Evolutionary Ecology*, ed. B. Stonehouse & C. Perrins, pp. 239–51. Baltimore: University Park Press.

Stacey, P. B. & Ligon, J. D. (1987). Territory quality and dispersal options in the acorn woodpecker, and a challenge to the habitat saturation model of co-operative breeding. *Am. Nat.*, **130**, 654–76.

Stacey, P. B. & Ligon, J. D. (1990). The benefits of philopatry hypothesis for the evolution of cooperative breeding: variation in territory quality and group size effects. *Am. Nat.* (In press.)

Stacey, P. P. (1979). Habitat saturation and communal breeding in the acorn woodpecker. *Anim. Behav.*, **27**, 1153–66.

Taylor, C. E. & McGuire, M. T. (1988). Reciprocal altruism: 15 years later. *Ethol. Sociobiol.*, **9**, 67–72.

Trivers, R. L. (1971). The evolution of reciprocal altruism. *Q. Rev. Biol.*, **46**, 35–57.

Trivers, R. L. (1985). *Social Evolution*. Menlo Park: Benjamin Cummings.

Walters, J. R., Doerr, P. D. & Carter III, J. H. (1988). The cooperative breeding system of the red-cockaded woodpecker. *Ethology*, **78**, 275–305

Waltz, E. C. (1981). Reciprocal altruism and spite in gulls: a comment. *Am. Nat.*, **118**, 588–92.

Waser, P. M. (1988). Resources, philopatry, and social interactions among mammals. In *The Ecology of Social Behavior*, ed. C. N. Slobodchikoff, pp. 109–30. New York: Academic Press.

Wasser, P. M. & Jones, W. T. (1983). Natal philopatry among solitary mammals. *Q. Rev. Biol.*, **58**, 355–90.

Waser, S. K. (1982). Reciprocity and the trade-off between associate quality and relatedness. *Am. Nat.*, **119**, 720–31.

West-Eberhard, M. J. (1975). The evolution of social behavior by kin selection. *Q. Rev. Biol.*, **50**, 1–33.

Wiley, R. H. & Rabenold, K. N. (1984). The evolution of cooperative breeding by delayed reciprocity and queuing for favourable social positions. *Evol.*, **38**, 609–21.

Wilkinson, G. S. (1984). Reciprocal food sharing in vampire bats. *Nature*, **308**, 181–4.

Wilkinson, G. S. (1988). Reciprocal altruism in bats and other mammals. *Ethol. Sociobiol.*, **9**, 85–100.

Woolfenden, G. E. & Fitzpatrick, J. W. (1984). *The Florida Scrub Jay: Demography of a Cooperative-breeding Bird*. Princeton: Princeton University Press.

Zack, S. & Ligon, J. D. (1985a). Cooperative breeding in *Lanius* shrikes. I. Habitat and demography of two sympatric species. *Auk*, **102**, 754–65.

Zack, S. & Ligon, J. D. (1985b). Cooperative breeding in *Lanius* shrikes. II. Maintenance of group-living in a nonsaturated habitat. *Auk*, **102**, 766–73.

3

Kinship and fellowship in ants and social wasps

Pierre Jaisson

Introduction

The complex societies of wasps, ants and bees constitute a special problem for the Darwinian theory of evolution. This did not escape the attention of Charles Darwin. He considered these insects to represent 'one special difficulty, which at first appeared to me insuperable, and actually fatal to my whole theory' (*The Origin of Species*, Chap. VII, p. 236). If we accept that natural selection operates at the level of individual organisms, it is difficult to comprehend just how it could occur in species like those of eusocial insects, where the majority of individuals are almost always excluded from reproduction because of morphological specializations which generally involve sterility. Darwin conceived of an explanation which allowed him to reconcile the existence of insect societies with his theory of natural selection ('This difficulty, though appearing insuperable, is lessened, or, as I believe, disappears, when it is remembered that selection may be applied to the family, as well as to the individual . . .': ibid., p. 237). This interpretation led eventually to a significant step forward, when an analogy was drawn between the multicellular organism and the insect society, which was conceptualized as a 'superorganism' (Wheeler, 1911; Emerson, 1952). After being neglected for some years, this theory has recently been rehabilitated (Lumsden, 1982; Jaisson, 1985; Wilson, 1985). This mode of thinking helps to resolve the difficulties in understanding the functioning of natural selection in insect societies. It suggests that the social group, as a unit, can be influenced by selective pressures, and that there is a relationship between the reproductive caste and the sterile worker caste comparable to that between the soma and the germ plasm of a multicellular organism.

A second important advance furthering the application of Darwinian thinking to the evolution of insect societies occurred in 1964 when W. D. Hamilton proposed the concept of inclusive fitness. This emphasizes the relationship between altruistic behaviour by workers, and its effect on the spread of genomes similar to that of the altruist. This kinship theory (so termed by Trivers & Hare, 1976) has strong affinities with the superorganism concept (the notion of inclusive fitness could, for example, be used, as an analogy, to explain the evolution of the somatic cells which constitute a multicellular organism). According to Hamilton, kin selection provides an explanation for the previously perceived multiple evolution of eusociality in Hymenoptera, in which the haplodiploid sex determination mechanism, and the long-term storage of sperm by females, serve to increase the levels of inclusive fitness, because the offspring of a single mother share high levels of genetic relatedness.

All this is possible only if a mechanism exists for kin recognition, which would enable cohesiveness among the related members of each hymenopteran society (their genetic relatedness increases with a reduction in the number of egg-layers). Thus, it is not surprising that a progressive increase in studies involving kin recognition followed Hamilton's major contribution (see review by Waldman *et al.*, 1988).

Many years ago Fielde (1901; 1904) emphasized the importance of *colony odour* in recognition between the members of a single ant nest. With perceptiveness astonishing in hindsight she considered that this constituted 'a means for the recognition of blood relations'.

The superorganism concept allows analogy between the ability in hymenopteran societies to discriminate nestmates, and the immunological barrier which protects organisms in other animal species. This leads to the postulate that there exists in hymenopterans a genetic mechanism which produces a high degree of odour polymorphism, analogous to the major histocompatibility complex (MHC) of vertebrates (the involvement of which in recognition is gradually being documented, see chapter by Boyse, this volume). This polymorphism must be accompanied by sophisticated sensorial discrimination and learning if it is to function properly.

The empirical demonstration of a relationship between kinship and the nature of altruistic acts in social hymenopterans has been the subject matter of several reviews (Gadagkar, 1985; Jaisson,1985; Breed & Bennett, 1986, 1987; Gamboa *et al.*, 1986a; Isingrini & Lenoir, 1986). Thus, I will place emphasis here on the behavioural *mechanisms* which form the obligatory links between the two components of the above correlation. Furthermore, I will limit myself to ants and social wasps, since another chapter of this book considers the social bees (Getz, chapter 13).

Natural and experimental interspecific associations in ants and social wasps: recognition of phylogenetically distant cues

The behavioural mechanisms involved in the development of a recognition ability among individuals in social Hymenoptera were first studied in ants by using interspecific mixed societies as a model (Forel, 1874;

Fielde, 1903). I have returned to this model of mixed societies in order to study the behavioural mechanisms operating during worker ontogenesis and involved with development of the ability to recognize nestmate pupae (Jaisson, 1972, 1973, 1975a,b), as well as adults (Jaisson, 1980).

Artificially-mixed nests of ants offer the advantage that natural dummies can be used, which allows cross-fostering experiments, as was done in the classical studies on *imprinting* in vertebrates. Thus, it is not surprising that some analogous processes can be revealed in ants, which confirm the hypotheses of Wallis (1963) and Wilson (1971): imprinting-like phenomena are involved in species recognition. The same types of phenomena have subsequently been found for intraspecific colony recognition in ants and social wasps.

Natural models

In actual fact, interspecific mixed colonies which are based on the host/parasite relationship do exist in nature. These involve perfect social integration between two species. This situation is found mainly in two categories of social parasitism:

> *inquilinism*, where the parasitic species invades nests of another species and then exploits the worker force of the host. This permanent phenomenon exists in both wasps and ants. Inquilinism has evolved many times and is found throughout these groups. In social wasps, the 'cuckoo strategy', as it is often called, has been documented in the European species of *Polistes* (de Beaumont & Matthey, 1945; Scheven, 1958) and *Vespula* (de Beaumont, 1958). In these, the worker caste has disappeared, and the queens enter young host colonies, where they behave as usurpers. Often the queen of the host species is expelled or killed; sometimes (in some *Polistes*) she remains to become integrated with the worker force and to participate in the rearing of new sexuals of the parasitic species. In ants there are numerous examples in the higher subfamilies (Formicinae and Myrmicinae, especially), and one example in the archaic subfamily Myrmeciinae (see description of *Myrmecia inquilina*, Douglas & Brown, 1959).
>
> *slave-making (or dulosis)*, exists only in ants (first described by Huber, 1810), where it seems to occur only in two subfamilies, Formicinae and Myrmicinae (see Table 3.1). This restricted distribution suggests that slave-making is more difficult to evolve than inquilinism, and that it requires intermediate steps. This is not surprising, because it is easier to select mechanisms which integrate one parasitic individual into the social group of a host species (as in the large numbers of myrmecophilous and termitophilous arthropods), than to assimilate individuals as slaves in the foreign societies

Table 3.1. *The ant genera in which slave-making occurs in all or some species, listed with the genera which provide their slaves.*

Subfamily	Slave-making genera	Parasitised corresponding genera
Formicines	*Raptiformica*	*Serviformica*
	Polyergus	*Serviformica*
	Rossomyrmex	*Proformica*
Myrmicine	*Epimyrma*	*Leptothorax*
	Harpagoxenus	*Leptothorax*
	Leptothorax	*Leptothorax*
	Strongylognathus	*Tetramorium*

of a slave-making parasitic species. This behavioural trait essentially involves the process of early learning, which is normally associated with nestmate recognition in the host. For that purpose, not only must the slave-making species have its own specialized attributes (essentially the behavioural repertoire necessary for raiding and capturing slaves), but it must also exploit to its own advantage the mechanisms of behavioural development which have evolved in species with the preadapted capacities to becoming slaves. This is what I have demonstrated in the relationship between *Raptiformica sanguinea* and *Serviformica fusca* (Jaisson, 1973, 1975a,b).

Mature *Serviformica* workers collected from their natal colony cannot recognize as nestmates the pupae of *Raptiformica* which are offered to them. For these workers, the pupae have only trophic significance: they are eaten. On the contrary, when *Serviformica* workers are observed in a *Raptiformica* nest, they care for the slave-maker pupae, which are recognized as nestmates. In fact, the integration of *Serviformica*, like that of other host species, into the slave-maker's colony is absolute: altruistic acts are directed toward alien nestmate pupae, and after emergence the enslaved workers act to increase the inclusive fitness of the slave-maker, to which they are not genetically related (*Serviformica* queens are not tolerated in the mixed colonies). Finally, in this type of social parasitism, the slave-making species increase their own fitness by exploiting an allospecific altruistic force. This benefit is substantial, so that, in the extreme cases of dulosis (*Polyergus*) there has been a loss of almost all altruistic acts towards conspecific brood.

The occurrence of interspecific altruism is probably the fundamental trait in the evolution of slave-making in ants. Following a period during which the anthropocentric viewpoint of entomologists led them to imagining a supposed co-operation of the slaves with their masters' raids, different authors have attempted to discover the underlying mechanisms. An important observation by the Italian myrmecologist Carlo Emery (1909) proposed a tentative explanation. Emery emphasized the close phylogenetic relationships which existed between dulotic ants and their slaves, and concluded that in each case the two probably had a common evolutionary history. The example of *Leptothorax duloticus* (Wilson, 1975), which captures slaves of a congeneric species, has confirmed this hypothesis.

A second, proximate, level of explanation has become possible more recently, at least for the formicines, with the discovery of early learning phenomena involved in the development of recognition of the slave-maker's brood by the enslaved workers. In fact, young callow workers of *S. fusca* isolated from their mother colony and reared in groups of 50 to 100 in the presence of *R. sanguinea* pupae, are known to adopt the latter, and then to show a strong tendency to behave altruistically only towards cocoons of this type, provided that the contact has been maintained during the first two weeks of imaginal life (Jaisson, 1975b). This time-period is relatively short because *S. fusca* workers can live for more than 18 months. In this situation, it is possible for *Serviformica* even to prefer tending cocoons of *Raptiformica*, rather than conspecific cocoons (the studies on *S. cunicularia* by Le Moli & Mori, 1985 and 1987, have led to similar results).

This tending behaviour of *Serviformica* workers towards allospecific brood does not seem to be associated with any special characteristic of the *Raptiformica* pupae, because it is seen also when cocoons of *Formica polyctena* are used in adoption experiments. In the field, this related species is neither parasitic nor enslaved. Once they have been in contact with *F. polyctena* cocoons, the exposed *Serviformica* workers adopt the familiar alien pupae. Furthermore, the recognition of *F. polyctena* cocoons is even more marked when, within the two weeks following eclosion, callow workers of *F. polyctena* are added to the group (Jaisson, 1975b). This suggests that in naturally-mixed colonies of slave-maker ants, the mere presence of parasitic ants has a stimulating effect on the process of recognition of the allospecific brood by the enslaved workers.

A question arises concerning species such as *R. sanguinea* which, in contrast to *Polyergus*, remain capable of tending their own brood: how is it that the parasite is not tricked by its own strategy and proceeds to tend also the pupae of its enslaved species? The answer is simple: *Raptiformica* do not learn to recognize brood during the first few days of adult life. Orientation towards conspecific pupae is spontaneous and appears early: even isolation with the brood of a foreign species during 45 days following eclosion does not cancel or change the strong preference for conspecific brood. There are three likely hypotheses: (1) innate recognition, (2) pre-imaginal learning, or (3) post-eclosion self-phenotype learning. Among these, pre-imaginal learning is not likely because *Raptiformica* larvae are also exposed to stimulation from the enslaved species. It is very difficult to differentiate experimentally between the two other hypotheses.

However, a certain plasticity could occur in other aspects of the behaviour of formicine slave-making species. Indeed, it has been reported that, while several different species of *Serviformica* can be available and used as slaves by colonies in a population of the slave-maker, a single species is generally found in each particular nest of *Raptiformica* (Jaisson, 1985) or of *Polyergus* (Goodloe *et al.*, 1987) which demonstrates host-specificity in the raiding behaviour of each slave-maker colony. This suggests the existence of some imprinting by the slave-maker species, with respect to the orientation of its raids towards *Serviformica* colonies.

The ability to recognize brood, even that of a very remote kin, and to accept as nestmates adults of a different species through early learning processes, explains also why *Raptiformica* or *Polyergus* selectively carry off the pupae (sometimes also the young callows) of *Serviformica* species. The probable reason they do not carry off the larvae is because they would need to be fed in the slave-maker nest.

Much research is still necessary for a proper understanding of these phenomena, but the formicines are better known than the myrmicines in these respects. Nonetheless, an important step forward was the recent study of the development of brood recognition in the Myrmicinae ants *L. ambiguus* and *L. longispinosus*, two species which are commonly enslaved by *Harpagoxenus americanus* and *L. duloticus*: Hare & Alloway (1987) have found that workers from *L. ambiguus* or *L. longispinosus* which are reared during the 10 days following eclosion, either with brood from the other species, or isolated without any brood, will tend larvae of both species. In return, if the callow workers are reared in early imaginal life with conspecific larvae, they develop a strong preference for that kind of brood. Although the phenomenon of plasticity in brood discrimination differs in its details from that found in *Serviformica*, an important common feature is that susceptibility to slave-making exists in both subfamilies, because of the evolution of early learning processes related to nestmate recognition.

Experimental interspecific associations of ants as a model for the understanding of the mechanisms and cues involved in nestmate recognition

Mixed colonies involving adult ants

The behavioural plasticity of certain ants during the early stage of imaginal life, which is exploited by slave-maker species to parasitize their hosts, has also been used by various researchers following Forel (1874). As I emphasized elsewhere (Jaisson, 1980), artificial mixed colonies of ants are a useful model for the study of behavioural development and plasticity. In particular, this approach can serve to clarify the mechanisms involved in nestmate recognition, through the manipulation of the types of nestmates and of the periods of contact between species.

When appropriate species are chosen there seems to be no limit to the number which can be brought together within an mixed colony. Indeed, I was able to bring together six different species of Formicinae and Myrmicinae. However, success in the establishment of an experimental mixed colony involves two rules (Jaisson, 1980), between which there is summation:

    1 use individuals soon after eclosion (i.e. young callow workers)
    2 choose species which are phylogenetically related

These two rules have been confirmed by Errard (1984a, 1986a) in a series of systematic experiments. She observed that it was particularly easy to establish mixed colonies using the workers of two species of *Camponotus* (*C. senex* and *C. abdominalis*). Mixing between the individuals occurs immediately if they are brought together on the day of eclosion; it occurs after four days if they are brought together four days after eclosion; and after five days if the ants are six-days old when they are brought together. We see that the optimal condition is when the individuals are youngest, but until they are six-days old (as adults) it is still possible to obtain a successful result, even though it takes longer to happen. In contrast, when Errard used *C. senex* (Formicinae) and *Pseudomyrmex ferruginea* (Pseudomyrmicinae), which are two phylogenetically-distant species, she showed that the establishment of a mixed society was possible only when the workers were less than 24 hours old; and to be certain of success it was preferable to use individuals within three hours of eclosion. It is thus possible to operate earlier when the subject species are phylogenetically distant, since there is summation between the two above rules.

Thereafter, Errard achieved all the possible combinations between 11 species of ants belonging or not to either the same genus or the same subfamily. In this way, three situations were possible: (1) two congeneric species, (2) two species from different genera, and (3) two species from two different subfamilies. Since the ants had similar ages when they were brought together, it was found that the minimum level of agonistic interactions occurs in situation (1) and the maximum in situation (3), with an intermediate level in situation (2).

These experiments suggest that the origin of the odour cues involved in recognition is related to phylogeny, and thus to the part of the genome which is common to the different species. This means that callow workers accept as nestmates individuals with a chemical visa which shares a common component with that of their own species. It is logical to think that, coarsely, the less closely related two species are, the smaller will be the common component of their chemical visa. This interpretation would mean that workers elaborate the recognition template of their adult nestmates through a progressive mechanism of associative learning which takes place in the four days after eclosion. The elaboration of a complex template corresponding to the odour of the group would have as its beginning an innate template (possibly learned during larval life), the basic features of which would be shared in part with a range of species or genera. The eventual association between familiar odours, detected ever since emergence, and the unfamiliar odours which characterize the social group, would lead progressively to the recognition of unfamiliar nestmates. In this hypothesis about the progressive construction of a template during the first hours or the first days of adult life, the last features to be learned by the callow worker could be those which specifically characterize the group to which it belongs. This interpretation makes it possible to understand why the *sensitive periods*, which are evident during attempts to establish a mixed colony, have durations which are inversely related to phylogenetic distance (as the common component between innate templates gets smaller, it becomes necessary to start earlier).

In agreement with the above, the use of experimental mixed colonies also allows one to suggest that the characteristic odour of a group is *progressively constructed* (Fielde, 1905; Errard, 1986b,c). For example, in Errard's experiments, 10 callow workers of *Manica rubida* (Myrmicinae) and 10 callow workers of *F. selysi* (Formicinae) are brought together during the 24 hours following eclosion, and kept for 15 days. These mixed groups are then broken up for a period of 8 to 90 days, and then brought together again. In some cases the individuals which knew each other following eclosion are brought together (known recombination), in others they are brought together with allospecific individuals from a different group of origin (unknown recombination). Although only one colony of each species was used to constitute these groups, Errard has shown that recognition was more easily redeveloped in the case of a known recombination, than in those of unknown recombination. However, the agonistic interactions increased in frequency and intensity when the duration of the separation was increased. For example, aggressive responses of *F. selysi* workers after 90 days of separation are nearly twice as numerous as those after only 15 days of separation. They are also twice as numerous in the case of an unknown recombination than a known recombination. This last result supports the idea that a group odour is learned, and varies with the experimental group, even when only one original nest of each species is used for the experiment, and all the groups are given the same diet.

The decrease in recognition ability as a function of duration of separation can be interpreted as the progressive decline of memory in ants. But it can also be explained as a loss in the cuticular compounds which constitute the chemical visa of the group. Indeed, using gas chromatographic analysis, Errard & Jallon (1987) showed that cuticular hydrocarbons (which are thought to function during recognition of myrmecophilous beetles by their hosts (Vander Meer & Wojcik, 1982), and between conspecific colonies: see below) are exchanged between callow workers of the two species during the first eight days of being together. Thus, the cuticle of *Manica* workers becomes coated with hydrocarbons of *Formica* and vice versa. More precisely, it has been found that hydrocarbons with a low molecular weight are the first to be transferred from one species to another, and those with a high molecular weight the last. Furthermore, when the two species are separated, the allospecific hydrocarbons are lost from the cuticle in an order inverse to that in which they arrived. Small-molecule hydrocarbons are more stable than large ones in their association with cuticular waxes. Consequently, the longer the separation, the larger is the degradation of the individual's chemical visa with respect to the template learned during the initial period, and the more marked are the agonistic interactions of the two species when they are reunited.

In the examples mentioned so far, only the workers were manipulated. However, several authors consider that the queen is of prime importance as the source of the colony odour, especially in monogynous species (see Wilson, 1971; Hölldobler & Wilson, 1977; Hölldobler & Michener, 1980). The occurrence of queen pheromones which prevent workers from reproducing is considered to be an argument in favour of that point of view. We will see later, in the case of conspecific social recognition, that the queen does not alone have that role, even in monogynous species. Furthermore, one of the known variables is seldom considered by authors comparing nestmate recognition in groups with and without a queen. It is, the queen's effect on the behaviour of workers, as revealed by the degree of intolerance or attraction towards a test individual. In other words, if a 'queenright group' of workers rejects a foreign individual, while the latter is tolerated by another group from the same colony which had been reared without a queen, one interpretation is that the queen is the source of the chemical visa distributed to the workers, but another interpretation may be that the queen enhances the intolerance of workers. In *F. polyctena* for example, preference for the one type of brood is more marked when the workers learned the brood recognition template in the presence of a queen than when they were reared in an orphaned group (Jaisson, 1975a).

Carlin & Hölldobler (1983, 1986) also used interspecific ant nests as a model to analyze nestmate recognition, but only in the context of queen presence. By setting up a number of small mixed colonies each with two out of six species of *Camponotus*, they showed that workers reared from larvae adopted by a foreign queen do not recognize their unfamiliar sisters which remained in the original stock colony. This result is unaffected by the ratio between the number of allospecific ants and the number of conspecific ants in the mixed colony (i.e. the lack of kin recognition persists even if the queen is the only allospecific individual). It is also unaffected by the species to which the queen belongs, which these authors consider to be proof that workers which have lived in mixed colonies do not rely on a species-specific odour, but rather on an individual odour which originates from the queen.

The experiments of Carlin & Hölldobler, like those of Errard, demonstrate that the behaviour of an ant worker in a mixed colony, in which she comes into contact with unfamiliar individuals, results from information which she learned in her adoptive group. The chemical visa from which her recognition template is learned is partly (according to Errard) or almost exclusively (according to Carlin & Hölldobler) composed of exogenous chemical factors. In small mixed colonies of *Camponotus* the queen is definitely the main source of recognition substances ('discriminators' as termed by Hölldobler & Michener 1980).

If we consider that mixed interspecific colonies offer a valid model for the study of nestmate recognition, the results of Errard on associations of *Manica* and *Formica* support the gestalt odour hypothesis championed by Crozier & Dix (1979) and Crozier (1987), according to which colony odour is a cocktail which is shared by all nestmates and which results from the contribution of each individual.

Interspecific ant colonies involving mixed adults and brood

For many years it has been known that it is easier to transfer a conspecific brood from one colony to another than to transfer adults. In addition, it was discovered that mature adults of various species could be led to adopt allospecific brood (Plateaux, 1960). However, when the adult ants are callow, it is possible to make them adopt brood belonging to different genera, particularly within the subfamilies Formicinae and Myrmicinae (see review by Jaisson, 1985).

Once again phylogenetic distance is important in the search for an optimal mix. Thus, callow workers of *F. polyctena* learn to recognize cocoons of other species of *Formica* more easily than cocoons of *Camponotus*, and cocoons of *Camponotus* are more easily recognized than cocoons of *Lasius*. In contrast, it is impossible to achieve the adoption by *F. polyctena* of pupae from Ecitoninae or from Myrmicinae (Jaisson, 1974, 1975b and unpublished data). This relationship between the affinity for brood and phylogeny of the species involved has been proposed and successfully used as a taxonomic tool by Rosengren & Cherix (1981).

It was shown in the genus *Formica* (*F. polyctena*, *F. rufa* and *F. lugubris*) that the mechanism by which an adult worker learns to recognize familiar brood is similar to imprinting (Jaisson, 1972 to 1975a,b; Le Moli & Passetti, 1977, 1978; Le Moli & Mori, 1982). The latter has been known for a long time in vertebrates, where it is involved in the recognition of sexual partners and offspring. Indeed, in *Formica*, callow workers must have been in contact with cocoons during the first few days of their imaginal life if they are to be capable of recognizing and tending brood, i.e. to serve as nurses. In the absence of this early learning, any cocoons that are presented to such a worker, including conspecific cocoons, are neglected or treated as food. The sensitive period involved lasts from one to two weeks, depending on the particular experimental conditions (Jaisson & Fresneau, 1978). Judging from these results it seems that *Formica* is much more restricted than the myrmicine *Leptothorax* (see above), in which allospecific early learning induces greater tolerances, including those directed towards unfamiliar conspecific brood (Hare & Alloway, 1987).

The possibility of exchanging brood between colonies within the same species, and that of sometimes being able to lure adult workers by getting them to adopt allospecific brood, has stimulated numerous researchers to try to identify the brood pheromones which underlie the possibility of species-specific recognition. However, in a recent article, Morel & Vander Meer (1988) have critically reviewed these studies and, on the basis of valid methodological considerations, reached the conclusion that nobody has really found a brood pheromone, or even brood-specific chemicals. These authors suggest alternative hypotheses which involve brood behaviour (for example the rewards given to the adult nurse workers by the larvae), morphology, cuticular hydrocarbons (colony-recognition compounds) and learning. But these alternative hypotheses fail to explain why, in *F. polyctena* for example, stable species-specific brood recognition occurs even when the brood originates from different colonies (which eliminates the effect of colony odour), even when pupae which have been previously killed by freezing are used (thus excluding brood behaviour), and even when cocoon shape has been modified (Jaisson, 1975a,b). Such results (together with the influence of the phylogenetic variable, with respect to the feasibility of constituting mixed interspecific colonies involving callow adults) support the existence of brood pheromones. However, it is possible that in the Myrmicinae, where they have been studied, brood pheromones are less likely to be species-specific than in Formicinae, where they have not yet been researched.

Recognition between members of conspecific colonies

In ants, as in wasps, studies on nestmate recognition in pure colonies have accumulated considerably over the last few years, thus following the trend of work on kin recognition in other taxa (Waldman *et al.*, 1988). Table 3.2 gives a list of the species of ants and wasps in which nestmate discrimination has been demonstrated following a quantitative study, and shows that nestmate discrimination between workers has been found in Polistinae wasps (two species of *Polistes* and one of *Ropalidia*) and in Vespinae wasps (one species each of *Dolichovespula* and *Vespula*). In ants, it has been studied in the primitive subfamily Ponerinae (*Neoponera, Odontomachus* and *Rhytidoponera*) and in the higher subfamilies (Formicinae: *Camponotus, Cataglyphis, Lasius, Serviformica* and *Formica*; Myrmicinae: *Acromyrmex, Atta, Leptothorax, Myrmica, Novomessor, Pristomyrmex* and *Solenopsis*; Pseudomyrmicinae: *Pseudomyrmex*; Dolichoderinae: *Tapinoma*). It is present as much in monogynous species with a high intra-nest relatedness (e.g. *Lasius* and *Camponotus*) as in species with a greatly reduced relatedness due to the presence of many reproductive individuals (e.g. *Rhytidoponera* and *Pristomyrmex*). In the archaic ants of the genus *Myrmecia*, Haskins & Haskins (1950) reported that nestmate recognition was distinct: foreign cocoons are accepted, but after eclosion callow workers are discriminated as aliens and killed. These data have not been confirmed by Crosland (1989c) who used four species of *Myrmecia* and was able not only to achieve the acceptance of foreign conspecific cocoons, but also the acceptance of callow workers emerging from them. Recently Jaisson & Taylor (unpublished data) obtained the same result in *Myrmecia nigriceps*. However, if the callow workers (aged from one to five days) are removed from their foster colony for two hours and then returned, they are no longer recognized: but are attacked and killed. This rejection, which does not exist towards callow nestmate workers, demonstrates that adoption is precarious. It is possible that the adopted callow in isolation continues to produce the chemical visa corresponding to her natal colony while at the same time she also loses the chemicals (corresponding to the chemical visa of her adoptive colony) adsorbed on her cuticle following the imaginal moult. This interpretation is in agreement with the model studied by Errard & Jallon (1987) on allospecific mixed colonies.

As seen in Table 3.2, the behavioural test commonly used by authors has been acceptance versus rejection. Sometimes researchers were concerned to refine this test by constructing a hierarchy of rejection behaviours (Pfennig *et al.*, 1983a on *Polistes*; Carlin & Hölldobler, 1986 on *Camponotus*).

In most instances when brood recognition was studied, research results paralleled those obtained with adults. In wasps, intolerance towards foreign brood is evidenced by its destruction (Klahn & Gamboa, 1983), while in ants the response is generally less dramatic, and foreign brood is less well-cared for than nestmate brood (e.g. Isingrini *et al.*, 1985). This greater tolerance of ants for foreign brood is strongest in *Rhytidoponera confusa*, where it was impossible, despite careful observation, to reveal any discrimination of related *versus* unrelated larvae, although discrimination is present among adults (Crosland, 1988). These results agree with my (1985) hypothesis, according to which species-specific compounds play a greater role in brood recognition than in recognition between adults.

Finally, no nestmate recognition has been evidenced in some other ant species: foreign conspecific individuals can thus enter their societies. This is

Table 3.2. *Species of ants (a) and social wasps (b) where nestmate recognition has been demonstrated. The taxa are listed by subfamily, in alphabetic order. Workers = mature workers; L = learning (presence indicated by +).*

(a)

Subfamily	Species	Stages recognized	Behavioural test	L	Authors
Dolichoderinae	*Tapinoma erraticum*	larvae	brood retrieving		Meudec 1978
Formicinae	*Cataglyphis cursor*	callows	altruistic interactions	+	Lenoir *et al.*, 1982
		larvae	brood tending	+	Isingrini *et al.*, 1985, Isingrini, 1987
		queen	altruistic interactions		Berton *et al.*, 1990
	Camponotus abdominalis	larvae	brood tending	+	Errard, 1984b
		workers	behavioural interactions	+	Errard, 1984b
		queen	behavioural interactions		Errard, 1984b
	C. floridanus	workers	acceptance vs. rejection	+	Carlin & Hölldobler, 1987a
		workers	acceptance vs. rejection	+	Morel, 1988,
		callows	acceptance vs. rejection	+	Morel & Blum, 1988
	C. lateralis	workers	acceptance vs. rejection		Provost, 1979
	C. pennsylvanicus	workers	acceptance vs. rejection	+	Carlin & Hölldobler, 1986
	C. rufipes	workers	behavioural interactions		Jaffé & Sanchez, 1984
	C. vagus	workers callows	acceptance vs. rejection	+	Morel, 1983
		larvae	brood retrieving		Bonavita-Cougourdan *et al.*, 1988

	Species	Caste	Measure		Reference
	Formica argentea	sexuals	acceptance vs. rejection		Bennett, 1988
	F. cunicularia	workers	acceptance vs. rejection	+	Le Moli & Mori, 1987
	F. fusca	workers	acceptance vs. rejection		Wallis, 1962
	F. lugubris	workers	acceptance vs. rejection	+	Le Moli & Parmigiani, 1982
	F. podzolica	sexuals	acceptance vs. rejection		Bennett, 1988
	F. polyctena	workers	acceptance vs. rejection		Lange, 1960, Mabelis, 1979
	Lasius niger	larvae	brood retrieving		Lenoir, 1981
Myrmicinae	*Acromyrmex landolti*	workers	acceptance vs. rejection		Jaffé & Navarro, 1985
	A. octospinosus	workers	behavioural interactions		Jutsum *et al.*, 1979
	Atta cephalotes	workers	acceptance vs. rejection		Jaffé, 1983
	Leptothorax ambiguus	workers	acceptance vs. rejection		Stuart, 1987b
	L. curvispinosus	workers	acceptance vs. rejection		Stuart, 1987a,b
	L. longispinosus	workers	acceptance vs. rejection		Stuart, 1987b
	Myrmica laevinodis	callows	acceptance vs. rejection		Jaisson, 1971
	M. ruginodis	workers	acceptance vs. rejection		Le Roux, 1980
	Novomessor albisetosus	workers	acceptance vs. rejection		Goldstein & Topoff, 1985
	Pristomyrmex pungens	workers	acceptance vs. rejection		Tsuji & Itô, 1986; Tsuji, 1988
	Solenopsis invicta	workers	behavioural interactions		Obin, 1986

Table 3.2. (a) *(cont.)*

Subfamily	Species	Stages recognized	Behavioural test	L	Authors
Ponerinae	*Neoponera apicalis*	workers	acceptance vs. rejection		Fresneau, 1980
	Odontomachus bauri	workers	acceptance vs. rejection		Jaffé & Marcuse, 1983
	Rhytidoponera confusa	workers	acceptance vs. rejection		Crosland, 1989a,b
	R. metallica	workers	acceptance vs. rejection		Haskins & Haskins, 1979
	R. sp. 12	workers	acceptance vs. rejection		Peeters, 1988
Pseudomyrmicinae	*Pseudomyrmex ferruginea*	workers	acceptance vs. rejection		Mintzer, 1982 Mintzer & Vinson, 1985
	P. termitarius	workers	acceptance vs. rejection		Jaffé *et al.*, 1986
	P. triplarinus	workers	acceptance vs. rejection		Jaffé *et al.*, 1986

Subfamily	Species	Stages recognized	Behavioural test	L	Authors
Polistinae	*Polistes carolina*	fall gynes	spatial tolerance	+	Pfennig *et al.*, 1983a
	P. exclamans	spring-foundresses	spatial proximity for overwintering		Allen *et al.*, 1982
	P. fuscatus	brood	acceptance vs. rejection	+	Klahn & Gamboa, 1983
		workers	spatial tolerance	+	Pfennig *et al.*, 1983b
		fall gynes	spatial tolerance	+	Shellman & Gamboa, 1982
		fall gynes	spatial tolerance and behavioural interactions		Pfennig *et al.*, 1983a,b, Gamboa *et al.*, 1986b, 1987, Gamboa 1988
		males	spatial tolerance and copulation	+	Ryan & Gamboa, 1986
		spring-foundresses	spatial tolerance and cooperation		Bornais *et al.*, 1983
		spring-foundresses	spatial proximity		Post & Jeanne, 1982
	P. metricus	spring-foundresses	spatial tolerance		Ross & Gamboa, 1981
Vespidae	*Dolichovespula maculata*	males	spatial tolerance		Ryan *et al.*, 1985
	Ropalidia marginata	females	behavioural interactions	+	Venkataraman *et al.*, 1988
	Vespula maculifrons	males	mate choice		Ross 1983

the case for example in *F. execta*, where it is possible to exchange adult ants between different colonies (Pisarski, 1972). Wilson (1971) and Isingrini & Lenoir (1986) present other examples of societies which exhibit tolerance towards foreign individuals. One of the most conspicuous and spectacular cases is that of the imported Argentine ant (*Iridomyrmex humilis*). Along the French and Italian rivieras, the whole population of this abundant ant constitutes one single super-colony.

It is difficult to suggest an evolutionary explanation for nestmate recognition. Wilson (1971), using data from Haskins & Haskins (1950) suggests that the loss of inter-colonial hostility in ants may be a secondary process typical of various higher species. However, some primitively eusocial bees are frequently tolerant (e.g. Plateaux-Quénu, 1962), and the most archaic known living ant, *Nothomyrmecia macrops*, exhibits no hostility between colonies (Taylor, 1978; Hölldobler & Taylor, 1983; Jaisson & Taylor, unpublished data).

The behavioural mechanisms and recognition cues involved in nestmate discrimination: the examples of *Polistes* (wasps) and *Camponotus* (ants)

Table 3.2 indicates that among wasps, the genus *Polistes* (with four species studied) is the best known with respect to nestmate recognition, and among ants it is the genus *Camponotus* (with six species studied).

In *Polistes*, the first conclusive proof of the involvement of a mechanism of early learning in nestmate recognition was given by Shellman & Gamboa (1982). These researchers compared the behavioural interactions and spacing patterns between three groups of fall (i.e. Autumn) gynes. The first group was made up of isolated gynes; the second of gynes which had been exposed to their nestmates, and the third of gynes previously exposed to both their natal nest and nestmates immediately after adult emergence. The period of exposure or isolation was altered from 15 to 115 days for different individuals. The results showed that only the gynes previously exposed to both nest and nestmates were able to discriminate between nestmates and non-nestmates. Exposure only to nestmates appears to provide as little information than total isolation. The same results were obtained independently (Pfenning *et al.*, 1983b) with workers of *P. fuscatus*.

A study by Pfennig *et al.* (1983a) with gynes of the same species showed that sole exposure to the natal nest (without the adult nestmates) was sufficient, and that nestmate recognition is established even when exposure lasted only one hour after adult emergence and a rest period of 36 hours on average preceded the recognition test. Furthermore, increasing the period of exposure apparently does not increase the strength of the nestmate recognition response. The same result was obtained with gynes of *P. carolina* exposed for only two hours (following adult emergence) to their natal nest. The existence of a very short critical period, relative to the life expectancy of the wasps, and the temporal stability of the information concerning the chemical characteristics of nestmates during this period, reveal that this process of ethogenesis (behavioural development) is analogous to the imprinting process involved during the elaboration of species-specific cocoon recognition by workers in *F. polyctena* (Jaisson, 1975a). In general, it is likely that imprinting is the most common process by which social hymenopterans elaborate their nestmate recognition ability.

Since Shellman & Gamboa (1982), the most common experimental approach has been that of 'triplet experiments', whereby two nestmates wasps are placed together with an unrelated wasp. Thereafter, the bioassay consists of evaluating nestmate recognition by measuring spatial proximity between one individual and each of the two others (i.e. number of minutes paired in a definite observation), and by quantifying the behavioural interactions between the wasps. Pfennig *et al.* (1983b) used a scale of ten defined interactions, which were then averaged, in order to obtain a 'tolerance value'. At the extremes of this scale are, on the one hand, the interaction which represents the weakest form of tolerance (*chasing*) and, on the other hand, that which represents strong tolerance (*prolonged or repeated mutual antennation*).

The method of triplets has been widely used in order to try and localize the source of the recognition cues. This source may be, *a priori*, the wasp herself (endogenous origin) or external (nestmates, nest, or the physical environment). In the external case, we need to hypothesize that the cues are adsorbed in the cuticular waxes of the individual. It is also possible that endogenous and exogenous odours combine.

With a thorough experimental study, Gamboa *et al.* (1986b) demonstrated the existence of both an environmental and an hereditary component in the odour labels involved in nestmate recognition. For example, unrelated non-nestmate gynes of *P. fuscatus*, reared separately in the same strictly-controlled laboratory environment, displayed a significantly higher tolerance than non-nestmate gynes coming from an uncontrolled field environment. This experiment revealed that wasps are able to adsorb and learn environmental odours involved in nestmate recognition. However, in the next experiment, these wasps displayed a greater tolerance towards nestmates when, after a period of isolation from their nest, they were introduced in triplets with a related and an unrelated wasp. They prefer to pair with the related individual, which indicates that there is a genetic component in nestmate recognition.

A possible interpretation is that the wasp recognizes her nestmates on the basis of either a nest odour learned by imprinting during exposure at an early age to the nest and/or brood, or an environmental odour (from the food or nest-construction material) learned by olfactory conditioning. In this hypothesis two distinct ethological phenomena (only the first of which is dependent on a critical period) would be involved in the dynamics of nestmate recognition. This would explain why wasps, which prefer to use environmental (exogenous) odour cues first, could also use inherited (endogenous) odour cues after a prolonged isolation from their nest had led them to forget their olfactory conditioning to the environment. It would also explain why there is no summation between the endogenous components and the exogenous cues (Gamboa *et al.*, 1986b; Gamboa, 1988), and why the wasp uses either one or the other ethological process.

Lastly, for recognition to be possible, a wasp needs only to perceive in a conspecific individual either the inherited odour (which she herself produces), or the environmental odour. The use of one of these cues precludes use of the other.

In further advancing our knowledge of genetic (or endogenous) cues Gamboa *et al.*, (1987) and Gamboa (1988) studied discrimination between nestmate and non-nestmate wasps which are more or less genetically related (sisters, aunt–niece and cousins).

By experimental manipulation of *P. fuscatus* in the field, Gamboa *et al.*, (1987) were able to rear non-nestmate gynes which were related as aunts and nieces. This made it possible to conduct an experiment in which fall gynes were arranged in triplets. Each experimental triplet included two nestmate sister gynes and one non-nestmate niece of the other two. In the control triplets the non-nestmate wasp was unrelated to the two nestmates. Observation of the interactions within triplets, and calculation of a mean tolerance value, as explained above, showed that the nestmates could discriminate the unrelated non-nestmate wasp: the mean tolerance value between the latter and the nestmate sisters was lower than that between the two nestmates. In contrast, within the experimental group there were no significant differences among the interactions – those between nestmate sisters, or those between the nestmates and their non-nestmates nieces. The tolerance values were in any case very close (8.21 ± 0.89 and 8.22 ± 0.76), which leads one to think that the non-nestmate nieces are effectively treated in the same way as nestmate sisters.

This experiment clearly demonstrates that the recognition cues of *P. fuscatus* include a strong genetic component, given that the nests from which nestmates and non-nestmates originate are sufficiently separated geographically to eliminate interference by an environmental effect. Once again we notice that there is no summation between the effects of the genetic and environmental components of colony odour; non-nestmates have both of these components while the nieces share only a genetic similarity with the nestmates. Nonetheless, the mean tolerance value is not higher between nestmate sisters than between non-nestmate nieces and nestmate sisters.

Another important point to emphasize is that the genetic relatedness between sisters is twice as big as that between aunts and nieces, while the mean tolerance value between sisters is identical with that between those sisters and their non-nestmate nieces. Despite some variations in the recognition cues the mean tolerance value does not vary.

With an experimental design adapted for field study, Gamboa (1988) compared the mean tolerance between sisters, first cousins and unrelated gynes. The experimental procedure consisted of inducing three to five females to fly out of their comb (by flushing them), and then removing the nest box containing the comb (the total number of adults in each of the colonies studied ranged from 6 to 40). Substituted in its place were either the same nest box (control), a nest box containing a sister colony (cousin treatment), or a nest box containing an unrelated colony (unrelated treatment). Generally, the flushed wasps took less than 10 minutes to return to their nest-site. Following this, interactions were recorded using the scale of Pfennig et al., (1983b), in order to determine the mean tolerance values for each group.

The results (obtained with 12 colonies in each group) showed that the resident females were equally intolerant towards either unrelated females or females cousins. The controls revealed, as expected, a high tolerance value (approximately double that obtained in the cousin and unrelated treatments).

If we attempt an overview of these recent results (being aware that the experimental contexts used by Gamboa et al., 1987 and Gamboa 1988 are not identical), and if we compare interactions between sisters, aunt–nieces, cousins and unrelated wasps, a critical equilibrium point appears to exist between aunt–nieces and cousins. This equilibrium point was termed the 'cue similarity threshold' by Gamboa et al., (1986a,b).

In 3 of 12 cases, cousins showed the same high tolerance value as the controls, while the unrelated gynes were poorly tolerated by the same resident colony. This indicates that in these cases cousins shared a critical minimum number of alleles, which affected recognition odours in the same way as in all of the aunt–niece interactions.

As far as their average genetic relatedness is concerned, aunt–nieces in Hymenoptera are related at one-half the level of sisters, but are twice as closely related as cousins. Behavioural responses, however, do not follow this progression, since acceptance is suddenly exhibited at a point between cousins and aunt–nieces. This is what the concept of cue similarity threshold is about.

Compared to the ants there is apparently less data on the chemical characteristics of the cuticle of *Polistes* with reference to nestmate recognition. Studies of the lipids involved, also focus on the material of the nest, which contributes to the chemical stimuli learned during imprinting (when the individual elaborates the template of her nestmates) (Espelie & Hermann, 1990; Espelie *et al.*, 1990). It is likely that more information on this subject will become available during the next few years.

The carpenter ants (genus *Camponotus*) have been well-studied with respect to intraspecific colony recognition. The first study, on the northern Mediterranean species *C. vagus*, was published in 1983 by L. Morel. In order to determine the behavioural mechanism responsible for the mutual recognition between callows and mature nestmate workers, Morel prevented individuals from emerging in their home nest. It is easy to remove cocoons just before eclosion because this species is black, and one can see through the cocoon walls that pupae are sufficiently pigmented to indicate that emergence is imminent (the movement of limbs by the pupa confirms that the time is appropriate for removal). After removal the cocoon is opened with forceps and the pharate callow worker is released from the pupal exuvium. Young adult ants which have been eclosed artificially in this way are placed together for periods ranging from a few hours to a few days (up to 15 days). Thereafter, when they are brought together with nurses or foragers from their natal colony, the callows display intense aggression, regardless of the behavioural subcaste of the mature workers. They are also attacked by the latter, and the foragers exhibit more intense aggression, than nurses. This reveals that interactions during emergence, and in the time period which directly follows emergence, are very important, and even indispensable for the callow to develop its ability to recognize nestmates and to acquire the chemical visa of her colony (there is no significant difference between callows a few hours old and those aged a few days). In other words, there is on the one hand an early learning process with a short critical period, and on the other adsorbtion of chemicals (which constitute the colony odour) onto the cuticle of callow workers. The same results were obtained later by Morel (1988) with *C. floridanus*. In this species, the aggressive interactions of callow workers are less intense than those of *C. vagus*. Information about nestmate recognition, which was acquired during emergence, was found to be retained after a month of separation, beginning only a few hours after emergence.

In the neotropical species *C. abdominalis* Errard (1984b) compared the behaviour within the same adoptive colony of callows, which emerged from previously-adopted heterocolonial cocoons, of heterocolonial callows introduced immediately after emergence in their natal colony, and of callows which had emerged experimentally (following Morel's techniques) and were then isolated for eight days before adoption. All of the adoptees originated from the same foreign colony. These different groups were compared among themselves and with control ants, which were the workers of the adoptive colony and were similar in age to that of the adoptees. The results (two to four observations per day over 20 days) showed that the control ants, as well as the callows which emerged in the adoptive colony, provided most of the care received by the brood. The controls, however, were the most active nurses. The frequency of interactions with the queen (antennal contacts, allogrooming, trophallaxis) separated the controls clearly from all the adopted individuals. The controls produced more interactions with their *kin mother*, than the three groups of adoptees, and no significant difference appeared among the latter in the levels of their interactions with their adoptive mother.

Furthermore, the groups of adopted ants exhibited among themselves more altruistic interactions than were shown by any of them towards the workers of the control group. It appears that, despite the introduction of callows in the adoptive colony, the monitoring of altruistic behaviours revealed a tendency to prefer nestmates. This result (obtained in *C. abdominalis*, and later confirmed in *C. floridanus* by Carlin & Hölldobler, 1987b), suggests the possibility that pre-imaginal learning affects the nature of altruistic behaviour, but not intraspecific aggression. The latter, in contrast, is determined by imprinting which occurs during and soon after emergence.

The origin of recognition cues was studied by Carlin & Hölldobler (1986) in small colonies of *C. pennsylvanicus*. In this study, the role of early learning of the imprinting type previously observed by other authors was confirmed. Here again the method of adoptions was used. Founding queens were led to adopt brood from one or two different colonies (the number of workers adopted was less than eight). After overwintering, the adopted workers from this brood were tested against unfamiliar sisters, and their aggressivity was recorded on a scale of seven possible aggressive behaviours (this allows calculation of the frequency distribution of aggression). The results showed that the adopted workers are highly aggressive towards unfamiliar sisters, regardless of their diet, and regardless of the proportion of adoptees in the conspecific colony (even when they constituted 100%). Therefore, the founding queen seems to be the main source of the recognition cue. When unrelated workers shared a common, switched queen, they did not show more aggression to one another than to familiar sisters. Finally, workers from two orphaned worker groups are able to recognize each other if they are kin (the major factor) and if they were fed with the same diet (the minor factor). All these results lead Carlin & Hölldobler to conclude that the most significant cues involved in colony odour originate from the queen, with worker cues less important, and substances obtained from the diet still less so. One can hypothesize that the low worker numbers is the reason why the larger founding queen is the source of the most powerful recognition labels. However, a study on *C. floridanus* by Carlin & Hölldobler (1987a) seems to show the opposite: with conspecific mixed colonies of about 190 workers they confirmed the results obtained with *C. pennsylvanicus*, but only when the adoptive queen exhibited a high level of reproductive activity, measured by ovarian development and brood production. When the queen had less well-developed ovaries the workers were able to exceed her influence in production of recognition labels. In this case the workers may be less aggressive towards unfamiliar sisters than towards unfamiliar unrelated workers.

In a recent study on orphaned groups of *C. floridanus* workers, Morel & Blum (1988) showed that callow workers reared by nurses from an alien colony recognize as nestmates the unfamiliar sisters of their nurses, rather than their own unfamiliar sisters. This confirms that, in the absence of a queen, callow workers elaborate their template for nestmate recognition based upon the chemical visa of the workers which tend them following eclosion, rather than on the basis of their own odour. Conversely, foragers recognize as nestmates foreign callows reared by their sisters. However, these same foragers attacked their callow sisters when they were tended by unrelated nurses. When the foragers are brought together with unfamiliar related callows or unfamiliar unrelated callows, both reared by nurses unrelated to the foragers, however, the foragers are more aggressive towards unrelated callows. To interpret this result we must assume that chemical cues circulate from one individual to another within the colony, as observed by Errard & Jallon (1987) in interspecific mixed colonies. This introduces a further question about the nature of the colonial chemical visa in *Camponotus* – what are the chemicals involved?

A partial answer is obtained from a study on *C. vagus* by Bonavita-Cougourdan & Clement (1986), who used as dummies ants killed by freezing, deprived of their odour by washing with pentane, and then painted with an extract derived from alien workers. When these dummies were presented to their sisters they released strong aggressive behaviour. If the extract had been prepared from nestmate workers no aggressive behaviour was recorded. These researchers established a correlation between the behavioural response to the dummy and the gas chromatographic profile of the hydrocarbon cocktail of the extracts (which probably originated from the cuticular waxes). However, the acceptance of dummies comprising alien workers painted with extracts from workers of the colony tested remains to be demonstrated, and substances other than hydrocarbons might be present in the pentane or hexane extracts. Nonetheless, these results strongly suggest that cuticular hydrocarbons are intimately involved in nestmate recognition in *C. vagus*.

The same conclusion was reached by Morel *et al.*, (1988) in *C. floridanus*, using a different extraction technique and conducting a multivariate analysis of the 15 major gas chromatographic peaks obtained from the extracts. Similarly, they clearly showed that the callow workers obtain their recognition labels from their nurses, during emergence and/or the 12 hours that follow: newborn individuals have quite a different hydrocarbon profile, when compared to mature workers from the same colony.

In order to consolidate current knowledge about the main features of the

Table 3.3. *Synopsis of the information presently available to interpret how a* Camponotus *callow worker recognizes, and is recognized by, her nestmates.* ON = *a naive callow ant less than 12-hours old (who was artificially eclosed by the experimenter, and isolated from her colony during several following hours);* OC = *a control callow ant, less than 12-hours old (eclosed in an ant nest);* C = *a control callow ant more than 4-days old.*

	Experimental data	Interpretations
V I S A	*ON* is attacked more by workers of an unrelated colony than by related workers.	*ON* has an odour which is characteristic of the group to which she is related. This odour can be acquired either at a pre-imaginal stage, or produced endogenously.
	ON is attacked less by workers from an alien colony than her own *OC* sisters.	*ON* is more acceptable to an alien colony because she is less strongly odouriferous and/or she has an acceptance pheromone which elicits tolerance by adults, including non-nestmates
	Foragers recognize *C* reared with mature workers sharing the phenotype of the foragers.	Odours transferred from nurses to the callows they rear contribute to the formation of the chemical visa on the cuticle.
T E M P L A T E	(1) *ON* attacks her unfamiliar sisters. (2) *C* reared by unrelated nurses recognizes as nestmates the sisters of her nurses rather than her own, unfamiliar, sisters.	*ON/C* learns to recognize her nestmates and is able to generalize from her early experience.
	C isolated from her mature nurses for 30 days after emergence is still able to recognize her nestmates when she is brought back into either a related or an adoptive colony.	The learning process involved in kin recognition permits a long-term memorization of the colony odour template. This process belongs to the family of imprinting-like phenomena.

dynamic processes leading to the integration of a newborn callow worker into its colony, the major experimental results and their conclusions are shown in Table 3.3 (this is based mostly on data obtained from *C. floridanus*).

Conclusion

Despite numerous studies, the difficulties involved in determining the exact nature of colony odour suggest the existence of a subtle chemical cocktail. This contrasts with the relative simplicity of other pheromones.

This complexity results from evolutionary imperatives found throughout hymenopteran societies: the necessity for closure of the society, and the need for development of intolerance by its members toward those of other societies, whether conspecifics or not. Recent studies of polistine wasps and higher ants lead to the conclusion that members of the colony are the principal source of colony odour (i.e. workers and/or queens), and that environmental cues may sometimes also be involved.

Each individual is potentially the source of chemical cues which constitute the colony odour. If an individual is the source of odours for its nestmates (either by direct contact, or via the nest), there are no *a priori* reasons why it should not also be a source of odours for itself. It is thus likely that the chemical visa of an individual includes some self-produced labels.

Future studies will no doubt consider the relationship between the production of colony odour and age polyethism. Indeed, it would not be surprising, given that workers produce components of the colony odour, if it were discovered that the synthesis and release of these could vary quantitatively with age. In *Polistes*, where the nest material itself seems to be a major source of the colony odour, individuals which are more active in nest construction could be more influential in adding secretions to the paper-maché, if these are involved in recognition. In *Camponotus*, where the nurses play an important part in the 'initiation' of callows, it is not impossible that this behavioural subcaste is more active in the production of colony odour than others.

It is important in hymenopteran societies that aggressive interactions are directed in a precise manner, so that altruistic behaviour does not benefit genomes which are distant from those of the colony. However, it is also essential that aggression should only be beyond the error limit which is necessary to allow both for the variability of individual visas within the colony (which is greater with an increase in the number of egg layers), and to prevent interference from foreign odours (e.g. a forager may be contaminated by its prey or during a fight with a conspecific alien). Between these two constraints lies the *Rubicon* which separates familiar from alien. This interpretation is compatible with the 'cue similarity threshold concept' of Gamboa *et al.*, (1986a,b).

I am of the opinion that, when two adult individuals tolerate each other, it is more a consequence of factors which they shared than those which are different. In contrast, intolerance is probably affected more by the perception of differences, when the shared components of templates are not important enough to be significant. Much additional work is necessary to confirm or deny this interpretation.

General discussion: kinship, fellowship, and the importance of ethological mechanisms during development

The sociobiological approach currently deals only with the correlations between the two extremities of the continuum which extends from genome to the altruistic behaviour. This often gives rise to learned mathematical interpretations. The drawback of this approach is that the intermediate steps in the continuum can be obscured, i.e. at the level of the physiological, ethological and ecological mechanisms which are involved in the conflict between tolerance and intolerance, or in that between indifference and active preference (the first alternative does not necessarily imply the latter).

Knowledge of the mechanisms must not be neglected. On the one hand it is inherently important, and on the other hand it sometimes demonstrates that the two extremities which are referred to above are not always part of the same continuum.

When one sums up the studies of mechanisms involved in ethogenesis (i.e. the development of an individual's behaviour), one realizes that learning processes have been discovered almost every time they had been previously hypothesized in nestmate recognition in ants and social wasps. It can be objected that this is not fundamental, because evolution has preferred the plasticity of early learning, relative to the rigidity and the risk involved in a strictly genetic determinism. However, early learning is not always available to orientate altruism towards individuals of the same kin. This is clear in interspecific mixed colonies, where the relationship between host and parasite orientates the altruistic behaviour of the host species to benefit a different, allospecific, genome. This is also clear in colonies of some non-parasitic species, as shown for example in a recent study by Venkataraman *et al.*, (1988) on the polistine wasp *Ropalidia marginata*. Multiple mating is the rule in this species, and individuals are unable to discriminate between full and half sisters; however, they need to learn to recognize their nestmates through a process of early experience. This is one of the many examples discussed here which demonstrate that the ethogenesis of nestmate recognition is based directly on spatial proximity of the individuals involved, rather than on their kinship. Of course, there are many cases in which there is correspondence between kinship and the familiarity (which I have termed *fellowship*; Jaisson 1985, 1987) which results when individuals (whose ethogenesis is in progress) live together with their older nestmates. It is on the basis of their fellowship that the individuals thereafter orientate their altruistic behaviour, not on the basis of the genetic relatedness (see Fig. 3.1). In the naturally-mixed colonies of slave-maker ants, the successive chain of events shown in Fig. 3.1 can operate even with the altruistic individuals having contrasting genetic interests from those of the beneficiaries of the altruism.

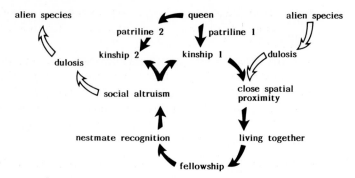

Fig. 3.1 Diagram illustrating (1) the steps connecting kinship status (genetic bonds between individuals) and fellowship status (bonds built through early experience processes), and (2) the consequences of fellowship on the orientation of altruistic behaviours. The dark arrows indicate the typical pattern in a colony with a single queen (who might have been mated more than once), the light arrows indicate modifications to this typical situation in instances when slave-making occurs (modified from Jaisson, 1987).

Regardless of what one believes the function of fellowship to be, whether this leads to the spread of the particular genetic formula of the individuals displaying altruism or not, the major characteristic on which the entire dynamic of nestmate recognition is based remains the existence of sensitive periods. These allow the ant or social wasp early in life to elaborate the template of social partners. The timing and duration of the sensitive periods clearly varies from one species to another, and, even within one species, it can be influenced by the experimental methods (Jaisson & Fresneau, 1978). In the wasp *Polistes*, the sensitive period occurs in newly-emerged individuals and lasts only a few hours following the exit by the callow from its cell. In ants it occurs in general during emergence and the following few days (usually the first week after emergence). However, in the formicine *Cataglyphis cursor*, a pre-imaginal sensitive period occurs during larval life (Isingrini *et al.*, 1985, 1986), and thus nestmate preference is already displayed by the newly-emerged callow worker. In this case, the young ant tends unfamiliar unrelated larvae (sisters of the nurses which cared for her when she was a larva) rather than sister larvae. The pupal instar, in *C. cursor* (and *Camponotus floridanus* (Carlin & Schwartz, 1989)), could be a stage during which the social experiences of larvae are consolidated in the memory of the future adult.

Early learning during sensitive periods is not the only characteristic of the immature stages in the life of ants or wasps. Additional characteristics are associated with the integration process. I have suggested five of these which might be typical of callow workers, the duration of which varies, each relative to the others, and from one species to another (Jaisson, 1985):

- tolerance and absence (or low levels) of aggressive behaviour
- production of acceptance pheromones which inhibit adult aggression
- attractiveness (possibly an effect of the acceptance pheromones)
- that adsorbtion of the colony odour must be possible, and efficient enough for prompt acquisition of the chemical visa
- reduced mobility, which ensures that odours from outside the nest will not be acquired, and that those which are relevant to nestmate recognition will be clearly perceived

There remains much scope for researchers who wish to study these characteristics, which are so vital as contributors to the integration of young individuals into the fellowship of the colony, and thus central to our understanding of the societies of hymenopterous insects, which have been justifiably 'much admired by man' (Joseph Banks, H.M.S. *Endeavour* journal, 1896).

Acknowledgments

I wish to thank Christian Peeters and Robert W. Taylor for improving the English text.

References

Allen J. L., Schulze-Kellman, K. & Gamboa, G. J. (1982). Clumping patterns during overwintering in the paper wasp, *Polistes exclamans*: effects of relatedness. *J. Kans. Entomol. Soc.*, **55**, 97–100.

Beaumont, J. de (1958). Le parasitisme social chez les guêpes et les bourdons. *Mitt. Schweiz. Entomol. Gesell.*, **31**, 168–76.

Beaumont, J. de & Matthey, R. (1945). Observations sur les *Polistes* parasites de la Suisse. *Bull. Soc. Vaud. Sc. Nat.*, **62**, 439–54.

Bennett, B. (1988). Discrimination of nestmate and non-nestmate sexuals by ants (Hymenoptera: Formicidae). *Insectes Soc.*, **35**, 82–91.

Berton, F., Lenoir, A., Leroux, A.M. & Leroux, G. (1990). Reconnaissance et attractivité de la reine de *Cataglyphis cursor* (Hymenoptera Formicidae). *Actes Coll. Insectes Soc.* (In press.)

Bonavita-Cougourdan, A. & Clément, J.-L. (1986). Processus de reconnaissance chez la fourmi *Camponotus vagus* Scop. *Bull. S.F.E.C.A.*, **1**, 49–55.

Bonavita-Cougourdan, A., Clément, J.-L. & Lange, C. (1988). Reconnaissance des larves chez la fourmi *Camponotus vagus* Scop. Phénotypes larvaires des spectres d'hydrocarbures cuticulaires. *C. R. Acad. Sci. Paris, D.*, **306**, 299–305.

Bornais, K. M., Larch, C. M., Gamboa, G. J. & Daily, R. B. (1983). Nestmate discrimination among laboratory overwintered foundresses of the paper wasp, *Polistes fuscatus* (Hymenoptera: Vespidae). *Can. Entomol.*, **115**, 655–8.

Breed, M. D. & Bennett, B. (1986). Colony member discrimination. In *The Individual and Society*, ed. L. Passera & J.-P. Lachaud, pp. 21–30. Toulouse: Privat.

Breed, M. D. & Bennett, B. (1987). Kin recognition in highly eusocial insects. In *Kin Recognition in Animals*, ed. D. J. C. Fletcher & C. D. Michener, pp. 243–85. New York: John Wiley & Sons.

Carlin, N. & Hölldobler, B. (1983). Nestmate and kin recognition in interspecific mixed colonies of ants. *Science*, **222**, 1027–9.

Carlin, N. & Hölldobler, B. (1986). The kin recognition system of carpenter ants (*Camponotus* spp.). I. Hierarchical cues in small colonies. *Behav. Ecol. Sociobiol.*, **19**, 123–34.

Carlin, N. & Hölldobler, B. (1987a). The kin recognition system of carpenter ants (*Camponotus* spp.). II. Larger colonies. *Behav. Ecol. Sociobiol.*, **20**, 209–17.

Carlin, N. & Hölldobler, B. (1987b). The kin recognition system of carpenter ants (*Camponotus* spp.) III. Within-colony discrimination. *Behav. Ecol. Sociobiol.*, **20**, 219–27.

Carlin, N. F. & Schwartz, P. H. (1989). Pre-imaginal experience and nestmate brood recognition in the carpenter ant, *Camponotus floridanus*. *Anim. Behav.*, **35**, 89–95.

Crosland, M. W. J. (1988). Inability to discriminate between related and unrelated larvae in the ant *Rhytidoponera confusa* (Hymenoptera; Formicidae). *Ann. Entomol. Soc. Am.*, **81**, 844–50.

Crosland, M. W. J. (1989a). Kin recognition in the ant *Rhytidoponera confusa*. I. Environmental odour. *Anim. Behav.*, **37**, 912–19.

Crosland, M. W. J. (1989b). Kin recognition in the ant *Rhytidoponera confusa*. II. Gestalt odour. *Anim. Behav.*, **37**, 920–6.

Crosland, M. W. J. (1989c). Intraspecific aggression in the primitive ant genus *Myrmecia. Insectes Soc.*, **36**, 161–72.

Crozier, R. H. (1987). Genetic aspects of kin recognition: concepts, models, and synthesis. In *Kin Recognition in Animals*, ed. D. J. C. Fletcher & C. D. Michener, pp. 55–73. New York: John Wiley & Sons.

Crozier, R. H. & Dix, M. W. (1979). Analysis of two genetic models for the innate components of colony odor in social hymenoptera. *Behav. Ecol. Sociobiol.*, **4**, 217–24.

Darwin, C. (1859). *On the Origin of Species*. London: John Murray.

Douglas, A. & Brown, W. L. (1959). *Myrmecia inquilina* new species: the first parasite among the lower ants. *Insects Soc.*, **6**, 13–19.

Emerson, A. E. (1952). The superorganismic aspects of the society. In *Structure et Physiologie des Sociétés Animales*, ed. P.-P. Crasse, pp. 333–53. Paris: Colloque International 34, C.N.R.S.

Emery, C. (1909). Uber der ursprung der dulotischen, parasitischen und myrmekophilen ameisen. *Biol. Zentralbl.*, **29**, 352–62.

Errard, C. (1984a). Evolution, en fonction de l'âge, des relations sociales dans les colonies mixtes hétérospécifiques chez les fourmis des genres *Camponotus* et *Pseudomyrmex*. *Insectes Soc.*, **31**, 185–98.

Errard, C. (1984b). Influence des stimulations sociales précoces sur l'intégration sociale de l'adulte de *Camponotus abdominalis* (Hymenoptera: Formicidae). In: *Processus d'Acquisition Précoce. Les Communications*, ed. A. de Haro & X. Espadaler, pp. 71–80. Barcelona: Universidad Autónoma Press.

Errard, C. (1986a). Interactions biotope-phylogenèse sur la tolérance interspécifique chez les fourmis. *Actes Coll. Insectes Soc.*, **3**, 143–52.

Errard, C. (1986b). Artificial mixed colonies: a model for the investigation of colony odour in ants. In *The Individual and Society*, ed. L. Passera & J.-P. Lachaud, pp. 55–66. Toulouse: Privat.

Errard, C. (1986c). Role of early experience in mixed colony odor recognition in the ants *Manica rubida* and *Formica selysi*. *Ethology*, 72, 243–9.

Errard, C. & Jallon, J.-M. (1987). An investigation of the development of chemical factors in ants intrasociety recognition. In *Chemistry and Biology of Social Insects*, ed. J. Eder & H. Rembold, p. 478. München: Peperny Verlag.

Espelie, K. E. & Hermann, H. R. (1990). Surface lipids of the social wasp *Polistes annularis* (L.) and its nest and nest pedicel. *J. Chem. Ecol.*, 16, 1841–52.

Espelie, K. E., Wenzel, J. W. & Chang, G. (1990). Surface lipids of social wasp *Polistes metricus* Say and its nest and nest pedicel and their relation to nestmate recognition. *J. Chem. Ecol.*, 16, 2229–41.

Fielde, A. M. (1901). Further study of an ant. *Proc. Acad. Sc. Philad.*, 53, 521–44.

Fielde, A. M. (1903). Artificial mixed nests of ants. *Biol. Bull.*, 5, 320–5.

Fielde, A. M. (1904). Power of recognition among ants. *Biol. Bull. Mar. Biol. Lab.*, 7, 227–50.

Fielde, A. M. (1905). The progressive odor of ants. *Biol. Bull.*, 10, 1–16.

Forel, A. (1874). Les fourmis de la Suisse. *Nouv. Mém. Soc. Helv. Sc. Nat.*, 26, 1–447.

Fresneau, D. (1980). Fermeture des sociétés et marquage territorial chez des fourmis du genre *Neoponera*. *Biol. Ecol. Médit.*, 7, 205–6.

Gadagkar, R. (1985). Kin recognition in social insects and other animals. A review of recent findings and a consideration of their relevance for the theory of kin selection. *Proc. Indian Acad. Sci. (Anim. Sci.)*, 94, 587–621.

Gamboa, G. J. (1988). Sister, aunt–niece, and cousin recognition by social wasps. *Behav. Genet.*, 18, 409–23.

Gamboa, G. J., Klahn, J. E., Parman, A. O. & Ryan, R. E. (1987). Discrimination between nestmate and non-nestmate kin by social wasps (*Polistes fuscatus*, Hymenoptera: Vespidae). *Behav. Ecol. Sociobiol.*, 21, 125–8.

Gamboa, G. J., Reeve, H. K. & Pfenning, D. W. (1986a). The evolution and ontogeny of nestmate recognition in social wasps. *Ann. Rev. Entomol.*, 31, 431–54.

Gamboa, G. J., Reeve, H. K., Ferguson, I. D. & Wacker, T. L. (1986b.) Nestmate recognition in social wasps: the origin and acquisition of recognition odours. *Anim. Behav.*, 34, 685–95.

Goldstein, M. H. & Topoff, H. (1985). Reaction of the ant *Novomessor albisetosus* to intruders in the nest arena. *Insectes Soc.*, 32, 173–85.

Goodloe, L., Sanwald, R. & Topoff, H. (1987). Host specificity in raiding behavior of the slave-making ant *Polyergus lucidus*. *Psyche*, 94, 39–44.

Hamilton, W. D. (1964). The genetical evolution of social behaviour. *J. Theor. Biol.*, 7, 1–52.

Hare, J. F. & Alloway, T. M. (1987). Early learning and brood discrimination in leptothoracine ants (Hymenoptera: Formicidae). *Anim. Behav.*, 35, 1720–4.

Haskins, C. P. & Haskins, E. F. (1950). Notes on the biology and social behavior of the archaic ponerine ants of the genera *Myrmecia* and *Promyrmecia*. *Ann. Entomol. Soc. Am.*, 43, 461–91.

Haskins, C. P. & Haskins, E. F. (1979). Worker compatibilities within and between populations of *Rhytidoponera metallica*. *Psyche*, 86, 299–312.

Hölldobler, B. & Michener, C. D. (1980). Mechanisms of identification and discrimination in social Hymenoptera. In *Evolution of Social Behaviour: Hypothesis and Empirical Tests*, ed. H. Markl, pp. 35–58. Weinheim: Verlag Chemie.

Hölldobler, B. & Taylor, R. W. (1983). A behavioral study of the primitive ant *Nothomyrmecia macrops* Clark. *Insectes Soc.*, **3**, 384–401.

Hölldobler, B. & Wilson, E. O. (1977). The number of queens: an important trait in ant evolution. *Naturwiss.*, **64**, 8–15.

Huber, P. (1810). *Recherches sur les Moeurs des Fourmis Indigènes*. Paris: J. J. Paschoud.

Isingrini, M. (1987). La reconnaissance coloniale des larves chez la fourmi *Cataglyphis cursor* (Hymenoptera, Formicidae). *Insectes Soc.*, **34**, 20–7.

Isingrini, M. & Lenoir, A. (1986). La reconnaissance coloniale chez les hyménoptères sociaux. *Ann. Biol.*, **25**, 219–54.

Isingrini, M., Lenoir, A. & Jaisson, P. (1985). Preimaginal learning as a basis of colony-brood recognition in the ant *Cataglyphis cursor*. *Proc. Natl. Acad. Sci. USA*, **82**, 8545–7.

Jaffé, K. (1983). Chemical communication system in the ant *Atta cephalotes*. In *Social Insects in the Tropics* ed. P. Jaisson, pp. 165–80. Villetaneuse: Presses de l'Université Paris-Nord.

Jaffé, K., Lopez, M. E. & Aragort, W. (1986). On the agonistic communication systems of the ants *Pseudomyrmex termitarius* and *P. triplarinus*. *Insectes Soc.*, **33**, 105–17.

Jaffé, K. & Marcuse, M. (1983). Individual recognition and territorial behaviour in the ant *Odontomachus bauri*. *Insectes Soc.*, **30**, 466–81.

Jaffé, K. & Navarro, J. G. (1985). Communicación química entre obreras de la hormiga cortadora de grama *Acromyrmex landolti*. *Rev. Bras. Entomol.*, **28**, 351–62.

Jaffé, K. & Sanchez, C. (1984). Nestmate recognition and territorial behaviour in the ant *Camponotus rufipes*. *Insectes Soc.*, **31**, 302–15.

Jaisson, P. (1971). Expériences sur l'agressivité chez les fourmis. *C. R. Acad. Sci. Paris, D*, **273**, 2320–3.

Jaisson, P. (1972). Note préliminaire sur l'ontogenèse du comportement de soin au couvain chez la jeune fourmi rousse (*Formica polyctena* Först.): rôle d'un mécanisme probable d'imprégnation. *C. R. Acad. Sci. Paris, D*, **275**, 2721–3.

Jaisson, P. (1973). L'imprégnation dans l'ontogenèse du comportement de soins aux cocons chez les formicines. In *Proc. VIIth Int. Conf. IUSSI*, pp. 176–80, London.

Jaisson, P. (1974). Etude du développement des comportements de soins aux cocons chez la jeune fourmi rousse (*Formica polyctena* Först) élevée en milieu précoce hétérospécifique. *C. R. Acad. Sci. Paris, D*, **279**, 1205–7.

Jaisson, P. (1975a). L'imprégnation dans l'ontogenèse du comportement de soin aux cocons chez la jeune fourmi rousse (*Formica polyctena* Först.). *Behaviour*, **52**, 1–37.

Jaisson, P. (1975b). *Contribution à l'étude de l'expérience précoce chez les fourmis*. Université Pierre-et-Marie Curie, Paris: Thèse Doct. Etat.

Jaisson, P. (1980). Les colonies mixtes plurispécifiques: un modèle pour l'étude des fourmis. *Biol. Ecol. Médit.*, **7**, 163–6.

Jaisson, P. (1985). Social behaviour. In *Comprehensive Insect Physiology, Biochemistry and Pharmacology*, ed. G. A. Kerkut & L. I. Gilbert, vol. 9, pp. 673–94.

Jaisson, P. (1987). The construction of fellowship between nestmates in social hymenoptera. In *From Individual to Collective Behaviour in Social Insects*, ed J. M. Pasteels & J.-L. Deneubourg, Experientia Suppl., vol. 54, pp. 313–31. Basel: Birkhauser Verlag.

Jaisson, P. & Fresneau, D. (1978). The sensitivity and responsiveness of ants to their cocoons in relation to age and methods of measurement. *Anim. Behav.*, **26**, 1064–71.

Jutsum, A. R., Saunders, T. S. & Cherrett, J. M. (1979). Intraspecific aggression in the leaf-cutting ant *Acromyrmex octospinosus*. *Anim. Behav.*, **27**, 839–44.

Klahn, J. E. & Gamboa, G. J. (1983). Social wasps: discrimination between kin and nonkin brood. *Science*, **221**, 482–4.

Lange, R. (1960). Über die futterwitergabe swischen angehörigen verschiedener waldameisenstaaten. *Z. Tierpsychol.*, **17**, 389–401.

Le Moli, F. & Mori, A. (1982). Early learning and cocoon nursing behaviour in the red wood ant *Formica lugubris* Zett. (Hymenoptera: Formicidae) *Boll. Zool.*, **49**, 93–7.

Le Moli, F. & Mori, A. (1985). The influence of the early experience of worker ants on enslavement. *Anim. Behav.*, **33**, 1384–7.

Le Moli, F. & Mori, A. (1987). The problem of enslaved ant species: origin and behavior. In *From Individual to Collective Behaviour in Social Insects*, ed. J. M. Pasteels & J.-L. Deneubourg, Experientia Suppl., vol. 54, pp. 333–63. Basel: Birkhauser Verlag.

Le Moli, F. & Parmigiani, S. (1982). Intraspecific combat in the red wood ant (*Formica lugubris* Zett.). *Aggress. Behav.*, **8**, 145–8.

Le Moli, F. & Passetti, M. (1977). The effect of early learning on recognition, acceptance and care of cocoons of the ant *Formica rufa* L. *Atti Soc. Ital. Sci. Nat. Mus. Civ. Stor. Nat. Milano*, **118**, 49–64.

Le Moli, F. & Passetti, M. (1978). Olfactory learning phenomena and cocoon nursing behaviour in the ant *Formica rufa* L. *Boll. Zool.*, **45**, 389–97.

Lenoir, A. (1981). Brood retrieving in the ant *Lasius niger* L. *Sociobiology*, **6**, 153–78.

Lenoir, A., Isingrini, M. & Nowbahari, M. (1982). Le comportement d'ouvrières de *Cataglyphis cursor* introduites dans une colonie étrangère de la même espèce (Hym. Formicidae). In *La Communication chez les Sociétés d'Insectes*, ed. A. de Haro & X. Espadaler), pp. 107–14. Barcelona: Universidad Autónoma Press.

Le Roux, A. M. (1980). Possibilités de réintégration dans leur groupe d'origine d'individus ayant subi une période d'isolement ou un changement de milieu social (*Myrmica ruginodis* Nyl.). *Biol. Ecol. Médit.*, **7**, 203–4.

Lumsden, C. J. (1982). The social regulation of physical caste: the superorganism revived. *J. Theor. Biol.*, **95**, 749–81.

Mabelis, A. A. (1979). Wood ant wars. *Nether. J. Zool.*, **23**, 451–620.

Meudec, M. (1978). Response to and transport of brood by workers of *Tapinoma erraticum* (Formicidae, Dolichoderinae) during nest disturbance. *Behav. Proc.*, **3**, 199–209.

Mintzer, A. (1982). Nestmate recognition and incompatibility between colonies of the acacia ant *Pseudomyrmex ferruginea*. *Behav. Ecol. Sociobiol.*, **10**, 165–8.

Mintzer, A. & Vinson, S. B. (1985). Kinship and incompatibility between colonies of the acacia ant *Pseudomyrmex ferruginea*. *Behav. Ecol. Sociobiol.*, **17**, 75–8.

Morel, L. (1983). Relation entre comportement agressif et privation sociale précoce chez les jeunes immatures de la fourmi *Camponotus vagus* Scop. (Hymenoptera:

92   P. Jaisson

Formicidae). *C. R. Acad. Sci. Paris, D*, **296**, 449–52.

Morel, L. (1988). Ontogenèse de la reconnaissance des membres de la société chez *Camponotus floridanus* (Hymenoptera: Formicidae). Rôle de l'expérience sociale précoce. *Biol. Behav.*, **13**, 59–72.

Morel, L. & Blum, M. S. (1988). Nestmate recognition in *Camponotus floridanus* callow worker ants: are sisters or nestmates recognized? *Anim. Behav.*, **36**, 718–25.

Morel, L. & Vander Meer, R. K. (1988). Do ant brood pheromones exist? *Ann. Entomol. Soc. Am.*, **81**, 705–10.

Morel, L., Vander Meer, R. K. & Lavine, B. K. (1988). Ontogeny of nestmate recognition cues in the red carpenter ant (*Camponotus floridanus*). *Behav. Ecol. Sociobiol.*, **22**, 175–83.

Obin, M. S. (1986). Nestmate recognition cues in laboratory and field colonies of *Solenopsis invicta* Buren (Hymenoptera: Formicidae). *J. Chem. Ecol.*, **12**, 1965–75.

Peeters, C. P. (1988). Nestmate discrimination in a ponerine ant (*Rhytidoponera* sp. 12) without a queen caste and with a low intra-nest relatedness. *Insectes Soc.*, **35**, 34–46.

Pfennig, D. W., Gamboa, G. J., Reeve, H. K., Shellman Reeve, J. & Ferguson, I. D. (1983a). The mechanism of nestmate discrimination in social wasps (*Polistes*, Hymenoptera: Vespidae). *Behav. Ecol. Sociobiol.*, **13**, 299–305.

Pfennig, D. W., Reeve, H. K. & Shellman, J. S. (1983b). A learned component of nestmate discrimination in workers of a social wasp, *Polistes fuscatus* (Hymenoptera: Vespidae). *Anim. Behav.*, **31**, 412–16.

Pisarski, B. (1972). La structure des colonies polycaliques de *Formica* (*Coptoformica*) *exsecta* Nyl. *Ekol. Polsk.*, **20**, 111–16.

Plateaux, L. (1960). Adoptions expérimentales de larves entre des fourmis de genres différents: *Leptothorax nylanderi* Förster et *Solenopsis fugax* Latreille. *Insectes Soc.*, **7**, 163–70.

Plateaux-Quénu, C. (1962). Biology of *Halictus marginatus* Brullé. *J. Apicult. Res.*, **1**, 41–52.

Post, D. C. & Jeanne, R. L. (1982). Recognition of former nestmates during colony founding by the social wasp *Polistes fuscatus* (Hymenoptera: Vespidae). *Behav. Ecol. Sociobiol.*, **11**, 283–5.

Provost, E. (1979). Etude de la fermeture de la sociéte de fourmis chez diverses espèces de *Leptothorax* et chez *Camponotus lateralis* (Hyménoptères Formicidae). *C. R. Acad. Sci. Paris, D*, **288**, 429–32.

Rosengren, R. & Cherix, D. (1981). The pupa-carrying test as a taxonomic tool in the *Formica rufa* group. In *Biosystematics of Social Insects*, ed. P. E. Howse & J.-L. Clément, pp. 263–81, London: Academic Press.

Ross, K. G. (1983). Laboratory studies of the mating biology of the eastern yellow jacket *Vespula maculifrons* (Hymenoptera: Vespidae). *J. Kans. Entomol. Soc.*, **56**, 523–37.

Ross, N. M. & Gamboa, G. J. (1981). Nestmate discrimination in social wasps (*Polistes metricus*, Hymenoptera: Vespidae). *Behav. Ecol. Sociobiol.*, **9**, 163–5.

Ryan, R. E., Cornell, T. C. & Gamboa, G. J. (1985). Nestmate recognition in the bald-faced hornet, *Dolichovespula maculata* (Hymenoptera: Vespidae). *Z. Tierpsychol.*, **69**, 19–26.

Ryan, R. E. & Gamboa, G. J. (1986). Nestmate recognition between males and gynes of the social wasp *Polistes fuscatus* (Hymenoptera: Vespidae). *Ann. Entomol. Soc. Am.*, **79**, 572–5.

Scheven, J. (1958). Beitrag zur biologie der schmarotzerfeldwespen *Sulcolpolistes atrimandibularis* Zimm., *S. semenovi* F. Morawitz und *S. sulcifer* Zimm. *Insectes Soc.*, **5**, 409–37.

Shellman, J. S. & Gamboa, G. J. (1982). Nestmate discrimination in social wasps: the role of exposure to nest and nestmates (*Polistes fuscatus*, Hymenoptera: Vespidae). *Behav. Ecol. Sociobiol.*, **11**, 51–3.

Stuart, R. J. (1987a). Individual workers produce colony-specific nestmate recognition cues in the ant *Leptothorax curvispinosus*. *Anim. Behav.*, **35**, 1062–9.

Stuart, R. J. (1987b). Nestmate recognition in leptothoracine ants: testing Fielde's progressive odor hypothesis. *Ethology*, **76**, 116–23.

Taylor, R. W. (1978). *Nothomyrmecia macrops*: a living fossil ant rediscovered. *Science*, **201**, 979–85.

Trivers, R. L. & Hare, H. (1976). Haplodiploidy and the evolution of the social insects. *Science*, **191**, 249–63.

Tsuji, K. (1988). Inter-colonial incompatibility and aggressive interactions in *Pristamyrex pungens* (Hymenoptera: Formicidae). *J. Ethol.*, **6**, 77–81.

Tsuji, K. & Itô, Y. (1986). Territoriality in a queenless ant, *Pristomyrmex pungens*. *Appl. Ent. Zool.*, **21**, 377–81.

Vander Meer, R. K. & Wojcik, D. P. (1982). Chemical mimicry in the myrmecophilous beetle *Myrmecaphodius excavaticollis*. *Science*, **218**, 806–8.

Venkataraman, A. B. Swarnalata, V. B., Nair, P. & Gadagkar, R. (1988). The mechanism of nestmate discrimination in the tropical social wasp *Ropalidia marginata* and its implications for the evolution of sociality. *Behav. Ecol. Sociobiol.*, **23**, 271–9.

Waldman, B., Frumhoff, P. C., & Scherman, P. W. (1988). Problems of kin recognition. *Trends Ecol. Evol.*, **3**, 8–13.

Wallis, D. I. (1962). Aggressive behaviour in the ant *Formica fusca*. *Anim. Behav.*, **10**, 267–74.

Wallis, D. I. (1963). A comparison of the response to aggressive behaviour in two species of ants, *Formica fusca* and *Formica sanguinea*. *Anim. Behav.*, **11**, 164–71.

Wheeler, W. M. (1911). The ant colony as an organism. *J. Morphol.*, **22**, 307–25.

Wilson, E. O. (1971). *The Insect Societies*. Cambridge: Harvard University Press.

Wilson, E. O. (1975). *Leptothorax duloticus* and the beginnings of slavery in ants. *Evolution*, **29**, 108–19.

Wilson, E. O. (1985). The principles of caste evolution. In *Experimental Behavioral Ecology and Sociobiology*, ed. B. Hölldobler & M. Lindauer, Forschur. Zool. 31., pp. 307–24. Stuttgart: Gustav Fischer Verlag.

4

Successes and failures of parent–offspring recognition in animals

Michael D. Beecher

Introduction

Animals show both impressive feats and surprising failures of recognition. On the one hand, we have the Mexican free-tailed bat mother finding her young in a maternity cave of a million bats. On the other we have the redwinged blackbird parent failing to eject from its nest the conspicuously different eggs of a brood parasitic cowbird. From the evolutionary point of view, the failures of recognition are, on the face of it, much more difficult to explain than the successes. Any treatment of the evolution of recognition systems, therefore, must consider both faces of recognition – the failures as well as the successes. This is the perspective I take in this paper, drawing my examples from the parent–offspring context.

A source of conceptual confusion in the analysis of 'failures' of recognition has been the tendency of investigators to equate 'recognition' with 'discrimination'. Typically, recognition is operationally defined in terms of discrimination, an animal being said to recognize a particular individual or class of individuals if it discriminates that individual or class from others (e.g. discriminates its offspring from other, unrelated young). For successes of recognition, this operational definition is perfectly reasonable. A failure of discrimination, however, could imply either that (1) discrimination is not possible, a true failure of recognition, or (2) discrimination is not adaptive in this circumstance. In the latter case, the animal may well recognize the difference between two classes of individuals but treat them similarly. In this case, additional tests are required to determine if recognition is in fact possible. Throughout this paper I will use the term 'recognition' in a broad, theoretical sense and the term 'discrimination' to refer to specific contexts. Thus, an animal 'recognizes' class A individuals if it discriminates A from non-A, in at least some contexts, but it need not do so in all.

The general recognition model

General perspective

I treat recognition as a communication problem and as a decision problem. The communication perspective gives the individual to be recognized (the potential *sender*) equal billing with the recognizer (the *receiver*). This emphasis is relatively unusual for discussions of recognition; indeed, the term 'recognition' may predispose to a receiver-biased perspective. Viewing recognition as a decision problem opens up the approaches and concepts of decision theory (e.g. Luce & Raffia, 1957; Green & Swets, 1966; McNamara & Houston, 1980). I have developed different aspects of the model to be described here in several papers (Beecher, 1982; Beecher, 1989; Beecher, unpub. data).

Recognition requires the receiver to decide whether or not the sender belongs to the recognition class (conspecific, offspring, mate, neighbour, etc.). To this end the sender may provide identity or *signature* cues. It need not do so (in which case 'sender' is a bit of a misnomer), nor need it necessarily signal 'honestly'. But in any case the receiver is clearly favoured to extract whatever signature information is available. The receiver must *perceive* the signature cues, if they are present, integrate them with whatever contextual evidence is available (e.g. the presence of the sender in the home nest), make its *decision*, and take appropriate *action* (e.g. evict the individual from the nest). In this view, recognition is viewed as having four distinct components: identification (the providing of signature cues by the sender), perception, decision and action by the receiver. All four components are potential targets of natural selection. Thus recognition should be viewed as an outcome of adaptations of identification (signature), perceptual, decision and behavioural mechanisms. 'Adaptation' here means shaped by natural selection for the function specified (Williams, 1966).

Most discussions of recognition in animal behaviour have focused on the question of capability: *can* the animal under study recognize a particular class of animals? Thus, the implicit focus has been on identification and perception (and generally the two processes are not separated out). Yet, in fact, a failure of the animal to discriminate may be traced to any of the four components of recognition, to (a) inadequate signature information, (b) an inability to perceive signature information that is present, (c) a decision not to discriminate in this situation, which in fact may be the optimal decision, or (d) a lack of the appropriate discrimination behaviour (e.g. the receiver may be unable to eject a brood parasite egg from the nest).

Examples in the parent–offspring context

Using the context of parent–offspring recognition in birds as examples, four general recognition cases will be distinguished. In the *detection* case the parent at the nest asks, is this individual my offspring or not? Here the default assumption is that any young in the nest are offspring, and the parent seeks to detect the relatively rare event of an intruder in the nest. In the *simple choice* case, the parent has the additional information that not all the young in the nest are its offspring (it may nevertheless choose to accept them all). The *forced choice* case contains an additional constraint that the parent must make a hard choice between the young (i.e. cannot accept them all). The *search* case takes place away from the nest and is the inverse of the detection problem: which one of these many individuals is my offspring?

In the following discussion, the bank swallow (*Riparia riparia*, the conspecific sand martin in Europe) will be used to illustrate these four cases (see Beecher *et al.*, 1981a). I believe that the arguments are quite general, however, and should apply to a wide variety of recognition cases beyond the parent–offspring context.

Case 1: detection

Imagine a parent at the nest evaluating a particular chick as to relatedness (the 'focal' chick of this discussion). We suppose that in this species unrelated chicks sometimes enter nests and so it is possible that the focal chick is an interloper (though it is not likely). The detection case is defined by the absence of any *contextual evidence* as to the possible presence of an interloper. The parent, therefore, must evaluate the chick on its own merits, so to speak. If there is external evidence of an intruder (e.g. the parent has seen an intruder enter the nest, or counts an extra chick), then we have the simple choice case (see below). As an example, bank swallows live in large, dense colonies and it is not unusual for newly-fledged birds to fly into alien burrows. As these intrusions typically happen while the parents are away from the nest, the interlopers are not immediately detected by parents. Moreover, the physical nature of the burrow impedes counting chicks or side-by-side comparisons. Thus an example such as this conforms reasonably well to our detection model.

The parent must decide if the focal chick is its own, in which case it will accept it, or if the chick is alien, in which case it will reject it. *Acceptance* and *rejection* refer to decisions or actions which irreversibly increase (when correct) or decrease (when incorrect) the parent's reproductive success via effects on the survival probability of its offspring. For example, a bank swallow parent rejects an alien chick either by forcibly evicting the interloper from the nest, or by consistently ignoring (not feeding) it; in the latter case, the chick eventually leaves. We do not use 'rejection' to refer to the short-term action of not feeding on a particular trip to the nest, since this can be made up without loss on subsequent trips. 'Acceptance' similarly refers to a decision having long-term consequences. Because acceptance is the 'default' condition at the nest, we know it primarily as the absence of rejection. Nevertheless, experiments in several species have revealed a 'positive' acceptance process, in which the parents learn the signature characteristics (distinctive odours, calls, visual markings, etc.) of their young, and subsequently accept only individuals with those (or very similar) traits. In addition to this 'signature imprinting', parents typically also track the development of their young and so can detect an interloper that is obviously too old or too young. In the detection case, we are interested in acceptance or rejection that occurs on the basis of direct evidence provided by the chick, either via signatures or via the chick's general appearance, demeanour or behaviour (e.g. is it the right age, does it act nervous or strangely?).

Consider the case where the focal chick is the parent's offspring. If the parent accepts rather than rejects the chick it gets net benefit W. In the case where the chick is an interloper, if the parent accepts rather than rejects the chick it gets net cost $-K$. Thus if p_0 is the probability that a given chick in the nest is the parent's own, and $p_1 = 1 - p_0$ is the probability that a given chick is alien, by the expected value criterion a parent should accept the chick if

$$\frac{p_0}{p_1} \geq \frac{K}{W} \tag{1}$$

Otherwise it should reject the chick. I refer to K/W as the 'payoff function'. The net benefit of a correct acceptance W represents the increment in the fitness of the focal chick summed with the fitness decrements to all other offspring of the parent, present and future. Because we assume that the parent has already chosen the optimum clutch size, W is necessarily positive. The cost of an incorrect acceptance K represents the decrements in the fitnesses of all other offspring of the parent present and future, due to the loss of parental care (mostly lost feedings to chicks in the nest). So long as the alien chick takes not more than its pro rata share of parental care, K will usually be less than W. That is, of the two types of errors, accepting an alien chick will usually be less costly than rejecting your own chick. However, in some circumstances K will be greater than W. To take a case where this is so, consider a brood parasite chick with a voracious appetite that takes enough food for several of the parent's own chicks. Clearly acceptance of the brood parasite chick (costs several offspring) is more costly than rejection of a single offspring (costs that one less the fitness increments to the remaining offspring).

How does the receiver evaluate p_0/p_1? In our decision model, the receiver is viewed as a Bayesian decision maker, choosing between the hypothesis h_0 that the chick is its offspring and the hypothesis h_1 that it is not. We suppose that the parent makes two independent probability estimates concerning the alternative hypotheses, and combines these to arrive at its best estimate of the true state of affairs. First it estimates the *prior probability* that h_0 (or h_1) is true. In this model, the prior probability is derived from information *external* to the sender, given by the structure of the situation. For a parent evaluating a potential offspring, the relevant prior information might include the following. The chick is in the parent's nest, therefore it very likely is the parent's. How likely will depend on the rate of mix-up at the nest in this population of the species. Evaluation of this probability can depend on information this particular parent has obtained – for example, it may have observed that it is in a very large colony, thus mix-ups are more likely – and on its 'innate' knowledge of the situation (see the discussion of the 'innateness of priors' in Staddon, 1980). That is, in the evolutionary past of this species, certain recognition events (e.g. unrelated young sneaking into the nest) have happened with certain average frequencies, and natural selection has shaped decision rules in parents, which reflect these probabilities. To consider another 'external' aspect of the situation which would set the prior probabilities, the parent may observe that there are two chicks in the nest where before there had been only one, hence the prior probability that a given chick is its own is 0.5 (before it was much closer to 1); in this case we have the simple choice problem (see below).

The second, independent estimate of the probability of the alternative hypotheses can be derived from whatever additional information the *sender* supplies during the trial. In general, the major sources of information are the sender's general appearance, its behaviour and its *signature* traits. We assign this evidence a value (X). We treat this evidence as provided by the sender, with no implication that the information is purposefully provided, or that the sender will necessarily benefit. The receiver then estimates the probability of X given each of the two alternative hypotheses, i.e. $p(X/h_0)$ and $p(X/h_1)$. The receiver can then revise its estimate of the *posterior probabilities* $p(h_0/X)$ and $p(h_1/X)$ by Bayes' Rule

$$p(h_0/X) = \frac{p(h_0)\, p(X/h_0)}{p(X)}$$

and

$$p(h_1/X) = \frac{p(h_1)\, p(X/h_1)}{p(X)}$$

These posterior probabilities are the p_0 and p_1 of equation 1. That is,

$$\frac{p_0}{p_1} = \frac{p(h_0)\cdot p(X/h_0)}{p(h_1)\cdot p(X/h_1)} \tag{2}$$

If we substitute them into equation 1 we have that the receiver should accept h_0 if

$$\frac{p(h_0)\cdot p(X/h_0)}{p(h_1)\cdot p(X/h_1)} \geq \frac{K}{W} \tag{3}$$

Otherwise it should reject h_0. I will refer to this as a *likelihood test*. To simplify some formulas, I will let $\Phi = p_0/p_1$, the posterior odds, $\Omega = p(h_0)/p(h_1)$, the prior odds, $L = p(X/h_0)/p(X/h_1)$, the *likelihood ratio*, and $v = (K/W)$, the *pay-off function* (this will vary from situation to situation). Thus equation 2 becomes

$$\Phi = \Omega \cdot L \qquad (4)$$

Furthermore, I will refer to the values of Φ and L just satisfying equation 3 as Φ^* and L^*, the *critical values* of L and Φ respectively, i.e.

$$\Phi^* = v \text{ and } L^* = \frac{v}{\Omega} \qquad (5)$$

Note that for each pair of hypotheses we have two each of the quantities Φ, Ω and L, e.g. $\Omega_0 = p(h_0)/p(h_1)$ and $\Omega_1 = p(h_1)/p(h_0)$, with $\Omega_0 = 1/\Omega_1$. Usually it is not necessary to designate which prior odds, likelihood ratio or posterior odds one is referring to; above we have been referring to L_0 etc.

Two conclusions can be drawn about Case 1 recognition. First, when little or no evidence can be gleaned from the focal chick, blanket acceptance by the parent is favoured. This is primarily because Ω invariably is much greater than 1 (a parent is much more likely to find offspring than unrelated chicks in its nest, but also because v is generally less than 1 (the cost of rejecting your own chick is greater than the cost of accepting an alien chick). Under these circumstances, equation 3 will always be satisfied, and acceptance favoured, unless L_1 is quite large. Hence, even in species in which there is a need for recognition, parental discrimination will not be favoured unless there is very strong evidence as to identity.

Second, by the same reasoning, in species with little need for recognition at the nest (probably the majority of species), the prior odds will overwhelmingly favour h_0, and will offset even strong evidence as to identity.

In summary, non-discrimination will be favoured (1) in species where recognition is seldom required, and (2) also in species where parents would benefit by recognition but cannot extract strong identity information. I will apply these generalizations to specific cases below.

Case 2: simple choice

Consider a parent returning to the nest, where it has left a single chick, and finding two chicks. Suppose that the parent can count, and does. Then it knows one of the chicks is not its own, but is uncertain as to which chick this is. Had it seen the interloper arrive, it might know which one to evict. Since it has not, it must carry out the Bayesian calculation of equation 3. This situation is different in one important respect from the detection problem, the problem the parent has faced on previous trips to the nest. When just one chick was in the nest (the normal state of affairs), the prior odds overwhelmingly favoured h_0 and the parent might not even have bothered to evaluate L. Now with two chicks at the nest, the prior odds favouring one hypothesis over the other are weak (in the symmetrical case they equal 1). Thus the parent is left with L as the major (in the symmetrical case, sole) basis for a judgement.

This simple choice case, unlike the forced choice case to be considered next, does not require a *hard* choice by the parent, and is not fundamentally different from the detection case (except that the prior odds are equal, or much closer, to 1). In particular, the parent can choose to accept *both* chicks. The parent follows the same decision rule as above, for each chick. That is, the parent evaluates Φ for each chick and if each is greater than v, it accepts both chicks. Nevertheless, the change in prior odds has a dramatic effect on the expression of recognition: A 'suspicious' L (say $\ll 1$) that previously could not have offset the high prior odds in favour of h_0, can easily overcome it now.

Case 3: forced choice

This case is similar to the simple choice situation but with the additional constraint that the parent *must* choose between the two young (or two groups of young). A natural circumstance we see in blank swallows occurs when the separation between two burrows breaks down (through erosion or a partial bank collapse) so that it is ambiguous as to which nest is which. A common experimental circumstance consists of providing the parent with a choice between young in a nest just to the left of the old one and a nest just to the right. In both cases, if the two nests contain eggs or very young chicks, the parent must choose between them; it cannot simultaneously incubate both clutches, or brood both sets of chicks. Most situations contain elements of both simple and forced choice. For example if the two nests contain older young, the parent can in fact choose to feed both. Nevertheless, even if the parent splits its care between the two, the split can favour one over the other.

The two competing hypotheses in the (pure) forced choice case are h_{01} that chick A is the parent's offspring, and chick B is not, and h_{10} that A is and B is not. Thus the parent is favoured to accept h_{01} if $\Phi_{01} \geq \Phi_{10}$. Otherwise it should accept h_{10}. Note that the key difference between simple choice and forced choice is not that the alternative hypotheses are mutually exclusive (for they are so in both cases) but that the parent's decisions with respect to the two young (or groups of young) are mutually exclusive – the parent is constrained – in forced choice.

Note that in the forced choice case the pay-off function has been removed from the decision rule, leaving the prior odds and the likelihood ratio as the only variables affecting discrimination. Indeed, many experiments set the prior odds at 1 (e.g. the choice is between young in nests symmetrically placed either side of the original nest); in this case the likelihood ratio is the sole variable affecting discrimination. We have seen that in the detection problem, both the prior odds and the pay-off function bias the parent to conservative, non-discriminative behaviour (accept h_0). In the simple choice situation, and even more so in the forced choice situation, on the other hand, the parent is biased toward discrimination. A parent that is non-discriminating in the detection case may become quite discriminating in the choice case. Especially in the forced choice case, a relatively small difference between chick signatures or behaviour will be used by the parent as a basis for discrimination.

A clear prediction from this argument concerns tests of recognition in animals (field or laboratory): simultaneous recognition tests (one offspring compared with one unrelated young with contextual evidence removed, e.g. both tested on neutral ground) should be more sensitive tests of recognition than chick substitution tests (one or more alien young substituted for one or more offspring). The simultaneous test is analogous to the forced choice case while the substitution test is analogous to the detection case. Evidence on this point will be considered in a later section.

Case 4: search

In many species, recognition becomes a significant problem when the young leave the nest. The most extreme case occurs in those species where the young remain in a large communal group ('creche') while the parents hunt for food. The parent returns to the creche to feed its young, who it must locate in this very large group.

The creche is essentially the inverse of the nest situation. In both cases the parent returns to a location where there are N individuals, one or more of whom is its offspring. At the nest, however, N is small and all or most of the young are offspring, whereas at the creche N is large and only one or a few of the young are offspring (we will assume one throughout the rest of the discussion). Thus the prior odds make acceptance the default assumption at the nest, and rejection the default assumption in the creche.

We assume that once in the creche the parent checks signatures of individuals until it finds one that is 'acceptable', and then it feeds that individual. The problem for the parent is that in this very large group there may be other young with signatures similar to that of its offspring. The larger the group, the greater this danger. If there are N individuals in the creche, then the prior probability of the offspring is $1/N$ and the prior probability of an alien chick is $(N-1)/N$. Thus the prior odds are $1/(N-1)$. Thus if searching for young is to be favoured, strong signature evidence is needed to offset the unfavourable prior odds. That is equation 5 becomes

$$L^* = (N-1)\cdot v \tag{6}$$

It is not realistic to suppose that the parent compares signatures from all N individuals in the creche. Therefore, we assume that the parent has some rule of thumb as to what is an 'acceptable' signature, and that it searches until it finds an individual whose signature falls in this range. To analyze this situation, we require a signature model. The signature model is presented in detail later (see section 'The senders perspective'), and I present here only the features relevant to the search decision (Beecher, 1989; Beecher, unpub. data).

We assume that an individual's signature is not constant, but varies from trial to trial. Variation may originate within the sender, as typically occurs with acoustic signals, or during transmission of the signal through the environment, or within the perceptual system of the receiver. We refer to all such variation as *within-sender variation* (σ^2_W), and assume that it is normally distributed. The greater part of the variation in signatures is *between-sender variation*, (σ^2_B). It is assumed that the sources of between-individual and within-individual variation are independent, hence the total variation $\sigma^2_T = \sigma^2_B + \sigma^2_W$. The key point is that discrimination among individuals will become sharper, the larger the ratio σ_T/σ_W.

We conceive of the searcher as setting up an acceptance region $\mu_i \pm k\sigma_W$ where μ_i is the mean of the within-distribution for the target individual (which is different for every individual), σ_W is the standard deviation of the within-distribution (which is the same for every individual) and k is the number of standard deviation steps defining the acceptance region. We assume that the receiver sets k sufficiently large that its offspring's X_i is almost always within the acceptance region and hence the parent will rarely reject its offspring, should it find it. For the ensuing discussion we assume that the parent sets k at 1.96, which for a normally distributed population gives an acceptance probability of 0.95 (the results are not particularly sensitive to the value of k however, Beecher, unpub. data). The major problem for the parent, therefore, is not that of rejecting its offspring on finding it, but of finding an alien chick whose X_i falls within the acceptance region before it finds its offspring. That is, the parent is concerned with the Bayesian posterior odds

$$\Phi = \frac{\text{prior prob (offspring)} \cdot \text{prob } (X_i \text{ within acceptance region/offspring})}{\text{prior prob (alien)} \cdot \text{prob } (X_i \text{ within acceptance region/alien})}$$

As noted above the prior odds $\Omega = 1/(N-1)$. Elsewhere I show that the likelihood ratio $L \approx \sigma_T/\sigma_W$ (Beecher, unpub. data). Hence

$$\Phi = \frac{1}{N-1} \cdot \frac{\sigma_T}{\sigma_W} \tag{7}$$

approximately. Thus in the search context, the odds of the parent making a correct feeding decision on a given visit are inversely related to the number of individuals in the group and directly related to the ratio of total variation to within-sender variation. The parent should search in the creche if

$$\frac{\sigma_T}{\sigma_W} \geq (N-1)\cdot v \tag{8}$$

Concerning the pay-off function, in the creche the relevant decision for the parent is which chick to feed on this trip to the creche. If the parent delivers the food to its own chick, it gets B, but any feeding, correct or incorrect, costs C. Here B is the increment in the offspring's fitness due to the feeding and C is the decrement in the fitnesses of other offspring of the parent, present and future, due to the feeding. In the creche situation, there is no rejection action comparable to ejecting the young from the nest. Indeed, it will not generally pay the parent to waste time and energy rejecting alien young (say clearly unrelated young that approach it and beg), since the parent is not likely to encounter that young again. This is quite different from the situation in the nest, where an alien young fed on one trial will very likely be there on subsequent trials, therefore making active rejection of this chick beneficial. Thus equation 8 becomes

$$\frac{\sigma_T}{\sigma_W} \geq (N-1)\cdot\frac{C}{B-C} \tag{9}$$

Combining equations 7 and 9 and remembering that $\Phi = p_0/p_1$, we see that the parent should continue feeding in the creche so long as

$$p_0 B \geq C \tag{10}$$

As $B > C$, we have the surprising conclusion that it pays the parent to continue feeding in the face of relatively high probability of misdirected feedings, provided that the benefit devalued by the probability of a correct feeding is greater than the cost of a feeding. Needless to say, it pays the parent to discriminate at higher levels than this minimum, but the key point is that we should not expect to find perfect or even near-perfect recognition in the creche. We return to this prediction of the model below.

Summary

In the four general recognition contexts, the decision function can be described in terms of three variables, the likelihood ratio L of the evidence given the two hypotheses, the prior odds Ω of the two hypotheses (L and Ω can be combined as the posterior odds Φ), and the pay-off function v (a function of the benefits and costs). The basic decision is accept if $\Omega \cdot L \geq v$. The pay-off functions and the prior odds are given. The parent can manipulate the prior odds only indirectly. By increasing its attendance at the nest, for example, the parent might greatly reduce the frequency of interlopers. Similarly, by keeping its young out of the creche, the parent can forego dealing with the unfavourable prior odds of this situation. However, increased attendance at the nest or opting out of the creche will carry their own significant costs, and presumably the parent has already chosen the better alternative. The likelihood ratio, on the other hand, is a variable potentially under the control of the parent and the young (receiver and sender). The sender can provide better signals as to identity, and the receiver can improve its attention to and perception of identity cues and other predictors of relatedness.

The discussion is summarized in Table 4.1. In the *detection* case both the prior odds ($\Omega \gg 1$) and the pay-off function ($v < 1$) favour acceptance (non-discrimination), hence strong evidence must be extracted for parental discrimination to pay. In the *simple choice* case, external evidence drops the prior odds to close to 1 and in the *forced choice* case, the parent is constrained to choose among the young as well. This effectively removes the pay-off function as a factor, leaving the likelihood ratio as the major basis for the decision. Thus in this case even weak evidence will be sufficient for discrimination to pay. In the *search* case, the prior odds favour rejection and the parent should opt out of the situation unless it can extract strong signature information. For this reason, the very existence of the creche situation in a species is prima facie evidence for good signatures. Nevertheless, a fairly high level of mistakes can be tolerated, since v in this case is generally well below 1.

Parent–offspring recognition at the nest: predictions of the model

Prior odds favour blanket acceptance in most but not all cases

In most species, most of the time, presence in the home nest is strong contextual evidence that the young there are in fact the parent's offspring. That is to say, the prior odds of a particular young being unrelated are exceptionally small, often zero. In these circumstances, natural selection might favour a simple decision rule such as, 'if the young is in my nest, accept it'. In a minority of species, however, presence in the nest is not a totally reliable predictor of relatedness. For example, as mentioned above, bank swallow young fly between nests for a few days before totally fledging (Beecher *et al.*, 1981a). In these cases, while the prior odds favour the hypothesis that the young in the nest are your own, it is hardly certain that they are, and the parent will be favoured to extract information about relatedness from the chicks themselves (signature information).

As an example, consider intraspecific cross-fostering experiments (reviews in Falls, 1982; Colgan, 1983). This vast body of studies can be summarized by saying that young cross-fostered between nests are invariably accepted by foster parents, the only exceptions occurring in species where young move

Table 4.1. *Summary of major parent-offspring recognition contexts (a) and decision (b) rules.*

(a)

Case 1. Detection. Typical context: at the nest. Receiver evaluates one of N individuals, who may or may not be its offspring. No external evidence as to prior odds (Ω). Decision rule is accept individual if $L_i > v/\Omega$. Because Ω strongly favours h_0, parent is biased toward acceptance.

Case 2. Simple choice. Typical context: at the nest. External evidence indicates that only $M < N$ of the individuals are offspring, and therefore sets prior odds at near 1 (in symmetrical 2-individual case, $\Omega = 1$ exactly). Parent *not* constrained to choose among the young (can accept them all). Then the decision rule is as in Case 1, but the strong acceptance bias has been removed.

Case 3. Forced choice. As in 2, but here parent is constrained to choose among young. In the 2-choice, 2-individual case, the parent must choose between $h_{0,1}$ that individual A is its offspring and individual B is not, and $h_{1,0}$ that B is and A is not. The decision rule in the 2-choice, 2-individual case is accept A if $\Omega_{01} L_{01} > \Omega_{10} L_{10}$. (Note that $L_{01} = 1/L_{10}$.) Thus the pay-off function v is irrelevant in this case. In the symmetrical case ($\Omega_{01} = \Omega_{10}$) the likelihood ratio becomes the sole determinant of the decision, thus this situation is the most sensitive test for discrimination by signature cues.

Case 4. Search. Typical context: in the creche. N individuals (N large). Receiver searches until it finds individual whose likelihood ratio exceeds L^*. Decision rule same as in Case 1. Prior odds here bias toward rejection. Very good signatures required here to offset unfavourable prior odds.

(b)

	Detection	Simple choice	Forced choice	Search
Ω	very large	≈ 1	≈ 1	very small
v	$\dfrac{K}{W}$	$\dfrac{K}{W}$	1	$\dfrac{C}{B-C}$
Rule	$L_i \geqslant v/\Omega$	$L_i \geqslant v/\Omega$	$\Omega_{01} L_{01} \geqslant \Omega_{10} L_{10}$	$L_i \geqslant v/\Omega$
Bias	acceptance	$\approx$ even	even	rejection

between nests or where large numbers of young intermingle on leaving the nest. For example, we have done cross-fostering experiments on three species of swallow – barn swallows (*Hirundo rustica*), northern rough-winged swallows (*Stelgidopteryx serripenis*) and bank swallows (*Riparia riparia*). Of these, only the colonial bank swallow gave clear evidence of recognition (Beecher *et al.*, 1981b). In comparable experiments with the noncolonial rough-winged swallows and barn swallows, on the other hand, exchanged and sham-exchanged birds were accepted equally (Medvin & Beecher, 1986; Beecher & Beecher, unpub. data; see also, Hoogland & Sherman, 1976). This contrast is understandable in that only in bank swallows does extensive mix-up of young occur (outside of the nest, and to a lesser extent in the nest). Even in bank swallows, discrimination of own from alien at the nest occurs only when the fostering is carried out late, at about the time young fly from the nest (indeed it is difficult to do the experiment at this time because the transferred young may fly from the nest immediately).

When prior odds swamp signature evidence

In Michigan, colonies of hundreds of bank swallows typically have one or two rough-winged swallow pairs nesting on the periphery. This situation was used to carry out interspecific cross-fostering experiments (Beecher, 1981; Beecher & Beecher, unpub. data). Normally such between-species exchanges fail because the chicks are poorly adapted to the heterospecific nest environment. In this case, however, both species are nesting in precisely the same habitat. Moreover, the diet of the two species is highly similar. In these exchanges, a single bank swallow was added to a rough-winged swallow brood, or a single rough-winged swallow to a bank swallow brood.

It was found that rough-winged swallows added to bank swallow nests were typically rejected whereas bank swallows added to roughwing nests were invariably accepted. I should add that we also have seen rough-winged swallow parents feeding bank swallow chicks that had flown into their nest. What these interspecific transfers tell us that the intraspecific transfers cannot, is that rough-winged swallow parents do not accept alien chicks purely because they cannot discriminate their own from alien chicks. Bank swallow chicks are clearly different from rough-winged swallow chicks (visually and acoustically) and we have seen rough-winged parents do visible double-takes before feeding a bank swallow chick.

This inference is supported by a second type of cross-fostering experiment we carried out in which we exchanged rough-winged swallow and bank swallow broods from adjacent or close burrows (Beecher & Beecher, unpub. data). Note that this experiment is essentially the forced choice paradigm discussed above. We observed that both sets of parents would shortly begin to feed their chicks at the new location. Although the behaviour of the two sets of parents in this situation cannot be treated as independent (which is why we generally didn't use this very convenient design), rough-winged swallow parents were clearly attracted to the calls of their young. These experiments suggest that when forced to make a hard choice, rough-winged swallow parents can indeed discriminate conspecific from heterospecific chicks.

I offer the following interpretation of these interspecific cross-fostering experiments. The two species normally employ different decision rules when confronted with discrimination between own and alien young. In bank swallows, parents are often confronted with such discriminations, and have been selected to base their decisions on individually distinctive cues. In rough-winged swallows, where such discriminations are almost never required, the criterion of 'feed any chick you find in your nest' has been a generally reliable, conservative rule. A chick's presence in the home nest is, of course, strictly circumstantial evidence as to its relatedness, but in rough-winged swallows it is a virtually fail-safe criterion. In bank swallows, however, it is an unreliable predictor of relatedness, at least for chicks near flying age. According to this hypothesis, in the chick addition test, on finding a bank swallow chick in its nest, a rough-winged swallow parent gives priority to the chick's location in the home nest over its unusual appearance and sound and so accepts it. In the close interchange test, however, the rough-winged swallow parent is confronted not only with an entire brood of transplanted heterospecific chicks in its nest, but also its own brood of chicks calling at the mouth of a nest close by. It cannot tend to both broods of chicks. In this case, the very large difference in calls and physical appearance is pitted against the very small difference in location (prior odds), and the parent gives priority to calls and appearance.

The rough-winged swallow example illustrates a point that probably applies to any non-colonial nesting species: a parent returning to its nest should assume that the young there are its own. That is, the prior odds overwhelmingly favour h_0. In this case, there will have been no selection for attending to evidence indicating an alien chick is in the nest, or for discriminating between chicks in the face of such evidence, for such an unusual event simply has not been a part of the species' selective background. (In the case of rough-winged swallows, nesting in bank swallow colonies is likely to be a relatively recent event; it is quite uncommon in most of North America.)

The difference between bank swallows and rough-winged swallows in these interspecific fostering experiments parallels the difference between those passerine species that accept and those that reject cowbird eggs (Rothstein, 1982). The difference between cowbird and host eggs is quite conspicuous, and acceptance of cowbird eggs probably reflects a decision rule (possibly adaptive, see Rohwer & Spaw, 1988) or a lack of appropriate eviction behaviour (Rothstein, 1982), rather than a perceptual inability.

Forced choice tests are more sensitive than substitution tests

It has been noted by several investigators that 'simultaneous discrimination' tests are more sensitive than 'sequential discrimination' tests of recognition (Evans, 1970; Beer, 1979; Shugart, 1977). The key to this difference, I would suggest, is not the timing, but the fact that simultaneous tests are forced choice tests, with the offspring pitted against unrelated young. The sequential tests are usually substitution tests at the nest, in which the offspring are removed and replaced with unrelated young. I have suggested that the parent should assume that the young it finds at the home nest are its own, provided there is no conflicting evidence emanating from adjacent nests.

There are only a few direct comparisons of substitution and forced choice tests of parents recognizing young. One is the rough-winged swallow and bank swallow interspecific cross-fostering experiments just described. The best such study, however, has been done by Shugart (1977) on caspian terns. Shugart found that caspian tern parents will accept young substituted for their own in the first week of life, yet when given a choice between their own and alien young, in nest scrapes either side of the original nest, they will unfailingly choose their own. He got essentially the same results in egg-fostering experiments in this species (Shugart, 1987).

The tern case is somewhat different from the rough-winged swallow case discussed above, for terns normally nest in dense colonies and young may move between nests if disturbed. The early development of signature cues in caspian terns has probably been favoured because of the potential for mix-ups at the nest. If a tern parent finds its nest deserted, it will start searching, and it can use the signature cues to choose among chicks. If instead of an empty nest, it finds two or more chicks in or around its nest, it will use the signature cues to choose among the chicks (analogous to the simultaneous discrimination test). But it will never have to deal with the situation of its chick disappearing out of sight and earshot, neatly replaced with a new chick (which describes the chick substitution test). Experimenters moving young between widely-separated nests has not been part of the species' selective history. To the contrary, in this situation selection will have favoured tern parents that go with the prior odds which support h_0, rather than the signature evidence, which supports h_1, provided there is no other evidence favouring h_1 (e.g. signs of disturbance, the parent's true offspring calling from nearby, and so on).

Parent–offspring recognition in the creche: do high error rates occur?

Our decision model points out that a parent may be favoured to seek out its young in the creche even under circumstances where recognition frequently fails. The only requirement is that the net gain from correct feedings devalued by their probability be greater than the cost of misdirected feeding devalued by their probability. Needless to say, the more discriminating parent is favoured by natural selection, but even a poorly-discriminating parent is favoured to keep on trying in the creche so long as its success rate exceeds that set by the minimum level of the likelihood function.

There are very few studies that have tracked known parent–offspring pairs in the creche and measured recognition success. Most of these are fragmentary. For example, in a bank swallow creche of about 100 individuals, we observed 18 feedings of marked young by marked parents: in every case the adult was the parent of the chick (Beecher *et al.*, 1981a). Although this was about the average creche size for bank swallows in our study area, they do come much bigger, and observational studies of these larger creches would make for a better test of this hypothesis.

Far and away the best test of the idea to date is McCracken's (1984) study of Mexican free-tailed bats (*Tadarida brasiliensis mexicana*). The 'maternity caves' of these bats may contain up to a million young with 4000 bats per square metre. The effective size of the creche is much smaller than this, however, for the pups will remain in a relatively circumscribed area to which the mother homes on returning to the creche. McCracken & Gustin (1987) estimate the effective creche size as about 1500 bats. McCracken (1984) tested 167 mother–pup pairs for heteromorphic blood allozymes and estimated that 17 per cent of the mothers were nursing pups that could not be their offspring. On the one hand, this is very good recognition considering the magnitude of the problem. On the other hand, this is a relatively high error rate, especially in light of the high cost of lactation.

Parent–offspring recognition in the creche: quantitative predictions

Signature adaptation hypothesis

Our decision model predicts that in species with strong selection for parent–offspring recognition, such as occurs in large creches, there will be selection for strong signature information. Elsewhere I have discussed how selection might act on individuals so as to shape signature systems that have high information value (Beecher, 1982; 1988). Our decision model predicts that the target of this selection will be σ_T/σ_W (Eq. 8): genetic changes which increase between-individual signature variation or decrease within-individual signature variation will be favoured. I will refer to this as the *signature adaptation* hypothesis.

With this hypothesis as the impetus, I have developed a model and method for measuring the information value of signature systems (Beecher, 1989). I will first sketch out the major features of the model and then turn to tests of the hypothesis.

The signature model and information measure

Identification and recognition are inherently quantitative concepts, and can be readily analyzed from the perspective of information theory (Shannon & Weaver, 1949; Quastler, 1958; Wilson, 1975; Hailman, 1977). Two information measures are defined, the first describing the recognition needs of the animal and the second describing the recognition capabilities of the signature system.

H_0 is the initial or inherent uncertainty as to identity within the group of N individuals. It is defined purely in terms of the number requiring identification:

$$H_0 = \log N \tag{11}$$

where the log here and throughout is to the base 2 and is measured in bits (here bits/individual). H_0 is the minimum number of binary decisions the recognizer would need to narrow the search down to the target individual if all individuals were unambiguously identified.

H_S is the potential signature information present in the entire signature system. H_S is a more difficult concept than H_0, and requires a signature model. The model is basically the analysis of variance (ANOVA) model II or random effects model (e.g. Sokal & Rohlf, 1981). This model has been used many times in the past in analyzing signature traits, although this has never been explicitly recognized. Rather, the model is implicit in the many analyses that have used either linear discriminant functions to classify individuals or have carried out ANOVAs on these data.

Consider a single variable trait, such as the duration of a call. By the model a particular observation, X_{ij}, is assumed to be composed of two independent components: a component B_i, reflecting true differences between individuals, and a 'within-individual' or 'error' component, W_{ij}, i.e.

$$X_{ij} = B_i + W_{ij}$$

assuming that the means are zero. Because B_i and W_{ij} are independent, the variances have the simple relationship

$$\sigma^2_T = \sigma^2_B + \sigma^2_W$$

where σ^2_T is the total variance in X and σ^2_B and σ^2_W are the variances in B and W, respectively.

H_S is then defined as the amount of information needed to reduce the total uncertainty to the within-individual uncertainty, which turns out to be

$$H_S = \log \frac{\sigma_T}{\sigma_W} \qquad (12)$$

so long as B and W are distributed similarly. Note the following properties: (1) signature information increases directly with σ_B (or σ_T,) and inversely with σ_W. (2) $H_S = 0$ when $\sigma_B = 0$. (3) H_S is an absolute measure with a non-arbitrary zero, the unit of measure being the within-individual uncertainty. The original units of measurement are immaterial. The amount of signature information conveyed by the amount of dark feathering on the face can be compared, for example, with that conveyed by the average frequency of a call. An additional feature of the model is that since it is formally identical to the ANOVA model II, we immediately have an appropriate statistical test for the presence of signature information.

Invariably signature traits are multivariate (e.g. a call might differ with respect to duration, peak frequency, modulation rate, etc.). The information measure can be generalized to the multivariate case by submitting the data to a principal components analysis and carrying out the analysis on the resulting, uncorrelated variables. Then the total information H_S is simply the sum of the information H_i in each of the independent variables, and

$$H_S = \sum H_i = \sum \log \frac{\sigma_{Ti}}{\sigma_{Wi}} \qquad (13)$$

where σ_i is the standard deviation of the ith principal component. Equation 13 is the information measure used in the call analyses described below.

Connecting the signature and decision models

Earlier we developed the prediction of our decision model concerning the value $L^* = \sigma_T/\sigma_W$ necessary for parents to attempt recognition in the creche. If we take the log of both sides of equation 9 and note the definitions of H_S and H_0 given above (Eq. 11 and 12) we see that at L^*

$$H_S^* = H_0 + \log \left[\frac{C}{B-C} \right] \qquad (14)$$

approximately. Note that for $B > 2C$, H_S^* can actually be less than H_0. For low-error recognition, however, H_S will have to be considerably larger than H_S^*. It is not clear exactly what one should predict in a given case. Clearly we should find at least that $H_S(\text{obs.}) \geq H_S^*$. Since presumably there is selection for recognition better than the bare minimum necessary to keep searching in the creche, our prediction should be based on some criterion concerning the posterior odds of a correct acceptance, which in this instance is (from Eq. 7)

$$\log \frac{p_0}{p_1} = H_S - H_0 \tag{15}$$

approximately. Thus if we set our criterion probability correct level at some low level p_0^{**} ($<p_0^*$), our criterion H_S^{**} will be

$$H_S^{**} = H_0 + \log \frac{p_0^{**}}{p_1^{**}} \tag{16}$$

To take an example, suppose $N = 101$ an $B/C = 3$. Then by equations 9–12, $L^* (\sigma_T/\sigma_W) = 50, p_0^* = .33, H_0 = 6.66$ bits and $H_S^* = 5.04$ bits. If we set p_0^{**} at 0.95, however, then $H_S^{**} = 10.9$ bits. Thus about 6 bits worth of additional information is needed in the signature system to move discrimination from the minimum acceptable (33%) to the 95 per cent correct acceptance level. Note that H_S^* and H_S^{**} are not strictly comparable as the former requires an estimate of v and the later an estimate of p_0^{**}.

Testing the signature adaptation hypothesis

In this section I will turn to an explicit test of the signature adaptation hypothesis. The hypothesis predicts that if we compare several closely-related species, which vary with respect to selection pressure for recognition, signature information will be greater in the species requiring recognition. To test this hypothesis we studied four North American swallows. The study species were chosen to give us two colonial-noncolonial pairs. The first pair, mentioned earlier, is the colonial bank swallow and noncolonial northern rough-winged swallow. The second pair is the colonial cliff swallow (*Hirundo pyrrhonata*) and noncolonial barn swallow (*Hirundo rustica*, the conspecific 'swallow' in Europe). This second pair is particularly interesting, as the two species are especially closely related – they are congenerics and hybridize on occasion (Martin, 1980). For the purposes of this discussion, the key difference between these species is that creches are common and often large in bank swallows and cliff swallows, but do not occur in the two comparison species. Reciprocal calling between parents and fledglings is conspicuous in all four species. It is clearly a begging and contact call in all four species. Our hypothesis was that this call had been further modified for a signature function in bank swallows and cliff swallows. We thus predicted that we would measure a greater information capacity in the calls of these two species relative to the comparison species.

We did cross-fostering experiments and playback studies on all four species (Beecher *et al.*, 1981a,b, 1986; Stoddard & Beecher, 1983; Medvin & Beecher, 1986). These studies provide background for the call analysis studies. The cross-fostering and playback experiments were done on nests where young were close to fledging; some of the young had already taken some trips to and from the nest, or were on the verge of doing so. The results of both the cross-fostering and playback experiments indicated that parent–

offspring recognition is well developed in the colonial bank swallow and cliff swallow but absent or weak in the noncolonial rough-winged swallow and barn swallow. Our failure to find parental recognition in cross-fostering experiments on barn swallows and rough-winged swallows counters the argument that their failure to recognize in the playback experiment was due to their normal use of non-vocal cues. In addition, circumstantial evidence suggests that barn swallows do not use visual cues (essentially the only other source of cues for individual recognition in birds): cliff swallow chicks show marked individual variation in face colour pattern, while barn swallow chicks show no such variation. Although we have not investigated whether cliff swallow parents use this visual variation for recognition, this species difference is opposite to that expected if barn swallows use the visual modality rather than the acoustic modality; additionally, we know of no case in which visual recognition has been shown in birds where the visual variation is not conspicuous to the eye of the human observer (the same cannot be said for the acoustic modality). Neither bank swallow nor rough-winged swallow chicks show facial variation.

The earlier discussion points out, however, that the 'failures' of recognition seen in the barn swallows and rough-winged swallows may reflect decision rule or perceptual adaptations, and not signature adaptations. Thus, to test the signature adaptation hypothesis, the calls must be analyzed directly. To analyze the relative information capacities of chick calls for each of the four species, we used the information measures described above (Beecher, 1982; Beecher & Beecher, unpub. data; Medvin, Stoddard & Beecher, unpub. data). The measured information capacity of the calls systems were 8.74 bits for cliff swallows, 4.57 bits for barn swallows, 10.2 bits for bank swallows and 3.83 bits for rough-winged swallows. Thus, as predicted, the information capacity of the signature calls of the creching cliff swallow and bank swallow is greater than for that of the non-creching barn swallow and rough-winged swallow. Referring back to the various equations relating prior odds (or creche size) and L, the 6.37 bit difference between bank swallows and rough-winged swallows means that about 83 times more individuals can be identified, to the same degree of precision, with the bank swallow signature system. Similarly, the 4.17 bit difference between cliff swallow and barn swallow signature systems implies that the former is approximately 18 times better than the latter.

Jouventin (1982) has done a similar analysis using a less-sophisticated quantitative approach. Comparing several species of penguin varying in degree of coloniality, he has shown that the signature calls of the more colonial species are more complex. His results parallel ours, and support the signature adaptation hypothesis.

To check whether our call analysis truly reflected the perception of the birds, we have trained laboratory-reared cliff swallows and barn swallows to discriminate among the calls of different individuals of each species in an operant conditioning paradigm (Beecher *et al.*, 1989; Loesche *et al.*, unpub. data). We found that individuals of both species learned to discriminate cliff swallow call pairs more readily than barn swallow call pairs. This study supports the notion that cliff swallow chick calls are more individually distinctive.

A double filter?

As indicated earlier, we have few good quantitative studies of recognition in the creche (indeed McCracken, 1984 is the only one). We do have fragmentary evidence and a number of anecdotal reports to suggest that in many creching species, recognition is very accurate. Yet it is difficult to see from the call analysis data how parents could do as well as they evidently do. For example, from the small field study I mentioned earlier, it appears that bank swallow recognition is very good indeed. That this discrimination was better than that which McCracken observed for the free-tailed bats, however, may merely reflect the difference in the creche size in the two cases – an order of magnitude difference.

Both bank swallows and cliff swallows, in the populations we have observed, often have to deal with creches in the range of 100–500 individuals. If we set the probability correct level at 0.95, then by equation 15, H_S** should be in the range 10.9–13.2 bits. By our call measurements, both bank swallows and cliff swallows fall below that predicted range. Because the 0.95 is an arbitrary choice, and because our measured H_Ss may underestimate the true value, we should not attach too much significance to this discrepancy. In addition, cliff swallows (but not bank swallows) have a second set of potential signature cues, a conspicuous variation in chick facial coloration (Stoddard & Beecher, 1983). Nevertheless, it seems quite possible that precise recognition in very large creches may well require greater information capacities than is found in these chick signature systems. This leads me to suspect that parents may need a mechanism to 'filter' out senders and reduce the creche size down to more manageable proportions. Two observations of parent–offspring interactions in the creche suggest a possible filter mechanism.

The first observation is that in bank swallows and cliff swallows (and many other creching species), parent–offspring reunions are typically preceded by reciprocal calling. Thus if chicks recognize the calls of their parents, and respond only to those calls or similar ones, this immediately and drastically reduces the effective creche size: the parent need only discriminate among the relatively small number of young that respond to their calls (this small number is the effective creche size). This would imply selection on parent signatures as well as chick signatures. Recognition of parents by young has been demonstrated innumerable times, and reciprocal recognition is probably the rule in species where recognition occurs. For example, recognition of parent calls by chicks is well developed in bank swallows (Sieber, 1985) and cliff swallows (Beecher *et al.*, 1985).

The major theoretical problem with this mechanism is sender honesty. The parent benefits only by feeding its own young, but young benefit by any feeding, regardless of the relatedness of the provider. Therefore, if responding to an unrelated adult's call increases its probability of getting a free meal then it will be favoured to do so. It has been noted many times that the creche presents an ideal opportunity for low-cost free-loading by the young.

A second observation of typical creche behaviour suggests a way out of this dilemma, at least for some species. 'Feeding chases' were first reported in the adelie penguin (Spurr, 1975; Thompson, 1981): on arriving at the creche, the parent calls and is accosted by a number of calling chicks. The parent leads the chicks off on a chase, and the number of beggars decreases, usually down to one, who the parent ultimately feeds. We have seen similar events in bank swallow creches. Perhaps the major function of these feeding chases is to discourage free-loading by imposing a sufficient cost on it. An additional function of the chase is the opportunity it offers the parent to sample the calls of the chasing chicks (and vice versa).

The sender's perspective

General considerations

The decision model, emphasized in this chapter, takes the receiver's perspective: it tells us what the receiver should do given certain partial information and certain consequences. By itself, it provides little insight into the sender's perspective. The sender figures in the argument as the source of the evidence X the receiver uses in computing likelihood ratios. In the creche situation, X becomes essentially synonymous with the signature (phenotypic traits adapted for signature function). In the nest, however, the sources of this evidence are multiple. In particular, an interloper can 'unwittingly' provide a variety of evidence that betrays it: it may be older and clearly larger than the offspring in the nest, it may be nervous, its demeanour may be otherwise unusual, it may be hanging out at the edge of the nest rather than in it, and so on. Much of the so-called parental recognition of alien chicks in gulls appears to involve such indirect cues (e.g. Miller & Emlen, 1975). In fact, cross-fostering experiments are unreliable as indicators of parental recognition of signature traits of young since the parents often react to the disturbed behaviour of chicks introduced into the nest.

It seems clear that selection will act on young confronting foster parents (whether in the nest or the creche) to suppress behaviours that give them away. It should also act on signature mechanisms so that alien young either

(a) withhold signatures (crypticity), (b) provide 'generic' signatures that are less distinctive and so less obviously alien, or (c) mimic the signatures of the parent's offspring. I will consider those possibilities in several contexts below.

Honest signatures in the creche

As suggested earlier, the existence of the creche in a species is almost prima facie evidence for honest signatures. Although our search model suggests that given a high B/C ratio, parents should stick with the creche in the face of relatively high error rates, it is obvious that reliable signatures are precisely what permit even imperfect recognition in the face of the strongly unfavourable prior odds. Yet it is precisely in this situation, where no contextual evidence is present, that we should most expect young to deceive.

Signatures probably work in the creche because parents make them a requirement for feeding (no signature, no food) and because they are essentially 'unbluffable'. If the potential free-loader cannot withhold its signature, its only hope is that its signature is a generic signature or a signature that mimics the signature of the parent's offspring. In the creche, however, the youngster will not know the signature of the approaching parent's offspring, and so the latter strategy cannot work. (It might work at the nest, however; see below.) Nor is it likely that a generic signature could work either, because the nonparental feeds a chick gains by resembling the young in the creche, would be offset by the parental feeds it loses to these other chicks. In short, so long as a signature is a precondition for feeding in the creche, it does not appear that there is any feasible way for the young to cheat. It can certainly beg and take any food offered, but the usual routes to free meals in the creche must be chance similarity to other young coupled with the good fortune of being in the right place at the right time.

When offspring should not signal identity

I have suggested that there are situations, such as the chick substitution test, in which the parent is favoured to disregard potential signature evidence, i.e. favoured not to discriminate, and other situations, such as the forced choice discrimination, in which the parent is favoured to utilize this same signature evidence because it constitutes the only basis for decision. In this section I suggest that there are other situations in which the parent is favoured to use signature evidence but the young are not favoured to provide it. This too will lead to a failure of recognition but for quite a different reason.

As more studies have examined both parental recognition of offspring and offspring recognition of parents, several cases have been found where young appear to recognize parents but not vice versa. Certainly any theory of kin recognition should suggest why parental recognition does not evolve in a situation where it is clearly possible and desirable for the parent. I will suggest that when the prior odds of equation 3 favour the hypothesis that the young in the nest are the parent's offspring, so that acceptance is the default assumption, then offspring may not stand to gain by providing reliable signatures.

In several gull species, parents fail to recognize their young until the young begin to fly, despite the fact that such recognition would appear to be useful: from a few days after hatching, young are mobile and may stray between nests in the colony. Moreover, we know that recognition is possible since young recognize their parents by voice shortly after hatching and parents do learn the calls of their young eventually, shortly before the young begin to fly at about a month of age (reviews in Beer, 1979; Shugart 1990). The oft-stated but erroneous generalization that herring gull parents recognize their chicks a few days after hatching is based on cross-fostering experiments in which recognition was expressed primarily by the chicks, and only secondarily by the parents. This asymmetry in recognition is surprising given the conventional argument that the onus of recognition is on the parent: parental care benefits a parent only if the recipient young is related, but it benefits the young no matter what their relationship (Beecher, 1981, 1982; Holmes & Sherman, 1983).

It has been noted for quite some time – the generalization may have been first made by Davies & Carrick (1962) – that in those species where parental recognition occurs, it develops just shortly before chick mobility. In ground-nesting gulls, there are essentially two such periods, first when the young begin to move about on the ground a few days after hatching, and second when the young begin to fly at about a month. The key point is that the chick does not need the parent to recognize it in order to feed it before fledging, for it will be fed merely by being in the nest. While a recognizable call might benefit the chick when lost away from the nest by facilitating parental searching, it may not be necessary if the parent calls and searches on finding a missing chick at the nest; if the chick recognizes the calling parent it can home to the parent and nest. Moreover, when away from the nest there is a danger to having a reliable signature, namely that unrelated parents may reject or attack you. Gull adults can easily injure or kill a chick. Moreover, an abandoned chick may be adopted if it makes it to another nest and acts as if it belongs there (e.g. Holley, 1984). Thus reliable signatures would appear to have a net cost to gull chicks. It is only when the young fledge, and the nest is abandoned, that, a reliable signature is required, for the young can no longer expect to receive parental care simply by being in a particular place. It is precisely in this circumstance that we would predict offspring to delay the development of a signature until it is required, i.e. until contextual evidence is lacking.

Intraspecific brood parasitism and genetic signatures

There is now a great deal of evidence to suggest that in many birds, parents cannot necessarily use location in the home nest as a completely reliable predictor of relatedness. A number of electrophoresis studies of passerine birds have shown that a fairly high fraction of offspring are unrelated to one or both of the putative parents (Gowaty & Karlin, 1984; Gavin & Bollinger, 1985; Westneat, 1987). These are two sources of parental uncertainty. First, extra-pair copulations are evidently much more common than was once realized. Second, 'egg-dumping' is likewise much more widespread than had been realized, and may even be a relatively common parental strategy in some colonial species (e.g. cliff swallows, Brown, 1984).

The parental uncertainty resulting from extra-pair copulations and egg-dumping should favour kin recognition by parents. The most plausible mechanism would be for the parent to match its signature to the young's signature; if signatures are genetically determined, then the goodness of the match would be predictive of the likelihood of kinship (Beecher, 1982). We know that kin recognition by self-matching is possible (for reviews see Holmes & Sherman, 1983; Blaustein et al., 1987; Waldman, 1987; Waldman et al., 1988). Moreover, it is likely that signature traits have a strong genetic component in the typical case. For example, Baker & Bailey (1987a) have done cross-fostering experiments to show that the characteristics of the separation calls of northern bobwhite quail are genetically determined (in other experiments, they have shown that these calls are individually recognized, Baker & Bailey, 1987b).

While this kin recognition mechanism is possible, and it is likely that many signature traits have a genetic component, the catch is that such identification would not benefit the young. The argument is similar to that given in the previous section. Consider a cliff swallow chick, that may or may not be related to the parent at that nest. Clearly the chick would not benefit by signalling that it is unrelated to the parent. If it were related, it would benefit by so signalling only if it were in a brood in which some other chick signalled that it is not related (in which case the parent could redirect care from the latter to the former). Presumably the chick is unaware of the circumstances of its entry into the nest, and so cannot evaluate whether or not it will gain or lose by an honest identity signal and adjust its signalling accordingly. Since the cost of signalling that it is unrelated would seem to be substantially greater than the benefit of signalling that it is, chicks would seem to be favoured not to signal relatedness.

In all the examples of possible self-matching discovered to date, the kin recognition has been between collateral kin, a case that may be fundamentally different from the parental uncertainty case. For example, Holmes & Sherman (1982, 1983) found that Belding's ground squirrels behave more altruistically toward littermate full sisters than towards littermate half-sisters, and they interpreted this as kin recognition by self-matching. The essential difference between a case such as this and the parental uncertainty case is that in the former the sender can benefit by a reliable signal (when it interacts with a more closely related kin) as well as lose by the signal (when it interacts with a less closely related kin). In the parental uncertainty case, however, the major effect of the signal for the sender is a cost (if the sender is unrelated to the receiver).

It is clear that in the parental uncertainty case we have a conflict of interest between parent and putative offspring with regard to the value of an honest signal of genetic identity. It is difficult to predict how this conflict of interest will be resolved in actual cases. The outcome would seem to hinge on which evolves first, (a) the use of self-matching by parents, or (b) the withholding of signature information, or providing of deceptive information, by young.

Consider the possible strategies available to a chick which hatches in a particular nest, but which may have arrived via egg dumping, and so is unrelated to both parents at the nest. Neither the chick nor the parents know how it arrived in the nest. We assume that selection pressure for signatures (e.g. a creche situation) occurred prior to selection pressure for kin recognition by self-matching (e.g. egg dumping). Thus the chick will develop an individually distinctive signature at some point; the chick has no choice in this, for parental care beyond this point becomes contingent on this signature. The default condition is for the parent to treat the chick as its own and learn its signature when it develops. However, if in the process of learning this signature, the parent detects a sufficiently poor match to its own signature, then the parent will reject the chick. Since the chick may be unrelated to the parents, it would like to have it both ways: to provide a signature that (a) is individually distinctive and allows the parent to discriminate it from alien chicks that have flown into the nest, or permits the parent to locate it in the creche, yet (b) contains minimal or even deceptive information about genetic identity. The problem for the chick, then, is that any adaptation which promotes function (b) may impede function (a). Thus for example, the chick cannot simply refuse to provide a signature (e.g. give a simple begging call instead of the individually distinctive calls that most colonial birds develop). Delaying the development of the signature (as suggested in the previous section) will only postpone detection, not prevent it. (The situation described in the previous section is different from the one discussed here. The problem for the gull chick is primarily that of not getting killed if it strays from the nest, whereas the problem of the cliff swallow chick is of getting itself adopted into the nest it finds itself in.) Releasing the signature from genetic control is an alternative strategy, but the simplest ways this might happen (loss of penetrance of, or reduction in polymorphism at the genetic signature loci) lead to loss of signature information value and again work against the primary function of the signature. Thus the necessary adaptation would seem to have to be one which maintains the information value of the signature (in the sense detailed in the previous section 'Parent–offspring recognition in the creche: quantitative predictions') while reducing its value as a kinship signal. A possible adaptation to this end, at least in the case of acoustic signals in birds, is imitation of the signature calls of the parents or siblings. This seems quite plausible in passerine birds, since song-learning by imitation is the rule in passerines, and call learning by imitation may not be unusual either. We have measured the calls of cliff swallow chicks and found very high correlations between nestmates (Medvin, Stoddard & Beecher unpub. data). High sibling correlations can reflect dominance variance as well as additive genetic variance, however, and until we do the appropriate cross-fostering experiment, we cannot evaluate the relative roles of genetic and experiential factors in signature development in this species.

What is the possibility that parents actually exercise this sort of discrimination? To date there is no evidence for kin recognition by self-matching in any bird where parental uncertainty is a potential problem. In fact, many cross-fostering experiments have shown that parents learn the signatures of the young while they are in the nest, usually shortly before fledging, and that parents will evidently learn those of unrelated foster chicks as readily as those of offspring. For example, we have done such cross-fostering experiments with bank swallows, a possible candidate species for parental uncertainty (as noted above, extensive egg dumping has been demonstrated in the other North American colonial swallow, the cliff swallow, Brown, 1984). We have found that not only are foster chicks accepted and cared for after fledging (if fostered before signatures develop), but that offspring that are fostered to another nest and happen to return to the home nest are rejected by their true parents!

Nevertheless, kin recognition in the parent–offspring context has not received the same intense examination as it has in the sibling context, and it would be rash to conclude that it cannot occur. Careful study of the problem is warranted.

Conclusions

I believe that the perspective of this chapter, which views recognition simultaneously as a communication problem and as a decision problem, provides some insights into parent–offspring recognition. In particular, it leads to the following conclusions.

1 Discrimination is not always beneficial for the parent. In particular, there will be circumstances in which the parent is reasonably certain that an individual is not its offspring but *is not certain enough* that it can afford to reject the individual. Thus certain

'failures' of recognition reflect conservative decision-making rather than perceptual inability.

2 As a corollary, if the investigator's goal is to evaluate the species' ability to discriminate between offspring and unrelated young, then a forced-choice test is the procedure of choice. If properly carried out, this method eliminates all factors affecting discrimination save information provided by the young themselves.

3 Recognition in the creche or similar situations requires strong signature information to overcome the unfavourable prior odds of finding the offspring. It is likely that signature adaptations are common in species where such large-scale recognition is required.

4 Signature adaptations in themselves may not be enough to ensure precise recognition. Reciprocal recognition (parent of offspring, offspring of parent), in conjunction with adaptations of parent and offspring signatures, may provide a 'double filter' permitting precise recognition. A major problem with this mechanism is that it requires 'honesty' on the part of offspring, who otherwise are favoured to take any feeding they can get, regardless of the relatedness of the donor. Such honesty may be maintained by feeding chases or other devices which impose a cost on 'free-loading'.

5 The argument above notwithstanding, parents can actually tolerate a fairly high level of errors (misdirected feedings) in the creche so long as the benefit of a feeding times the probability of a correct feeding exceeds its cost.

6 Offspring should signal individuality or genetic identity only if it benefits them. At the nest in particular, the 'default' condition is for the parent to feed the young, and so signatures do not necessarily benefit offspring. Where signatures are required in some other context (e.g. in the creche) and where there is a conflict of interest between parents and young, a simple prediction may not be possible. In this case, further empirical studies will probably prove more fruitful than further modelling.

Finally, I believe the communication/decision approach is quite applicable to other types of recognition in animals. For example, consider recognition of neighbours by song, as occurs in many species of songbirds (Falls, 1982). Although this ability is widespread, it appears to break down in species with song repertoires: most studies have found that the discrimination of the songs of neighbours and strangers is sharp in single-song and small-repertoire species, but poor or lacking altogether in medium- and large-repertoire species. It has been suggested by several workers that song repertoires degrade individual recognition by overloading the working memories of these birds: the birds can memorize only so many songs (see e.g.

Falls, 1982). We have argued elsewhere, however, that observed weak neighbour–stranger discrimination may reflect not an inability to discriminate between the songs of neighbours and strangers but a decision to treat neighbours and strangers as equally threatening. Careful forced-choice tests in the field on song sparrows (a species thought to discriminate neighbour song from stranger song weakly or not at all) showed clear neighbour–stranger discrimination, and laboratory experiments showed that the song memory capacities of these birds are impressively large. The difference between discriminating and non-discriminating species, therefore, may reflect variation in decision rules, appropriate to the differing ecologies, rather than variation in true recognition ability.

Acknowledgements

I thank my colleagues Inger Mornestam Beecher, Philip Stoddard, Mandy Medvin and Patricia Loesche. Much of the work was supported by grants from the National Science Foundation.

References

Baker, J. A. & Bailey, E. D. (1987a). Sources of phenotypic variation in the separation call of Northern bobwhite (*Colinus virginianus*). *Can. J. Zool.*, **65**, 1010–15.

Baker, J. A. & Bailey, E. D. (1987b). Auditory recognition of covey mates from separation calls in Northern bobwhite (*Colinus virginianus*). *Can. J. Zool.*, **65**, 1724–8.

Beecher, M. D. (1981). Development of parent–offspring recognition in birds. In *Development of Perception*, ed. R. K. Aslin, J. R. Alberts & M. R. Petersen, vol. 1. pp. 45–66. New York: Academic Press.

Beecher, M. D. (1982). Signature systems and kin recognition. *Am. Zool.*, **22**, 477–90.

Beecher, M. D. (1988). Kin recognition in birds. *Behav. Genet.*, **18**, 465–82.

Beecher, M. D. (1989). Signalling systems for individual recognition: an information theory approach. *Anim. Behav.*, **38**, 248–61.

Beecher, M. D., Beecher, I. M. & Lumpkin, S. (1981a). Parent–offspring recognition in bank swallows: I. Natural history. *Anim. Behav.*, **29**, 86–94.

Beecher, M. D., Beecher, I. M. & Hahn, S. (1981b). Parent–offspring recognition in bank swallows. II. Development and acoustic basis. *Anim. Behav.*, **29**, 95–101.

Beecher, M. D., Loesche, P., Stoddard, P. K. & Medvin, M. B. (1989). Individual recognition by voice in swallows: signal or perceptual adaptation? In *Comparative Psychology of Audition: Perceiving Complex Sounds*, ed. R. J. Dooling & S. H. Hulse, pp. 227–94. Hillsdale: Erlbaum.

Beecher, M. D., Medvin, M. B., Stoddard, P. K. & Loesche, P. (1986). Acoustic adaptations for parent–offspring recognition in swallows. *Exp. Biol.*, **45**, 179–93.

Beecher, M. D., Stoddard, P. K. & Loesche, P. (1985). Recognition of parents' voices by young cliff swallows. *Auk*, **102**, 600–5.

Beer, C. G. (1979). Vocal communication between laughing gull parents and chicks. *Behaviour*, **70**, 118–46.

Blaustein, A. R., Bekoff, M. & Daniels, T. J. (1987). Kin recognition in animals (excluding primates): empirical evidence. In *Kin Recognition in Animals*, ed. D. J. C. Fletcher & C. D. Mitchener, pp. 287–331. Chichester: John Wiley & Sons.

Brown, C. R. (1984). Laying eggs in a neighbour's nest; benefits and costs of colonial living in swallows. *Science*, **224**, 518–19.

Colgan, P. (1983). *Comparative Social Recognition*. New York: John Wiley.

Davies, S. & Carrick, R. (1962). On the ability of crested terns to recognize their own chicks. *Aust. J. Zool.*, **10**, 171–7.

Evans, R. M. (1970). Imprinting and the control of mobility in ring-billed gulls. *Anim. Behav. Monogr.*, **3**, 193–248.

Falls, J. B. (1982). Individual recognition by sounds in birds. In *Acoustic Communication in Birds*, vol. 2, ed. D. E. Kroodsma & E. H. Miller, pp. 237–78. New York: Academic Press.

Gavin, T. A. & Bollinger, E. K. (1985). Multiple paternity in a territorial passerine: the bobolink. *Auk*, **102**, 550–5.

Gowaty, P. A. & Karlin, A. A. (1984). Multiple maternity and paternity in single broods of apparently monogamous eastern bluebirds. *Behav. Ecol. Sociobiol.*, **15**, 91–5.

Green, D. M. & Swets, J. A. (1966). *Signal Detection Theory and Psychophysics*. New York: John Wiley.

Hailman, J. P. (1977). *Optical Signals: Animal Communication and Light*. Bloomington: Indiana University Press.

Holley, A. J. F. (1984). Adoption, parent–chick recognition and maladaptation in the herring gull. *Z. Tierpsychol.*, **64**, 9–14.

Holmes, W. G. & Sherman, P. W. (1982). The ontogeny of kin recognition in two species of ground squirrels. *Am. Zool.*, **22**, 491–517.

Holmes, W. G. & Sherman, P. W. (1983). Kin recognition in animals. *Am. Sci.*, **71**, 46–55.

Hoogland, J. L. & Sherman, P. W. (1976). Advantages and disadvantages of bank swallow coloniality. *Ecol. Monogr.*, **32**, 33–58.

Jouventin, P. (1982). *Visual and Vocal Signals in Penguins, their Evolution and Adaptive Characters*. Berlin: Verlag Paul Parey.

Luce, R. D. & Raffia, H. (1957). *Games and Decisions*. New York: John Wiley.

Martin, R. F. (1980). Analysis of hybridization between the Hirundinid genera *Hirundo* and *Petrochelidon* in Texas. *Auk*, **97**, 148–59.

McCracken, G. F. (1984). Communal nursing in Mexican free-tailed bat maternity colonies. *Science*, **223**, 1090–1.

McCracken, G. F. & Gustin, M. F. (1987). Batmom's daily nightmare. *Natural History*, **96**, (10), 66–73.

McNamara, J. M. & Houston, A. I. (1980). The application of statistical decision theory to animal behaviour. *J. Theor. Biol.*, **85**, 673–90.

Medvin, M. B. & Beecher, M. D. (1986). Parent–offspring recognition in the barn swallow. *Anim. Behav.*, **34**, 1627–39.

Miller, D. E. & Emlen, J. T., Jr. (1975). Individual chick recognition and family integrity in the ring-billed gull. *Behaviour*, **52**, 124–44.

Quastler, H. (1958). A primer on information theory. In *Symposium on Information Theory in Biology*, ed. H. P. Yockey & R. L. Platzman, pp. 3–48. New York: Pergammon.

Rohwer, S. & Spaw, C. D. (1988). Evolutionary lag versus bill-size constraints: a comparative study of acceptance of cowbird eggs by old hosts. *Evol. Ecol.*, **2**, 27–36.

Rothstein, S. I. (1982). Success and failures in avian egg and nestling recognition with comments on the utility of optimality reasoning. *Am. Zool.*, 22, 547–60.

Shannon, C. E. & Weaver, W. (1949). *The Mathematical Theory of Communication.* Urbana: University of Illinois.

Shugart, G. W. (1977). The development of chick recognition by adult caspian terns. *Proc. Colonial Waterbird Group*, **1**, 110–17.

Shugart, G. W. (1987). Individual clutch recognition by Caspian terns, *Sterna caspia. Anim. Behav.*, **35**, 1563–5.

Shugart, G. W. (1990). Experimental analysis of parent–offspring recognition in gulls. *Anim. Behav.* (In press.)

Sieber, O. (1985). Individual recognition of parental calls by bank swallow chicks (*Riparia riparia*). *Anim. Behav.*, **33**, 107–16.

Sokal, R. R. & Rohlf, F. J. (1981). *Biometry*, 2nd edn. San Francisco: Freeman.

Spurr, E. B. (1975). Behaviour of the adelie penguin chick. *Condor*, **77**, 272–80.

Staddon, J. E. R. (1980). *Adaptive Behaviour and Learning.* Cambridge: Cambridge University Press.

Stoddard, P. K. & Beecher, M. D. (1983). Parental recognition of offspring in the cliff swallow. *Auk*, **100**, 795–9.

Thompson, D. H. (1981). Feeding chases in the adelie penguin. In *Terrestrial Biology III. Antarctic Res. Series 30.* Washington, DC: American Geophysics Union.

Waldman, B. (1987). Mechanisms of kin recognition. *J. Theor. Biol.*, **128**, 159–85.

Waldman, B., Frumhoff, P. C. & Sherman, P. W. (1988). Problems of kin recognition. *Trends Ecol. Evol.*, **3**, 8–13.

Westneat, D. F. (1987). Extra-pair fertilizations in a predominantly monogamous bird: genetic evidence. *Anim. Behav.*, **35**, 877–86.

Williams, G. C. (1966). *Adaptation and Natural Selection.* New Jersey: Princeton University Press.

Wilson, E. O. (1975). *Sociobiology.* Cambridge: Harvard University Press.

5

Kinship, kin discrimination and mate choice

C. J. Barnard and Peter Aldhous

Non-random mating can be an important cause of evolutionary change within populations (see, e.g. Partridge, 1983). While several factors can lead to non-randomness in mating – dispersal patterns, intrasexual competition, mating preferences, etc. – it is the possible role of mate choice in differential mating success that has excited the greatest interest among evolutionary biologists. In part this is because, while mate choice is at the heart of part of Darwin's (1859, 1871) original theory of sexual selection, attempts to model the evolution of mating preferences have proved contentious and tests of models often equivocal (Read, 1990; Bateson, 1983; Bradbury & Andersson, 1987). Nevertheless, there is no shortage of suggestions as to the criteria on which mating preferences might be based (e.g. Halliday, 1978; Hamilton & Zuk, 1982; Bateson, 1983) and, in some cases, convincing empirical support has been forthcoming (e.g. Semler, 1971; Andersson, 1982; Majerus, 1986).

A number of lines of argument point to the degree of relatedness between potential mates as a criterion in mate choice (Bateson, 1983, 1988; Smith, 1979; Shields, 1982, 1983; Partridge, 1983). The degree to which individuals of sexually reproducing species outbreed and therefore mate with others of differing genotype is likely to have important consequences for their reproductive success, mainly through the effects of dispersal costs, changes in the level of homozygosity and indirect fitness (e.g. Smith, 1979; Bateson, 1983; Partridge, 1983). It seems likely that extremes of inbreeding or outbreeding will incur both advantages and disadvantages for individual reproductive success and that an optimal balance between them might be expected under selection. Where the balance falls, however, is likely to vary both between populations, depending on, for example, levels of dispersal and mate competition (Maynard Smith, 1978; Smith, 1979) and between the sexes, especially where there are sex differences in mating opportunity (Smith, 1979; Sutherland, 1985). Degree of relatedness may thus be important in choosing a mate and, where there is appreciable variation in degree of relatedness between individuals, some form of kin discrimination (see Holmes & Sherman, 1983; Hepper, 1986; Fletcher & Michener, 1987; Barnard, 1990 and below) provides an obvious mechanism of optimising choice.

Empirical studies of a number of taxonomic groups provide some evidence for kin bias in sexual interaction, though often only at the level of odour preferences or spending more time with certain classes of individual (e.g. Sade, 1968, Wells, 1987, Murray & Smith, 1983 (humans and other primates); Gilder & Slater, 1978, Ågren, 1984, Dewsbury, 1982, Winn & Vestal, 1986, Barnard & Fitzsimons, 1988 (rodents); Bateson, 1978, 1982 (birds); Greenberg, 1982, Smith, 1983, Simmons, 1989 (insects)). In three cases, studies have demonstrated a correlation between kin bias and measures of reproductive success (Bateson, 1988; Barnard & Fitzsimons, 1989; Keane, 1990). However, while it is easy to envisage advantages and disadvantages to inbreeding or outbreeding, and thus a selection pressure for kin discrimination as a source of kin bias in mate choice, optimal inbreeding/outbreeding is only one factor which may lead to kin bias; correlations between kinship and mate choice may arise for reasons that have nothing at all to do with kin discrimination, though determining empirically whether kinship effects are due to discrimination or to some other factor may be difficult (see also Barnard, 1990). Nevertheless, despite the empirical difficulties, it is important to be clear about how mate choice criteria may give rise to kin bias in mating and to caution against concluding that kin bias, even where it is demonstrably based on mating preference, is necessarily due to kin discrimination. In this chapter, therefore, we examine the potential role of kinship in mate choice and the conditions under which kin bias is and is not due to kin discrimination.

Kinship and kin discrimination

'Kinship' denotes the sharing of alleles and thus heritable phenotypic characters between individuals by common descent. It is a special case of genetic similarity where the probability of individuals sharing an allele at a particular locus is determined by their distance apart on the path of common descent (their coefficient of relatedness) (see also Grafen, 1990).

It follows, therefore, that 'kin discrimination' refers to discrimination between individuals for the purpose of biasing responses with respect to those individuals sharing ancestry with the discriminator. Following Byers & Bekoff (1986), Waldman *et al.* (1988) and Barnard (1990), we distinguish between kin *recognition*, which is an (externally, at least) unobservable neural process, and kin *discrimination*, which is the differential response towards kin resulting from kin recognition. Where discrimination occurs, therefore, it must be the result of recognition, but an absence of discrimination does not necessarily mean an absence of recognition. As Waldman *et al.* (1988) have pointed out, kin discrimination involves an interaction between three components: (a) a label, which is a genetic or environmental cue allowing the kinship status of an individual to be determined; (b) a template, which is an internal model (learned or genetically encoded) of kinship characteristics against which labels can be compared and (c) decision rules, which evaluate the fit between label(s) and template(s) and determine whether any resulting recognition is translated into discrimination (see Aldhous, 1989a and Barnard, 1990, for fuller discussions). Waldman *et al.* (1988) proposed (a)–(c) in the context of kin discrimination based on the cues borne by individuals themselves (what Hepper, 1986 and Barnard, 1990, call discrimination by conspecific cues). However, 'labels' and 'templates' could also incorporate the spatial/temporal cues used in discrimination by non-conspecific cues. We thus extend the use of Waldman *et al.*'s components to cover all forms of kin discrimination. When talking about the evolution of kin discrimination, we envisage alleles which influence any or all of these components (e.g. what is regarded as a label, what is included in the template, the form of decision rules) and thus the nature and extent of discrimination. Following Barnard (1990), we refer to these as kin discrimination alleles. However, there may also be discrimination alleles which use criteria other than indices of kinship to influence inter- and intrasexual interactions (see below) and some of these may cause kin bias. Nevertheless, since they do not discriminate on the basis of kinship, we should not wish to call them kin discrimination alleles.

Indirect allele co-bearer discrimination

Kin discrimination alleles may use phenotypic similarities between their bearer and its kin or non-conspecific cues correlating with kinship as a means of biasing social responses in favour of individuals carrying copies of themselves. The advantage of kin discrimination thus lies in the indirect component of inclusive fitness accruing to the discrimination allele. In this case, it is genetic similarity only at the discrimination allele locus which is important; where it is used, similarity at other loci, and thus kinship, acts simply as a guide to discrimination allele co-possession. Kin discrimination here is what Barnard (1990) refers to as indirect co-bearer discrimination, to be contrasted with direct co-bearer or 'green beard'-type discrimination in which similarity at the discrimination allele is perceived directly (using phenotypic cues produced by the discrimination allele itself) and kinship, as defined above, is not involved. However, as Barnard (1990) points out, selection on discrimination mechanisms is likely to blur the distinction between direct and indirect co-bearer discrimination at the level of behaviour.

Kin discrimination for genetic similarity

While apparent genetic similarity between individuals may be used by kin discrimination alleles to identify likely co-bearers, genetic similarity itself may be the object of discrimination and contribute to the fitness of discrimination alleles distinguishing similarity directly and not through kinship. Genetic complementarity (the bringing together of adaptively complementary genotypes) between mates (e.g. Halliday, 1978), increased competition between genetically similar individuals with similar resource requirements (e.g. Ellestrand & Antonovics, 1985; Jasieński *et al.*, 1988; Barnard 1990), aggregation in times of danger (e.g. Bertram, 1978) and so on, may provide selection pressures for discriminating on the basis of overall genetic similarity (see Blaustein *et al.*, 1987). Since kinship implies characters shared through recent common ancestry, both allele co-possession and overall genetic similarity naturally arise through that route. However, it is important to stress that both can arise other than through common ancestry. The discrimination of genetic similarity for either purpose, therefore, does not necessarily imply kin discrimination or result in kin bias (but see later).

Although we have talked in terms of using indices of genetic similarity in kin discrimination, such indices may be very indirect. In most cases, similarity is likely to be gauged by phenotypic characteristics of the individual (conspecific cues; see above) but it is possible in certain circumstances that less direct cues such as spatial proximity or timing of encounter (non-conspecific cues) may be used (e.g. Hepper, 1986; Elwood & Ostermeyer, 1984). Indeed, such cues may be the most reliable guides to kinship if the probability of detectable phenotypic traits being shared between unrelated individuals is high. However, we do not intend to discuss mechanisms of kin discrimination in detail here; these have recently been reviewed by several authors (see e.g. Waldman, 1987, Waldman *et al.*, 1988, Barnard 1990). The purpose of this discussion is to explore the ways in which kin bias in mating success might arise and the extent to which it might reflect kin discrimination in mating preferences. We deal first with the potential causes of kin bias in mating success that do not involve mate choice and then with the role, if any, of kinship *per se* in kin-biased mating preferences.

Kinship and non-random mating without mate choice

Mating preferences are one cause of non-random mating, but non-randomness in mating alone does not necessarily imply mating preferences. Differences between individuals in patterns of mobility and dispersal or intrasexual competitive ability, for example, may bias access to mates and lead to non-random mating. There are several ways that such factors could lead to kin bias in mating success.

Behavioural similarity among kin

Close kin share alleles and thus the phenotypic characters for which they code. This means that relatives may share behavioural responses in given situations which lead to kin bias in behaviour. In some species, individuals disperse from natal areas in kin groups (e.g. Bygott *et al.*, 1979; Packer, 1979) and there is evidence that, in some cases, dispersal distance may be characteristic of sibships (Keppie, 1980). In such populations, and in those where there is little or no dispersal or there are other reasons, such as shared microclimatic preferences, why kin might associate, kin bias in mating may arise simply because behavioural responses lead to kinship structuring of the population. Indeed, the reason dispersal in some species tends to be in kinship groups may be because close kin share emigration thresholds (say, in response to increasing population density or food shortage).

Even where kin groups mix, however, behavioural and other phenotypic similarities between close relatives could bias mating success. Access to mates may be determined largely by success in intrasexual competition (e.g. LeCroy, 1981; Borgia, 1985). If close relatives share characteristics that affect competitive ability (Dewsbury, 1990), such as body size, hormone levels, motivational thresholds and so on, access to mates may be biased towards certain kin groups because individuals within them share high competitive ability. Behavioural similarities between kin may also influence the outcome of intersexual interactions. Barnard (1990) argues that the appearance of kin discrimination in dyadic encounters between animals (e.g. Kareem & Barnard, 1982, 1986; Barnard & Fitzsimons, 1988; Aldhous, 1989b) could in some cases be due to individuals sharing behavioural characteristics with kin and differing in these characteristics from non-kin. If close kin have, say, similar thresholds of response to aggressive or investigatory approaches, and unrelated individuals have very different ones, patterns of social interaction between related and unrelated individuals may be very different for reasons that have nothing to do with kin discrimination. While different physiological processes are likely to be involved in male and female sexual arousal, a similar argument can be applied to intersexual interactions where close kin share arousal thresholds. There is some evidence that differences between interacting dyads of different degrees of relatedness and age and sex composition in mice are attributable to pair rather than individual effects (Kareem & Barnard, 1986). If shared thresholds are low and mating speed high, there may be a mating success bias in favour of close kin; if thresholds are high, the opposite may be true. Tendencies to inbreed or outbreed in particular instances could, therefore, arise in the absence of any active preference/avoidance. Assuming thresholds to be random with respect to kinship groups across the population as a whole, however, kin bias in mating success as a result of shared thresholds is likely to be of only local importance and is unlikely to lead to general inbreeding or outbreeding.

Intrasexual competition and kin bias through discrimination

While shared characters within intrasexual kin groups may bias access to mates as argued above, bias could also occur through discrimination in intrasexual competition. If, for whatever reason (see Grafen, 1990; Barnard, 1990 and below), there is differential aggression towards competitors along kinship lines, bias in favour of or against close relatives in access to mates may occur. Several experiments have demonstrated kin bias or bias in genetic similarity in intrasexual aggressive or aggression-related behaviours which could result in a bias in access (Kareem & Barnard, 1982; Kareem, 1983; Aldhous, 1989a; Everitt *et al.*, 1991). However, it is also clear that any bias in access may be affected by independent bias in the receptivity of potential mates (Barnard & Fitzsimons, 1988) and that patterns of intrasexual discrimination may differ depending on experience of potential mate availability (Aldhous, 1989a).

There are at least two ways in which the fitness of kin discrimination alleles might be enhanced by intrasexual discrimination. The first is through the indirect component of inclusive fitness. Depending on the level of local mate competition (see Smith, 1979), indirect fitness advantages may accrue from reducing competition with closely related individuals. In some cases, this may extend as far as co-operation between kin in mate acquisition (Watts & Stokes, 1971) or even sharing mates with relatives (Maynard Smith & Ridpath, 1972). In these cases, the importance of kinship is as a guide to co-possession of the discrimination allele(s). The second is in reducing competition between individuals with similar genotypes and thus potentially similar resource requirements (e.g. Levene, 1953; Williams, 1975; Ellestrand & Antonovics, 1985; Jasieński, 1988). Where competition for local resources is intense, it may pay individuals to discriminate against close kin at the mate acquisition stage if the offspring of relatives impose severer competition on their own in the next generation. In many social mammals, breeding females suppress reproduction among other females which are closely related (e.g. Payman & Swanson, 1980; Vandenburgh & Coppola, 1986). In some cases, females are simply less likely to breed if close kin are present (Armitage, 1986); in others, kin competition among females extends to overt aggression and even infanticide (Hoogland, 1985; Armitage, 1986).

Kinship and mating preferences

While kin bias in mating success may arise as a result of intraspecific kinship effects, mating preferences within one or both sexes may also cause kin bias. This may work in conjunction (positively or negatively) with intrasexual bias or it may be the principal or only cause of bias. Intrasexually, kin bias through mating preference could arise in a number of different ways. These all involve discrimination of the degree of genetic similarity between potential mates but differ in the mechanism and function of discrimination.

Kinship and similarity as indirect guides to mate quality

The most familiar and widely-quoted argument for a relationship between kinship and mating preference concerns the costs and benefits of inbreeding/outbreeding (see Introduction) and the role of kinship preferences in optimizing the degree of inbreeding. Optimal inbreeding/outbreeding arguments imply discrimination on the basis of either phenotypic characters (moderately different from those of the discrimination allele bearer or its close relatives) which are encoded at loci other than that of the discrimination allele or non-genetic cues (e.g. location or environmentally-conferred odours) that happen to correlate with relatedness. The spread of the discrimination allele is determined by its consequences at the other loci with which it shares resultant offspring and which determine the offspring's survivorship and reproductive success. The discrimination allele thus 'hitchhikes' (Maynard Smith, 1978) on the back of its consequences elsewhere in the genome.

If bearer survivorship and reproductive success is enhanced by gene complementarity between mates at particular loci (as implied by the notion of optimal inbreeding/outbreeding), then direct discrimination of cues for complementary alleles at those loci may be favoured (see below). Since several different loci, and degrees of complementarity at those loci, may be involved, direct discrimination is likely to correlate with overall genetic similarity between potential mates and thus with kinship. The degree of relatedness between mates, however, is here an incidental consequence of the correlation between kinship and genetic similarity; kin discrimination *per se* is not occurring. On the other hand, kinship may be used as an indirect guide to potential complementarity in which case kin discrimination, perhaps based on shared characters that are unconnected with those having consequences for bearer survival and reproductive success (including, for example spatial location or familiarity during early development), may be occurring.

While discrimination on the basis of overall phenotypic (and thus potentially genetic) similarity seems an obvious mechanism of achieving complementarity, there are good reasons for caution in suggesting that matching across large numbers of loci might be used. Some mechanism of kin discrimination, perhaps using non-conspecific cues, is likely to be a simpler way of ensuring an appropriate degree of similarity than making detailed phenotypic comparisons. However, such multiple character comparisons might occur where a high degree of genetic similarity between mates is favoured in highly outbred populations in which most individuals are unrelated. Even here, though, the probable scarcity of individuals meeting the matching requirements is likely to impose high mate searching costs and render mating with kin, using the above indirect cues, a more cost-effective alternative (see Grafen, 1990).

A more feasible alternative to matching across large numbers of loci is to match for rare alleles at one or a few loci which are likely to be shared only by close kin (Grafen, 1990). Highly polymorphic loci such as those involved in histocompatibility systems (e.g. Klein, 1976, 1979; Grosberg, 1988) are ideal candidates and accumulating evidence suggests they may be used in phenotypic matching by potential mates. Grosberg & Quinn's (1986) study of larval settlement in the tunicate *Botryllus schlosseri* is a nice example.

In a series of experiments, Grosberg & Quinn (1986) found that settlement patterns in larval *B. schlosseri* on homogeneous plates appeared to be influenced by relatedness. Groups of siblings settled non-randomly, tending to clump together, while groups of unrelated larvae settled at random. As colonies grow, adjacent ones may meet and fuse, forming a chimaera in which intimate contact extends to sharing a blood vascular system. Fusion appears to be adaptive for both members of the chimaera, in that colony survivorship increases with colony size, larger colonies show some evidence of reproducing earlier and chimaeras have different physiological attributes from either component colony, thus potentially increasing the range of environmental tolerance (Buss, 1982). Whether or not colonies fuse depends on their genotypes at a single histocompatibility locus. If they share the same allele at the locus, colonies fuse; if they do not share alleles, fusion does not occur. The rarity of any given allele at the histocompatibility locus means that fusions are almost certain to be between close kin, though certainty is not absolute because fusions would occur between unrelated colonies that just happened to share histocompatibility alleles and fail to occur between related colonies that happened not to share alleles. An effective mechanism of kin bias is likely to be important in colony fusion decisions because co-operative chimaera formation is potentially cheatable. By fusing with kin, co-operators reduce the risk of being cheated as kin are likely to share alleles for co-operation as well as the histocompatibility marker. Alternatively, 'exploitation' between kin may simply be a nepotistic division of labour. Grafen (1990) discusses the role of cheating in maintaining polymorphism at the histocompatibility marker locus.

The highly specific historecognition system in *B. schlosseri* bears some similarity to the long familiar self-incompatibility mechanisms in plant pollination in which pollen growth is checked if the haploid pollen carries either of the two alleles present at a particular incompatibility locus in the diploid tissue of the stigma (e.g. Emerson, 1940). In this case, therefore, differences rather than similarity at crucial loci are favoured.

The *Botryllus* and pollen incompatibility mechanisms are just two examples of what may be a widespread role of histocompatibility differences in mate choice. Since histocompatibility systems such as those in tunicates and the major histocompatibility complex (MHC) in mammals are generally highly polymorphic, some of these may reflect kin discrimination through rare marker alleles. However, as we shall argue below, histocompatibility differences may form the basis for mating preferences for other reasons which incidentally lead to kin bias. Distinguishing experimentally between kin discrimination and other mechanisms leading to kin bias in mating is likely to be difficult, but there may be *a priori* reasons for favouring one or other explanation in particular cases. For example, the risk of cheating in chimaera formation by *B. schlosseri* provides a reason for expecting kin discrimination (Grafen, 1990).

Similarity as a direct guide to mate quality

The above mechanisms are concerned with achieving a degree of similarity in genotype using phenotypic similarities which themselves may have no consequences for the fitness of discrimination alleles. The matching of marker cues (overall phenotypic similarity, shared rare alleles, etc.) is used as a guide to likely similarity at loci other than those coding for the marker(s) where matching has consequences for offspring survivorship and reproductive potential, and thus discrimination allele fitness. However, it may be that matching of discriminated cues themselves has fitness consequences for the discrimination allele. The role of allelic differences in the MHC and related loci in mating preferences among mammals (e.g. Yamazaki *et al.*, 1976, 1978; Yamaguchi *et al.*, 1978) may be a good example. While differences in highly polymorphic MHC alleles may simply be used as marker cues for desirable degrees of similarity elsewhere in the genome, they may have important fitness consequences for discrimination alleles in their own right. Since the MHC is concerned with the self/non-self recognition involved in resistance to disease, judicious choice of mate MHC genotype could confer enhanced disease resistance on offspring. Outbreeding with respect to MHC alleles as suggested by the work of Yamazaki *et al.* (1976, 1978) with mice, will result in heterozygosity and diversity among offspring which may enable them to respond immunologically to a wider range of antigens (see Beauchamp *et al.*, 1985). Such benefits in disease resistance may provide the selection pressure not only for observed preferences for outbreeding with respect to the MHC but also for the effects of MHC differences on pregnancy blocking (Yamazaki *et al.*, 1983). Although there is no direct evidence for it, MHC differences may be instrumental in the acceleration of puberty in male and female mice reared with less closely related adults of the opposite sex (Lendrem, 1985, 1986). Such discrimination may result in kin bias simply because kin are less likely to differ at MHC loci than unrelated individuals. This is true whether it is overall differences across loci or differences at a particular locus that are important.

At first sight, recent immunological studies appear to bear out the MHC heterozygosity argument. Comparisons of the ability of two inbred strains of laboratory mice (of different MHC type) and their F_1 hybrids to clear intestinal nematode infections showed that hybrids cleared infections more rapidly than mice from either parental strain (Robinson *et al.*, 1989). While one interpretation of this is that MHC heterozygosity in the hybrids enhanced their immune response to infection, MHC heterozygosity in the experiment by Robinson *et al.* was correlated with heterozygosity at other loci which may have influenced responses. Furthermore, the enhanced response by hybrids was not evident in crosses between other inbred strains (Wahid *et al.*, 1989). Although these caveats do not rule out offspring MHC heterozygosity as a potential criterion in mating preferences, they suggest that MHC characteristics may be used instead, or in addition, to indicate degrees of genetic similarity elsewhere in the genome, as discussed in the previous section. The relative importance of the 'MHC heterozygosity' and 'genetic similarity indicator' functions has been the subject of some recent experiments with house mice.

Using semi-natural populations of mice (*Mus domesticus*) (wild mice crossed with inbred strains of different MHC genotype), Potts *et al.* (1990) found that females preferred males of dissimilar MHC type. Moreover, MHC-heterozygous males were more successful in establishing territories, suggesting a fitness advantage to alleles for MHC-based disassortative mating. However, as in the experiment by Robinson *et al.*, heterozygosity at MHC loci correlated with heterozygosity in the genome as a whole. When the experiment was repeated using a population in which the correlation had been eliminated using a selective breeding program, the apparent fitness consequences of MHC heterozygosity disappeared. Potts *et al.* (1990) interpret their results as indicating that MHC-determined cues are important as indicators of kinship in mating decisions by females rather than as a guide to offspring MHC heterozygosity *per se* (though they could simply indicate the likelihood of specific genetic differences elsewhere in the genome). While this may be the case, however, it again does not rule out offspring MHC heterozygosity as a mate choice criterion. Increasing the selection pressure from pathogenic organisms might shift preferences towards MHC heterozygosity criteria.

As an important aside, it is worth pointing out that a high degree of polymorphism *per se* does not necessarily mean that discrimination will be based on specific alleles. For example, while the polymorphic t-allele complex in house mice has also been implicated in kin discrimination (see Lenington, *et al.*, 1988), discrimination in this case appears to be based on broad groupings of mutants within the complex which reduces the effective diversity of alternative genotypes in mating preferences. A large number of t mutations have been identified in the complex (a large segment of chromosome 17) of which most are recessive lethals causing death during foetal development in homozygotes. A few are termed 'semilethal'; homozygous 'semilethals' may survive but all males are sterile. Despite the intense selection against these deleterious alleles, about 25 per cent of wild mice are heterozygous for a t-mutation (Bennett, 1978, Lenington *et al.*, 1988) because t-alleles are associated with segregation distortion effects in hetero-zygous males (Bennett, 1975). While mice appear to have mating preferences based on the possession of t-alleles by potential mates (e.g. Lenington 1983; Lenington & Egid, 1985), preferences seem to be determined by whether a potential mate carries alleles belonging to the broad categories of lethal or semilethal rather than the unique identity of the allele itself. Unlike the situation with the MHC, therefore, the t-allele discrimination system is unlikely to be used in kin discrimination.

Kinship and co-bearer discrimination

In the previous section, we argued that kin bias in mating preference could arise through discrimination for characters that correlate with kinship or through kinship being used as a guide to the possession of characters enhancing the fitness of discrimination alleles. In these cases, the fitness of discrimination alleles is influenced by the alleles with which they subse-quently share bodies as a result of discrimination. A second route through which discrimination alleles can increase their fitness is by discriminating and favouring co-possessors. The probability of a discrimination allele being present in the offspring of a female mating with, say, her own father is 0.75, since the allele will be passed on with a probability of 0.5 from the female herself and 0.25 from the resultant offspring's father/grandfather (Smith, 1979). Incestuous mating will thus be favoured as long as kin selected increases in the fitness of the discrimination allele outweigh any decrease due to inbreeding depression (Harvey & Read, 1988) at other loci (Smith, 1979) and the fitness advantage of being passed on through non-incestuous matings (Dawkins, 1979). A high degree of mate competition among members of one sex may thereby select for inbreeding by the other (Smith, 1979; Parker, 1983). In principle, co-bearer discrimination could be achieved directly through a green beard-type mechanism or indirectly through kin discrimination (see above and Barnard, 1990). If discrimination alleles do not advertise their presence by a recognizable phenotypic marker, kin discrimination (indirect co-bearer discrimination) at some level becomes the most parsimonious mechanism for increasing assortatative mating between allele co-bearers. However, fitness increases arising from direct co-bearer discrimination may be sufficiently great to favour the evolution of phenoty-pic markers if they arise (Barnard, 1990). If the discrimination allele already codes for both recognition and discrimination processes (Byers & Bekoff,

1986; Barnard, 1990), the additional encoding of a marker character would result in a green beard-type discrimination system. While a green beard system is unlikely to arise *de novo*, therefore, it might arise by the more piecemeal pathway of adding a marker to an existing indirect co-bearer discrimination system.

Although selection for co-bearer discrimination might be expected to result in direct discrimination mechanisms, discrimination may still be based in part on phenotypic characters encoded at other loci so that direct co-bearer discrimination may in practice be indistinguishable from kin discrimination or direct discrimination for similarity. This could arise where phenotypic characters encoded elsewhere correlate with survivorship and reproductive success among resultant offspring and thus constitute additional indices of parental value. Phenotypic characters other than those of the discrimination allele may therefore be used in mate choice in two ways: (a) as kinship cues indicating the probability of allele sharing or degree of genetic similarity or (b) cues to the parental value of known allele co-bearers.

Variation in preferences based on kinship

So far, we have discussed a number of problems that need to be borne in mind when thinking about the role of kinship in non-random mating in general and mating preferences in particular. In many cases, kin bias in mating success may have nothing to do with kin discrimination but may instead be due to other causes of kin bias, the consequences of which are difficult to distinguish from those of discrimination. While these arguments caution against assuming too readily that discrimination is the cause of bias, the opposite problem could also arise: discrimination may be occuring but the decision rules determining its expression are complex and subtle so that it is expressed only under certain circumstances. Whether or not discrimination occurs, the degree of kin bias it produces when it does occur and the kinship classes it favours/penalizes may be conditional on a number of factors, some of which we now discuss.

Sex differences in mating preference

In cases where both sexes show mating preferences, there may be conflicts of interest between the sexes in preferred criteria (e.g. Halliday, 1978; Smith, 1979; Parker, 1983). Where kinship between potential mates is important in mating decisions, the sexes may differ in preferred degree of relatedness or even whether relatedness matters at all.

In their study of intersexual interactions between laboratory mice (*Mus musculus*), Barnard & Fitzsimons (1988) found that males and females differed in apparent kinship preferences. When males were given access to pots of soiled sawdust from females of different degrees of relatedness (full and half-siblings, first and second cousins and unrelated females), they spent more time investigating the sawdust of less closely related females. However, the effect of relatedness was apparent only when males had had previous experience of the females during rearing. There was no significant effect with sawdust from unfamiliar females. When males were given access to the females themselves (restrained in individual cages to prevent intrasexual interference), there was a pronounced bias towards spending time with and investigating second cousin females, but again only when females were familiar. Females showed no tendency to discriminate between males or their soiled sawdust on the basis of kinship, irrespective of familiarity. In freely mixing groups, there was still some bias in investigation by males towards second cousin females but no effect of relatedness on time spent together; furthermore, bias in investigatory behaviours occurred only with unfamiliar females. At the level of time allocation and investigatory interest, therefore, the form and degree of apparent kin discrimination depended on the sex of the subject, the cues available (odour, individual) for discrimination, the experience of subjects during rearing and the social context in which any choices were being made.

While females in Barnard & Fitzsimons' experiment showed no evidence of discrimination in time spent or investigation, there appeared to be some discrimination in proceptivity and receptivity towards investigating males when females showed pre-mounting sexual interest (an index of sexual arousal). In some cases, females showed differences in both towards males of different degrees of relatedness. Where there was a significant effect of relatedness, females appeared to prefer closely related (full or half-sibling) males. Where there was no significant effect of relatedness on female behaviour, most mounting occurred between full and half-siblings and second cousins. Overall, therefore, Barnard & Fitzsimons' results suggest different male and female kinship preferences in mating, with female preferences exerting a stronger influence when their sexual interest was high. Further experiments (Barnard & Fitzsimons, 1989), however, suggested a fitness advantage to outbreeding in terms of litter size with some evidence of a peak with second cousin matings. In addition, male offspring of second cousin matings tended to be more aggressive, thus possibly increasing their competitive ability. Why, then, should females apparently prefer to mate with more closely related males?

While the reason may be kin-related consequences for litter size and female lifetime reproductive success (Barnard & Fitzsimons (1988), for instance, suggest that producing and caring for the larger litters of outbred matings may reduce female longevity and/or ability to raise future offspring (see Fuchs, 1981, 1982)), the difference in preference may arise because females use kinship as a guide to some other factor influencing offspring survival and reproductive success rather than as a choice criterion in its own right. One obvious factor is male dominance (Barnard & Fitzsimons (1989). The social structure of wild house mouse colonies is very variable (Bronson, 1979), but most matings appear to be achieved by dominant individuals of both sexes (Wolff, 1985; Hurst, 1987) so that resident young animals are likely to be the offspring of dominants. This in turn means that, as long as physical and behavioural characters influencing dominance are heritable, kinship and potential dominance among available males are likely to be correlated. Kinship could, therefore, be used as a guide to male quality. Since females tend to breed within family groups (Lidicker, 1976; Baker, 1981), it could be argued that mating indiscriminately with any male would result in kin bias so that a mechanism of kin discrimination is unnecessary. However, although generally regarded as inbred (Reimer & Petras, 1967; Anderson, 1970, Selander, 1970), evidence suggests there may sometimes be appreciable gene flow between house mouse populations in the wild (see Baker, 1981) so that discrimination may be favoured. Alternatively, where mate competition within one sex is high, there may be inclusive fitness advantages to discrimination alleles from matings between close relatives (Smith, 1979; but see Dawkins, 1979).

In other studies, sex differences in apparent kinship preferences have been less pronounced. Bateson (1982) found that, while both sexes of Japanese quail (*Coturnix coturnix japonica*) preferred intermediately related novel individuals (i.e. cousins rather than siblings or unrelated individuals) of the opposite sex, preference for first cousins was slightly stronger among females and preference for third cousins was stronger among males. In both sexes, though, first cousins were preferred over other individuals. Later experiments showed that mating between cousins reduced the age of first laying relative to matings between siblings or more distant relatives (Bateson, 1988). Furthermore, when the variability of Bateson's quail population was increased by outbreeding with new stock, there was a shift to earlier laying among sibling pairs. One interpretation of this is that, by increasing genetic variation within populations, the optimal degree of inbreeding had shifted towards closer relatives compared with the lower diversity, inbred populations (Bateson, 1988).

Population structure and kin bias

The kinship structuring of populations and their overall degree of inbreeding are likely to influence the degree and form of kin bias whether through discrimination or any other cause. At a trivial level, population viscosity and low dispersal from natal areas are likely to result in kin bias simply because most available mates are closely related. While it is still possible that selection will favour discrimination between kinship classes even within highly inbred populations, there are reasons for supposing that the degree of inbreeding will influence selection for kin preferences in a complex way. As Dawkins has pointed out (see McGregor, 1989), the more a population inbreeds, the less costly inbreeding is likely to become. This is because any deleterious genetic consequences of inbreeding may be removed by long-term inbreeding (see Partridge, 1983) and the genetic difference between mates (and thus the difference in fitness consequences of choosing one individual rather than another) is reduced. The costs and benefits of inbreeding may thus involve an element of frequency dependence and introduce instability in the optimal degree of inbreeding.

Where the development of discrimination depends on learning appropriate models, variation in population structure may impose local constraints on discrimination through its effects on the availability and label characteristics of models for template formation, thus leading to population differences in discrimination. Constraints on model availability may preclude later discrimination altogether. For example, there is evidence that female great tits (*Parus major*) use male song as a means of optimizing the degree of outbreeding (McGregor & Krebs, 1982a). Females tend to pair with males whose songs are moderately similar to those of their fathers (McGregor & Krebs, 1982a). While, in theory, it might pay females to refine their discrimination by also taking into account the songs of their brothers, the latter do not acquire their songs until the spring after they have fledged and dispersed when they learn them from territorial neighbours rather than their fathers (McGregor & Krebs, 1982b). Since females have no opportunity to learn their brothers' songs, it is not surprising that brother–sister (and also mother–son since, by the same token, mothers do not experience the songs of their sons) incest occurs at about chance levels, whereas father–daughter incest is rarer than expected (Greenwood *et al.*, 1978; see also van Tierderen & van Noordwijk, 1988).

As well as influencing the costs and benefits of inbreeding/outbreeding and the information on which choices of mate can be based, population structure may also influence kin bias through its effects of mate availability and mating competition. Whether or not degree of relatedness affects behavioural bias is likely to depend on the choices available at the time. As Barnard (1990) points out, if discrimination is based on absolute degrees of relatedness, so that what matters is whether an individual is a full sibling or a first cousin, then discrimination will not be apparent unless the appropriate kinship class is present. Moreover, the use of absolute degrees of relatedness will result in preferences for more closely related individuals in some cases and less closely related individuals in others as the range of available alternatives varies. Preferences may also be inconsistent with respect to relatedness if relatedness cannot be distinguished accurately and mistakes are made. Even where relative degree of relatedness is used, so that animals, for example, generally prefer the most or least closely related individuals of any available set, there may be thresholds of relatedness above or below which discrimination is not favoured by selection or is not possible with the mechanism being used (see e.g. Gamboa *et al.*, 1986); discrimination will then be apparent over only a limited range of kinship classes.

If population changes result in changes in the availability of mates generally, the consequences for mating competition may affect both intra- and intersexual kin bias in mating success. Intrasexually, severer competition may accentuate the effect of kin group differences in competitive ability (see earlier) so that access to mates is biased towards certain groups. On the other hand, it may reduce nepotistic intrasexual discrimination and thus bias due to kin discrimination. Intersexually, an increase in local mate competition is likely to increase selection for incestuous mating through its consequences for the inclusive fitness of discrimination alleles (see Smith, 1979 and earlier).

Motivational effects

While it may be possible to predict broad trends in kin bias due to discrimination in different contexts, differences in the motivational state of animals at the time of testing or observation may create variability in outcome. For example, studies which have shown intersexual kin preferences in female mice have used females which were in natural or induced oestrus (e.g. Winn & Vestal, 1986; Barnard & Fitzsimons, 1988). Hayashi & Kimura (1983), however, failed to show kin preferences in females which were not in oestrus. Motivational changes may similarly confound studies of intrasexual kin bias. Aldhous (1989a) found that male laboratory mice given experience of females for 48 hours and then tested in male dyads showed no evidence of the apparent kin discrimination in investigatory and passive contact behaviours previously shown in such dyads (Kareem & Barnard, 1982; Aldhous, 1989b), but did show discrimination in a different range of behaviours. Which behaviours showed evidence of discrimination was also affected by whether or not the male had copulated with the female (Aldhous, 1989a). The effect of experience with females was generally to increase levels of aggression and investigatory behaviours towards, and reduce levels of passive contact between, all males regardless of degree of relatedness. Isolating males prior to testing had similar effects on levels of behaviour but abolished kin discrimination (Aldhous, 1989a). Little attention has so far been paid to the effects of differences in motivational state on kin bias, but since both internal (e.g. hormone levels, 'confidence' in aggressive disputes) and external (e.g. mate availability) environmental factors are likely to be key elements in the decision rules governing kin discrimination, experiments which systematically vary these are likely to yield interesting results.

Phenotype-limited kin bias

Earlier we discussed the potential role of kin-correlated differences in competitive ability in intrasexual kin bias. However, competitive ability and other phenotypic differences between individuals may result in variation in kin bias *within* kinship classes. As in other areas of decision-making, optimal discrimination policies may vary with individual phenotype (Parker, 1982). For instance, both intra- and intersexual discrimination may be conditional on the competitive abilities of interactants or vary with their age and sex in different contexts (e.g. Sherman, 1977; Kareem & Barnard, 1986). Such context-specific and phenotype-limited decision rules for discrimination could lead to pronounced variation in access to mates and/or mating preferences within any given kinship class.

The phenotypes of discriminators are one potential source of variation in kin bias within kinship classes; phenotypic differences among discriminanda may be another. As discussed earlier, where mating preferences are based on kinship, whether for genetic similarity or co-bearer discrimination, discrimination may still include phenotypic characters encoded at loci other than those producing the cues used in matching or coding for the discrimination allele in the case of direct co-bearer discrimination because variation at other loci may well affect offspring survivorship and reproductive success. Individuals of a given kinship class may thus not be equivalent in terms of mate quality (see also Barnard, 1990).

When apparent kin discrimination may not be kin discrimination

In his recent discussion, Grafen (1990) points out that even where kin bias in behaviour is due to preferences for or discrimination against individuals of different degrees of relatedness, bias may not be due to kin discrimination in that kinship *per se* is not the object of discrimination. Grafen's point is that, in several detailed studies of what appears to be kin discrimination, kin bias could arise as a result of kinship cues being used as a guide to discrimination at other levels such as species or social group. Close kin may provide a model for species recognition simply because they are at hand at the appropriate stages in development and are a sufficiently reliable guide or because self-matching as a species identification mechanism biases responses to conspecifics which are most like self. If mating preferences hinge on species identification (e.g. Capranica *et al.*, 1973), kin bias may thus arise because of the rules used in identifying conspecifics.

Discrimination at levels other than kinship may result in apparent kin discrimination in other ways. Animals may recognize a number of familiar closely related conspecifics individually and behave differentially towards them on that basis. Unfamiliar relatives will share characteristics with the familiar relatives with a probability determined by their coefficient of relatedness. When these unfamiliar relatives are encountered they may thus be mistaken through stimulus generalization for one or other familiar relative (Waldman, 1987). What appears to be kin discrimination in fact arises simply by confusing individual identities. If the cost of mistaken identity is low or mistakes are likely to be rare in nature, there may be only weak selection against stimulus generalization for familiar characters. As Waldman (1987) points out, such a mistaken identity effect could give rise to a close correlation between social responses and degree of relatedness since the probability of making a mistake will decrease as the degree of relatedness (and thus phenotypic similarity) decreases.

Acknowledgements

We should like to thank Jane Hurst and Wayne Potts for constructive comments on the manuscript, Jerzy Behnke for various helpful discussions, Pat Bateson for sending us copies of papers and Wayne Potts for sending manuscripts and allowing us to refer to some as yet unpublished results.

References

Ågren, G. (1984). Incest avoidance and bonding between siblings in gerbils. *Behav. Ecol. Sociobiol.*, **14**, 161–9.

Aldhous, P. G. M. (1989a). *Mechanisms of Apparent Kin Discrimination in Male Mice*. Ph.D. Thesis, University of Nottingham.

Aldhous, P. G. M. (1989b). The effects of individual cross-fostering on the development of intrasexual kin discrimination in male laboratory mice, *Mus musculus* L. *Anim. Behav.*, **37**, 741–50.

Anderson, P. K. (1970). Ecological structure and gene flow in small mammals. *Symp. Zool. Soc. Lond.*, **26**, 299–325.

Andersson, M. (1982). Female choice selects for extreme tail length in a widowbird. *Nature*, **299**, 818–20.

Armitage, K. B. (1986). Marmot polygyny revisited: determinants of male and female reproductive strategies. In *Ecological Aspects of Social Evolution: Birds and Mammals*, ed. D. I. Rubenstein & G. R. Michener, pp. 303–31. Princeton. Princeton University Press.

Baker, A. E. M. (1981). Gene flow in house mice: behavior in a population cage. *Behav. Ecol. Sociobiol.*, **8**, 83–90.

Barnard, C. J. (1990). Kin recognition: problems, prospects and the evolution of discrimination systems. *Adv. Stud. Behav.* **19**, 29–81

Barnard, C. J. & Fitzsimons, J. (1988). Kin recognition and mate choice in mice: the effects of kinship, familiarity and social interference on intersexual interactions. *Anim. Behav.*, **36**, 1078–90.

Barnard, C. J. & Fitzsimons, J. (1989). Kin recognition and mate choice in mice: fitness consequences of mating with kin. *Anim. Behav.*, **38**, 35–40.

Bateson, P. P. G. (1978). Sexual imprinting and optimal outbreeding. *Nature* **273**, 659–60.

Bateson, P. P. G. (1982). Preferences for cousins in Japanese quail. *Nature* **295**, 236–7.

Bateson, P. P. G. (1983). Optimal outbreeding. In *Mate Choice*, ed. P. P. G. Bateson, pp. 257–77. Cambridge: Cambridge University Press.

Bateson, P. P. G. (1988). Preferences for close relations in Japanese quail. *Acta XIX Congr. Int. Orn.*, **1**, 961–72.

Beauchamp, G. K., Yamazaki, K. & Boyse, E. A. (1985). The chemosensory recognition of genetic individuality. *Sci. Am.*, **253**, 66–72.

Bennett, D. (1975). The T-locus of the mouse. *Cell*, **6**, 441–54.

Bennett, D. (1978). Population genetics of the T/t complex mutations. In *Origins of Inbred Mice*, ed. H. Morse, pp. 615–32. New York: Academic Press.

Bertram, B. C. R. (1978). Living in groups: predators and prey. In *Behavioural Ecology: An Evolutionary Approach*, ed. J. R. Krebs & N. B. Davies, pp. 64–96. Oxford: Blackwell.

Blaustein, A., Bekoff, M. & Daniels, T. J. (1987). Kin recognition in vertebrates (excluding primates). In *Kin Recognition in Animals*, ed. D. J. C. Fletcher & C. D. Michener, pp. 287–357. New York: John Wiley.

Borgia, G. (1985). Bower quality, number of decorations and mating success of male bower satin birds (*Ptilinorhynchus violaceus*): an experimental analysis. *Anim. Behav.*, **33**, 353–89.

Bradbury, J. W. & Andersson, M. (Eds). (1987). *Sexual Selection: Testing the Alternatives*. Berlin: Springer Verlag.

Bronson, F. H. (1979). The reproductive ecology of the house mouse. *Q. Rev. Biol.*, **54**, 265–99.

Buss, L. W. (1982). Somatic cell parasititism and the evolution of somatic tissue compatibility. *Proc. Nat. Acad. Sci., U.S.A.*, **79**, 5337–41.

Byers, J. A. & Bekoff, M.(1986). What does kin recognition mean? *Ethology*, **72**, 342–5.

Bygott, D., Bertram, B. C. R. & Hanby, J. P. (1979). Male lions in large coalitions gain reproductive advantages. *Nature*, **282**, 839–41.

Capranica, R. R., Frishkopf, L. S. & Nevo, E. (1973). Encoding of geographic dialects in the auditory system of the cricket frog. *Science*, **182**, 1272–5.

Darwin, C. (1859). *On the Origin of Species*. London: John Murray.

Darwin, C. (1871). *The Descent of Man and Selection in Relation to Sex*. London: John Murray.

Dawkins, R. (1979). Twelve misundertandings of kin selection. *Z. Tierpsychol.*, **51**, 184–200.

Dewsbury, D. A. (1982). Avoidance of incestuous breeding between siblings in two species of *Peromyscus* mice. *Biol. Behav.*, **7**, 157–69.

Dewsbury, D. A. (1990). Fathers and sons: genetic factors and social dominance in deer mice, *Peromyscus maniculatus*. *Anim. Behav.*, **39**, 284–9.

Ellestrand, N. & Antonovics, J. (1985). Experimental studies of the evolutionary significance of sexual reproduction. II. A test of the density-dependent selection hypothesis. *Evolution*, **39**, 657–66.

Elwood, R. W. & Ostermeyer, M. C. (1984). Does copulation inhibit infanticide in male rodents? *Anim. Behav.*, **32**, 293–4.

Emerson, S. H. (1940). Growth of incompatible pollen tubes in Oenothera. *Botan. Gaz.*, **101**, 890–911.

Everitt, J., Hurst, J. L., Ashworth, D. & Barnard, C. J. (1991). Aggressive behaviour among wild-caught house mice (*Mus domesticus* Rutty) correlates with a measure of genetic similarity using DNA fingerprinting. *Anim. Press.* (In Press).

Fletcher, D. J. C. & Michener, C. D. (Eds.) (1987). *Kin Recognition in Animals*. New York: John Wiley.

Fuchs, S. (1981). Consequences of premature weaning on the reproduction of mothers and offspring in laboratory mice. *Z. Tierpsychol.*, **55**, 19–32.

Fuchs, S. (1982). Optimality of parental investment: the influence of nursing on reproductive success of mother and female young house mice. *Behav. Ecol. Sociobiol.*, **10**, 39–51.

Gamboa, G. J., Reeve, H. K. & Pfennig, D. W. (1986). The evolution and ontogeny of nestmate recognition in social wasps. *Ann. Rev. Entomol.*, **31**, 431–54.

Gilder, P. M. & Slater, P. J. B. (1978). Interest of mice in conspecific odours is influenced by degree of kinship. *Nature*, **274**, 364–5.

Grafen, A. (1990). Do animals really recognize kin? *Anim. Behav.*, **39**, 42–54.

Greenberg, L. (1982). Persistent habituation to female odor by male sweat bees (*Lasioglossum zephyrum*). *J. Kans. Ent. Soc.*, **55**, 525–31.

Greenwood, P. J., Harvey, P. H. & Perrins, C. M. (1978). Inbreeding and dispersal in the great tit. *Nature*, **271**, 52–4.

Grosberg, R. K. (1988). The evolution of allorecognition specificity in clonal invertebrates. *Q. Rev. Biol.*, **63**, 377–412.

Grosberg, R. K. & Quinn, J. F. (1986). The genetic control and consequences of kin recognition by the larvae of a colonial marine invertebrate. *Nature*, **322**, 456–9.

Halliday, T. R. (1978). Sexual selection and mate choice. In *Behavioural Ecology: an Evolutionary Approach*, ed. J. R. Krebs & N. B. Davies, pp. 64–96. Oxford: Blackwell.

Hamilton, W. D. & Zuk, M. (1982). Heritable true fitness and bright birds: a role for parasites? *Science*, **218**, 384–7.

Harvey, P. H. & Read, A. F. (1988). Copulation genetics: when incest is not best. *Nature*, **336**, 514–15.

Hayashi, S. & Kimura, T. (1983). Degree of kinship as a factor regulating preferences among conspecifics in mice: *Anim. Behav.*, **31**, 81–5.

Hepper, P. G. (1986). Kin recognition: functions and mechanisms. *Biol. Rev.*, **61**, 63–93.

Holmes, W. G. & Sherman, P. W. (1983). Kin recognition in animals. *Am. Sci.*, **71**, 46–55.

Hoogland, J. L. (1985). Infanticide in prairie dogs: lactating females kill offspring of close kin. *Science*, **230**, 1037–40.

Hurst, J. L. (1987). Behavioural variation in wild house mice (*Mus domesticus* Rutty): a quantitative assessment of female social organization. *Anim. Behav.*, **35**, 1846–57.

Jasieński, M. (1988). Kinship ecology of competition: size hierarchies in kin and nonkin laboratory cohorts of tadpoles. *Oecologia*, **77**, 407–13.

Jasieński, M., Korzeniak, U. & Lomnicki, A. (1988). Ecology of kin and nonkin interactions in *Tribolium* beetles. *Behav. Ecol. Sociobiol.*, **22**, 277–84.

Kareem, A. M. (1983). Effect of increasing periods of familiarity on social interactions between male sibling mice. *Anim. Behav.*, **31**, 919–26.

Kareem, A. M. & Barnard, C. J. (1982). The importance of kinship and familiarity in social interactions between mice. *Anim. Behav.*, **30**, 594–601.

Kareem, A. M. & Barnard, C. J. (1986). Kin recognition in mice: age, sex and parental effects. *Anim. Behav.*, **34**, 1814–24.

Keane, B. (1990). The effects of relatedness on reproductive success and mate choice in the white-footed mouse (*Peromyscus leucopus*). *Anim. Behav.*, **39**, 264–73.

Keppie, D. M. (1980). Similarity of dispersal among sibling male spruce grouse. *Can. J. Zool.*, **58**, 2102–4.

Klein, J. (1986). *The Natural History of the Histocompatibility Complex*. New York: John Wiley.

Klein, J. (1979). The major histocompatibility complex of the mouse. *Science*, **203**, 516–21.

LeCroy, M. (1981). The genus *Paradisea* – display and evolution. *Amer. Mus.*, Nov. No. 2714.

Lendrem, D. W. (1985). Kinship affects puberty acceleration in mice (*Mus musculus*). *Behav. Ecol. Sociobiol.*, **17**, 397–9.

Lendrem, D. W. (1986). Kinship, pheromones and reproduction. In *The Individual and Society*, ed. L. Passera & J.-P. Lachaud, pp. 67–71. Toulouse: Privat.

Lenington, S. (1983). Social preferences for partners carrying 'good genes' in wild house mice. *Anim. Behav.*, **31**, 325–33.

Lenington, S. and Egid, K. (1985). Female discrimination of male odors correlated with male genotype at the T-locus: a response to T-locus or H-2 locus variability? *Behav. Genet.*, **15**, 53–67.

Lenington, S., Egid, K. & Williams, J. (1988). Analysis of a genetic recognition system in wild house mice. *Behav. Genet.*, **18**, 549–64.

Levene, H. (1953). Genetic equilibrium when more than one ecological niche is available. *Amer. Nat.*, **87**, 131–3.

Lidicker, W. Z. Jr. (1976). Social behaviour and density regulation in house mice living in large enclosures. *J. Anim. Ecol.*, **45**, 677–97.

Majerus, M. E. N. (1986). The genetics and evolution of female choice. *Trends Ecol. Evol.*, **1**, 3–7.

Maynard Smith, J. (1978). *The Evolution of Sex*. Cambridge: Cambridge University Press.

Maynard Smith, J. & Ridpath, M. G. (1972). Wife sharing in the Tasmanian native hen, *Tribonyx mortierii*: a case of kin selection? *Amer. Nat.*, **106**, 447–52.

McGregor, P. K. (1989). Bird song and kin recognition: Potential, constraints and evidence. *Ethol. Evol. Ecol.*, **1**, 123–7.

McGregor, P. K. & Krebs, J. R. (1982a). Song types in a population of great tits (*Parus major*): their distribution, abundance and acquisition by individuals. *Behaviour*, **79**, 126–52.

McGregor, P. K. & Krebs, J. R. (1982b). Mating and song types in the great tit. *Nature*, **297**, 60–1.

Murray, R. D. & Smith, E. O. (1983). The role of dominance and intrafamilial bonding in the avoidance of close inbreeding. *J. Human Evol.*, **12**, 481–6.

Packer, C. (1979). Inter-troop transfer and inbreeding avoidance in *Papio anubis*. *Anim. Behav.*, **27**, 1–36.

Parker, G. A. (1982). Phenotype-limited evolutionarily stable strategies. In *Current Problems in Sociobiology*. ed. King's College Sociobiology Group, pp. 173–201. Cambridge: Cambridge University Press.

Parker, G. A. (1983). Mate quality and mating decisions. In *Mate Choice*, ed. P. P. G. Bateson, pp. 141–64. Cambridge: Cambridge University Press.

Partridge, L. (1983). Non-random mating and offspring fitness. In *Mate Choice*, ed. P. P. G. Bateson, pp. 227–55. Cambridge: Cambridge University Press.

Payman, B. C. & Swanson, H. H. (1980). Social influence on sexual maturation and breeding in the female mongolian gerbil (*Meriones unguiculatus*). *Anim. Behav.*, **28**, 528–35.

Potts, W. K., Manning, C. J. & Wakeland, E. K. (1990). The maintenence of MHC genetic diversity: mating preferences, disease resistance and inbreeding depression. Abstract. *IV Int. Congr. Syst. Evol. Biol.*, Wisconsin.

Read, A. F. (1990). Parasites and the evolution of host sexual behaviour. In *Parasitism and Host Behaviour*, ed. C. J. Barnard & J. M. Behnke. London: Taylor and Francis. (In Press.)

Reimer, J. D. & Petras, M. L. (1967). Breeding structure of the house mouse, *Mus musculus*, in a population cage. *J. Mammal.*, **45**, 88–9.

Robinson, M., Wahid, F., Behnke, J. M. & Gilbert, F. S. (1989). Immunological relationship during primary infection with Heligosomoides polygyrus (*Nematospiroides dubius*): dose-dependent expulsion of adult worms. *Parasitology*, **98**, 115–24.

Sade, D. S. (1968). Inhibition of son–mother mating among free-ranging rhesus monkeys. *Sci. Psychol. Anal.*, **12**, 18–38.

Selander, R. K. (1970). Behavior and genetic variation in natural populations. *Am. Zool.*, **10**, 53–66.

Semler, D. E. (1971). Some aspects of adaptation in a polymorphism for breeding colours in the three-spined stickleback (*Gasterosteus aculeatus* L.). *J. Zool. Lond.*, **165**, 291–302.

Sherman, P. W. (1977). Nepotism and the evolution of alarm calls. *Science*, **197**, 1246–53.

Shields, W. M. (1982). *Philopatry, inbreeding and the Evolution of Sex*. Albany: State University of New York.

Shields, W. M. (1983). Optimal inbreeding and the evolution of philopatry. In *The Ecology of Animal Movement* ed. I. R. Swingland & P. J. Greenwood, pp. 132–59. Oxford: Clarendon Press.

Simmons, L. W. (1989). Kin recognition and its influence on mating preferences of the field cricket, *Gryllus bimaculatus* (de Geer). *Anim. Behav.*, **38**, 68–77.

Smith, B. H. (1983). Recognition of female kin by male bees through olfactory signals. *Proc. Nat. Acad. Sci., U.S.A.*, **80**, 4551–3.

Smith, R. H. (1979). On selection for inbreeding in polygynous animals. *Heredity*, **43**, 205–11.

Sutherland, W. J. (1985). Chance can produce a sex difference in variance of mating success and account for Bateman's data. *Anim. Behav.*, **33**, 1349–52.

Tierderen, P. H. van & Noordwijk, A. J. van (1988). Dispersal, kinship and inbreeding in an island population of the great tit. *J. Evol. Biol.*, **2**, 117–37.

Vandenburgh, J. G. & Coppola, D. M. (1986). The physiology and ecology of puberty modulation by primer pheromones. *Adv. Stud. Behav.*, **16**, 71–107.

Wahid, F. N., Robinson, M. & Behnke, J. M. (1989). Immunological relationships during primary infection with *Heligosomoides polygyrus* (*Nematospiroides dubius*): expulsion of adult worms from fast-responder syngeneic and hybrid strains of mice. *Parasitology*, **98**, 459–69.

Waldman, B. (1987). Mechanisms of kin recognition. *J. Theor. Biol.*, **128**, 159–85.

Waldman, B., Frumhoff, P. C. & Sherman, P. W. (1988). Problems of kin recognition. *Trends Ecol. Evol.*, **3**, 8–13.

Watts, C. R. & Stokes, A. W. (1971). The social order of turkeys. *Sci. Am.*, **224**, 112–18.

Wells, P. A. (1987). Kin recognition in humans. In *Kin Recognition in Animals* ed. D. J. C. Fletcher & C. D. Michener, pp. 395–415. New York: John Wiley.

Williams, G. C. (1975). *Sex and Evolution*. Princeton: Princeton University Press.

Winn, B. E. & Vestal, B. M. (1986). Kin recognition and choice of males by wild female house mice (*Mus musculus*). *J. Comp. Psychol.*, **1**, 72–5.

Wolff, R. J. (1985). Mating behaviour and female choice: their relation to social structure in wild caught house mice (*Mus musculus*) housed in a semi-natural environment. *J. Zool. Lond.*, **207**, 43–51.

Yamaguchi, M., Yamazaki, K. & Boyse, E. A. (1978). Mating preference tests with the recombinant congenic strain BALB.HTG. *Immunogenetics*, **6**, 261–4.

Yamazaki, K., Beauchamp, G. K., Wysocki, C. J., Bard, J., Thomas, L. & Boyse, E. A. (1983). Recognition of H-2 types in relation to the blocking of pregnancy in mice. *Science*, **221**, 186–8.

Yamazaki, K., Boyse, E. A., Mike, V., Thaler, H. T., Mathieson, B. J., Abbot, J., Boyse, J., Zayas, Z. A. & Thomas, L. (1976). Control of mating preferences in mice by genes in the major histocompatibility complex. *J. Exp. Med.*, **144**, 1324–35.

Yamazaki, K., Yamaguchi, M., Andrews, P. W., Peake, B. & Boyse, E. A. (1978). Mating preference of F2 segregants of crosses between MHC-congenic mouse strains. *Immunogenetics*, **6**, 253–9.

6

Genetic components of kin recognition in mammals

Edward A. Boyse, Gary K. Beauchamp, Kunio Yamazaki and
Judith Bard

In practice, although not in theory, the subject of this chapter – the genetic determination of body scents that distinguish one individual from another individual of the same mammalian species – is fairly new. Systematic work on this topic was made possible by certain incidental observations made by animal technicians responsible for deriving and maintaining congenic mouse strains in a special facility for that purpose at Sloan Kettering Institute, in New York, USA.

Studies on the major histocompatibility complex (MHC)

In general, mice of an inbred strain are genetically identical to one another. Mice of an inbred congenic strain are likewise identical with one another and differ from a selected standard inbred strain only in the vicinity of a particular gene or gene complex. This discrete genetic difference between a pair of congenic strains, meaning a standard inbred strain and its congenic companion strain, is achieved by crossing two inbred strains and then serially back-crossing to the selected inbred parental strain for many generations with selection for an allelic genetic trait of interest, derived from the opposite parental strain, in each generation. Any difference that distinguishes a pair of congenic strains, provided that this is shown to be genetic by appropriate segregation tests, must be due to the selected gene or a gene in that vicinity, i.e. within the small segment of donor chromosome carried over together with the selected gene (Boyse, 1977; Foster *et al.*, 1981).

The observations referred to above were made with a pair of congenic strains differing at the MHC, a group of linked genes that is represented in all vertebrates and called H-2 in the mouse and HLA in man (see Klein, 1986). For reasons that have nothing to do with the studies described here, certain cages contained a pair of breeders of the same inbred strain together with a second female belonging to an MHC congenic strain and so genetically identical to the first female except for the MHC. The technicians observed that the male and female of different MHC type paid greater attention to one another and tended to nest together to the relative exclusion of the female whose MHC type was the same as the male's.

To authenticate the male's apparent distinction of the MHC types of the females as 'self' or 'non-self' with respect to the male's own MHC type, and to determine any significance to a vital aspect of reproduction, namely mating itself, the mating preference test was devised.

The suitability of the terms 'self' and 'non-self' in this context is questioned later in this chapter.

Spontaneous mating preference

As in all studies mentioned in this review, investigation of MHC-selective mating began with the assembly of panels of mice representing the genotypes of interest, matched for age and individually numbered for systematic use in such a manner that a different pair of mice would be used for comparison of their odours in successive tests. In testing for mating preference, three panels were assembled, patterned on the circumstances of the original observation: a panel of inbred males, a panel of syngeneic females and a panel of MHC-congenic females. For each test, a female in oestrus was selected from each female panel and the two females were caged with one selected member of the male panel. The trio was watched until successful copulation, verified by a vaginal plug, had occurred with one of the females, at which time the receptivity of the second female was verified by mating with another male (not belonging to the male test panel).

With two exceptions there was a highly significant bias towards mating between the H-2-disparate male and female rather than between the syngeneic male and female (Yamazaki *et al.*, 1976), i.e. a *prima facie* preference of the male to mate with a female of non-self H-2 type. In the first exceptional instance, the male's preference for females of a different MHC type was observed when the mouse panels were isolated in separate quarters housing no other mice (Yamazaki *et al.*, 1978), but no preference was evident when the panels were maintained in the usual quarters housing other mouse strains (Yamazaki *et al.*, 1976, 1978); such communal conditions favour interchange of aerially dispersed sex pheromones and doubtless other chemosensory signals. The second exceptional instance involved the same pair of congenic mice, communally housed as usual, but the male mouse panel was of the other (alternative) MHC type. In this case there was plainly selection for the MHC-identical female (Yamazaki *et al.*, 1976); this is the only instance of self-preference we have seen with congenic strains whose entire MHC genotypes are of distant origin, which may suggest that one of the MHC types involved in this case, or the general genetic background on which it is expressed, is peculiar in some unknown respect. Latterly, our studies of mating preference have been conducted with mouse panels maintained in environmental isolation, by the use of individually ventilated cage racks.

Learned MHC discrimination

Several aspects of genetically determined chemosensory identity have been studied by the use of a specially designed Y-maze (see Yamazaki *et al.*, 1979). In this apparatus, air propelled by a fan through two chambers containing mice or other materials whose odours are to be compared conveys their odours to the two arms of the maze. The mouse-in-training or trained mouse employed for testing, deprived of water beforehand, runs the maze and is rewarded with a drop of water dispensed mechanically on the mouse's entry to the arm scented with whichever of the odour sources had been designated for reward reinforcement for that particular mouse. The criteria employed in validating a distinction between two alternative odour sources include a highly significant concordance score, generally 80 per cent or more, observed not only in rewarded runs with the familiar odour sources, but also in unrewarded blind trials of newly encountered odour sources that duplicate the genetic constitution of the familiar odour sources used in training (Yamaguchi *et al.*, 1981).

In this way urine was shown to be as potent a source of MHC odour-types as the intact mouse (Yamaguchi *et al.*, 1981) and replaced the intact mouse as an odour source in most subsequent studies.

It appears that any mouse, male or female, can be trained to distinguish the MHC odour-type of any other mouse, male or female, regardless of whether either MHC type is selfsame or both are non-self.

Influence of MHC odour-types on maintenance of pregnancy; neuroendocrine responses

If an isolated female mouse in pre-implantation pregnancy is exposed to the scent of a novel male, of a strain different from that of the stud male, the risk of abortion is greatly raised (Bruce, 1960). This phenomenon, called pregnancy block, is averted by administering hormones that sustain implantation (Dominic, 1966). Pregnancy block provided experimental access to a second phase of reproduction in addition to mating, namely the period of early embryogenesis and implantation.

The mice assembled for these studies comprised two panels of males of congenic strains having equivalent sexual experience and differing genetically only in their MHC types, as sources of alternative MHC odour-types, and a panel of females of an unrelated inbred strain whose MHC type differed from that of both male panels.

The female was mated to a male of one of the male panels and later placed on one side of a cage divided by a perforated screen, with a male or the urine of a male on the other side. This second male was either the original stud male again, or a genetically identical male of the same male panel, or a male of the MHC congenic panel, genetically identical to the stud male except for the MHC. In the first two instances, involving no MHC disparity between stud and second males, there was no increased incidence of terminated pregnancy. Only in the third circumstance, involving MHC disparity between stud and second male, was the incidence of terminated pregnancy raised regardless of which male panel was assigned to the stud versus secondary role. Clearly the chemosensory recognition of a male of unfamiliar MHC odour-type, operating through neuroendocrine channels, has a harmful influence on the maintenance of pregnancy (Yamazaki *et al.*, 1983a).

Interspecies recognition of MHC odour-types

Rats can distinguish odour-types of MHC-congenic mice (Beauchamp *et al.*, 1985) as well as of MHC-congenic rats (Brown *et al.*, 1987). Apparently some human subjects can distinguish MHC odour-types of congenic mice (Gilbert *et al.*, 1986). The formidable task of determining the extent to which MHC (HLA) odour-types of man expressed on the varied genetic background of a freely segregating population can be distinguished by human subjects or by other mammals has not yet been tackled. Use of the rat in sensing odour-types among other species has the advantage of a totally automated and computer-programmed olfactometer adapted to the rat for this purpose (Beauchamp *et al.*, 1985).

Multiple components of the MHC odour-type

The extended MHC of the mouse occupies ≈ 2 cM of a total haploid genome of $\approx 1\,600$ cM and can accommodate perhaps 50 genes (see Carroll *et al.*, 1987). MHC antibodies identify different genes within the complex, permitting the identification of MHC recombinant mice from which recombinant congenic strains can be derived, differing genetically in only a part of the MHC. Several such recombinant congenic strains, representing three main divisions of the MHC, were all shown to be discriminated either in the Y-maze or mating preference test systems (Yamaguchi *et al.*, 1978; Andrews & Boyse, 1978; Yamazaki *et al.*, 1982), signifying that at least three loci within the MHC autonomously specify an odour-type. Thus a complete MHC odour-type is compounded of at least three independent odour-types arising from genes of different sections of the MHC.

Involvement of a class I MHC gene

The MHC includes a family of up to 30 'class I' genes, not all necessarily functional, that uniquely characterize the MHC throughout its length and encode transmembrane glycoproteins whose outer domains bear distinctive MHC cell-surface antigens (Klein, 1986).

With congenic strains that differ with respect to a given trait, as is true also of any trait that exhibits linkage with a recognized gene, a question which must be settled is whether the observed trait is determined by the recognized gene itself or whether the recognized gene is merely a fortuitously linked marker for an unknown determining gene in the same neighbourhood. The usual method of resolving this issue is to determine whether mutation of the recognized gene alters the trait under study. If so, then the recognized gene must be the determining gene.

This criterion has been met for the gene H-2K, a member of the class I gene family. The mutant class I gene H-2K^{bm1} differs from the non-mutant H-2K^b gene in base substitutions corresponding to a difference in three amino acids of the protein backbone of the class I H-2K glycoprotein coded within a stretch of 13 nucleotides (Zeff et al., 1985). The critical finding was that mutant H-2K^{bm1} mice can be distinguished from non-mutant H-2K^b mice in the Y-maze (Yamazaki et al., 1983b), albeit after longer training than is required for distinction of general MHC disparity (Boyse et al., 1987). Furthermore, spontaneous distinction of H-2K^{bm1} and H-2K^b odour-types was evident in the pregnancy block system, described above, in which maintenance of pregnancy was prejudiced by the scent of a male of H-2K^{bm1} genotype if the stud male was H-2K^b, and vice versa (Yamazaki et al., 1986a). Although H-2K is the only known gene so far shown to be implicated in odour-type determination, one may infer that other genes of this class I MHC family are similarly competent.

Extreme diversity of MHC genotypes
Although not all species display it, extreme diversity of the MHC is probably characteristic of such populous and social species as mouse and man. So much so that the number of potential MHC types, comprising two MHC sets in each diploid individual, might theoretically exceed the population of a given species. There are several sources of this diversity. As regards class I MHC genes, H-2K and H-2D each have at least 50 and perhaps many more alleles, generated in part by a mechanism by which one class I gene confers part of its sequence on another, and H-2K^b is said to be the most mutable of all known genes (Klein, 1986).

The great diversity of the MHC implies an evolutionary investment in mechanisms that promote it, one of which is mating preference that favours outbreeding and MHC disparity. Inbreeding is fostered particularly in animals like mice which may generate a large social population comprising many generations in one season even from a single pair. Possible explanations of why MHC diversity should be advantageous centre on the best comprehended functions of class I and other genes of the region, which are immunological. For instance, the hypothesis of viral mimicry supposes that a virus may use its capacity for genetic variation to alter the constitution of its glycoprotein envelope to mimic an MHC glycoprotein, thus simulating self and avoiding immune recognition and response. Structural diversification of MHC glycoproteins may be viewed as a counter to this viral gambit in that it is calculated to deny access of such a mutant virus to the population at large. Again, diversification of class II immune response (Ir) genes, mostly unique to the MHC and commonly co-dominant with respect to various antigens, is calculated to confer upon a species a corresponding wide range of specific immune responses to antigens of various pathogens. These are cogent arguments since adaptive acquired immunity, which is highly developed only in vertebrates, is vital to their survival and its absence is lethal to individuals; recessive mutations causing defective expression of class I or class II genes are responsible for congenital immune deficiency syndromes in humans (for references see Lee, 1989) and other mammals.

Cellular origin of MHC odourants

Having found that urine rivals the intact mouse as a source of MHC odours, we wondered whether the kidney might be the sole generator of MHC-related odourants. This was studied by making chimeras in which the entire haematopoietic and lymphoid systems of a lethally irradiated mouse are replaced from stem cells of administered bone marrow, in this context from an MHC congenic donor. In the Y-maze, such chimeras were shown to have acquired an MHC odour-type characteristic of the bone marrow donor (Yamazaki *et al.*, 1985). Thus haematopoietic cells of the bone marrow, blood, spleen and lymphoid organs are one source of odourants composing MHC odour-types. Nothing is yet known of the contributions of other cells and organs.

Odour-types of MHC heterozygotes

As a result of inbreeding, all inbred and inbred congenic mice are homozygous for the MHC as for all their genes. Two studies in the Y-maze, using F_1 (genetically uniform) hybrid progeny of a cross between MHC congenic strains, bear on the odour-type of MHC heterozygotes. First, mice trained to distinguish either parental strain from the MHC-heterozygous F_1 hybrid in the Y-maze could as readily distinguish between the two parental homozygous strains (Yamazaki *et al.*, 1984). Thus the odour-type of an MHC heterozygote includes dominant constituents typical of each homozygote odour-type. Secondly, the urine of heterozygotes is readily distinguished in the Y-maze from an equal mixture of half quantities of urines of the two parental homozygotes (Yamazaki *et al.*, 1984), which would not be the case if MHC odour-types were determined simply by co-dominant genes with alleles for distinguishable odourants. Thus the odour-type of an MHC heterozygote is in part typical of each parental homozygote odour-type and in part distinctive of the heterozygous state *per se*.

Nature of MHC-related odours and odourants

Except for one report of statistically significant differences between immature (but unaccountably not between mature) MHC-congenic females with respect to profiles of urinary volatiles observed in high-resolution chromatography (Schwende *et al.*, 1984) extensive use of conventional physico-chemical methods such as gas chromatography and mass spectroscopy that are commonly applied to olfactants have not revealed significant characteristics distinguishing urinary constituents of MHC congenic mice (personal unpublished data). Hypotheses of odour determination, in no way mutually exclusive, rest on various considerations. Degraded moieties of MHC molecules reported to occur in urine (Singh *et al.*, 1987), might conceivably constitute structurally distinctive odourants. Or such genetically variable moieties might act as carriers of variable binding affinity for odourous metabolites voided in urine, thereby formulating distinctive compound odours (defined as odours whose perception depends solely on differing relative proportions of the same range of constitutive odourants, assumed in this case to be odourous metabolites (Boyse, 1985)). Still further removed from direct MHC specification is the idea that the numerous anatomical and physiological variations that have been traced to MHC genetic variation might be responsible for distinctive compound odours. These developmental variations include differences in the relative sizes of organs and cell sets within individual mice. Thus distinctive compound odours might be determined by collective differences in the output of odourants by different cell lineages (referenced and discussed by Boyse *et al.*, 1983). Also, bacteria are a notable source of odourants and differences in commensal flora might be governed by the extensive variation of immune response class II genes which inhabit the MHC (Boyse, 1986).

A salient theme uniting all these proposals is the extraordinary theoretical potentiality implicit in combinatorial variation of odourant concentrations for composing compound odours of great diversity according to genotype (Boyse, 1986).

Chemosensory imprinting; self versus non-self (?): exposure history

Since all definitive studies of MHC-selective mouse mating preference have involved a choice by the male between females of the male's own (self) MHC odour-type and another (non-self) odour-type, a question arises concerning whether the more usual spontaneous preference for non-self is pre-ordained, as for instance by differential recognition alleles expressed by the male, or whether the preference is determined by exposure history, by familial/environmental imprinting. The latter alternative is strongly supported by the results of foster nursing. When inbred males that would normally express a non-self preference are reared in a non-self MHC familial environment, by transfer at birth to congenic foster parents of a non-self

MHC type, the mating preference of these males at maturity is the reverse, and they now preferentially select the self-MHC, i.e. the MHC type of which they had no exogenous familial experience from birth to weaning (Yamazaki *et al.*, 1988). This being so, the terms 'self' and 'non-self', denoting in this context genotypic self, are not entirely appropriate, at least in regard to mating choices of males of inbred strains in cases where the natural choice favours a disparate MHC type. Inbreeding results in genetic uniformity for the MHC, meaning that parents and progeny share the same (self) MHC type, and we now see, from the foster-nursing studies, that the ostensible choice of inbred males for a non-self MHC type is in reality a choice for a non-familial MHC type. Thus the terms 'self' and 'non-self' in this context should be replaced by 'familial' and 'non-familial', since the self MHC type *per se* is not the prime referent whereby preference is established.

Male and female initiatives

A further question of interest in studies of kin recognition is whether the mating choice is primarily the prerogative of the male rather than the female. The reversal of preference by foster nursing does not decisively prove this, because although the only experimental variable in that study was the rearing of males by parents of non-self MHC type, it is conceivable that this experience altered the scent of the male, providing a basis for choice by the female. However, that alternative explanation was largely excluded by determining that such males could not be distinguished in the Y-maze from control males reared by foster parents of self MHC genotype (Yamazaki *et al.*, 1988). This is a rigorous test because mice can be trained in the Y-maze to distinguish a difference in odour-type due to fine mutation of a single MHC gene, as noted above. Clearly the choice is primarily the male's in the particular circumstances of mating tests employed in our studies.

Recognition and differential response of females to MHC odour-types of males is evident from Y-maze data and from the fact that maintenance of pregnancy is prejudiced by the scent of a second male of unfamiliar MHC, as noted above, even if the genetic difference is merely mutation of single MHC gene. No question of self versus non-self is involved in the latter studies because the MHC types of stud and second males were both different from the female's. No matter which of the two alternative male MHC types was experienced first, via the stud male, pregnancy was subsequently compromised by the odour-type of the alternative male. A parallel temporal relation applies to the male mating preferences observed in the foster-nursing experiments described above, because, as noted, whichever of the two alternative MHC types was experienced first, in the postnatal period, became the less favoured of the two types presented subsequently in preference tests.

It has yet to be seen whether there is any innate genetically programmed MHC-selective mating preference independent of familial experience.

The rest of the genome

The remainder of the genome has been examined only with respect to contributions to odour-types that can be distinguished by mice trained in the Y-maze or rats trained in the automated olfactometer. No reproductive behavioural studies of the mating preference or pregnancy block sort have been conducted.

Autosomes

As outlined above, MHC congenic strains are derived by crossing two inbred strains of different MHC types and serially backcrossing to one of them while selecting for the MHC type of the other strain in each succeeding backcross generation. Thus the congenic strain and its partner strain come to exhibit different MHC genotypes on an otherwise identical genotypic background. Comparison of the congenic strain with the other inbred strain of the original cross constitutes the reverse circumstance, namely identity of MHC type on a background of the same overall genomic dissimilarity exhibited by the unrelated inbred strains of the original cross. Such comparisons in the Y-maze or rat olfactometer signify that by the criterion of relatively similar ease of training the entire autosomal genome exclusive of the MHC may be as potent a source of odour-type as the MHC.

It is not feasible to test intact autosomes separately, chromosome by chromosome, but mouse strains congenic for loci other than the MHC provide a means to test autosomal disparities which average very approximately say one-tenth of a chromosome. A few such congenic strains, one including a difference in the immunoglobulin light-chain *kappa* gene have been tested in the Y-maze. None were decisively distinguishable. Since the genetic differences between these congenic strains represent only a small part of the non-MHC autosomal genome it is not clear whether the odour-type efficacy of the non-MHC autosomal genome as a whole is due to a few scattered relatively potent loci or to many marginally effective loci. In any event there is no known gene complex which approaches the MHC in diversification (Klein, 1986). The diversification of immunoglobulin and T-cell receptor genes is not relevant here because it is largely somatic and to that extent not transmitted in the germ line.

The sex chromosomes

Unlike autosomes, which freely recombine in meiosis, the X and Y sex chromosome pair is not generally subject to mutual crossing-over. When two inbred strains are crossed, the female progeny (XX) are all genetically identical whichever way the cross is made. The male progeny (XY) also are genetically identical to one another except for their sex chromosomal constitution, which depends on which way the cross was made, i.e. on which parent contributed the X and which the Y. Thus, reciprocal hybrid males differ genetically in their X and Y chromosomes but are otherwise genetically identical. Reciprocal hybrid males were distinguished by trained mice in the Y-maze, indicating that the X and/or Y chromosomes contribute to odour-type (Yamazaki *et al.*, 1986b). Trained mice also distinguished a pair of congenic strains derived by introducing a different Y chromosome into a standard inbred strain, although only after relatively long training, showing that the Y chromosome contributes to odour-type (Yamazaki *et al.*, 1986b). A further comparison, made possible by the joint use of reciprocal hybrid males and the Y congenic strain, indicated that the X chromosome also contributes to odour-type, more strongly than the Y according to comparative ease of training (Yamazaki *et al.*, 1986b). As in the case of the non-MHC autosomal genome, it cannot be said whether the X chromosome (> 100 cM) carries one or a few potent loci or several marginally effective loci.

Mitochondria and other maternal factors

The statement above, that reciprocal hybrid females are genetically identical, applies only to the mendelian genome, for they differ in maternally inherited factors, notably mitochondria, which are exclusively transmitted by the mother. The possibility that mitochondria might contribute to odour-type is emphasized by the observation that a mitochondrial gene appears to govern expression of a certain class I gene of the MHC (Fischer Lindahl *et al.*, 1983). However, Y-maze testing entirely failed to distinguish reciprocal hybrid females. Nevertheless, more definitive exclusion of mitochondrial genes as contributors to odour-type would require comparison of reciprocal hybrid females of a cross between strains known to differ in governance of the pertinent reputed class I gene, which has not yet been undertaken.

Summary

The major histocompatibility complex (MHC), found in all verte-brates, comprises in the mouse about 50 linked genes that jointly are so highly polymorphic that they constitute a source of germ-line genetic variation exceeding that of any other gene complex. Each of the three main divisions of the MHC confers on each mouse an idiotypic self-marking MHC odour-type. Some 25–30 class I genes, specifying transmembrane glycopro-teins of the cell surface, occupy these three divisions of the MHC and uniquely characterize this section of the genome. Mutation of a single identified class I gene sufficed to determine an idiotypic MHC odour-type. Thus the immense genotypic diversity implicit in the MHC may be reflected in a similarly vast repertoire of MHC odour-types subserving kin and non-kin recognition.

Behavioural responses of males to perception of MHC odour-types are illustrated in the context of mating by a usual preference for females of a different MHC type. This propensity is due to familial imprinting, since it is reversed to favour the self MHC type when males are reared by MHC-congenic foster parents of a non-self MHC type. Thus, at least in the case of inbred males, the self MHC type is not a prime referent whereby preference of males is determined. Since the prime determining factor is in fact exposure to familial MHC types, whether self or non-self, male MHC preferences previously described in terms of 'self' versus 'non-self' are now more aptly denoted by the terms 'familial' versus 'non-familial'. That behavioural MHC mate choice by the male is determined by familial experience prior to weaning is in keeping with most work on kin recognition (Blaustein, 1983; Waldman, 1987) and accords with the advanced refinements of the mouse's behavioural repertoire.

Behavioural responses of females to the perception of MHC odour-types were observed in neuroendocrine changes disposing to termination of pregnancy resulting from incidental exposure to an MHC odour-type different from that of the stud male, both perceived MHC types being in this case different from that of the responding female. Again the response is seen to be conditioned by MHC exposure history, aptly described as familial in the sense that female and stud male composed a family.

The rest of the autosomal genome as a whole, exclusive of the MHC, rivals the MHC in capacity to constitute a strong odour-type, not necessarily so diverse in repertoire, but no individually effective autosomal genomic region other than the MHC has been identified. Each of the sex chromosomes, the X more obviously than the Y, also contributes significantly to odour-type. No contribution by mitochondria or other maternally transmitted factors has been evident.

Thus, the MHC together with unknown autosomal and sex chromosomal genes, offers a potentially inexhaustible store of information for the generation of individual odour-types that distinguish genetic kin from more distantly related and unrelated individuals, relationships which must in general reflect degrees of familial and genealogical relatedness. Behavioural consequences of the perception of MHC odour-types, in the context of kin and non-kin recognition, observed in males or females, have connotations in the settings of altruism or mate choice.

On the whole, the data support the hypothesis that a natural preference of males to mate with females whose MHC types differ from their own and from familial MHC types, and perhaps also neuroendocrine effects consequent on perception of MHC odour-types in other reproductive contexts, serve to promote outbreeding and its attendant advantages.

Acknowledgement

Support from the National Institutes of Health, the National Science Foundation and the American Cancer Society is acknowledged.

References

Andrews, P. W. & Boyse, E. A. (1978). Mapping of an H-2-linked gene that influences mating preferences in mice. *Immunogenetics*, **6**, 265–8.

Beauchamp, G. K., Yamazaki, K., Wysocki, C. J., Slotnick, B. M., Thomas, L. & Boyse, E. A. (1985). Chemosensory recognition of mouse major histocompatibility types by another species. *Proc. Natl. Acad. Sci., USA.*, **82**, 4186–8.

Blaustein, A. R. (1983). Kin recognition mechanisms: Phenotypic matching or recognition alleles. *Am. Nat.*, **121**, 749–54.

Boyse, E. A. (1977). The increasing value of congenic mice in biomedical research. *Lab. Animal Sci.*, **27**, 771–81.

Boyse, E. A. (1985). Class I genes: What do they all do? In *Cell Biology of the Major Histocompatibility Complex*, ed. B. Pernis & H. Vogel, pp. 219–22. New York: Academic Press.

Boyse, E. A. (1986). HLA and the chemical senses. *Human Immunol.*, **15**, 391–5.

Boyse, E. A., Beauchamp, G. K. & Yamazaki, K. (1983). The sensory perception of genotypic polymorphism of the major histocompatibility complex and other genes: Some physiological and phylogenetic implications. *Human Immunol.*, **6**, 177–83.

Boyse, E. A., Beauchamp, G. K. & Yamazaki, K. (1987). The genetics of body scent. *Trends Genet.*, **3**, 97–102.

Brown, R. E., Singh, P. B. & Roser, B. (1987). The major histocompatibility complex and the chemosensory recognition of individuality in rats. *Physiol. Behav.*, **40**, 65–73.

Bruce, H. (1960). A block to pregnancy in the mouse caused by proximity of strange males. *J. Reprod. Fertil.*, **1**, 96–103.

Carroll, M. C., Katzman, P., Alicot, E. M., Koller, B. H., Geraghty, D. E., Orr, H. T., Strominger, J. L. & Spies, T. (1987). Linkage map of the human major histocompatibility complex including the tumor necrosis factor genes. *Proc. Natl. Acad. Sci., USA*, **84**, 8535–9.

Dominic, C. J. (1966). Observations on the reproductive pheromones of mice. II. Neuroendocrine mechanisms involved in the olfactory block to pregnancy. *J. Reprod. Fertil.*, **11**, 415–21.

Fischer Lindahl, K., Hausmann, B. & Chapman, V. M. (1983). A new H-2-linked class I gene whose expression depends on a maternally inherited factor. *Nature*, **306**, 383–5.

Foster, H. L., Small, J. D. & Fox, J. G. (Eds). (1981). *The Mouse in Biomedical Research*, vol I. New York: Academic Press.

Gilbert, A. N., Yamazaki, K., Beauchamp, G. K. & Thomas, L. (1986). Olfactory discrimination of mouse strains (*Mus musculus*) and major histocompatibility types by humans. *J. Comp. Psychol.*, **100**, 262–5.

Klein, J. (1986). *Natural History of the Major Histocompatibility Complex*. New York: J. Wiley & Sons.

Lee, J. S. (1989). Regulation of HLA class II gene expression. In *Immunobiology of HLA*, vol. 1, *Histocompatability Testing*, ed. B. Dupont, pp. 49–62. New York: Springer Verlag.

Schwende, F. J., Jorgenson, J. W. & Novotny, M. (1984). Possible chemical basis for histocompatibility-related mating preference in mice. *J. Chem. Ecol.*, **10**, 1603–15.

Singh, P. B., Brown, R. E. & Roser, B. (1987). MHC antigens in urine as olfactory recognition cues. *Nature*, **327**, 161–4.

Waldman, B. (1987). Mechanisms of kin recognition. *J. Theor. Biol.*, **128**, 159–85.

Yamaguchi, M., Yamazaki, K., Beauchamp, G. K., Bard, J., Thomas, L. & Boyse, E. A. (1981). Distinctive urinary odours governed by the major histocompatibility locus of the mouse. *Proc. Natl. Acad. Sci., USA.*, **78**, 5817–20.

Yamaguchi, M., Yamazaki, K. & Boyse, E. A. (1978). Mating preference tests with the recombinant congenic strain BALB.HTG. *Immunogenetics*, **6**, 261–4.

Yamazaki, K., Beauchamp, G. K., Bard, J., Thomas, L. & Boyse, E. A. (1982). Chemosensory recognition of phenotypes determined by the Tla and H-2K regions of chromosome 17 of the mouse. *Proc. Natl. Acad. Sci., USA*, **79**, 7828–31.

Yamazaki, K., Beauchamp, G. K., Egorov, I. K., Bard, J., Thomas, L. & Boyse, E. A. (1983b). Sensory distinction between H-2^b and H-2^{bm1} mutant mice. *Proc. Natl. Acad. Sci.*, **80**, 5685–8.

Yamazaki, K., Beauchamp, G. K., Kupniewski, D., Bard, J., Thomas, L. & Boyse, E. A. (1988). Familial imprinting determines H-2 selective mating preferences. *Science*, **240**, 1331–2.

Yamazaki, K., Beauchamp, G. K., Matsuzaki, O., Bard, J., Thomas, L. & Boyse, E. A. (1986b). Participation of the murine X and Y chromosomes in genetically determined chemosensory identity. *Proc. Natl. Acad. Sci., USA.*, **83**, 4438–40.

Yamazaki, K., Beauchamp, G. K., Matsuzaki, O., Kupniewski, D., Bard, J., Thomas, L. & Boyse, E. A. (1986a). Influence of a genetic difference confined to mutation of H-2K on the incidence of pregnancy block in mice. *Proc. Natl. Acad. Sci., USA.*, **83**, 740–1.

Yamazaki, K., Beauchamp, G. K., Thomas, L. & Boyse, E. A. (1984). Chemosensory identity of H-2 heterozygotes. *J. Mol. Cell. Immunol.*, **1**, 79–82.

Yamazaki, K., Beauchamp, G. K., Thomas, L. & Boyse, E. A. (1985). The hematopoietic system is a source of odorants that distinguish major histocompatibility types. *J. Exp. Med.*, **162**, 1377–80.

Yamazaki, K., Beauchamp, G. K., Wysocki, C. J., Bard, J., Thomas, L. & Boyse, E. A. (1983a). Recognition of types in relation to the blocking of pregnancy in mice. *Science*, **221**, 186–8.

Yamazaki, K., Boyse, E. A., Mike, V., Thaler, H. T., Mathieson, B. J., Abbott, J., Boyse, J., Zayas, Z. A. & Thomas, L. (1976). Control of mating preferences in mice by genes in the major histocompatibility complex. *J. Exp. Med.*, **144**, 1324–35.

Yamazaki, K., Yamaguchi, M., Andrews, P. W., Peake, B. & Boyse, E. A. (1978). Mating preferences of F$_2$ segregants of crosses between MHC-congenic mouse strains. *Immunogenetics*, **6**, 253–9.

Yamazaki, K., Yamaguchi, M., Baranoski, L., Bard. J., Boyse, E. A. & Thomas, L. (1979). Recognition among mice: Evidence from the use of a Y-maze differentially scented by congenic mice of different major histocompatibility types. *J. Exp. Med.*, **150**, 755–60.

Zeff, R. A., Geier, S. S., Gopas, J., Geliebter, J., Schulze, D. H., Pease, L. R., Pfaffenbach, G. M., Pontarotti, P., Mashimo, H., McGovern, D. A. & Nathenson, S. G. (1985). Mutants of the murine major histocompatibility complex: Structural analysis of in vivo and in vitro H-2K^b variants. In *Cell Biology of the Major Histocompatibility Complex*, ed. B. Pernis & H. Vogel, pp. 41–9. New York: Academic Press.

7

Kin recognition in amphibians

Bruce Waldman

Introduction

Kinship underlies many facets of vertebrate sociality. Yet vertebrate societies rarely consist exclusively of kin. Typically, individuals interact both with kin and non-kin in numerous and often unpredictable contexts. Clear advantages accrue to those individuals that discriminate behaviourally among conspecifics based on their genetic relatedness. The ability to recognize relatives permits individuals (1) to favour their kin, thereby enhancing their inclusive fitness (Hamilton, 1964; Grafen, 1985), and (2) to avoid incest or even to choose an optimally related mate (Shields, 1982; Bateson, 1983), potentially increasing their own reproductive success. Possible benefits of kin discrimination are easily enumerated in birds and mammals (reviewed in Waldman, 1988), so their kin recognition abilities (e.g. Holmes & Sherman, 1982; Hepper, 1983; Beecher, 1988) should not be surprising. Less is known about the social life of fishes, amphibians, and reptiles, but they also may recognize their kin. Some fishes and larval amphibians, for example, recognize their siblings and preferentially school with them (e.g. Waldman, 1982b; O'Hara & Blaustein, 1985; Quinn & Busack, 1985; Van Havre & FitzGerald, 1988; Olsén, 1989). Similarly, iguana hatchlings associate in social groups with their siblings rather than with unrelated individuals (Werner et al., 1987).

Mechanisms of kin recognition have been more thoroughly examined in amphibians than in other vertebrates. Indeed, information on the ontogeny and sensory bases of kin recognition comparable in detail to that known about social insects (e.g. Gamboa et al., 1986; Hölldobler & Carlin, 1987; Michener & Smith, 1987) and some invertebrates (Linsenmair, 1987) is currently available only for larval amphibians. Warm-blooded vertebrates can be unwieldy as subjects for ontogenetic studies in the laboratory. The kin recognition systems of amphibians, by contrast, are highly tractable. Experimental procedures for manipulating embryonic development, borrowed from developmental biology and immunology (e.g. Rugh, 1962; McKinnell, 1978), can serve as powerful tools for probing the mechanisms underlying discrimination. Many amphibians, especially frogs, can recognize their kin, but mechanisms vary considerably among species. An ecological basis for this diversity is suggested by the observation that interspecific differences in kin recognition mechanisms sometimes correspond to differences in life-history characteristics (Waldman, 1984; Blaustein, 1988; Fishwild *et al.*, 1990). Finally, larval amphibians have proven model organisms for the investigation of population and community structure (Wilbur, 1984), and field studies provide a framework for the examination of the functional significance of kin discrimination (Waldman, 1982b; Blaustein & O'Hara, 1987). Still, amphibians may seem curious subjects for studies of kin recognition, if only because so little is known about their social behaviour.

The evolution of kin recognition systems is not predicated on behavioural complexity. Rather, mechanisms to detect genetic similarity are common to most living organisms. Even animals with seemingly limited behavioural repertoires can discriminate between their genetic relatives and other conspecifics when they come in contact. Consider, for example, sponges and coelenterates, marine invertebrates that might be considered 'primitive' in that they lack a central nervous system or distinct organs (see Grosberg, 1988, for a review of allorecognition in marine invertebrates). Sponges fuse with clonemates and close genetic relatives, but grafts between genetically dissimilar individuals fail (Neigel & Avise, 1985). Branching corals that touch fuse only if they are genetically similar (Neigel & Avise, 1983). Sea anemones segregate into groups of genetically similar individuals when mixed (Francis, 1973a, 1976), and genetically dissimilar individuals act aggressively toward one another (Francis, 1973b, 1988; Ayre, 1982,1987). Plants, too, respond differentially to genetically similar and dissimilar individuals, as manifested in their growth rates (Williams *et al.*, 1983; Wilson *et al.*, 1987; Tonsor, 1989) and in their susceptibility to predators (Schmitt & Antonovics, 1986). Self-incompatibility mechanisms are common in plants, and these not only serve as a barrier to self-fertilization but also may promote some degree of outbreeding (Uyenoyama 1988; Waser, 1991).

Studies of the mechanisms underlying kin recognition are necessarily reductionist in character. Analyses of potentially simpler recognition systems thus are expected to proceed more rapidly at first than those of more complex systems. In addition, these studies are likely to yield paradigms that foster experimentation on other organisms. Kin recognition systems of simple and complex organisms may be more closely linked than they first appear. Kin recognition appears to parallel self/non-self recognition at the cellular level (Alexander, 1979; Brown, 1983). Indeed, systems effecting self/non-self recognition may have initially evolved to maintain the genetic integrity of individuals and their component tissues (see Buss, 1982). As a by-product, they may have conferred on individuals unique chemical labels (Thomas, 1975). Products of self/non-self recognition processes presumably are further honed by natural selection to promote kin recognition. Evidence is accumulating that odours associated with histocompatibility genes influence kin discrimination responses by organisms as diverse as sea squirts (Grosberg & Quinn, 1986, 1988) and rodents (Beauchamp *et al.*, 1986; Brown *et al.*, 1987; Yamazaki *et al.*, 1988; Egid & Brown, 1989). Although the possibility that histocompatibility antigens or their derivatives serve as recognition cues in amphibians has yet to be examined, levels of genetic polymorphism in the major histocompatibility complex of anurans (Du Pasquier *et al.*, 1975, 1989; Roux & Volpe, 1975; Flajnik *et al.*, 1985) correspond to those characteristic of the mouse H-2 (Klein, 1986), rat RT1 (Butcher & Howard, 1986), and human HLA (Dausset, 1981). The problem of how animals recognize their kin, while central to many issues in behavioural ecology, has direct implications for other disciplines as well.

Kin recognition as communication

Kin recognition systems are, in the broadest sense, vehicles of social communication (Waldman, 1986a). Organisms can be identifiable at multiple organizational levels: for example, by species, population, and social group (Hölldobler & Carlin, 1987). The traits individuals express, the signals they produce, and their behaviour patterns all serve as cues by which they can be categorized by others. Many cues associated with individuals are correlated with or predictive of their kinship identities. Such cues potentially communicate information concerning their bearer's genetic relatedness to conspecifics. The perception and concomitant evaluation of these cues constitute kin recognition, even if recognition leads to no discernible behavioural response (Waldman, 1987).

Communication systems promote the transmission of information among individuals, although not necessarily for their mutual benefit. Views as to what constitutes communication vary widely (see Halliday & Slater, 1983), but the basic dynamics are agreed upon. Social organisms generally act both as senders and receivers. They give off signals, and they perceive signals given off by other individuals. The mechanisms by which signal production and detection occur can be experimentally studied, as can the manner in which the signals decay in time and space as they are propagated through the environment.

What qualities characterize a signal? A mouse moving through the underbrush inadvertently makes rustling noises that may alert a hunting owl, but it certainly is not intentionally signalling its presence (Marler, 1967). Nor do the red spots covering the skin of a human infected with the measles necessarily represent a signal intended to reduce social contacts and thus stave off the spread of the virus. Yet they may ultimately have this effect. Detection mechanisms may evolve even in the absence of any specific signalling systems.

Kinship can be communicated through a variety of cues. For example, morphological characteristics such as body shape or colour patterns may reflect particular genotypes. Behavioural tendencies expressed by individuals – even their propensities to occupy certain territories or home ranges – can reveal information about their probable genetic relatedness to neighbours. But these traits lack the properties usually ascribed to signals (Green & Marler, 1979). Vocalizations, chemical secretions, and visual displays (which may emphasize distinctive body patterns) are discrete cues. They are generated by specialized structural or behavioural adaptations, and they can be turned on and off. General morphological characters, while they may fully encode kinship identity, continuously broadcast this information. Despite this difference, the components of a kin recognition system are much the same as those of any other communication system. Most importantly, experimental approaches to the study of animal communication are directly applicable to analyses of kin recognition.

The utility of analyzing kin recognition systems from the perspective of communication becomes clear when reviewing the schemes that have been put forth to classify mechanisms of kin recognition. Recognition is generally attributed to various forms of social learning ('association' or 'familiarity'), to direct comparisons of genetically-correlated phenotypic traits ('phenotype matching'), or to hypothetical genetic units that generate a phenotypic marker and the ability to recognize this marker in others ('recognition alleles') (Holmes & Sherman, 1982; Blaustein, 1983). Yet distinctions between these fade when examining the components of kin recognition systems (Waldman, 1983, 1985b, 1987). For kinship identity to be communicated by any of these processes, individuals (*senders*) must express 'labels', possibly including behaviours, morphological traits, or signals they actively produce, that provide information concerning their genotype. These cues are perceived and analyzed by the senders' conspecifics (*receivers*), which may compare perceived labels with those stored in a model 'template'.

Empirical inquiries into kin recognition systems need to address (1) attributes of the labels, (2) attributes of the templates, and (3) factors that regulate behavioural discrimination. First, the extent to which individuals' labels, either separately or taken together, correlate with their kinship identities needs to be assessed. Behavioural experimentation on individuals with known genealogies may prove useful in this regard, but the characterization and identification of the labels represents a principal goal. Studies of the genetic and ontogenetic mechanisms by which labels arise can clarify the basis of the genotype/phenotype correlation. In some cases labels may be at least in part genetically based, so that genotypic differences translate predictably into phenotypic differences (Greenberg 1979, 1988; Hepper, 1987). But often labels are environmentally influenced (Gamboa *et al.*, 1986; Breed *et al.*, 1988; Porter *et al.*, 1989), and family differences in label characteristics thus are more susceptible to environmental perturbations.

Labels can encode information concerning the sender's kinship identity, but kin recognition cannot occur unless receivers have some frame of reference, with which to compare the labels. Templates are less easily isolated and identified than are labels. Sensory neurons or portions of the central nervous system presumably are specialized as templates effecting behavioural responses, but non-specialized cells may also differentially respond to labels. Templates, incorporating expected label characteristics, may be genetically determined, socially learned from those conspecifics encountered during particular life stages, or selectively learned. In the last case, individuals may be genetically predisposed to most rapidly learn phenotypes of related conspecifics, thus introducing some genetic constraints into the learning process. Templates may be acquired by experience with conspecifics (possibly indirectly, e.g. by contact with nests or territorial markings), or by experience with one's own labels.

Recognition can occur if features of the labels exactly match those of the templates, or if some proportion of labels matches those of the templates. Often, related individuals share many labels, either because the labels are genetically determined or because they are influenced by environmental effects experienced in common by family members. The proportion of matching labels then may be used as a measure of the degree of genetic relatedness (Getz, 1981; Lacy & Sherman, 1983; Crozier, 1987). Genetically-correlated labels, however, are not a prerequisite for all forms of kin recognition. Family members potentially could develop random individual signatures (label profiles), reflecting neither genetic effects nor genetically-correlated environmental influences. If individuals learn each other's signatures in ecological contexts in which they are predictably found in kin groups, they presumably would recognize one another in subsequent encounters, even in unfamiliar circumstances.

When kin are recognized based on traits they express or signals they give off, kin recognition is termed *direct*. The sender's genetic identity is encoded in the labels themselves, whether they be vocalizations, visual displays, or chemical cues. In the absence of such direct cues, however, kin recognition is still possible. Recognition may be elicited by contextual features that are reliably associated with kin. For example, when the location of a nest serves as a good predictor for finding offspring, parents may not recognize the nestlings themselves, but rather characteristics of the nest. *Indirect* kin recognition may be evidenced when individuals' responses toward conspecifics are contingent on the circumstances in which they meet. Under conditions in which dispersal is limited and populations are highly structured ('viscous'), the genetic relatedness among individuals tends to decrease with the distance between their home areas. Strangers then are more likely to be non-kin than are familiar individuals. The efficacy of indirect recognition mechanisms varies considerably, but close relatives can be identified most readily by making use of indirect cues together with direct cues (Waldman, 1987).

Even if an individual recognizes a conspecific as its relative, this does not imply that it will act differentially toward that relative. Natural selection should favour kin discrimination only when its benefits exceed its costs, measured in terms of inclusive fitness. The expression of kin recognition abilities is thus expected to be context-dependent (Waldman, 1988; Reeve, 1989). Most behavioural studies of kin recognition fail to distinguish between two confounded factors: the ability of study subjects to recognize their kin, and their propensity to demonstrate these abilities under the conditions in which they are observed. A major challenge of research into kin-recognition systems is to identify the circumstances in which discrimination occurs, and to predict how mechanisms vary in response to changing social and ecological conditions (Waldman, 1987). A survey of the contexts in which kin discrimination may occur thus provides a good starting point for the examination of kin-recognition systems.

Contexts of kin recognition

Parental care

Among birds and mammals, kin recognition abilities are perhaps most frequently manifested in parental care (Hepper, 1986; Waldman, 1988). Parental care is less common in amphibians, but a remarkable diversity of mechanisms for nurturing and protecting young is evident in the reproductive tactics of frogs, salamanders, and caecilians (reviewed in Salthe & Mecham, 1974; Duellman & Trueb, 1986).

Parental care is widespread among salamanders (urodeles), but only during the earliest stages of development. Guarding and attendance cease when embryos hatch from their egg capsules and larvae begin feeding. Salamanders deposit their eggs in nest sites in ponds, lakes, streams, or on land. Species that oviposit in streams and in terrestrial habitats are most apt to provide protection for their offspring, and their clutches generally consist of fewer but larger eggs (Nussbaum, 1985, 1987). Several functions have been proposed for parental care: it may reduce cannibalism or interspecific oophagy, lessen susceptibility to fungal infestation, or promote normal development and hatching.

Although many accounts of nest attendance are anecdotal, several trends emerge. In those species in which fertilization is external, males provide parental care, whereas in those species in which fertilization is internal, females provide parental care (Nussbaum, 1985). Internal fertilization is the more common reproductive strategy in salamanders; females pick up spermatophores either directly from courting males or those that males have deposited on the substrate. Spermatozoa are commonly stored in the female's spermatheca for months or years before fertilization (references in Salthe & Mecham, 1974), so unless males can enforce some mechanism of sperm precedence, their certainty of paternity may be low (see Halliday & Verrell, 1984; Houck & Schwenk, 1984; Houck *et al.*, 1985). The correlation of male parental care with external fertilization and female parental care with internal fertilization suggests that parents identify their own young indirectly by contextual cues. Similar patterns emerge in some anurans that provide parental care (reviewed in Gross & Shine, 1981).

Salamanders recognize their eggs by a variety of cues. Brood recognition mechanisms of mountain dusky salamanders (*Desmognathus ochrophaeus*) have been studied in detail (Tilley, 1972; Forester, 1979; Forester *et al.*, 1983). First, *D. ochrophaeus* females evidently recognize the site in which they have laid their eggs, and they orient repeatedly with respect to it. Forester (1979) displaced brooding females 2 m and found that most returned to their own nest, in many cases moving past other unattended clutches. Homing abilities have been well documented in many salamanders (Twitty *et al.*, 1964; Landreth & Ferguson, 1967; Madison, 1969), and since brood parasitism has not been documented in amphibians, such behaviour constitutes a reliable indirect means of identifying kin. Second, laboratory studies suggest that *D. ochrophaeus* females recognize cues emanating from their eggs. Females selected their own progeny from a group of three unattended clutches that were housed in glass containers (Forester, 1979). These discriminatory responses can be elicited by chemical cues alone (Forester *et al.*, 1983; also see discussion in Juterbock, 1987). Under natural conditions, direct recognition of cues emanating from the brood probably supplements indirect identification of the brood by their presence in a recognizable nest. Experimentally, females can be induced to brood a clutch of another female (for example, if the clutches are switched; see Tilley, 1972; Forester, 1979), but adoptions are probably infrequent in nature because individuals home quite precisely. Any adoptions that occur are likely to be of clutches of close relatives, and would be favoured by kin selection (Forester, 1979).

Parental care is rare among frogs and toads (anurans), but species that spawn in terrestrial habitats frequently attend their eggs or larvae and sometimes transport them from the oviposition site to water (reviewed in Salthe & Mecham, 1974; Lamotte & Lescure, 1977; McDiarmid, 1978; Wells, 1981). Because virtually all frogs (but see Townsend *et al.*, 1981; Grandison

& Ashe, 1983) fertilize their eggs externally, both parents usually should be able to identify their spawn with certainty. When parents maintain contact with their eggs and larvae after oviposition, parent/offspring recognition effectively can occur based on differential responses to simple contextual cues. Progeny may be confined to a nest or burrow, they may be located in restricted areas within a pond or stream, or they may be carried on the parent's body or within its body cavities. Parental attendance of offspring thus makes possible, with varied degrees of certitude, continued parent/ offspring recognition as larvae mature. Contextual recognition of offspring would be ineffective, of course, should females mate repeatedly within a short time frame, with multiple males at a particular site. Indeed, cuckoldry has been documented in some species (e.g., *Dendrobates arboreus*, Myers *et al.*, 1984). Under these circumstances it is unclear whether attending males discriminate between their own offspring and those of other males.

No studies have been conducted on the abilities of anurans to recognize their own brood, but several largely anecdotal reports raise the possibility that indirect recognition mechanisms may frequently be effective. Male poison-arrow frogs (*D. auratus*) appear to recognize their own eggs by spatial cues, and repeatedly moisten these clutches. Wells (1978) noted that brooding males normally orient in a straight line to their oviposition site, but when he disturbed the leaves surrounding the clutch, males could no longer find their eggs. *Hyla rosenbergi* males defend their nest from conspecifics when their offspring are present (Kluge, 1981). Captive *D. pumilio* males cannibalize conspecific eggs fertilized by other males (Weygoldt, 1980). Males may avoid cannibalizing their own eggs by recalling their locations.

Tadpoles of some frogs are attended by their parents, who may guard the larvae from potential predators, feed them, or lead them from place to place. Female *Leptodactylus ocellatus* defend their eggs after depositing them in a foam nest. Upon hatching, females closely follow their schooling tadpoles as they move around the pond, and they may fend off potential predators of the tadpoles (Vaz-Ferreira & Gehrau, 1975). *Pyxicephalus adspersus* males similarly have been observed attending and defending tadpole schools (Balinksy & Balinksy, 1954; Rose, 1956). Kok *et al.* (1989) observed males, possibly parents, digging channels to give tadpoles access to water. However, males seem not to discriminate among clutches or even species (Poynton, 1957), and reports of guarding are disputed (Wager, 1965; Lambiris, 1971). Schools of *L. bolivianus* tadpoles are attended by their female parents. The parents repeatedly make 'pumping' movements which create waves in the water that may serve as a signal directing tadpoles to particular areas of the pond (Wells & Bard, 1988). *D. pumilio* females feed their tadpoles abortive eggs (Weygoldt, 1980). Females carry tadpoles to water-filled bromeliad leaf axils, and place only one larva in each leaf. That female parents minimally recall the locations in which they have deposited their larvae is suggested by the observation that individuals repeatedly and specifically feed tadpoles they have deposited in particular leaf axils, and fail to feed larvae of other females in adjacent axils (Weygoldt, 1980). Obligate oophagy may represent a form of direct parental feeding in many species with arboreal tadpoles (see Lanoo *et al.*, 1987, for other examples).

Schooling behaviour

Tadpoles of many anuran amphibians aggregate in groups, much like the schools and shoals formed by some fishes (see Pitcher, 1986). Not all larval anurans school, but social aggregations have been noted in numerous species (at least 10 families) and in diverse habitats (reviewed in Lescure, 1968; Wassersug, 1973). Schools may be highly polarized (i.e. individuals orient in a common direction, maintain constant spacing, and appear to be co-ordinated in their movements), as in some midwater planktivores (e.g. *Xenopus laevis*; *Phrynomerus annectans*) (van Dijk, 1972; Katz *et al.*, 1981; Wassersug *et al.*, 1981). Or they may be largely stationary and non-polarized, such as the schools typical of toad (*Bufo*) and spadefoot toad (*Scaphiopus*) tadpoles, which feed on the bottom of shallow ponds in groups that can range from tens to hundreds to many thousands of individuals (Bragg, 1965; Beiswenger, 1975).

Although movements of tadpoles in non-polarized schools seem unco-ordinated, the schools are often tightly packed and individuals frequently come into contact with one another. Non-polarized schools of feeding tadpoles sometimes change into polarized schools as individuals move between localities (Savage, 1952; Beiswenger, 1975; Test & McCann, 1976; Griffiths *et al.*, 1988). Some tadpoles aggregate in densely packed 'balls', which although non-polarized, appear highly organized. For example, a ball of red toad (*Schismaderma* [= *Bufo*] *carens*) tadpoles, just 15 to 23 cm in diameter, consists of thousands of individuals which move around 'like an army of soldiers on the march' (Power, 1926; also see Wager, 1965; van Dijk, 1972). Similar schools are sometimes formed by *Rhinophrynus dorsalis*, *H. geographica*, and *Osteocephalus taurinus* tadpoles (Starrett, 1960; Stuart, 1961; Duellman & Lescure, 1973; Duellman, 1978; Caldwell, 1989).

Schooling in anuran larvae takes many forms, and non-social aggregations sometimes are difficult to discern from social schools. Tadpoles may come together because they respond in similar ways to environmental factors (food, light, temperature, oxygen tension, substrate characteristics) rather than to one another (e.g. Carpenter, 1953; Brattstrom, 1962; Wiens, 1972; Punzo, 1976; Beiswenger, 1977; Noland & Ultsch, 1981; O'Hara, 1981; Dunlap & Scatterfield, 1982; Griffiths, 1985; Wollmuth *et al.*, 1987). Yet once aggregations form, individuals necessarily interact with one another, either by co-operating or competing for available resources. Social attraction is readily evident in laboratory tests of some species, as individuals approach one another when separated by clear plastic barriers (e.g. Wassersug & Hessler, 1971; Wassersug, 1973; O'Hara, 1981; cf. Foster & McDiarmid, 1982), or as they associate in statistically contagious distributions in uniform environments (Waldman, 1986a).

Some observers have noted that tadpoles tend to associate with similarly-sized individuals, sorting themselves into separate schools (e.g. *Phyllomedusa vaillanti*, Branch, 1983; *H. geographica*, Caldwell, 1989; K. D. Wells, pers. comm.), or orienting more closely to individuals similar in size to themselves within schools (*Bufo woodhousei*, Breden *et al.*, 1982). These observations prompt speculation that, in some instances, schools represent sibling cohorts formed from individual egg clutches (e.g. Branch, 1983). *H. geographica* tadpoles formed schools largely with their same-sized siblings when mixed together, soon after hatching, in a new pond (Caldwell, 1989). Yet even within schools, tadpoles are highly variable in size (see Altig & Christensen, 1981; Breden *et al.*, 1982). In other cases, tadpoles are constrained from moving away from egg deposition sites. If each pair oviposits in a different site, schools that form must consist of siblings. For example, Smith frogs (*H. faber*) spawn in clay nests that they construct and, after hatching, tadpoles form schools within these nests (Martins & Haddad, 1988).

In reviewing schooling in larval anurans, Wassersug (1973) contrasted the schooling patterns of midwater planktivores ('*Xenopus*-mode') with those of substrate-dwelling species ('*Bufo*-mode'), and suggested that the two patterns differ not only in form but also in function. *Bufo* tadpoles aggregate in open areas of pools and ponds where the tadpoles' dark body coloration tends to stand out from the lighter substrate. Schooling enhances this conspicuousness. *Bufo* tadpoles are distasteful and possibly toxic to a variety of their predators (Voris & Bacon, 1966; Heusser, 1971; Cooke, 1974; Dawson, 1982; Kruse & Stone, 1984; Brodie & Formanowicz, 1987; Kats *et al.*, 1988). These traits are shared by larvae of some other genera that have somewhat different schooling behaviours (Caldwell, 1989, and references therein). *Xenopus* tadpoles and other midwater planktivores, on the other hand, are thought to be cryptic and often palatable (e.g. see Walters, 1975; Kats *et al.*, 1988). The conspicuousness and unpalatability of *Bufo* tadpoles together point to the possibility that their schooling serves as an aposematic signal.

The evolution of aposematic traits may involve an element of apparent altruism (Fisher, 1958; Waldman & Adler, 1979; Harvey *et al.*, 1982; but see Guilford, 1988 and references therein). After learning characteristics of distasteful prey, predators might avoid feeding on phenotypically similar individuals or those nearby. In being preyed on by naive predators, then, individuals bearing alleles for these traits effectively sacrifice themselves for the benefit of others with whom they are aggregated. Alleles for aposematic traits are expected to spread most quickly when individuals associate in groups with their siblings (Fisher, 1958). This functional interpretation of *Bufo* behaviour generates the prediction that schools represent sibling groups (see Wassersug, 1973), and additionally that tadpoles recognize siblings to maintain these schools (Waldman & Adler, 1979; Waldman, 1982b).

Experimental studies subsequently confirmed that American toad (*Bufo americanus*) tadpoles do indeed recognize and preferentially associate with their siblings both in the laboratory (Waldman & Adler, 1979; Waldman, 1981; Dawson, 1982) and in the field (Waldman, 1982b). Amplectant pairs of toads were collected in the field, and were transported to the laboratory, where they spawned in laboratory containers. Upon hatching, tadpoles were reared in tanks with their siblings. Subsequently, pairs of sibships were distinctively marked with vital dyes and were mixed in an indoor laboratory pool. Tested at various developmental stages, from soon after becoming free-swimming to metamorphic climax, *B. americanus* tadpoles assorted with their siblings, and formed groups preferentially with their siblings. Individuals were significantly closer to sibling than non-sibling neighbours, and this result was consistent in numerous tests of different sibships from several localities (Waldman & Adler, 1979; Waldman, 1981). The aggregations formed did not always resemble natural schools, but densities of tadpoles in the laboratory pool were much lower than those typically found in nature.

When Waldman (1982b) mixed together larger numbers of *B. americanus* tadpoles, again colour-marked to indicate their sibship identities, the tadpoles sorted out and formed schools upon release in natural ponds. Densities approached those found in nature, and the schools were virtually indistinguishable from those observed in undisturbed wild populations (Fig. 7.1). Most schools consisted predominantly of siblings, indicating that school formation was non-random (Fig. 7.2). Different sibships showed no evidence of preferring different microhabitats (Waldman, 1982b), so that schooling selectively with siblings is attributable to tadpoles' reactions to one another rather than to their possible preferences for different regions of the pond. In control tests, siblings were dyed different colours and were then released into ponds. When all were siblings, tadpoles never sorted themselves into different schools based just on colour (Waldman, 1982b). Control tests in the laboratory also failed to reveal discrimination among siblings marked with different colours (Waldman & Adler, 1979; Waldman, 1981).

These results suggest that *B. americanus* tadpoles form schools with siblings in nature. Yet not all schools are likely to consist just of siblings. *B. americanus* tadpoles sometimes swarm together in large schools which consist of more individuals than could possibly hatch from a single clutch, but at other times these large schools subdivide into smaller schools that

Fig. 7.1 Schools of American toad (*Bufo americanus*) tadpoles in field experiments. Two well-defined schools are apparent along the pond's edge, and these consist largely of different sibling groups. (After Waldman, 1982.)

might indeed represent sibling cohorts (Beiswenger, 1972, 1975; Waldman, 1982b). *B. americanus* tadpoles evidently recognize kin by a direct rather than an indirect mechanism, and this permits them to selectively associate with kin in appropriate ecological circumstances despite their fluid social system.

Kin association tendencies shown by *B. americanus* tadpoles in laboratory pools seem to reflect their schooling propensities in natural conditions. The ability of tadpoles to recognize their kin also has been demonstrated using other behavioural assays. For example, Dawson (1982) recorded the time spent by individual *B. americanus* tadpoles on either side of a porcelain tray: near their siblings within an enclosure on one side, and near non-siblings within an enclosure on the other side. Test subjects associated preferentially with their siblings under varied conditions (Dawson, 1982). Similar testing protocols have been used to document kin recognition abilities in larvae of many anuran species (Cascades frogs, *Rana cascadae*, O'Hara & Blaustein, 1981; Blaustein & O'Hara, 1981, 1982a,b, 1983; red-legged frogs, *R. aurora*, Blaustein & O'Hara, 1986; wood frogs, *R. sylvatica*, Cornell *et al.*, 1989; western toads, *Bufo boreas*, O'Hara & Blaustein, 1982; common toads, *B.*

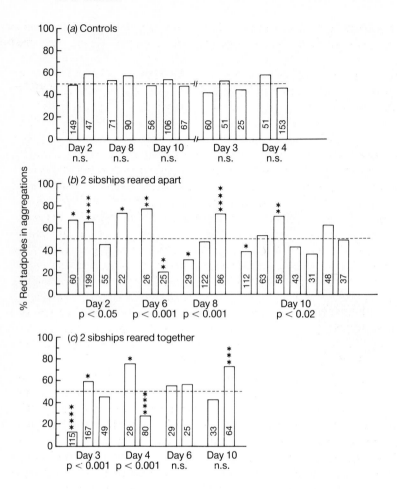

Fig. 7.2 Schooling preferences of *Bufo americanus* tadpoles in typical field tests. Tadpoles were reared in outdoor laboratory enclosures, either just with their siblings (*a, b*) or with siblings and non-siblings (*c*). In each test, 2000 tadpoles were dyed red or blue, and were then released into a natural pond from which other larvae had been removed. Schools were sampled repeatedly on days following the release. Each school that formed is represented by a bar, and the length of the bar denotes the proportion of red-marked individuals in the school. If tadpoles joined schools randomly, schools should consist of half red- and half blue-marked individuals (dotted lines). (*a*) In control tests, sibships were randomly divided into two parts, each half was marked a different colour, and tadpoles were released together. Results of two control tests are shown. No schools differed significantly from random expectations. (*b*) Different sibships initially reared apart, but marked differently and released together, formed numerous schools that departed significantly

bufo, Kiseleva, 1989; Mexican spadefoot toads, *Scaphiopus multiplicatus*, Pfennig, 1990; African clawed frogs, *Xenopus laevis*, Waldman, unpub. data).

Spatial affinity manifested in laboratory choice tests may or may not be associated with schooling tendencies. *R. aurora* tadpoles preferentially associate with kin during early developmental stages, but schooling behaviour has been rarely observed in this species (O'Hara & Blaustein, 1985; Blaustein *et al.*, 1987; Blaustein, pers. comm.). *R. sylvatica* tadpoles associate with their siblings in a choice tank (Cornell *et al.*, 1989), and tend to swim significantly closer to siblings than to non-siblings in laboratory pools (Waldman, 1984). In nature, however, *R. sylvatica* tadpoles are generally asocial and rarely school; under those limited conditions in which tadpoles are found in polarized schools, these are too large to possibly represent sibling groups (see Waldman, 1984; Fig. 7.3). Indeed, analyses of spatial distributions of *R. sylvatica* tadpoles in laboratory pools suggest that they space themselves randomly with respect to their siblings, but actively avoid non-siblings (Waldman, 1986a). While this may indicate that in ponds, *R. sylvatica* tadpoles show greater spatial proximity to siblings than non-siblings, these results do not implicate kin recognition as a component of schooling. Laboratory assays of kin discriminatory responses may demonstrate individuals' propensities to avoid interactions with competitors rather than any predilection to school with relatives.

Although choice tests reveal less about schooling tendencies than do analyses of spatial clumping, their use simplifies behavioural testing, facilitates more rapid data collection, and makes possible more powerful experimental designs. These techniques have been most fully exploited in analyses of the kin recognition system of *R. cascadae* tadpoles. In numerous laboratory experiments, *R. cascadae* tadpoles within compartmentalized testing tanks were shown to associate with siblings in preference to non-siblings (Blaustein & O'Hara, 1981, 1982a,b, 1983; O'Hara & Blaustein, 1981). In the field, *R. cascadae* tadpoles often associate in 'loose groups' in which nearest neighbours are on average 13 cm from one another (O'Hara, 1981). Denser aggregations (in which tadpoles are in physical contact or within 2 cm of one another) were observed only infrequently by O'Hara (1981) but are described

from random in sibship composition. (*c*) Different sibships, initially reared together, similarly formed numerous schools that differed significantly from random, although not on every day. Numbers of individuals in each school are denoted within bars. Mark-colour comparisons for each school were evaluated by a two-tailed binomial test (*$P < 0.05$; **$P < 0.01$; ***$P < 0.001$; ****$P < 0.0001$). Differences in mark-colour composition among schools on each sampling date were evaluated by chi-square contingency table analyses. (After Waldman, 1982b.)

Fig. 7.3 Wood frog (*Rana sylvatica*) tadpoles in a streaming school
formation. Wood frog tadpoles are rarely found in schools, and both
field and laboratory observations suggest that individuals tend to actively
avoid one another. This school, consisting of most tadpoles in the pond,
was observed near Dryden, New York. When the stream stopped,
tadpoles began feeding on vegetation in extremely dense non-polarized
schools. Densities in schools approached 160 000 individuals per cubic
meter, considerably larger than typical clutch sizes. These schools thus do
not consist of exclusive sibling groups. (See Waldman, 1984.)

as 'typical' in later reports (O'Hara & Blaustein, 1985; Blaustein, *et al.* 1987).
These aggregations generally consist of between 4 and 40 individuals
(O'Hara, 1981; O'Hara & Blaustein, 1985), although they may be larger.
Group sizes of up to 100 individuals were mentioned by Blaustein *et al.*
(1987), but Wollmuth *et al.* (1987) observed aggregations with similar
behavioural characteristics consisting of more than 5000 individuals layered
three to five deep. Using much the same procedures as Waldman (1982b),
O'Hara & Blaustein (1985) found that laboratory-reared *R. cascadae*
tadpoles showed a statistical bias to aggregate with their siblings when
released into natural ponds (Fig. 7.4). The large schools observed by
Wollmuth *et al.* (1987) cannot represent sibling cohorts, however, as typical
clutch sizes range from 300 to 500 eggs (Nussbaum *et al.*, 1983).

   Species with schooling, non-schooling, and even asocial larvae show
tendencies to associate with kin in laboratory choice devices. In retrospect,
Wassersug's (1973) speculation that behavioural traits of '*Bufo*-mode'

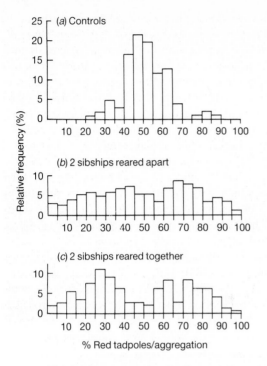

Fig. 7.4 Frequency distributions of the percentage of sibling Cascades frog (*Rana cascadae*) tadpoles marked red in aggregations sampled in field experiments. Using procedures similar to those of Waldman (1982b), all individuals were dyed red or blue prior to release into natural ponds. (*a*) In control tests, in which all tadpoles were siblings, most schools consisted of roughly half red and half blue individuals. (*b, c*) In experimental tests in which two sibships, marked red or blue, were released, many schools consisted predominantly of members of one or the other sibship. (After O'Hara & Blaustein, 1985.)

species – their conspicuousness, gregariousness and aposematism – arose by kin selection seems fortuitous, for many larval anurans that recognize their kin lack these characteristics. Kin recognition may not be a universal trait, however. Tadpoles of some species, tested using established laboratory protocols, fail to discriminate between siblings and non-siblings. Tadpoles of the western spotted frog, *R. pretiosa*, and the Pacific tree frog, *H. regilla*, reared under a variety of regimens, show no preference to associate with their siblings rather than non-siblings (O'Hara & Blaustein, 1988). Similarly, tadpoles of the leopard frog, *R. pipiens*, and the spring peeper, *Pseudacris crucifer*, spend roughly equal amounts of time near siblings and non-siblings in laboratory tests (Fishwild *et al.*, 1990).

Blaustein (1988) argues that interspecific variation in kin recognition abilities revealed by laboratory tests is predictable based on knowledge of the ecology and life-history strategies of these species (also see Blaustein *et al.*, 1987). Thus *R. pretiosa* and *H. regilla* tadpoles 'form aggregations only intermittently and there is little opportunity for siblings to interact' in nature (Blaustein, 1988). *R. aurora* tadpoles express sibling-association preferences only during early developmental stages, but this is consistent with their presumed lack of opportunity to associate with their kin beyond early developmental stages. *R. aurora* tadpoles apparently do not aggregate in the wild, and they disperse from oviposition sites and mix freely with members of other clutches (Blaustein & O'Hara, 1986; Blaustein *et al.*, 1987). Similarly, the propensity of *B. boreas* larvae, under natural conditions, to form extremely large schools, in which they presumably lack opportunities to learn and respond to their siblings, might account for their failure to retain kin-association preferences in the laboratory after exposure to non-siblings (O'Hara & Blaustein, 1982; Blaustein *et al.*, 1987; Blaustein, 1988). In other habitats, however, *B. boreas* tadpoles, like *B. americanus* tadpoles, form smaller schools (Nussbaum *et al.*, 1983), and tests of the kin recognition abilities of subjects from these populations would be illuminating. Fishwild *et al.* (1990) note that unlike species that have been shown to preferentially associate with kin in laboratory tests, larvae of neither *P. crucifer* nor *R. pipiens* are known to form schools in the wild. They suggest that the cryptic coloration and large size of larvae of these species, respectively, may confer anti-predator advantages comparable to those achieved by schooling. The arguments of Blaustein (1988) and Fishwild *et al.* (1990) seem to presuppose that kin recognition is selected primarily in the context of larval schooling, but this is yet to be firmly established.

Territoriality

Organisms that repeatedly come into contact over time may habituate to each others' presence so that they reduce conflicts over resources. Many examples of 'dear enemy' recognition among occupants of neighbouring territories have been documented in vertebrates, including amphibians. Territorial male bullfrogs, *R. catesbeiana*, discriminate between calls of neighbours and strangers, and respond more strongly to strangers (Davis, 1987). As pointed out by Madison (1975), several aspects of amphibian life-histories suggest the possibility that the ability to distinguish neighbours from non-neighbours at times is functionally equivalent to kin recognition. Many amphibians are philopatric or disperse only over limited distances, so genetic relatives may be spatially clumped (reviewed in Waldman & McKinnon, 1991).

Although territoriality has been studied less in amphibians than in other vertebrates, mechanisms by which neighbours recognize one another have been extensively studied in terrestrial lungless salamanders (family Plethodontidae) (reviewed in Jaeger, 1986). Tested prior to the onset of breeding, in a laboratory device, both male and female *Plethodon jordani* oriented preferentially to the odours of their neighbours rather than non-neighbours (Madison, 1975). Thus, information concerning individuals' genotypes seems to be encoded in the chemical cues they release into the air. Jaeger and his students have assayed neighbour recognition by measuring the rates at which individuals tap their nasolabial cirri (chemoreceptive organs) against the substrate. Their results suggest that *P. cinereus* adults can recognize their own odours from those of other conspecifics (Tristram, 1977), and can discriminate between odours of familiar and unfamiliar conspecifics (McGavin, 1978; Jaeger & Gergits, 1979), possibly through chemical cues present in faecal pellets (Jaeger *et al.*, 1986; Horne & Jaeger, 1988; also Simon & Madison, 1984).

Plethodontid salamanders exhibit intraspecific aggression (Arnold, 1976; Jaeger 1984), and individuals may retreat to refuges in dry periods where they establish territories to defend limited food resources (Jaeger *et al.*, 1982). Because salamanders that are excluded from territories may be forced to retreat underground where they are less likely to capture prey (Fraser, 1976), owners of superior territories might be selected to share resources with close relatives or to act less agonistically toward related neighbours. Although the role of kinship in regulating intraspecific competition in plethodontid salamanders is at present largely unknown, investigation of their kin recognition abilities seems warranted. Adult ambystomatid salamanders (*Ambystoma maculatum*) show patterns of intraspecific aggression similar in many respects to that shown by plethodontids (Ducey & Ritsema, 1988). Recently, Walls & Roudebush (1991) found that larval marbled salamanders, *A. opacum*, were significantly less aggressive toward their siblings than toward non-siblings in laboratory tests. Discrimination of kin was apparent whether or not opponents were familiar. Consequently levels of cannibalism among ambystomatid larvae (see Collins & Cheek, 1983) may be reduced among siblings as compared with non-siblings (Walls & Roudebush, 1991). After metamorphosis discrimination is based more on familiarity than kinship *per se*. (Walls, 1991).

Mating

Amphibian breeding populations can represent highly structured demes. Many amphibians are thought to return to their natal pond to breed (e.g. Oldham, 1966; Heusser, 1969; Ewert, 1969; Grubb, 1973). Homing tendencies have been experimentally verified in several species (e.g. Bogert, 1947; Jameson, 1957; Madison, 1969; Barthalmus & Bellis, 1972). In a detailed four-year study of Fowler's toad, *B. woodhousei fowleri*, Breden (1987) found that 73 per cent of metamorphosed individuals bred for the first time in their natal pond, and most migrants bred only a short distance away.

Wood frogs, *R. sylvatica*, also return to pools from which they have metamorphosed to breed, and show remarkable discrimination even between nearby ponds (Berven & Grudzien, 1990). Why amphibians return to their natal ponds is unknown. Homing tendencies may be favoured if the quality of breeding habitats tends to be consistent between generations: an individual's survival in a particular pond provides some assurance that its offspring will fare well there. Alternatively, philopatry may serve to maintain specific genetic adaptations to local environmental conditions (Shields, 1982).

If neither sex disperses, relatives may frequently encounter one another as potential mates. Matings between siblings are less likely, but are still possible, if sexes first breed at different ages (see Duellman & Trueb, 1986, for a summary of data on age at sexual maturity). Outbreeding, like inbreeding, may have potential costs, and by returning to their natal ponds, individuals may maximize the likelihood that they can obtain an optimally related mate (Bateson, 1982, 1983; Shields, 1982). Under these conditions, however, individuals that recognize and avoid mating with their siblings might be selected over those that mate indiscriminately.

The possibility that larval kin recognition abilities are retained after metamorphosis (Waldman, 1982b), perhaps through sexual maturity, suggests a means by which individuals might avoid incest. *R. cascadae* froglets, tested in a laboratory apparatus 4–12 or 39–47 days after metamorphosis, recognized and preferentially associated with siblings rather than non-siblings (Blaustein *et al.*, 1984). Metamorphosed *R. sylvatica* froglets, tested at comparable developmental stages in an apparatus virtually identical to that used by Blaustein *et al.* (1984), failed to discriminate between siblings and non-siblings (Waldman, 1989). *R. sylvatica* froglets, however, tested within 24 hours of metamorphosis in a smaller choice tank, spent significantly more time near a compartment housing either their siblings or paternal half-siblings than one housing non-siblings (Cornell *et al.*, 1989). These divergent results may reflect ontogenetic shifts in behavioural preferences, differences in experimental design, or variation among populations. Whether kin recognition following metamorphosis has any function is also unclear; larval discrimination tendencies might be retained only for a short time after metamorphosis and may be selectively neutral during this period (Cornell *et al.*, 1989).

While no studies of the kin recognition abilities of adult amphibians have been conducted to date, genetic studies on breeding populations provide some basis for inferring whether discrimination based on kinship might occur. Examining the underlying source of heterozygote deficiency observed in *B. americanus* breeding populations (Guttman & Wilson, 1973), Christein *et al.* (1979) found no evidence of non-random mating by genotype based on electrophoretic analyses of blood proteins of pairs they collected in amplexus. Analyses of DNA extracted from mitochondria (mtDNA) of individuals collected from *B. americanus* breeding populations located just 1 km apart suggest that they are genetically distinct, as some unique haplotypes are represented in each population and overall distributions of haplotypes significantly differ (Waldman *et al.*, in preparation). Because mtDNA is matrilineally inherited, the results indicate some level of philopatry, at least among females. Yet mating with close relatives is exceedingly rare. Based on analyses of restriction fragment length polymorphisms of the mitochondrial genome, only one of 86 pairs collected in amplexus were genetically similar to one another and thus could possibly be siblings (they might instead be more distant matrilineal relatives). One plausible explanation for these data is that individuals recognize close relatives and actively avoid mating with them, although similar breeding patterns might be expected if males dispersed between ponds, a hypothesis currently being examined (see Waldman & McKinnon, 1991).

Mechanisms of kin recognition

Indirect means of identifying kin may be effective in some contexts, especially in parental care. Yet the ability of many amphibians to identify their relatives appears not to be limited to particular settings. Discrimination of kin by the perception and evaluation of the cues they express (direct kin recognition) has been demonstrated in numerous experimental studies. This work has two main goals: the delineation (1) of the ontogeny of kin recognition, and (2) of the proximate sensory mechanisms by which kin are recognized. To address these issues, both the means by which recognition labels develop and are expressed – and the developmental pathways by which templates become specified so that they match labels – need to be examined. Because of the ease with which amphibian larvae can be manipulated for investigations of developmental and sensory mechanisms, they have been adopted as focal study organisms by several laboratories. At present, we know considerably more regarding how tadpoles of frogs and toads recognize their kin than why they do so.

Ontogeny of kin recognition

The development of kin recognition abilities in amphibians can be studied by rearing larvae under varied regimens and assaying the subsequent behavioural responses of tadpoles to siblings and non-siblings (Table 7.1). The first studies of kin discrimination by anuran larvae involved simple rearing protocols: *B. americanus* tadpoles were raised in sibling groups from the time of oviposition until four to six weeks after hatching, at which time they were mixed with non-siblings in a laboratory pool. Tadpoles quickly assorted into sibling clumps (Waldman & Adler, 1979). *R. sylvatica* tadpoles also associated with familiar siblings in preference to unfamiliar non-siblings

Table 7.1. *Ontogeny of kin recognition abilities of larval anurans*

Discrimination type	Familiar sibs/unfamiliar non-sibs	Familiar sibs/familiar non-sibs		Unfamiliar sibs/unfamiliar non-sibs		Unfamiliar sibs/familiar non-sibs	Previously familiar sibs/currently familiar non-sibs	Currently familiar sibs/previously familiar non-sibs
Rearing conditions	Sib groups	Mixed groups	Social isolation	Sib groups	Mixed groups	Mixed groups	Sib groups first, then mixed groups	Mixed groups first, then sib groups
Species (References)								
Bufo americanus (5,13,14,16,18)	yes	no (lab) yes (field)	yes	—	—	yes	yes	no
Bufo boreas (9)	yes	—	—	—	no	no*	no	—
Bufo bufo (7)	yes	—	—	—	—	—	—	—
Hyla regilla (11)	no	—	—	—	no	—	—	—
Pseudacris crucifer (6)	no	—	—	—	—	—	—	—
Rana aurora (3)	yes**	—	no	—	no	no*	—	—
Rana cascadae (1,2,8,10)	yes	no (lab) yes (field)	yes	yes	yes	yes*	—	—
Rana pipiens (6)	no	—	—	—	—	—	—	—
Rana pretiosa (11)	no	—	no	—	no	—	—	—

Rana sylvatica (4,6,15)	yes	yes	—	—	—	—
Scaphiopus multiplicatus (12)	—	—	yes	—	—	—
Xenopus laevis (17)	—	—	yes	—	—	—

Notes:

All results are from laboratory experiments except where indicated. 'Sibs' denotes full siblings. Testing protocols vary; see text.

* Test subjects previously were reared with members of one other family (familiar non-siblings). Subjects had not been exposed, however, to those individuals with which they were tested (members of the familiar family, reared apart).

** Early developmental stages; later stages do not discriminate between siblings and non-siblings.

Sources: (1) Blaustein & O'Hara, 1981; (2) Blaustein & O'Hara, 1983; (3) Blaustein & O'Hara, 1986; (4) Cornell *et al.*, 1989; (5) Dawson, 1982; (6) Fishwild *et al.*, 1990; (7) Kiseleva, 1989; (8) O'Hara & Blaustein, 1981; (9) O'Hara & Blaustein, 1982; (10) O'Hara & Blaustein, 1985; (11) O'Hara & Blaustein, 1988; (12) Pfennig, 1990; (13) Waldman, 1981; (14) Waldman, 1982b; (15) Waldman, 1984; (16) Waldman, 1986b; (17) Waldman, unpub. data; (18) Waldman & Adler, 1979.

Table 7.2. *Discrimination between familiar and unfamiliar full siblings.*

Species (References)	Rearing conditions	
	Sib groups	Mixed groups
Bufo americanus (2,4)	no	no*
Rana sylvatica (1,3)	no	yes*

Note:
* Unfamiliar siblings were reared in different mixed-group tanks, and each tank comprised a different set of sibships.
Sources: (1) Cornell *et. al.*, 1989; (2) Waldman, 1981; (3) Waldman, 1984; (4) Waldman, 1985b.

in laboratory tanks (Waldman, 1984; Cornell *et al.*, 1989). Similar results were obtained on tadpoles of *R. cascadae*, *R. aurora* and *B. boreas* (O'Hara & Blaustein, 1981, 1982; Blaustein & O'Hara, 1986) tested in laboratory choice tanks. In each case, tadpoles reared in sibling groups spent more time near compartments holding their familiar siblings than near those holding unfamiliar non-siblings. In all these tests, familiarity is confounded with relatedness, so tadpoles may be responding simply to those individuals with which they have been reared rather than to their relatives *per se*. However, tadpoles reared in sibling groups failed to discriminate between familiar siblings and unfamiliar siblings (*B. americanus*, Waldman, 1981; *R. sylvatica*, Cornell *et al.*, 1989; see Table 7.2). Thus behavioural discrimination involves the assessment of cues associated with genetic relatedness, rather than differential responses simply to familiar individuals.

Social isolation experiments

The necessary and sufficient social conditions for the development of kin recognition can be established by comparing the behavioural responses of socially deprived individuals with those of socially reared individuals. If, for example, individuals reared in social isolation show similar behavioural tendencies to those reared in sibling groups, experience with conspecifics may not be crucial to the development of normal sibling recognition abilities. With anuran amphibians, this experiment is simple. Embryos can be separated from one another shortly after fertilization and each is then reared in a separate dish so that individuals are visually and chemically isolated from conspecifics. Placed together with unfamiliar siblings and unfamiliar non-siblings in a laboratory pool for the first time at five to seven weeks of age, *B. americanus* tadpoles that had been reared in isolation assorted into sibling clumps, just as did tadpoles reared with their siblings (Waldman, 1981, 1986a). *R. cascadae* tadpoles, similarly reared in separate bowls from embryonic stages, also discriminated between unfamiliar siblings and unfamiliar non-siblings when tested in a choice tank (Blaustein & O'Hara, 1981). *R. aurora* isolates, on the other hand, did not discriminate between siblings and non-siblings – but even *R. aurora* tadpoles reared in sibling groups failed to associate with kin at the age tested (Blaustein & O'Hara, 1986).

What do these results mean? Evidence that individuals reared in social isolation can discriminate, on first encounter, relatives from non-relatives might constitute evidence of a 'genetic recognition system' (Blaustein & O'Hara, 1981), and may point to a role of special recognition alleles in effecting phenotypic labels and template components that match these labels. Yet even animals reared under conditions in which they are restricted from interacting with conspecifics presumably do experience some facets of their own phenotype. These traits, if genetically determined, are likely to be shared by conspecifics in proportion to their degree of relatedness, and thus they can communicate information concerning kinship between unfamiliar individuals (see discussions by Getz, 1981; Waldman, 1981; Holmes & Sherman, 1982; Blaustein, 1983; Hepper, 1987; Crozier, 1987). Maternally inherited or environmentally acquired cues may also permit unfamiliar kin to recognize one another (Gamboa *et al.*, 1986; Waldman, 1987).

From an ecological perspective, however, interpretation of experiments on subjects reared in social isolation must be tempered by the realization that none of the species that have been tested to date typically spawns by scattering single eggs throughout a pond. *B. americanus* deposits their eggs in strings, and tadpoles are likely to be surrounded by their siblings from the earliest developmental stages (see Waldman, 1981). *R. cascadae* egg masses are globular and are frequently deposited in large, communal clumps with 30 to 70 other pairs (Sype, 1975; Briggs, 1987; O'Hara & Blaustein, 1981, give 4 to 8 as typical clump sizes), so tadpoles are likely to be exposed both to siblings and non-siblings soon after hatching (see Blaustein & O'Hara, 1981). Thus, social cues that are available in the larval environment conceivably might influence the ontogeny of kin recognition even though recognition mechanisms appear to develop properly in the absence of such cues.

Social enrichment experiments

To resolve social influences on the ontogeny of kin recognition, larvae can be reared in socially enriched environments. *B. americanus* tadpoles raised with siblings and non-siblings together in special flow tanks, from shortly after fertilization until testing three to five weeks later, failed to discriminate between familiar siblings and familiar non-siblings, as shown by their nearest-neighbour relationships in laboratory pools (Waldman, 1981). They did, however, orient toward unfamiliar siblings rather than familiar non-siblings when tested in a Y-maze apparatus (Waldman, 1986b). In the rearing tanks, water was rapidly pumped among mesh compartments holding each sibship, so that tadpoles were continuously exposed to water-borne cues of their siblings and those of seven other sibships. These conditions are unnatural, however *B. americanus* rarely spawns in pools with rapid currents, so larvae typically are not exposed to labels of non-siblings at such high concentrations. Indeed, when *B. americanus* tadpoles were reared in still rather than flowing water, separated from non-siblings by the same mesh, they subsequently did discriminate between familiar siblings and familiar non-siblings in field tests (Waldman, 1982b).

The laboratory results suggest that at least some components of the recognition template of developing *B. americanus* larvae can be influenced by their social environment. In contrast, tadpoles reared in isolation have only their own cues available to learn, and they thus effectively form a 'self-template' (Waldman, 1981). Cues learned from oneself may be sufficient to recognize unfamiliar kin, but as stimuli they may nonetheless lack the strength of socially acquired cues. Whereas socially reared *B. americanus* tadpoles are found in spatially contagious distributions in laboratory pools, isolates distribute themselves randomly with respect to siblings but appear to avoid non-siblings (Waldman, 1986a). More generally, isolates often fail to develop normal propensities to aggregate with conspecifics (Shaw, 1961; Williams, 1976; Foster & McDiarmid, 1982). Socially reared *B. americanus* larvae may use self-templates too, perhaps to a lesser extent than do isolates. This might account for their preference for unfamiliar siblings rather than familiar non-siblings.

An inability to discriminate between siblings and non-siblings might indicate a breakdown in the specificity of the template if different labels cannot be distinguished. Alternatively, apparent recognition failures might indicate a breakdown in the specificity of labels: two or more sibling groups might develop similar or identical labels. Reduced label specificity may result when social interactions cause label components to converge, for example by chemical exchange, social mimicry, or the use of common nesting or feeding resources.

Although examples of label convergence abound, especially in social insects (Carlin & Hölldobler, 1986, 1987; Stuart, 1988; and references therein) and crustaceans (Linsenmair, 1987), additional experimental work suggests that the failure of *B. americanus* tadpoles to discriminate between familiar siblings and non-siblings is due to social influences on the development of the template and not the labels. Groups of *B. americanus* siblings that were reared in varied social environments – exposed to completely different family mixtures – continued to recognize and associate with unfamiliar siblings in the same manner as if they were familiar (Waldman, 1985b). If the labels of individuals change as a function of the genetic makeup of their social group, subjects should discriminate between familiar siblings, that share their labels, and unfamiliar siblings, that express different labels. In laboratory tests, subjects failed to discriminate between these groups (Table 7.2, see Waldman, 1985b).

Decrements in kin preferences expressed by individuals reared in mixed-sibship groups cannot necessarily be attributed to changes in their templates. *R. cascadae* tadpoles, reared in tanks together with siblings and separated by a mesh partition from members of one other sibship, subsequently failed to discriminate between these familiar siblings and non-siblings in laboratory tests (O'Hara & Blaustein, 1981). Yet *R. cascadae* tadpoles reared under these same conditions associated preferentially with unfamiliar siblings rather than unfamiliar non-siblings (Blaustein & O'Hara, 1981). Even tadpoles raised only with non-siblings showed the same preference for unfamiliar siblings (Blaustein & O'Hara, 1983). When reared in mixed-sibship groups, *R. cascadae* tadpoles can recognize siblings that they have not previously encountered, but apparently they cannot discriminate between the siblings and non-siblings with which they have been reared.

Social modulation of recognition labels may account for the apparent inability of tadpoles to recognize differences between siblings and non-siblings reared together. Although the templates of *R. cascadae* tadpoles appear not to be changed by the social regimen in which they are reared, their labels must be environmentally influenced. O'Hara & Blaustein (1981) suggest that 'tadpoles reared in mixed sibling groups might retain the "odour" of the composite group'. As pointed out by Waldman (1984, 1987), the apparent lack of integrity of the labelling system when different sibling groups come in contact suggests *a priori* that in natural conditions, *R. cascadae* tadpoles should not discriminate between siblings and non-siblings (which mix at hatching). That they sometimes do (O'Hara & Blaustein, 1985) suggests some artificiality of the laboratory testing procedure.

The kin recognition system of *R. sylvatica* tadpoles is intermediate in several respects to those of *B. americanus* and *R. cascadae* tadpoles. Like these species, *R. sylvatica* tadpoles, reared in sibling groups, associated preferentially with familiar siblings rather than to unfamiliar non-siblings (Waldman, 1984), and unfamiliar siblings in preference to unfamiliar non-siblings (Cornell *et al.*, 1989). Unlike *B. americanus* and *R. cascadae*, however, *R. sylvatica* tadpoles, reared in tanks in which they were exposed both to siblings and non-siblings, subsequently discriminated between these familiar individuals, and associated preferentially with their siblings (Waldman, 1984). Spawning *R. sylvatica*, like *R. cascadae*, typically deposit egg masses in communal clumps (see Waldman, 1982a, and references therein), and upon hatching, tadpoles interact both with siblings and non-siblings.

Although field tests have yet to be conducted on *R. sylvatica* larvae, results of these laboratory experiments suggest that in natural conditions they should be able to discriminate between siblings and non-siblings.

Variation in results of the *R. sylvatica* and *R. cascadae* tests cannot be readily explained by differences in the rearing protocols. *R. sylvatica* tadpoles reared in mixed groups were housed in special flow tanks, identical to those used in the *B. americanus* tests (Waldman, 1981, 1985b). Water was actively pumped among mesh compartments in these tanks, whereas O'Hara & Blaustein (1981) used an air bubbler to circulate water between the two mesh compartments of their rearing tank. Thus *R. sylvatica* tadpoles had at least as much – indeed probably more – contact with waterborne labels of siblings and non-siblings as did the *R. cascadae* tadpoles studied by O'Hara & Blaustein (1981). Given that they had been thoroughly exposed to the cues of non-siblings, the ability of *R. sylvatica* tadpoles to discriminate between familiar siblings and familiar non-siblings contrasts strikingly with the apparent inability of *R. cascadae* tadpoles to discriminate between these same classes in laboratory tests.

Both the recognition templates and the labels by which *R. sylvatica* kin are identified seem resilient to perturbation by the larval social environment. *R. sylvatica* tadpoles evidently learn their siblings' traits more readily than those of non-siblings that they encounter during development. That labels borne by sibling groups reared together remain discernible is self-evident; otherwise tadpoles would not be able to recognize differences between them. But just as the recognition labels of *R. cascadae* tadpoles reared together appear to converge due to their social interactions, so too – to a lesser extent – do those of *R. sylvatica* tadpoles. Waldman (1984) split *R. sylvatica* clutches into portions, and raised each in a different flow tank housing largely non-overlapping sets of sibships. Tadpoles reared under these conditions associated with the siblings with which they had been reared in preference to unfamiliar siblings that had been reared in a different flow tank. Thus, the tadpoles' recognition labels appear to be influenced in part by their social environment. If labels of non-siblings reared together converge, labels of siblings reared apart should diverge, exactly as these results suggest. Because tadpoles can discriminate between familiar siblings and familiar non-siblings, any shifts in their label characteristics (1) must be incomplete so that different sibships do not appear identical, and (2) must be accompanied by a concomitant revision of the template to ensure continued label/template matching. Together with Cornell *et al.*'s (1989) finding that *R. sylvatica* tadpoles can discriminate between unfamiliar siblings and unfamiliar non-siblings, the results raise the possibility that *R. sylvatica* recognition labels incorporate both genetic and non-genetic components.

Sensitive periods

The kin recognition system of *Rana sylvatica* tadpoles shows unusual plasticity, as both labels and templates can change, probably throughout larval development. In contrast, *B. americanus* larvae form a template based upon the labels that they encounter during a more narrow time window: within two weeks of hatching. After this period, the template appears 'crystallized' and is highly resistant to further modification. *B. americanus* tadpoles that were initially reared in sibling groups, and were then transferred to flow tanks containing mixed groups, subsequently assorted preferentially with their siblings (Waldman, 1981). But if their early experience included both siblings and non-siblings, *B. americanus* tadpoles later failed to discriminate between their siblings and non-siblings – even though their most recent experience had been restricted to siblings (Waldman, 1981). The development of the *B. americanus* kin recognition system thus bears some resemblance to imprinting (see Bateson, 1979) in that behavioural preferences are shaped during a critical or sensitive period.

The sensitive period for the formation of a recognition template in *B. americanus* roughly corresponds to the stage at which larvae hatch and become free-swimming. If pairs spawn in non-contiguous locations within a pond, as usually occurs in temporary pools, tadpoles will be surrounded by their siblings during this period, and individuals will learn their siblings' traits. Should pairs spawn communally and intertwine their eggs with those of other pairs, as they sometimes do especially in denser breeding aggregations within permanent ponds, the tadpoles might not recognize kin by this learning process. However, the template might be acquired prior to or immediately after hatching, still permitting tadpoles to recognize their kin (Waldman, 1981). Alternatively, the acquisition of templates might be guided by a genetically based system of stimulus filtering, so that labels of genetic relatives might appear more salient during development, even in mixed groups (Waldman, 1986b). These hypotheses have not been experimentally tested. Other work described below suggests, however, that maternal cues play an important role in the ontogeny of kin recognition in this species, and these cues would be present during early developmental stages prior to hatching.

The possibility that kin-recognition templates may be imprinted has not been specifically examined in other anurans, but some evidence argues against such mechanisms. Like *B. americanus* tadpoles, *B. boreas* tadpoles that had been reared with their siblings developed a preference to associate with siblings, and tadpoles reared with siblings and non-siblings failed to discriminate between these groups (O'Hara & Blaustein, 1982). *B. boreas* tadpoles reared only with their siblings for 75 days, and then exposed to mixed groups for short periods (2 to 6 days), showed no evidence of discriminating between unfamiliar siblings and unfamiliar non-siblings (O'Hara & Blaustein, 1982). Thus the recognition templates of *B. boreas* tadpoles appear to be rapidly modified following exposure to additional sibling groups. O'Hara & Blaustein (1982) report that these toads spawn in large breeding aggregations in ponds and lakes, and that egg masses are often clumped so that different sibships become intertwined. Under these circumstances, tadpoles presumably interact with siblings and non-siblings soon after hatching and have little opportunity to form sibling groups. As in *B. americanus*, however, association preferences of *B. boreas* tadpoles appear not to be determined solely by social experience, since individuals reared just with non-siblings did not develop a preference for these non-siblings over their own unfamiliar siblings (O'Hara & Blaustein, 1982).

Maternal factors

Labels may be derived from a variety of sources, including gene products, environmental cues, and maternal factors. If templates become crystallized during a sensitive period, as are those of *B. americanus* larvae, labels also need to remain constant as individuals mature. Should labels change, they would no longer match the template already formed. Although labels that incorporate environmental cues are expected to fluctuate with time more than genetically based cues, environmental labels, acquired at particular ontogenetic stages, sometimes may be retained and expressed later in life (see Gamboa *et al.*, 1986). Maternal factors, especially prevalent in the larval environment, might also be incorporated into labels, and may influence the ontogeny of templates.

Maternal factors influence the development of the kin recognition systems of anuran amphibians (Table 7.3). This was first illustrated in experiments on *B. americanus* tadpoles that were reared in social isolation from eggs for five to seven weeks, and were then placed together with their full siblings and either their maternal or paternal half-siblings in a laboratory pool. Analyses of the spatial distributions of the tadpoles in the pool suggest that they discriminated between full siblings and paternal half-siblings, but not between full siblings and maternal half-siblings (Waldman, 1981). Waldman (1981) hypothesized that female parents pass to their progeny maternal labels, possibly in the egg jelly or extrachromosomal DNA.

R. cascadae tadpoles also demonstrate maternal biases in their kin association preferences, measured in a choice tank (Blaustein & O'Hara, 1982b). *R. cascadae* tadpoles reared in sibling groups or in social isolation associated with maternal half-siblings in preference to paternal half-siblings, but were able to distinguish full siblings from both maternal and paternal half-siblings (Blaustein & O'Hara, 1982b). To test the hypothesis that recognition labels are determined by the egg jelly, Blaustein & O'Hara

Table 7.3. *Discrimination of half-siblings.*

| | Discrimination type | | | | |
| Species (References) | Half-sibs/Full-sibs | | Half-sibs/Non-sibs | | Maternal half-sibs/ Paternal half-sibs |
	Maternal	Paternal	Maternal	Paternal	
*Bufo americanus** (3)	no	yes	—	—	—
*Rana cascadae*** (1)	yes	yes	yes	yes	yes (maternal preference)
*Rana sylvatica*** (2)	—	—	—	yes	—

Notes:
* Individuals reared in social isolation. The failure of tadpoles to discriminate between maternal half-sibs and full sibs suggests a maternal basis to recognition cues.
** Individuals reared in sib groups, but stimulus groups are unfamiliar to test subjects (one replicate each of *Rana cascadae* maternal half-sib/full sib and paternal half-sib/non-sib tests reared in social isolation).
Sources: (1) Blaustein & O'Hara, 1982b; (2) Cornell *et. al.*, 1989; (3) Waldman, 1981.

(1982b) removed most of the jelly from larvae at hatching, and reared one group of individuals without jelly and another group in the presence of jelly from a second female. Both experimental groups subsequently discriminated between full siblings and paternal half-siblings. In choice tanks, *R. sylvatica* tadpoles discriminated between unfamiliar paternal half-siblings and unfamiliar non-siblings, both prior to and immediately after metamorphosis (Cornell *et al.*, 1989). Recognition labels of both species thus show evidence of a substantial genetic basis in addition to any maternal component.

Genetic factors

Genetic influences on recognition labels are implicated by experiments demonstrating recognition of unfamiliar paternal half-siblings from unfamiliar non-siblings (*R. cascadae*, Blaustein & O'Hara, 1982b; *R. sylvatica*, Cornell *et al.*, 1989). Paternal half-siblings share half the male parent's genes, but not the other factors that maternal half-siblings share (e.g. cytoplasm, egg yolk, egg jelly, or oviposition site), so these results provide strong evidence of a genetic component to kin recognition. Possibly through self-learning, recognition templates also are genetically influenced in these cases. The failure to find discrimination between paternal half-siblings and either full siblings or non-siblings does not necessarily imply, however, the absence of genetic components to kin recognition. As discussed earlier, the expression of kin recognition is expected to be context-dependent (Waldman, 1988; Reeve, 1989), and if interactions between paternal half-siblings are uncommon in natural conditions, discrimination may not have been selected. *B. americanus* tadpoles appear to respond to paternal half-siblings as non-siblings (Waldman, 1981). Observations on the mating behaviour of this species demonstrate that few males mate with more than one female (Howard, 1988), so paternal half-siblings must be quite rare. Genetic influences on the kin recognition system of *B. americanus* larvae may be apparent, nonetheless, in other discrimination tests.

Most studies of kin recognition focus on differential responses of individuals towards related and non-related members of their own population. The examination of kin discrimination tendencies expressed when members of different populations interact can serve as an additional means for evaluating the genetic components of recognition systems. Levels of genetic variability within and among populations are determined in large part by characteristics of the population structure. Evidence of philopatry in amphibians, reviewed earlier, suggests genetic partitioning among populations. Nearby *B. americanus* breeding populations, for example, are genetically quite different (Waldman & McKinnon, 1991). In these circumstances, recognition labels, to the extent they are genetically specified, should vary more greatly among members of different populations than within populations.

To investigate this possibility, Waldman (1981) obtained amplectant *B. americanus* pairs from several ponds, some up to 13 km apart, and allowed the pairs to spawn under controlled laboratory conditions. Tadpoles were reared in sibling groups or in social isolation using standardized laboratory protocols, and subsequently they were placed together in a laboratory pool with members of another sibling group. For those tadpoles that were reared with their siblings, association patterns varied depending on whether the two sibling groups were obtained from pairs collected in the same pond or from different ponds. Distances between non-siblings were substantially greater, and thus differences between sibling and non-sibling nearest-neighbour distances were significantly greater, when the different sibships were from distant ponds. These results are consistent with the hypothesis that recognition labels are at least in part genetically determined. Differences in label characteristics might also arise from influences of different environmental factors on the individuals collected in separate locations. Maternal effects thus may be confounded with genetic effects. Even though larvae were all reared under the same laboratory conditions, labels passed by parents to their progeny might be influenced by dietary factors or microbial fauna (Waldman, 1981), both of which should vary more between ponds than within ponds.

Environmental factors

To the extent that individuals' tendencies to choose particular habitats are correlated with their genotypes, they may incidentally associate with their kin. For example, if members of a sibling cohort predictably and regularly return to their hatching site, sibling schools may result. Kin recognition is then indirect, rather than direct, because individuals learn and respond to environmental cues (e.g. Wiens, 1972; Punzo, 1976; Dunlap & Scatterfield, 1982) rather than the labels or phenotypic traits of conspecifics. While differential behavioural responses to environmental cues may be favoured in the context of habitat selection, they may also promote nepotism or inbreeding avoidance. Once kin associate, for whatever reason, they can learn each other's trait and later can potentially recognize their relatives in other contexts. Preferences to associate with kin, that have been demonstrated in most experimental studies of anuran larvae, are attributable to direct rather than indirect mechanisms because environmental gradients are controlled in laboratory tests. But these findings do not exclude the possibility that in natural conditions, indirect mechanisms also play a role in effecting discrimination.

The ontogeny of direct kin recognition systems can be influenced by extrinsic environmental cues, just as it can be influenced by maternal factors. Environmental cues may be incorporated into recognition labels, and this in turn may affect the development of templates. Behavioural preferences also may be directly induced by exposure to these cues. Hepper & Waldman (unpub. data) injected various odourants into *R. temporaria* and *R. sylvatica* egg capsules soon after oviposition, and found that even 12 weeks later, larvae consistently oriented toward streams of the particular odour to which they had been exposed as embryos. For example, *R. sylvatica* larvae that had been injected with lemon extract oriented toward lemon in preference to orange, whereas larvae that had been injected with orange extract oriented toward orange in preference to lemon. While these cues are not typical of those found under natural conditions, Waldman (in preparation) found that *R. sylvatica* embryos inoculated with different algal cultures developed comparable preferences for odours associated with these cultures. In vernal pools, algae frequently grows commensally in *R. sylvatica* egg capsules, so the results suggest a natural context in which environmental cues experienced during early development might affect subsequent behavioural preferences. Whether these cues influence the labels expressed by tadpoles, as well as the templates by which they perceive these cues, is currently being studied.

Environmentally mediated preferences that are established during early development might become amplified if they cause sibships to partition available resources. Due to these preferences sibling groups might be stimulated to occupy different areas of a pond or to feed on particular foods.

Variation in the environmental cues to which sibships are exposed might be sufficient to impart distinctive properties to their recognition labels. Kin discrimination would then occur if individuals differentially interacted with conspecifics with which they shared these labels.

Experimental work on *Scaphiopus multiplicatus* tadpoles demonstrates how such a developmental scenario can come into play. Pfennig (1990) split groups of *S. multiplicatus* siblings into two treatments. One treatment was fed frozen brine shrimp (*Artemia*) and the other a commercial tadpole chow. Tested in a laboratory choice apparatus, tadpoles preferred unfamiliar siblings to unfamiliar non-siblings when all were fed on the same food. Subjects showed stronger preferences for water in which their familiar diet was present, however, than for water containing the other diet. Moreover, dietary factors appear to be more important than genetically determined cues, as tadpoles oriented toward non-siblings that had been fed their own diet in preference to siblings that had been fed the other diet. Pfennig (1990) interprets these data as evidence of philopatry, and suggests, based on field data showing greatest larval survivorship within enclosures at the natal site, that apparent kin recognition in this species is an epiphenomenon of preferences for particular habitats. Because siblings are more likely to experience similar environmental factors than are non-siblings, the results illustrate equally well a mechanism by which *S. multiplicatus* tadpoles might recognize their kin by environmentally influenced labels.

Sensory mechanisms

Anuran larvae perceive and respond to their environment, and to one another, through a variety of sensory modalities (e.g. see Katz *et al.*, 1981; Wassersug *et al.*, 1981) but kin discrimination occurs largely by chemosensory recognition. The association of close kin may also be facilitated by visual cues, but these cues appear not to encode specific information concerning kinship. The sensory basis of kin recognition abilities has been examined only in *B. americanus*, *B. bufo*, and *R. cascadae* tadpoles, however, and further work may reveal that other species recognize their kin by substantially different mechanisms.

The ability of *B. americanus* tadpoles to recognize their kin by chemical labels has been demonstrated using two different testing protocols. Dawson (1982) recorded the time spent by individuals on either side of a porcelain tray, near two compartments enclosed by visually opaque cheesecloth partitions. When five of the test subject's siblings were placed in one of the compartments and five non-siblings were placed in the other, subjects showed a significant tendency to swim toward their siblings. Subjects given a choice between siblings in one compartment and a second empty compartment also oriented toward their siblings, but individuals presented with non-siblings in one compartment and an empty compartment made no choice. In another series of experiments, Dawson replaced the cheesecloth partitions with ones made of aluminium screening so that subjects could visually interact with siblings and non-siblings in the end compartments. Under these conditions, test subjects did not discriminate between siblings and non-siblings. Subjects did orient, however, toward water previously conditioned by their siblings, that was intermittently introduced into one chamber, in preference to water conditioned by members of another sibling group, that was simultaneously introduced into the other chamber. In contrast, subjects showed no preference for water conditioned by siblings over 'blank' (unconditioned) water. *B. americanus* larvae thus tend to respond differentially to chemical cues given off by their siblings and non-siblings, but Dawson's failure to present numerical results to support his statistical conclusions makes some of the experiments difficult to evaluate.

Tested singly in a glass Y-maze apparatus, *B. americanus* tadpoles similarly oriented toward waterborne cues emanating from their siblings in preference to non-sibling cues (Waldman, 1985a). Subjects were simultaneously presented, through one arm of the maze, with water flowing from a container holding 25 siblings and, through the other arm, with water flowing from a container holding 25 members of a second sibship. Neither stimulus group was visible to test subjects, so as in Dawson's (1982) experiments, discrimination was evidently based on non-visual cues. To eliminate the possibility that kinship identity was being communicated through vocal, vibratory, or electrical signals, stimulus water was allowed to drip into intermediary reservoirs before entering the testing apparatus. Tadpoles nonetheless oriented toward their siblings in preference to non-siblings. They ceased to discriminate between siblings and non-siblings when their nares were blocked with a gelatinous paste, but later oriented toward their siblings when the plugs became dislodged. Subjects that were given a choice between water flowing from a container holding their siblings and blank water demonstrated no preference, whereas those presented with water flowing from a non-sibling container and blank water avoided non-siblings. Time spent oriented towards the stimulus groups was less, however, in these tests than in the others.

Dawson's (1982) results suggest that *B. americanus* tadpoles are attracted to labels associated with their siblings, as they approached their siblings in preference to an empty compartment. But subjects failed to demonstrate a preference for water conditioned by their siblings over blank water (Dawson, 1982). This may indicate that attraction to siblings is not entirely mediated by chemical cues. Indeed, Waldman's (1985a) experiments suggest that *B. americanus* tadpoles tend to avoid chemical labels secreted by non-siblings rather than approach labels secreted by siblings. Communication by vocal or vibratory cues was possible in Dawson's initial experiment, but not in these latter two experiments. The formation of sibling schools under natural conditions may result from visually mediated mutual attraction among conspecifics (without regard to genetic relationship) that is coupled with a negative, klinokinetic response to chemical factors released by non-siblings (Waldman, 1985a). In Dawson's testing apparatus, the visual component of the association response may have been stronger than the chemosensory component, leading to a lack of discrimination when visual cues were also available. Possibly the small size of his stimulus groups precluded the transmission of adequate concentrations of chemical cues.

Using a testing protocol similar to Dawson's (1982), Blaustein & O'Hara (1982a) determined that *R. cascadae* tadpoles also show a chemosensory response to cues released by their siblings. Time spent by test subjects near a partitioned compartment holding siblings on one side of a test tank was compared with time spent near a similar compartment holding non-siblings on the other side of the tank. When the partitions consisted of plastic mesh screens, allowing the diffusion of chemical cues, subjects spent significantly more time near their siblings than near non-siblings, regardless of whether visual cues were also available. When the partitions consisted of clear glass panes, allowing visual but no chemical communication, subjects showed no preference for either stimulus group. Hence, unlike Dawson's findings on *B. americanus* larvae, visual cues appear not to affect the sibling association response of *R. cascadae* larvae. Although Blaustein & O'Hara (1982a) did not control for other waterborne cues in their experimental apparatus, they monitored noise levels in their testing tank and failed to detect any sound production.

R. cascadae tadpoles appear to be attracted to chemical labels given off by their siblings rather than repulsed by non-siblings' labels. With plastic mesh partitions in their testing tank, Blaustein & O'Hara (1987) found that subjects spent more time near a compartment holding their siblings than near a compartment that contained no tadpoles. Subjects demonstrated no preference, however, for a compartment holding members of another sibling group over an empty compartment. These results are comparable to those obtained with *B. americanus* tadpoles when subjects were separated from their conspecifics by cheesecloth partitions that permitted communication by waterborne cues but no visual interactions (Dawson, 1982), but differ from those obtained when only chemical cues were available (Waldman, 1985a).

Recent experiments on *B. bufo* tadpoles suggest that sensory aspects of their recognition system are similar, in some respects, to those of *R. cascadae* and *B. americanus* tadpoles. Tested in a tank in which they were separated from their siblings and non-siblings by plastic mesh screening, young *B. bufo* larvae spent more time near a compartment holding their siblings than near one that contained non-siblings (Kiseleva, 1989). Test subjects showed no preference for a compartment holding non-siblings over an empty compartment. Kiseleva's (1989) data are difficult to interpret, however, as (1) the sibling preferences recorded were considerably weaker than those found in tests of other species, (2) the tendency to associate with siblings appeared to dissipate as tadpoles grew older, but this effect was confounded by a change in the number of individuals held in each compartment, and (3) tests for discrimination between siblings and blank water were conducted only on older tadpoles.

Whether kin discrimination is manifested principally as a tendency to approach siblings, or to avoid non-siblings, may depend on the precise ecological circumstances in which close kin associate (Waldman, 1988), and these factors generally have not been addressed in laboratory experiments. Analyses of the interplay between approach and avoidance tendencies might offer insight into the mechanisms by which kin are recognized. For example, depending on the extent of label variability within and among sibships, recognition of siblings by the rejection of non-siblings may represent a more efficient process than direct acceptance of siblings (see Getz, 1982; Waldman, 1987). Because tadpoles are likely to initially approach conspecifics based on visual cues (Wassersug & Hessler, 1971; Wassersug, 1973; O'Hara, 1981), either mechanism could lead to the formation of sibling schools under natural conditions (see discussions in Waldman, 1985a; Blaustein & O'Hara, 1987). Despite their demonstrable chemosensory avoidance of non-siblings in a Y-maze, socially reared *B. americanus* tadpoles swim closer to their siblings in a laboratory pool than would be expected if they moved randomly, but they distribute themselves randomly with respect to non-siblings (Waldman, 1986a). Indiscriminate visual attraction, coupled with the rejection of conspecifics bearing unfamiliar chemical labels, thus can lead to the formation of sibling aggregations.

Label characteristics

The properties of labels that communicate kinship identity have been subject only to cursory examination. In all the experiments just discussed, both stimulus animals and test subjects were removed from their rearing tanks and placed into clean water for tests. Subjects, therefore, responded to factors that were being secreted into the water at the time of testing rather than simply to familiar odours that may have been associated with their rearing environment.

B. americanus recognition labels probably lose their efficacy rapidly if they are not continually replenished in the medium. Water in which sibling groups had been housed for at least 24 hours failed to elicit a discrimination response if it was held for periods of 24–30 hours after the tadpoles were removed (Waldman, 1985a). More tests need to be done to quantify the decreased effectiveness of the stimulus over shorter time intervals. Dawson (1982) noted a decreased responsiveness with time when testing subjects' responses to water conditioned by sibling and non-sibling groups. This may indicate that labels diffuse rapidly, or that subjects become insensitive to them due to sensory accommodation. Boiling of water conditioned by a sibling group followed by reconstitution of the residue eliminates the effectiveness of the stimulus in attracting siblings (Dawson, 1982). For chemical labels to serve effectively as cues for kin discrimination, they should be relatively labile, should have fast fade-out times, and should be effective only over limited active space (Waldman, 1985a). Indeed *B. americanus* labels appear to have these characteristics. In natural conditions, *B. americanus* larvae typically swim with rapid undulatory tail movements (see Beiswenger, 1975). Although this behaviour might serve primarily to stir up food, the swimming patterns may direct the labels toward conspecifics so that they are more rapidly perceived.

Functional approaches to kin recognition

When individuals behaviourally discriminate between kin and non-kin, or between close and distant relatives, we infer that they recognize their kin. Sometimes kin discrimination tendencies evidenced in laboratory tests are inconsistent, even among organisms from single populations (for examples, see Waldman *et al.*, 1988). In other cases, behavioural responses toward kin and non-kin differ consistently, but the differences are small. For example, *R. cascadae* tadpoles, tested for 1200 sec in laboratory tanks, typically remain near siblings for only 50 to 120 sec longer than would be expected if they moved randomly (e.g. Blaustein & O'Hara, 1981; O'Hara & Blaustein, 1981; Blaustein & O'Hara, 1982b). Mean nearest-neighbour distances among *B. americanus* tadpoles released into indoor test pools range from 100 to 140 mm, and on average, tadpoles are 10 to 30 mm closer to siblings than to non-siblings (Waldman, 1981). Presuming that association with kin, specifically, is favoured by natural selection, we might expect tadpoles to more readily sort themselves into sibling groups under those ecological conditions in which selection acts. Indeed, under field conditions, kin-association propensities of both species are stronger (Waldman, 1982b; O'Hara & Blaustein, 1985) than in the corresponding laboratory tests.

The impetus to study kin recognition mechanisms stems ultimately from interest in the functional consequences of behavioural discrimination. Kin discrimination should be favoured in specific ecological and social circumstances (Waldman, 1988), and these conditions are rarely simulated in laboratory experiments. The identification of these conditions entails investigation of the costs and benefits of kin-directed behaviours. In that knowledge of the mechanisms by which organisms recognize their kin increases our ability to predict how their behavioural strategies may be influenced by kin selection or selection for mate choice, mechanistic studies provide a platform for evolutionary analyses of behaviour. But conversely, mechanisms are also influenced by natural selection. Kin recognition can be context-dependent: discrimination in different contexts may or may not come about by a common mechanism. It is unclear, for example, whether the same mechanism that permits tadpoles to preferentially associate with their siblings could come into play to prevent individuals from cannibalizing close relatives (Byers & Bekoff, 1986) – even though indiscriminate cannibals might suffer tremendous losses in inclusive fitness. Organisms may be able to recognize kin through multiple means, and may switch from one mechanism to another in response to ecological and social contingencies (Waldman, 1987). How these factors interact to effect recognition can be clarified only through studies that integrate functional and mechanistic perspectives.

Given the lack of compelling functional explanations for kin discrimination in amphibians, reports purporting to demonstrate their kin recognition abilities understandably elicit some consternation. Parental care, territorial maintenance, and incest avoidance all seem to provide obvious selective benefits, but the mechanisms by which amphibians might recognize kin in these contexts remain largely unexplored. And while the mechanisms by which larval anurans recognize kin have received much attention, the functions of kin discrimination in tadpoles are not understood. Indeed, the possibility that kin recognition in tadpoles represents an epiphenomenon cannot be disregarded (e.g. Waldman, 1984). Kin recognition might be a by-product of immune function, or it could simply be symptomatic of selection for species recognition (Grafen, 1990), to cite just two possibilities.

Why tadpoles recognize their kin has remained a puzzle not because their social behaviour is inherently simple, but because it has been scarcely studied! Several hypotheses regarding the role of kinship in mediating larval social interactions have been proposed, however, and these are now beginning to be tested. Tadpoles that associate with siblings may experience benefits in their abilities to grow, survive, and ultimately gain reproductive success. They may more efficiently utilize available resources, including food and space. They may more effectively deter and respond to predators. And their association with siblings during larval stages might be important after metamorphosis, in the ontogeny of mate choice, for example.

Growth and development

Many amphibians spawn in vernal pools. Because they periodically dry, the pools are devoid of some potential predators (e.g. fishes) of eggs and larvae. The benefits of life in a temporary pond are offset, however, by the risk of desiccation (e.g. Heyer *et al.*, 1975). To survive, larvae must metamorphose before pools dry. But if larvae metamorphose too soon, they may transform at a smaller size than conspecifics that delay metamorphosis.

Survivorship and reproductive success at maturity may be correlated with size at metamorphosis (Berven & Gill, 1983; Smith, 1987; Semlitsch et al., 1988), so that transforming too soon incurs risks comparable to those of transforming too late. Pools are patchy resources, both in space and time. How long a particular pool persists depends on numerous variables, but principally on precipitation, humidity, and temperature – factors that, for us at least, are notoriously difficult to predict. For inhabitants of these pools, determining how quickly to grow and when to emerge are non-trivial problems, and 'optimal' solutions may fluctuate with time.

By schooling with siblings, might individuals more effectively match their life-history patterns to available resources so that they grow more rapidly or metamorphose at a larger size? Many larval anurans are generalized feeders, often grazing on periphyton and detritus. Tadpoles that feed in groups, even with non-relatives, can collectively stir this material from the substrate so that food becomes more accessible to group members. These advantages are enhanced in kin groups. Should the net benefit be non-randomly distributed among group members, as when the largest individuals tend to monopolize resources, other individuals still stand to gain some benefit in terms of inclusive fitness. Beyond this, recent work suggests that anuran larvae regulate their own growth rates in response to those of their close relatives.

Growth rates of *B. americanus* tadpoles vary depending on whether experimental populations consist of kin or non-kin. In laboratory experiments (Waldman, unpub. data), *B. americanus* tadpoles were reared in either 'pure' or 'mixed' groups. Pure groups each consisted of 10 siblings, whereas mixed groups consisted of 10 individuals each from a different sibship. Groups were reared at two density levels: either crowded (in 1-litre containers) or uncrowded (in 10-litre containers). Equal numbers of individuals survived in pure and mixed groups. Under crowded conditions, however, tadpoles in pure groups varied more in size than did those in mixed groups. Some individuals in pure groups had grown to be quite large while others appeared to have ceased growing early in development. The largest individuals in pure groups were significantly larger than the largest individuals in mixed groups. Moreover, mixed groups lacked the small individuals that were found in pure groups. Under uncrowded conditions, variances in sizes within pure and mixed groups were comparable, and extreme size classes were absent from both groups. Indeed, individuals in pure groups in uncrowded environments were significantly larger in size, on average, than those reared in mixed groups.

Genetically similar individuals thus appear to diverge more in their growth trajectories than do genetically dissimilar individuals under intense competitive regimens. This effect is evident in the results of experiments conducted in natural ponds, as in laboratory experiments. Pure groups, consisting of 200 siblings, and mixed groups, consisting of 25 members of each of 8 sibships,

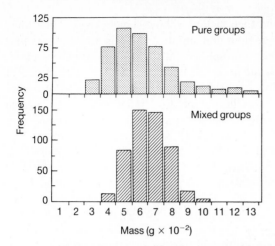

Fig. 7.5 Metamorphic masses of *Bufo americanus* toadlets emerging from field pens consisting of siblings (pure groups) or of groups of eight sibships (mixed groups). The eight sibships were represented in equal numbers in both the pure and mixed groups. Pens of each treatment were interspersed in the pond by a restricted randomization design to control for environmental heterogeneity. Distributions of metamorphic masses differed significantly between groups. Both the largest and the smallest individuals metamorphosed from pure groups. These results suggest that larval growth rates are influenced by the opportunity to school with siblings. (Waldman, unpub. data.)

were released into field enclosures (Waldman, unpub. data). Neither survivorship through metamorphosis, mean time to metamorphosis, nor mean mass at metamorphosis differed between pure and mixed groups, but toadlets emerging from pure pens were more variable in size than those from mixed pens (Fig. 7.5). Again, the largest individuals in pure groups were significantly larger than those in mixed groups. The greater size variance among individuals in pure groups reflects the rapid growth of a few large individuals, but at the apparent expense of a class of smaller, more slowly developing individuals. Contrasts in size variability between pure and mixed groups were particularly pronounced in pens with the highest mortality, presumably because of severe competition.

If *B. americanus* tadpoles school with siblings, they may be able to effectively adjust their growth rates to best utilize available resources. Within a particular sibling cohort, some individuals may initially grow more quickly than others, perhaps as a function of egg size (Crump, 1984; Kaplan & Cooper, 1984; Berven & Chadra, 1988; Williamson & Bull, 1989), and this size variation potentially could become amplified due to interference competition. In good seasons, resources may be sufficient for the entire brood to grow to an optimal size and then metamorphose. Under these conditions, siblings do better on average than non-siblings, and this may indicate that close kin co-operate for their mutual benefit. Co-operation may be manifested as decreased time spent in interference competition, and increased time spent foraging (possibly coupled with food sharing). Yet as resources become limiting and intraspecific competition intensifies, laggards face diminishing odds of surviving to metamorphosis, or of transforming at a viable size. Small individuals then may do best by restraining themselves from competing with their larger siblings, especially if by remaining in the school they can aid their siblings. Runts may, for example, shield their siblings from predators; they may increase their siblings' foraging efficiency by stirring up more food than they consume; or in the severest conditions, they may be cannibalized. Competitive restraint would not be expected, nor is it observed, among groups of non-kin, or when resources are sufficient to support the entire sibling group. Because of the patchiness of the environment, no one genotype is most fit under all conditions (Heyer, 1976). Runts are never eliminated by selection because particular genotypes potentially could develop into runts in some circumstances but into large robust individuals in others. Experiments on *R. sylvatica* larvae suggest that their growth rates are similarly influenced by the genetic relationship of group members and levels of resource availability (Waldman, unpub. data). While these data are consistent with a kin-selection model for schooling, as I have suggested, they also could be interpreted ostensibly as evidence of group selection to ensure genetic variation within populations under conditions of environmental stress (see discussion in Shvarts & Pyastolova, 1970).

Within ponds, larval anurans may be spatially segregated in kin groups even if they do not school. Whether this comes about by a behavioural kin recognition mechanism or because of non-random dispersal, co-operation among kin that enhances survivorship or accelerates growth should be strongly selected. Jasieński (1988) found that nearest-neighbour distances among *Bombina variegata* larvae were the same in genetically mixed and pure groups. Yet mean growth rates of the pure groups were significantly greater than those of the mixed groups, and individuals within pure groups were less variable in size than those within mixed groups (Jasieński, 1988). Similarly, *Pseudacris triseriata* tadpoles in pure groups grew larger than those in mixed groups, but only under conditions of high initial population density (Smith, 1990). In contrast, Shvarts & Pyastolova (1970) found that *Rana arvalis* tadpoles in mixed groups metamorphosed in higher numbers than those in pure groups. Travis (1980) found no difference in either the proportion of individuals metamorphosing or the size at metamorphosis of *Hyla gratiosa* tadpoles reared in mixed and pure groups, in competition with *H. femoralis* tadpoles; but variation within mixed groups was greater than that within pure groups. Apparent interspecific differences evident in these results are difficult to interpret because studies employed different competitive regimens. The results may indicate that kinship effects are differentially expressed in response to levels of intraspecific competition.

Effects of intraspecific competition on the growth of anuran larvae have been well documented, although the effects are complex and in some cases are still poorly understood (Brockelman, 1969; Wilbur & Collins, 1973; Wilbur, 1976; Steinwascher, 1978; Dash & Hota, 1980; Semlitsch & Caldwell, 1982; Woodward, 1987; Murray, 1990). Growth rates, manifested in larval body sizes, vary inversely with population densities, but slowly growing individuals tend to metamorphose at smaller sizes than their larger conspecifics. Crowding effects result both from exploitative competition, as individuals deplete available resources, and from interference competition, as they interact with one another to gain access to these resources. Individuals may interfere with one another actively by physically jostling (Savage, 1952; Gromko *et al.*, 1973; John & Fenster, 1975; Griffith, 1985), or by releasing diffusible chemical regulators into the water that inhibit the growth rates of smaller conspecifics (e.g. Richards, 1958, 1962; Rose, 1960; West, 1960; Akin, 1966; Licht, 1967; Steinwascher, 1979; but see Petranka, 1989a). Shvarts & Pyastolova (1970) found these factors acted with some specificity: *R. arvalis* tadpoles reared in water conditioned by non-siblings grew more quickly than those reared in water conditioned by their siblings. *B. americanus* tadpoles reared in social isolation also showed differential growth responses to waterborne cues released by siblings and non-siblings (Waldman, 1982b, 1986a). Exposure to siblings may either accelerate or inhibit growth, contingent possibly on the probabilities that individuals will successfully metamorphose (Waldman, 1986a).

Responses to predators

The characteristic conspicuousness and distastefulness of *Bufo* larvae were postulated to represent kin-selected responses to predation; consequently, early work on amphibian kin recognition focused on *Bufo* larvae (Waldman & Adler, 1979). Should predators fail to be deterred, however, tadpoles are not necessarily left defenceless. Upon being attacked by predators, some larval anurans warn nearby individuals by releasing an alarm pheromone into the water. The pheromone, stored in epidermal cells, is released only when the skin is injured and cannot be actively secreted into the water. Conspecifics that perceive this cue usually respond by fleeing and forming tighter schools. This form of alarm signalling, analogous to that found in some ostariophysan and gonorhynchiform fishes (Pfeiffer, 1973; Smith, 1977), is especially well developed in *Bufo* larvae (Eibl-Eibesfeldt, 1949; Hrbacek, 1950; Kulzer, 1954; Pfeiffer, 1966, 1974; Beiswenger, 1972; Hews & Blaustein, 1985; Hews, 1988; Petranka, 1989b). Comparable reactions to conspecific skin extracts are shown by larvae of other taxa (e.g.

Scaphiopus holbrooki, Richmond, 1947; *Discoglossus pictus*, Eibl-Eibesfeldt, 1962; *Rana heckscheri*, Altig & Christensen, 1981), but most show no response (e.g. *Xenopus laevis, Alytes obstetricans, Bombina variegata, H. arborea, R. temporaria, R. esculenta*; see Pfeiffer, 1966). Aquatic newts (*Trituris vulgaris* and *T. cristatus*) also respond with a fright reaction to skin extracts prepared from conspecifics (Margolis, 1986). Unlike anuran larvae, however, responses can be elicited even from uninjured newts, suggesting that individuals can secrete pheromones into the water.

The production and delivery of pheromones to warn conspecifics of danger from predators may be favoured by kin selection, especially if signalling individuals are unsuccessful in repelling predators (Williams, 1964), and wounds inflicted by predators are serious (Smith, 1977, 1982, 1986; Waldman & Adler, 1979). Certainly, if close relatives preferentially associate, signalling individuals should accrue inclusive-fitness benefits by inducing their kin to flee. Even if predators follow in pursuit, alerted tadpoles may be able to take evasive manoeuvers to decrease their own risk. Hews (1988) demonstrated that *B. boreas* tadpoles that had been exposed to conspecific skin extract were preyed on less frequently by dragonfly (*Aeshna umbrosa*) naiads than were tadpoles that had been exposed to extract prepared from heterospecifics. These results are attributable to changes in the behaviour of the tadpoles, not the predators (Hews, 1988), and they suggest that association in kin groups can have dramatic functional consequences.

Even in the absence of alarm pheromones, tadpoles conceivably might react differently toward predators when they are in kin groups than they would when interacting with non-relatives. Aggregating as a means of reducing the likelihood of being preyed on can be a purely selfish strategy (Hamilton, 1971), so there is no *a priori* reason to predict that members of kin groups co-ordinate their responses to predators to increase group survivorship. Yet given the consistent formation of kin groups revealed in experimental tests, detailed analyses of movement patterns of tadpoles in kin groups might very well reveal incidents of co-operation among group members that would not be expected in 'selfish herds'. Studies of the feeding strategies of individuals in kin and non-kin groups similarly might reveal behavioural co-operation among siblings not apparent among non-relatives. For example, predator detection might be enhanced in groups, so that individuals can forage with fewer interruptions while maintaining the same level of vigilance (Pulliam, 1973). Social co-operation in responding to predators can be evolutionarily stable among non-relatives, but its evolution is facilitated if individuals form kin groups (Axelrod & Hamilton, 1981).

Inbreeding and optimal outbreeding
Many amphibians are thought to be philopatric. If both sexes return to natal sites to breed, and population sizes are small, individuals that cannot

discriminate between kin and non-kin are likely to mate with their relatives. One possible function of kin recognition in amphibians may be incest avoidance (e.g. Blaustein *et al.*, 1984). Yet few data exist on the functional consequences of inbreeding, whether incestuous or between more distantly related kin.

Guttman & Wilson (1973) interpreted heterozygote deficiencies in breeding populations of toads (*B. americanus*) as resulting from negative heterosis. Pairs showed no evidence of assortative mating; indeed, individuals appeared to mate randomly by genotype (Christein *et al.*, 1979). Selection against heterozygotes has yet to be documented in amphibians (see Samollow, 1980). Rather, the scant evidence available suggests that heterozygosity confers advantages in terms of growth and survivorship. Sizes of larval tiger salamanders (*Ambystoma tigrinum*) collected from natural populations were positively correlated with their levels of heterozygosity at seven protein polymorphisms (Pierce & Mitton, 1982; Mitton *et al.*, 1986). After metamorphosis, heterozygotes also benefitted from a greater aerobic scope for activity (Mitton *et al.*, 1986). Samollow & Soulé (1983) found no evidence of differential survival of homozygotes and heterozygotes within larval *B. boreas* populations, but over winter, levels of heterozygosity (nine protein polymorphisms) among members of the population increased dramatically, indicating that heterozygotes survived in higher numbers. To the extent that inbreeding decreases levels of heterozygosity, it may impose a genetic cost on offspring in variable environments. On the other hand, higher levels of homozygosity that result from inbreeding may be associated with elevated fitness in more homogeneous environments (Dessauer *et al.*, 1975). Decreased levels of heterozygosity that are sometimes found in variable environments may reflect frequent extinctions and recolonizations rather than high levels of inbreeding (Shaffer & Breden, 1989).

Whether amphibians choose their mates in part based on their recognition of close kin is unclear, but so too are the functional consequences of such mate choice. The possible beneficial effects of inbreeding – including genetic adaptations to local environmental conditions, the maintenance of co-adapted gene complexes, and a reduction in risks due to limited dispersal – need to be balanced against its costs (Partridge, 1983). These include inbreeding depression associated with homozygosity, reduced fitness in patchy or variable environments, and increased susceptibility to parasites and pathogens (Hamilton, 1987). Philopatry coupled with active mechanisms for incest avoidance, such as kin recognition, in principle could allow individuals to strike an optimal balance between inbreeding and outbreeding (Shields, 1982; Bateson, 1983). Even if such an optimum exists, however, there is no reason to believe it remains fixed. Under some conditions close inbreeding may represent the best mating strategy, but in other cases individuals that outbreed may have higher fitness. If kin recognition is involved in mate choice, discrimination tendencies might fluctuate in response to local environmental conditions. Inbreeding may be favoured in stable environments, but outbreeding might prove more advantageous in disturbed populations or in stressed conditions.

Conclusions

Following the discovery 10 years ago that larval anurans recognize and preferentially associate with their siblings, much work has focused on the mechanisms by which tadpoles recognize their kin. The ontogeny of kin recognition abilities varies among species, but generally conforms to constraints imposed by larval ecology. Recognition systems of those species that deposit their egg masses communally, for example, appear less plastic than those of species with more variable oviposition strategies. Kinship labels and recognition templates of most species incorporate both environmental and genetic elements. Kin recognition in larval anurans is mediated by the transmission and perception and chemical cues.

Kin discrimination is often context-dependent, but the functional significance of kin association in larvae has only recently begun to be explored. Close kin may co-operate in responding to predators or in gaining access to food, but these effects have yet to be characterized. Kin recognition may also promote some level of outbreeding, or at least incest avoidance, although available data are equivocal on this point. A variety of studies point to a strong effect of kinship on larval growth rates. By associating with their siblings, tadpoles in some conditions may metamorphose more quickly or at a larger size than if they associated with non-kin. This may be extremely important in transient ephemeral pools in which some amphibians breed, but may be less important for species whose larvae inhabit permanent pools. To date, kin recognition abilities of species that breed in these more stable habitats have not been examined. If a major benefit of kin association is accelerating growth, kin recognition abilities may be less robust in these species.

Despite ample evidence that kin recognition influences the social behaviour of tadpoles, little is known about the kin recognition abilities of anurans in other contexts, such as territoriality or parental care. Even less is known about the kin recognition abilities of either urodeles or caecilians. In part this reflects a general scarcity of information on amphibian social behaviour. Given the large number of anurans that can recognize their kin, at least as larvae, the possibility that other amphibians also recognize their relatives merits consideration.

Acknowledgments

I thank A. R. Blaustein, G. J. Gamboa, C. J. Marshall, M. Schindlinger, K. Summers and R. J. Wassersug for comments on the manuscript, and K. D. Wells for making available an unpublished manuscript. My research has been supported by the

National Science Foundation (BSR-8717665, BNS-8820043, DIR-8901004 and BSR-9007760).

References

Akin, G. C. (1966). Self-inhibition of growth in *Rana pipiens* tadpoles. *Physiol. Zool.*, **39**, 341–56.

Alexander, R. D. (1979). *Darwinism and Human Affairs*. Seattle: University of Washington Press.

Altig, R. & Christensen, M. T. (1981). Behavioral characteristics of the tadpoles of *Rana heckscheri. J. Herpetol.*, **15**, 151–4.

Arnold, S. J. (1976). Sexual behavior, sexual interference and sexual defense in the salamanders *Ambystoma maculatum, Ambystoma tigrinum* and *Plethodon jordani. Z. Tierpsychol.*, **42**, 247–300.

Axelrod, R. & Hamilton, W. D. (1981). The evolution of co-operation. *Science*, **211**, 1390–6.

Ayre, D. J. (1982). Inter-genotype aggression in the solitary sea anemone *Actinia tenebrosa. Mar. Biol.*, **68**, 199–205.

Ayre, D. J. (1987). The formation of clonal territories in experimental populations of the sea anemone *Actinia tenebrosa. Biol. Bull.*, **172**, 178–86.

Balinsky, B. I. & Balinsky, J. B. (1954). On the breeding habits of the South African bullfrog, *Pyxicephalus adspersus. S. Afr. J. Sci.*, **51**, 55–8.

Barthalmus, G. T. & Bellis, E. D. (1972). Home range, homing and the homing mechanism of the salamander, *Desmognathus fuscus. Copeia*, **1972**, 632–42.

Bateson, P. (1979). How do sensitive periods arise and what are they for? *Anim. Behav.*, **27**, 470–86.

Bateson, P. (1982). Preferences for cousins in Japanese quail. *Nature*, **295**, 236–7.

Bateson, P. (1983). Optimal outbreeding. In *Mate Choice*, ed. P. Bateson, pp. 257–77. Cambridge: Cambridge University Press.

Beauchamp, G., Gilbert, A., Yamazaki, K. & Boyse, E. A. (1986). Genetic basis for individual discriminations: the major histocompatibility complex of the mouse. In *Chemical Signals in Vertebrates* 4. *Ecology, Evolution, and Comparative Biology*, ed. D. Duvall, D. Müller-Schwarze & R. M. Silverstein, pp. 413–22. New York: Plenum Press.

Beecher, M. D. (1988). Kin recognition in birds. *Behav. Genet.*, **8**, 465–82.

Beiswenger, R. E. (1972). Aggregative behavior of tadpoles of the American toad, *Bufo americanus*, in Michigan. Ph.D. Thesis, University of Michigan.

Beiswenger, R. E. (1975), Structure and function in aggregations of tadpoles of the American toad, *Bufo americanus. Herpetologica*, **31**, 222–23.

Beiswenger, R. E. (1977). Diel patterns of aggregative behavior in tadpoles of *Bufo americanus*, in relation to light and temperature. *Ecology*, **58**, 98–108.

Berven, K. A. & Chadra, B. G. (1988). The relationship among egg size, density and food level on larval development in the wood frog (*Rana sylvatica*). *Oecologia*, **75**, 67–72.

Berven, K. A. & Gill, D. E. (1983). Interpreting geographic variation in life history traits. *Am. Zool.*, **23**, 85–97.

Berven, K. A. & Grudzien T. A. (1990). Dispersal in the wood frog (*Rana sylvatica*): implications for genetic population structure. *Evolution.* (In Press.)

Blaustein, A. R. (1983). Kin recognition mechanisms: phenotypic matching or recognition alleles? *Am. Nat.*, **121**, 749–54.

Blaustein, A. R. (1988). Ecological correlates and potential functions of kin recognition and kin association in anuran larvae. *Behav. Genet.*, **18**, 449–64.

Blaustein, A. R., Bekoff, M. & Daniels, T. J. (1987). Kin recognition in vertebrates (excluding primates): empirical evidence. In *Kin Recognition in Animals*, ed. D. J. C. Fletcher & C. D. Michener, pp. 287–331. Chichester: John Wiley & Sons.

Blaustein, A. R. & O'Hara, R. K. (1981). Genetic control for sibling recognition? *Nature*, **290**, 246–248.

Blaustein, A. R. & O'Hara, R. K. (1982a). Kin recognition cues in *Rana cascadae* tadpoles. *Behav. Neural Biol.*, **36**, 77–87.

Blaustein, A. R. & O'Hara, R. K. (1982b). Kin recognition in *Rana cascadae* tadpoles: maternal and paternal effects. *Anim. Behav.*, **30**, 1151–7.

Blaustein, A. R. & O'Hara, R. K. (1983). Kin recognition in *Rana cascadae* tadpoles: effects of rearing with nonsiblings and varying the strength of the stimulus cues. *Behav. Neural Biol.*, **9**, 259–67.

Blaustein, A. R. & O'Hara, R. K. (1986). An investigation of kin recognition in red-legged frog (*Rana aurora*) tadpoles. *J. Zool., Lond. A*, **209**, 347–53.

Blaustein, A. R. & O'Hara, R. K. (1987). Aggregation behaviour in *Rana cascadae* tadpoles: association preferences among wild aggregations and responses to non-kin. *Anim. Behav.*, **35**, 1549–55.

Blaustein, A. R., O'Hara, R. K. & Olson, D. H. (1984). Kin preference behaviour is present after metamorphosis in *Rana cascadae* frogs. *Anim. Behav.*, **32**, 445–50.

Bogert, C. M. (1947). A field study of homing in the Carolina toad. *Am. Mus. Novit.*, **1355**, 1–24.

Bragg, A. N. (1965). *Gnomes of the Night. The Spadefoot Toads.* Philadelphia: University of Pennsylvania Press.

Branch, L. C. (1983). Social behavior of the tadpoles of *Phyllomedusa vaillanti*. *Copeia*, **1983**, 420–8.

Brattstrom, B. H. (1962). Thermal control of aggregation behavior in tadpoles. *Herpetologica*, **18**, 38–46.

Breden, F. (1987). The effect of post-metamorphic dispersal on the population genetic structure of Fowler's toad, *Bufo woodhousei fowleri*. *Copeia*, **1987**, 386–95.

Breden, F., Lum, A. & Wassersug, R. (1982). Body size and orientation in aggregates of toad tadpoles *Bufo woodhousei*. *Copeia*, **1982**, 672–80.

Breed, M. D., Williams, K. R. & Fewell, J. H. (1988). Comb wax mediates the acquisition of nest-mate recognition cues in honey bees. *Proc. Natl. Acad. Sci. USA*, **85**, 8766–9.

Briggs, J. L. Sr. (1987). Breeding biology of the Cascade frog, *Rana cascadae*, with comparisons to *R. aurora* and *R. pretiosa*. *Copeia*. **1987**, 241–5.

Brockelman, W. Y. (1969). An analysis of density effects and predation in *Bufo americanus* tadpoles. *Ecology*, **50**, 632–44.

Brodie, E. D. Jr. & Formanowicz, D. R. Jr. (1987). Antipredator mechanisms of larval anurans: protection of palatable individuals. *Herpetologica*, **43**, 369–73.

Brown, J. L. (1983). Some paradoxical goals of cells and organisms: the role of the MHC. In *Ethical Questions in Brain and Behavior*, ed D. W. Pfaff, pp. 111–24. New York: Springer-Verlag.

Brown, R. E., Singh, P. B. & Roser, B. (1987). The major histocompatibility complex and the chemosensory recognition of individuality in rats. *Physiol. Behav.*, **40**, 65–73.

Buss, L. W. (1982). Somatic cell parasitism and the evolution of somatic tissue compatibility. *Proc. Natl. Acad. Sci. USA*, **79**, 5337–41.

Butcher, G. W. & Howard, J. C. (1986). The MHC of the laboratory rat, *Rattus norvegicus*. In *Handbook of Experimental Immunology*, vol. 3, 4th edn., ed. by D. M. Weir, pp. 101.1–101.18. Oxford: Blackwell.

Byers, J. A. & Bekoff, M. (1986). What does 'kin recognition' mean? *Ethology*, **72**, 342–5.

Caldwell, J. P. (1989). Structure and behavior of *Hyla geographica* tadpole schools, with comments on classification of group behavior in tadpoles. *Copeia*, **1989**, 938–50.

Carlin, N. F. & Hölldobler, B. (1986). The kin recognition system of carpenter ants (*Camponotus* spp.). I. Hierarchial cues in small colonies. *Behav. Ecol. Sociobiol.*, **19**, 123–34.

Carlin, N. F. & Hölldobler, B. (1987). The kin recognition system of carpenter ants (*Camponotus* spp.). II. Larger colonies. *Behav. Ecol. Sociobiol.*, **20**, 209–17.

Carpenter, C. C. (1953). Aggregation behavior of tadpoles of *Rana p. pretiosa*. *Herpetologica*, **9**, 77–8.

Christein, D. R., Guttman, S. I. & Taylor, D. H. (1979). Heterozygote deficiencies in a breeding population of *Bufo americanus* (Bufonidae: Anura): the test of a hypothesis. *Copeia*, **1979**, 498–502.

Collins, J. P. & Cheek, J. E. (1983). Effect of food and density on development of typical and cannibalistic salamander larvae in *Ambystoma tigrinum nebulosum*. *Am. Zool.*, **23**, 77–84.

Cooke, A. S. (1974). Differential predation by newts on anuran tadpoles. *Brit. J. Herpetol.*, **5**, 386–90.

Cornell, T. J., Berven, K. A. & Gamboa, G. J. (1989). Kin recognition by tadpoles and froglets of the wood frog *Rana sylvatica*. *Oecologia*, **78**, 312–16.

Crozier, R. H. (1987). Genetic aspects of kin recognition: concepts, models, and synthesis. In *Kin Recognition in Animals*, ed. D. J. C. Fletcher & C. D. Michener, pp. 55–73. Chichester: John Wiley & Sons.

Crump, M. L. (1984). Intraclutch egg size variability in *Hyla crucifer* (Anura: Hylidae). *Copeia*, **1984**, 302–8.

Dash, M. C. & Hota, A. K. (1980). Density effects on the survival, growth rate, and metamorphosis of *Rana tigrina* tadpoles. *Ecology*, **61**, 1025–28.

Dausset, J. (1981). The major histocompatibility complex in man. *Science*, **213**, 1469–74.

Davis, M. S. (1987). Acoustically mediated neighbor recognition in the North American bullfrog. *Rana catesbeiana*. *Behav. Ecol. Sociobiol.*, **21**, 185–90.

Dawson. J. T. (1982). *Kin Recognition and Schooling in the American Toad* (Bufo americanus). Ph.D. Thesis, State University of New York, Albany.

Dessauer, H. C., Nevo, E. & Chaung, K.-C. (1975). High genetic variability in an ecologically variable vertebrate, *Bufo viridis*. *Biochem. Genet.*, **13**, 651–61.

Du Pasquier, L., Chardonnens, X. & Miggiano, V. C. (1975). A major histocompatibility complex in the toad *Xenopus laevis* (Daudin). *Immunogenetics*, **1**, 482–94.

Du Pasquier, L., Schwager, J. & Flajnik, M. F. (1989). The immune system of *Xenopus*. *Ann. Rev. Immunol.*, **7**, 251–75.

Ducey, P. K. & Ritsema, P. (1988). Intraspecific aggression and responses to marked substrates in *Ambystoma maculatum* (Caudata: Ambystomatidae). *Copeia*, **1988**, 1008–13.

Duellman, W. E. (1978). The biology of an equatorial herpetofauna in Amazonian Ecuador. *Misc. Publ. Univ. Kansas Mus. Nat. Hist.*, **65**, 1–352.

Duellman, W. E. & Lescure, J. (1973). Life history and ecology of the hylid frog *Osteocephalus taurinus*, with observations on larval behavior. *Occas. Pap. Mus. Nat. Hist. Univ. Kansas*, **13**, 1–12.

Duellman, W. E. & Trueb, L. (1986). *Biology of Amphibians*. New York: McGraw-Hill.

Dunlap, D. G. & Satterfield, C. K. (1982). Habitat selection in larval anurans: early experience and substrate pattern selection in *Rana pipiens*. *Develop. Psychobiol.*, **18**, 37–58.

Egid, K. & Brown, J. L. (1989). The major histocompatibility complex and female mating preferences in mice. *Anim. Behav.*, **38**, 548–50.

Eibl-Eibesfeldt, I. (1949). Über das Vorkommen von Schreckstoffen bei Erdkröten-quappen. *Experientia*, **5**, 236.

Eibl-Eibesfeldt, I. (1962). Die Verhaltensentwicklung des Krallenfrosches (*Xenopus laevis*) und des Scheibenzünglers (*Discoglossus pictus*) unter besonderer Berück-sichtigung der Beutefanghandlungen. *Z. Tierpsychol.*, **19**, 385–93.

Ewert, M. (1969). *Seasonal Movements of the Toads Bufo americanus and B. cognatus in Northwestern Minnesota*. Ph.D. Thesis, University of Minnesota.

Fisher, R. A. (1958). *The Genetical Theory of Natural Selection*. New York: Dover.

Fishwild, T. G., Schemidt, R. A., Jankens, K. M., Berven, K. A., Gamboa, G. J. & Richards, C. M. (1990). Sibling recognition by larval frogs (*Rana pipiens, R. sylvatica* and *Pseudacris crucifer*). *J. Herpetol.*, **24**, 40–4.

Flajnik, M. F., Du Pasquier, L. & Cohen, N. (1985). Immune responses of thymus/lymphocyte embryonic chimeras: studies on tolerance and major histocompati-bility restriction in *Xenopus*. *Eur. J. Immunol.*, **15**, 540–7.

Forester, D. C. (1979). Homing to the nest by female mountain dusky salamanders (*Desmognathus ochrophaeus*) with comments on the sensory modalities essential to clutch recognition. *Herpetologica*, **35**, 330–5.

Forester, D. C., Harrison, K. & McCall, L. (1983). The effects of isolation, the duration of breeding, and non-egg olfactory cues on clutch recognition by the salamander, *Desmognathus ochrophaeus*. *J. Herpetol.*, **17**, 308–14.

Foster, M. S. & McDiarmid, R. W. (1982). Study of aggregative behavior of *Rhinophrynus dorsalis* tadpoles: design and analysis. *Herpetologica*, **38**, 395–404.

Francis, L. (1973a). Clone specific segregation in the sea anemone *Anthopleura elegantissima*. *Biol. Bull.*, **144**, 64–72.

Francis, L. (1973b). Intraspecific aggression and its effect on the distribution of *Anthopleura elegantissima* and some related sea anemones. *Biol. Bull.*, **144**, 73–92.

Francis, L. (1976). Social organization within clones of the sea anemone *Anthopleura elegantissima*. *Biol. Bull.*, **150**, 361–76.

Francis, L. (1988). Cloning and aggression among sea anemones (Coelenterata: Actiniaria) of the rocky shore. *Biol. Bull.*, **174**, 241–53.

Fraser, D. F. (1976). Empirical evaluation of the hypothesis of food competition in salamanders of the genus *Plethodon*. *Ecology*, **57**, 459–71.

Gamboa, G. J., Reeve, H. K. & Pfennig, D. W. (1986). The evolution and ontogeny of nestmate recognition in social wasps. *Ann. Rev. Entomol.*, **31**, 431–54.

Getz, W. M. (1981). Genetically based kin recognition systems. *J. Theor. Biol.*, **92**, 209–26.

Getz, W. M. (1982). An analysis of learned kin recognition in Hymenoptera. *J. Theor. Biol.*, **99**, 585–97.

Grafen, A. (1985). A geometric view of relatedness. *Oxford Surv. Evol. Biol.*, **2**, 28–89.

Grafen, A. (1990). Do animals really recognize kin? *Anim. Behav.*, **39**, 42–54.

Grandison, A. G. C. & Ashe, S. (1983). The distribution, behavioural ecology and breeding strategy of the pygmy toad, *Mertensophryne micranotis* (Lov.). *Bull. Brit. Mus. Nat. Hist. (Zool.)*, **45**, 85–93.

Green, S. & Marler, P. (1979). The analysis of animal communication. In *Handbook of Behavioral Neurobiology*. Vol. 3, *Social Behavior and Communication*, ed. P. R. Marler & J. G. Vandenbergh, pp. 73–158. New York: Plenum Press.

Greenberg, L. (1979). Genetic component of bee odor in kin recognition. *Science*, **206**, 1095–7.

Greenberg, L. (1988). Kin recognition in the sweat bee, *Lasioglossum zephyrum*. *Behav. Genet.*, **18**, 425–38.

Griffith, D. C. (1985). Effects of density and environmental stimuli on aggregative behavior and growth in *Bufo americanus* tadpoles. M. S. Thesis, Michigan State University.

Griffiths, R. A. (1985). Diel pattern of movement and aggregation in tadpoles of the common frog, *Rana temporaria*. *Herpetol. J.*, **1**, 10–13.

Griffiths, R. A., Getliff, J. M. & Mylotte, V. J. (1988). Diel patterns of activity and vertical migration in tadpoles of the common toad, *Bufo bufo*. *Herpetol. J.*, **1**, 223–6.

Gromko, M. H., Mason, F. S. & Smith-Gill, S. J. (1973). Analysis of the crowding effect in *Rana pipiens* tadpoles. *J. Exp. Zool.*, **186**, 63–72.

Grosberg, R. K. (1988). The evolution of allorecognition specificity in clonal invertebrates. *Q. Rev. Biol.*, **63**, 377–412.

Grosberg, R. K. & Quinn, J. F. (1986). The genetic control and consequences of kin recognition by the larvae of a colonial marine invertebrate. *Nature*, **322**, 456–9.

Grosberg, R. K. & Quinn, J. F. (1988). The evolution of allorecognition specificity. In *Invertebrate Historecognition*, ed. R. K. Grosberg, D. Hedgecock & K. Nelson), pp. 157–67. New York: Plenum Press.

Gross, M. R. & Shine, R. (1981). Parental care and mode of fertilization in ectothermic vertebrates. *Evolution*, **35**, 775–93.

Grubb, J. C. (1973). Olfactory orientation in breeding Mexican toads, *Bufo valliceps*. *Copeia*, **1973**, 490–7.

Guilford, T. (1988). The evolution of conspicuous coloration. *Am Nat.*, **131**, S7–21.

Guttman, S. I. & Wilson, K. G. (1973). Genetic variation in the genus *Bufo* I. An extreme degree of transferrin and albumin polymorphism in a population of the American toad (*Bufo americanus*). *Biochem. Genet.*, **8**, 329–40.

Halliday, T. R. & Verrell, P. A. (1984). Sperm competition in amphibians. In *Sperm Competition and the Evolution of Animal Mating Systems*, ed. R. L. Smith, pp. 487–508. New York: Academic Press.

Halliday, T. R. & Slater, P. J. B. (Eds). (1983). *Animal Behaviour*. Vol. 2, *Communication*. New York: W. H. Freeman.

Hamilton, W. D. (1964). The genetical evolution of social behavior. I, II. *J. Theor. Biol.*, **7**, 1–52.

Hamilton, W. D. (1971). Geometry for the selfish herd. *J. Theor. Biol.*, **31**, 295–311.

Hamilton, W. D. (1987). Kinship, recognition, disease, and intelligence: constraints of social evolution. In *Animal Societies: Theories and Facts*, ed. Y. Itô, J. L. Brown & J. Kikkawa, pp. 81–102. Tokyo: Japan Science Society Press.

Harvey, P. H., Bull, J. J., Pemberton, M. & Paxton, R. J. (1982). The evolution of aposematic coloration in distasteful prey: a family model. *Am. Nat.*, **119**, 710–19.

Hepper, P. G. (1983). Sibling recognition in the rat. *Anim. Behav.*, **31**, 1177–91.

Hepper, P. G. (1986). Kin recognition: functions and mechanisms. A review. *Biol. Rev.*, **61**, 63–93.

Hepper, P. G. (1987). The discrimination of different degrees of relatedness in the rat: evidence for a genetic identifier? *Anim. Behav.*, **35**, 549–54.

Heusser, H. (1969). Die Lebensweise der Erdkröte, *Bufo bufo* (L.); Das Orientierungs-problem. *Rev. Suisse Zool.*, **76**, 443–518.

Heusser, H. (1971). Differenzierendes Kaulquappen-Fressen durch Molche. *Experientia*, **27**, 475.

Hews, D. K. (1988). Alarm response in larval western toads, *Bufo boreas*: release of larval chemicals by a natural predator and its effect on predator capture efficiency. *Anim. Behav.*, **36**, 125–33.

Hews, D. K. & Blaustein, A. R. (1985). An investigation of the alarm response in *Bufo boreas* and *Rana cascadae* tadpoles. *Behav. Neural Biol.*, **43**, 47–57.

Heyer, W. R. (1976). Studies in larval amphibian habitat partitioning. *Smithsonian Contrib. Zool.*, **242**, 1–27.

Heyer, W. R., McDiarmid, R. W. & Weigmann, D. L. (1975). Tadpoles, predation and pond habitats in the tropics. *Biotropica*, **7**, 100–11.

Hölldobler, B. & Carlin, N. F. (1987). Anonymity and specificity in the chemical communication signals of social insects. *J. Comp. Physiol. A*, **161**, 567–81.

Holmes, W. G. & Sherman, P. W. (1982). The ontogeny of kin recognition in two species of ground squirrels. *Am. Zool.*, **22**, 491–517.

Horne, E. A. & Jaeger, R. G. (1988). Territorial pheromones of female red-backed salamanders. *Ethology*, **78**, 143–52.

Houck, L. D. & Schwenk, K. (1984). The potential for long-term sperm competition in a plethodontid salamander. *Herpetologica*, **40**, 410–15.

Houck, L. D., Tilley, S. G. & Arnold, S. J. (1985). Sperm competition in a plethodontid salamander: preliminary results. *J. Herpetol.*, **19**, 420–3.

Howard, R. D. (1988). Sexual selection on male body size and mating behaviour in American toads, *Bufo americanus*. *Anim. Behav.* **36**, 1796–808.

Hrbacek, J. (1950). On the flight reaction of tadpoles of the common toad caused by chemical substances. *Experientia*, **6**, 100–1.

Jaeger, R. G. (1984). Agonistic behavior of the red-backed salamander. *Copeia*, **1984**, 309–14.

Jaeger, R. G. (1986). Pheromonal markers as territorial advertisement by terrestrial salamanders. In *Chemical Signals in Vertebrates* 4. *Ecology, Evolution, and Comparative Biology*, ed. D. Duvall, D. Muller-Schwarze & R. M. Silverstein, pp. 191–203. New York: Plenum Press.

Jaeger, R. G. & Gergits, W. F. (1979). Intra- and interspecific communication in salamanders through chemical signals on the substrate. *Anim. Behav.*, **27**, 150–6.

Jaeger, R. G., Goy, J. M., Tarver, M. & Márquez, C. E. (1986). Salamander territoriality: pheromonal markers as advertisement by males. *Anim. Behav.*, **34**, 860–4.

Jaeger, R. G., Kalvarsky, D. & Shimizu, N. (1982). Territorial behaviour of the red-backed salamander: expulsion of intruders. *Anim. Behav.*, **30**, 490–6.

Jameson, D. L. (1957). Population structure and homing responses in the Pacific tree frog. *Copeia*, **1957**, 221–8.

Jasieński, M. (1988). Kinship ecology of competition: size hierarchies in kin and nonkin laboratory cohorts of tadpoles. *Oecologia*, **77**, 407–13.

John, K. R. & Fenster, D. (1975). The effects of partitions on the growth rates of crowded *Rana pipiens* tadpoles. *Am. Midl. Nat.*, **93**, 123–30.

Juterbock, J. E. (1987). The nesting behavior of the dusky salamander, *Desmognathus fuscus*. II. Nest site tenacity and disturbance. *Herpetologica*, **43**, 361–8.

Kaplan, R. H. & Cooper, W. S. (1984). The evolution of developmental plasticity in reproductive characteristics: an application of the 'adaptive coin-flipping' principle. *Am. Nat.*, **123**, 393–410.

Kats, L. B., Petranka, J. W. & Sih, A. (1988). Antipredator defenses and the persistence of amphibian larvae with fishes. *Ecology*, **69**, 1865–70.

Katz, L. C., Potel, M. J. & Wassersug, R. J. (1981). Structure and mechanisms of schooling in tadpoles of the clawed frog, *Xenopus laevis*. *Anim. Behav.*, **29**, 20–33.

Kiseleva, E. I. (1989). Reactions of *Bufo bufo* L. toad tadpoles to chemical signals of individuals of the same and different species. *Zh. Obshch. Biol.*, **50**, 794–8.

Klein, J. (1986). *Natural History of the Major Histocompatibility Complex*. New York: John Wiley & Sons.

Kluge, A. G. (1981). The life history, social organization, and parental behavior of *Hyla rosenbergi* Boulenger, a nest-building gladiator frog. *Misc. Pub. Mus. Zool. Univ. Mich.*, **160**, 1–170.

Kok, D., du Preez, L. H. & Channing, A. (1989). Channel construction by the African bullfrog: another anuran parental care strategy. *J. Herpetol*, **23**, 435–7.

Kruse, K. C. & Stone, B. M. (1984). Largemouth bass (*Micropterus salmoides*) learn to avoid feeding on toad (*Bufo*) tadpoles. *Anim. Behav.*, **32**, 1035–9.

Kulzer, E. (1954). Untersuchungen über die Schreckreaktion der Erdkrötenkaulquappen (*Bufo bufo* L.). *Z. vergl. Physiol.*, **36**, 443–63.

Lacy, R. C. & Sherman, P. W. (1983). Kin recognition by phenotype matching. *Am. Nat.*, **121**, 489–512.

Lambiris, A. J. L. (1971). A note on 'parental care' in the bullfrog, *Pyxicephalus a. adspersus*. *J. Herp. Assoc. Africa*, **8**, 6.

Lamotte, M. & Lescure, J. (1977). Tendances adaptatives a l'affranchissement du milieu aquatique chez les amphibiens anoures. *Terre Vie*, **31**, 225–311.

Landreth, H. F. & Ferguson, D. E. (1967). Newts: sun-compass orientation. *Science*, **158**, 1459–61.

Lannoo, M. J., Townsend, D. S. & Wassersug, R. J. (1987). Larval life in the leaves: arboreal tadpoles types, with special attention to the morphology, ecology, and behavior of the oophagous *Osteopilus brunneus* (Hylidae) larva. *Fieldiana, Zool.*, **1381**, 1–31.

Lescure, J. (1968). Le comportement social des Batraciens. *Rev. Comp. Anim.*, **2**, 1–33.

Licht, L. E. (1967). Growth inhibition in crowded tadpoles: intraspecific and interspecific effects. *Ecology*, **48**, 736–45.

Linsenmair, K. E. (1987). Kin recognition in subsocial arthropods, in particular in the desert isopod *Hemilepistus reaumuri*. In *Kin Recognition in Animals*, ed. D. J. C. Fletcher & C. D. Michener, pp. 121–208. Chichester: John Wiley & Sons.

Madison, D. M. (1969). Homing behaviour of the red-cheeked salamander, *Plethodon jordani. Anim. Behav.*, **17**, 25–39.

Madison, D. M. (1975). Intraspecific odor preferences between salamanders of the same sex: dependence on season and proximity of residence. *Can. J. Zool.*, **53**, 1356–61.

Margolis, S. E. (1986). Behavioral reactions of newts to substances contained in the skin of conspecific individuals. *J. Evol. Biochem. Physiol.*, **21**, 174–8.

Marler, P. (1967). Animal communication signals. *Science*, **157**, 769–74.

Martins, M. & Haddad, C. F. B. (1988). Vocalizations and reproductive behaviour in the smith frog, *Hyla faber* Wied (Amphibia: Hylidae). *Amphibia-Reptilia*, **9**, 49–60.

McDiarmid, R. W. (1978). Evolution of parental care in frogs. In *The Development of Behavior: Comparative and Evolutionary Aspects*, ed. G. M. Burghardt & M. Bekoff, pp. 127–47. New York: Garland STPM Press.

McGavin, M. (1978). Recognition of conspecific odors by the salamander *Plethodon cinereus. Copeia*, **1978**, 356–8.

McKinnell, R. G. (1978). *Cloning. Nuclear Transplantation in Amphibia*. Minneapolis: University of Minnesota Press.

Michener, C. D. & Smith, B. H. (1987). Kin recognition in primitively eusocial insects. In *Kin Recognition in Animals*, ed. D. J. C. Fletcher & C. D. Michener, pp. 209–42. Chichester: John Wiley & Sons.

Mitton, J. B., Carey, C. & Kocher, T. D. (1986). The relation of enzyme heterozygosity to standard and active oxygen consumption and body size of tiger salamanders, *Ambystoma tigrinum. Physiol. Zool.*, **59**, 574–82.

Murray, D. L. (1990). The effects of food and density on growth and metamorphosis in larval wood frogs (*Rana sylvatica*) from central Labrador. *Can. J. Zool.* **68**, 1221–6.

Myers, C. W., Daly, J. W. & Martinez, V. (1984). An arboreal poison frog (*Dendrobates*) from western Panama. *Am. Mus. Novit.*, **2783**, 1–20.

Neigel, J. E. & Avise, J. C. (1983). Clonal diversity and population structure in a reef-building coral, *Acropora cervicornis*: self-recognition analysis and demographic interpretation. *Evolution*, **37**, 437–53.

Neigel, J. E. & Avise, J. C. (1985). The precision of histocompatibility response in clonal recognition in tropical marine sponges. *Evolution*, **39**, 724–32.

Noland, R. & Ultsch, G. R. (1981). The roles of temperature and dissolved oxygen in microhabitat selection by the tadpoles of a frog (*Rana pipiens*) and a toad (*Bufo terrestris*). *Copeia*, **1981**, 645–52.

Nussbaum, R. A. (1985). The evolution of parental care in salamanders. *Misc. Publ. Mus. Zool. Univ. Mich.*, **169**, 1–50.

Nussbaum, R. A. (1987). Parental care and egg size in salamanders: an examination of the safe harbor hypothesis. *Res. Popul. Ecol.*, **9**, 27–44.

Nussbaum, R. A., Brodie, E. D. Jr., & Storm, R. M. (1983). *Amphibians and Reptiles of the Pacific Northwest*. Moscow: University Press of Idaho.

O'Hara, R. K. (1981). Habitat selection behavior in three species of anuran larvae:

environmental cues, ontogeny, and adaptive significance. Ph.D. Thesis, Oregon State University.

O'Hara, R. K. & Blaustein, A. R. (1981). An investigation of sibling recognition in *Rana cascadae* tadpoles. *Anim. Behav.*, **29**, 1121–6.

O'Hara, R. K. & Blaustein, A. R. (1982). Kin preference behavior in *Bufo boreas* tadpoles. *Behav. Ecol. Sociobiol.*, **11**, 43–9.

O'Hara, R. K. & Blaustein, A. R. (1985). *Rana cascadae* tadpoles aggregate with siblings: an experimental field study. *Oecologia*, **67**, 44–51.

O'Hara, R. K. & Blaustein, A. R. (1988). *Hyla regilla* and *Rana pretiosa* tadpoles fail to display kin recognition behaviour. *Anim. Behav.*, **36**, 946–8.

Oldham, R. S. (1966). Spring movements in the American toad, *Bufo americanus*. *Can. J. Zool.*, **44**, 63–100.

Olsén, K. H. (1989). Sibling recognition in juvenile Arctic charr, *Salvelinus alpinus* (L.). *J. Fish Biol.*, **34**, 571–81.

Partridge, L. (1983). Non-random mating and offspring fitness. In *Mate Choice*, ed. P. Bateson, pp. 227–55. Cambridge: Cambridge University Press.

Petranka, J. W. (1989a). Chemical interference competition in tadpoles: does it occur outside laboratory aquaria? *Copeia*, **1989**, 921–30.

Petranka, J. W. (1989b). Response of toad tadpoles to conflicting chemical stimuli: predator avoidance versus 'optimal' foraging. *Herpetologica*, **45**, 283–92.

Pfeiffer, W. (1966). Die Verbreitung der Schreckreaktion bei Kaulquappen und die Herkunft des Schreckstoffes. *Z. vergl. Physiol.*, **52**, 79–98.

Pfeiffer, W. (1973). Alarm substances. *Experientia*, **19**, 113–23.

Pfeiffer, W. (1974). Pheromones in fish and amphibia. In *Pheromones*, ed. M. C. Birch, pp. 269–96. Amsterdam: North-Holland.

Pfennig, D. W. (1990). 'Kin recognition' among spadefoot toad tadpoles: a side-effect of habitat selection? *Evolution*, **44**, 785–98.

Pierce, B. A. & Mitton, J. B. (1982). Allozyme heterozygosity and growth in the tiger salamander, *Ambystoma tigrinum*. *J. Hered.*, **73**, 250–3.

Pitcher, T. J. (1986). The functions of shoaling behaviour. In *The Behaviour of Teleost Fishes*, ed. T. J. Pitcher, pp. 294–337. London: Croom Helm.

Porter, R. H., McFadyen-Ketchum, S. A. & King, G. A. (1989). Underlying bases of recognition signatures in spiny mice, *Acomys cahirinus*. *Anim. Behav.*, **37**, 638–44.

Power, J. H. (1926). Notes on the habits and life-histories of certain little-known Anura, with descriptions of the tadpoles. *Trans. Royal Soc. S. Africa*, **13**, 107–18.

Poynton, J. C. (1957). Bullfrog guardians. *Afr. Wildlife*, **11**, 80–1.

Pulliam, H. R. (1973). On the advantages of flocking. *J. Theor. Biol.*, **38**, 419–22.

Punzo, F. (1976). The effects of early experience on habitat selection in tadpoles of the Malayan painted frog, *Kaloula pulchra* (Anura: Microhylidae). *J. Bombay Nat. Hist. Soc.*, **73**, 270–7.

Quinn, T. P. & Busack, C. A. (1985). Chemosensory recognition of siblings in juvenile coho salmon (*Oncorhynchus kisutch*). *Anim. Behav.*, **33**, 51–6.

Reeve, H. K. (1989). The evolution of conspecific acceptance thresholds. *Am. Nat.*, **133**, 407–35.

Richards, C. M. (1958). The inhibition of growth in crowded *Rana pipiens* tadpoles. *Physiol. Zool.*, **31**, 138–51.

Richards, C. M. (1962). The control of tadpole growth by alga-like cells. *Physiol.*

Zool., **35**, 285–96.

Richmond, N. D. (1947). Life history of *Scaphiopus holbrookii holbrookii* (Harlan). Part I: Larval development and behavior. *Ecology*, **28**, 53–67.

Rose, S. M. (1960). A feedback mechanism of growth control in tadpoles. *Ecology*, **41**, 188–99.

Rose, W. (1956). Paternal care in batrachians. *Afr. Wildlife*, **10**, 257.

Roux, K. H. & Volpe, E. P. (1975). Evidence for a major histocompatibility complex in the leopard frog. *Immunogenetics*, **2**, 577–89.

Rugh, R. (1962). *Experimental Embryology*. 3rd edn. Minneapolis: Burgess.

Salthe, S. N. & Mecham, J. S. (1974). Reproductive and courtship patterns. In *Physiology of the Amphibia*, vol. 2, ed. B. Lofts, pp. 309–521. New York: Academic Press.

Samollow, P. B. (1980). Selective mortality and reproduction in a natural population of *Bufo boreas*. *Evolution*, **34**, 18–39.

Samollow, P. B. & Soulé, M. E. (1983). A case of stress related heterozygote superiority in nature. *Evolution*, **37**, 646–9.

Savage, R. M. (1952). Ecological, physiological and anatomical observations on some species of anuran tadpoles, *Proc. Zool. Soc. Lond.*, **122**, 467–514.

Schmitt, J. & Antonovics, J. (1986). Experimental studies of the evolutionary significance of sexual reproduction. IV. Effect of neighbor relatedness and aphid infestation on seedling performance. *Evolution*, **40**, 830–6.

Semlitsch, R. D. & Caldwell, J. P. (1982). Effects of density on growth, metamorphosis, and survivorship in tadpoles of *Scaphiopus holbrooki*. *Ecology*, **63**, 905–11.

Semlitsch, R. D., Scott, D. E. & Pechmann, J. H. K. (1988). Time and size at metamorphosis related to adult fitness in *Ambystoma talpoideum*. *Ecology*, **69**, 184–92.

Shaffer, H. B. & Breden, F. (1989). The relationship between allozyme variation and life history: non-transforming salamanders are less variable. *Copeia*, **1989**, 1016–23.

Shaw, E. (1961). The development of schooling in fishes. II. *Physiol. Zool.*, **34**, 263–72.

Shields, W. M. (1982). *Philopatry, Inbreeding, and the Evolution of Sex*. Albany: State University of New York Press.

Shvarts, S. S. & Pyastolova, O. A. (1970). Regulators of growth and development of amphibian larvae. I. Specificity of effects. *Sov. J. Ecol.*, **1**, 58–62.

Simon, G. S. & Madison, D. M. (1984). Individual recognition in salamanders: cloacal odours. *Anim. Behav.*, **32**, 1017–20.

Smith, D. C. (1987). Adult recruitment in chorus frogs: effects of size and date at metamorphosis. *Ecology*, **68**, 344–50.

Smith, D. C. (1990). Population structure and competition among kin in the chorus frog (*Pseudacris triseriata*). *Evolution*, **44**, 1529–41.

Smith, R. J. F. (1977). Chemical communication as adaptation: alarm substance of fish. In *Chemical Signals in Vertebrates*, ed. D. Müller-Schwarze & M. M. Mozell), pp. 303–20. New York: Plenum Press.

Smith, R. J. F. (1982). The adaptive significance of the alarm substance-fright reaction system. In *Chemoreception in Fishes*, ed. T. J. Hara, pp. 327–42. Amsterdam: Elsevier.

Smith, R. J. F. (1986). The evolution of chemical alarm signals in fishes. In *Chemical*

Signals in Vertebrates 4. *Ecology, Evolution, and Comparative Biology*, ed. D. Duvall, D. Muller-Schwarze & R. M. Silverstein, pp. 99–115. New York: Plenum Press.

Starrett, P. (1960). Descriptions of tadpoles of Middle American frogs. *Misc. Publ. Mus. Zool. Univ. Mich.*, **110**, 1–37.

Steinwascher, K. (1978). Interference and exploitation competition among tadpoles of *Rana utricularia*. *Ecology*, **59**, 1039–46.

Steinwascher, K. (1979). Host-parasite interaction as a potential population-regulating mechanism. *Ecology*, **60**, 884–90.

Stuart, L. C. (1961). Some observations on the natural history of tadpoles of *Rhinophrynus dorsalis* Dumeril and Bibron. *Herpetologica*, **17**, 73–9.

Stuart, R. J. (1988). Collective cues as a basis for nestmate recognition in polygynous leptothoracine ants. *Proc. Nat. Acad. Sci. USA*, **85**, 4572–5.

Sype, W. E. (1975). Breeding habits, embryonic thermal requirements and embryonic and larval development of the cascade frog, *Rana cascadae* Slater, Ph.D. Thesis, Oregon State University.

Test, F. H. & McCann, R. G. (1976). Foraging behavior of *Bufo americanus* tadpoles in response to high densities of micro-organisms. *Copeia*, **1976**, 576–8.

Thomas, L. (1975). Symbiosis as an immunologic problem. In *The Immune System and Infectious Diseases*. ed. E. Neter & F. Milgrom, pp. 2–11. Basel: Karger.

Tilley, S. G. (1972). Aspects of parental care and embryonic development in *Desmognathus ochrophaeus*. *Copeia*, **1972**, 532–40.

Tonsor, S. J. (1989). Relatedness and intraspecific competition in *Plantago lanceolata*. *Am. Nat.*, **134**, 897–906.

Townsend, D. S., Stewart, M. M., Pough, F. H. & Brussard, P. F. (1981). Internal fertilization in an oviparous frog. *Science*, **212**, 469–71.

Travis, J. (1980). Phenotypic variation and the outcome of interspecific competition in hylid tadpoles. *Evolution*, **34**, 40–50.

Tristram, D. A. (1977). Intraspecific olfactory communication in the terrestrial salamander *Plethodon cinereus*. *Copeia*, **1977**, 597–600.

Twitty, V., Grant, D. & Anderson, O. (1964). Long distance homing in the newt *Taricha rivularis*. *Proc. Nat. Acad. Sci. USA*, **51**, 51–8.

Uyenoyama, M. K. (1988). On the evolution of genetic incompatibility systems: incompatibility as a mechanism for the regulation of outcrossing distance. In *The Evolution of Sex*, ed. R. E. Michod & B. R. Levin, pp. 212–32. Sunderland, Mass: Sinauer Associates.

van Dijk, D. E. (1972). The behaviour of southern African anuran tadpoles with particular reference to their ecology and related external morphological features. *Zool. Afr.*, **7**, 49–55.

Van Havre, N. & FitzGerald, G. J. (1988). Shoaling and kin recognition in the threespine stickleback (*Gasterosteus aculeatus* L.). *Biol. Behav.*, **13**, 190–201.

Vaz-Ferreira, R. & Gehrau, A. (1975). Comportamiento epimeletico de la rana comun, *Leptodactylus ocellatus* (L.) (Amphibia, Leptodactylidae). I. Atencion de la cria y actividades alimentarias y agresivas relacionadas. *Physis B.*, **34**, 1–14.

Voris, H. K. & Bacon, J. P. Jr., (1966). Differential predation on tadpoles. *Copeia*, **1966**, 594–8.

Wager, V. A. (1965). *The Frogs of South Africa*. Capetown: Purnell and Sons.

Waldman, B. (1981). Sibling recognition in toad tadpoles: the role of experience. *Z.*

Tierpsychol., **56**, 341–58.

Waldman, B. (1982a). Adaptive significance of communal oviposition in wood frogs (*Rana sylvatica*). *Behav. Ecol. Sociobiol.*, **10**, 169–74.

Waldman, B. (1982b). Sibling association among schooling toad tadpoles: field evidence and implications. *Anim. Behav.*, **30**, 700–13.

Waldman, B. (1983). Kin recognition and sibling association in anuran amphibian larvae. Ph.D. Thesis, Cornell University.

Waldman, B. (1984). Kin recognition and sibling association among wood frog (*Rana sylvatica*) tadpoles. *Behav. Ecol. Sociobiol.*, **14**, 171–80.

Waldman, B. (1985a). Olfactory basis of kin recognition in toad tadpoles. *J. Comp. Physiol. A*, **156**, 565–77.

Waldman, B. (1985b). Sibling recognition in toad tadpoles: are kinship labels transferred among individuals? *Z. Tierpsychol.*, **68**, 41–57.

Waldman, B. (1986a). Chemical ecology of kin recognition in anuran amphibians. In *Chemical Signals in Vertebrates* 4. *Ecology, Evolution, and Comparative Biology*, ed. D. Duvall, D. Müller-Schwarze & D. M. Silverstein, pp. 225–42. New York: Plenum Press.

Waldman, B. (1986b). Preference for unfamiliar siblings over familiar non-siblings in American toad (*Bufo americanus*) tadpoles. *Anim. Behav.*, **34**, 48–53.

Waldman, B. (1987). Mechanisms of kin recognition. *J. Theor. Biol.*, **128**, 159–85.

Waldman, B. (1988). The ecology of kin recognition. *Ann. Rev. Ecol. Syst.*, **19**, 543–71.

Waldman, B. (1989). Do anuran larvae retain kin recognition abilities following metamorphosis? *Anim. Behav.*, **37**, 1055–8.

Waldman, B. & Adler, K. (1979). Toad tadpoles associate preferentially with siblings. *Nature*, **282**, 611–13.

Waldman, B., Frumhoff, P. C. & Sherman, P. W. (1988). Problems of kin recognition. *Trends Ecol. Evol.*, **3**, 8–13.

Waldman, B. & McKinnon, J. S. (1991). Inbreeding and outbreeding in fishes, amphibians and reptiles. In *The Natural History of Inbreeding and Outbreeding: Theoretical and Empirical Perspectives*, ed. N. W. Thornhill. Chicago: University of Chicago Press. (In Press.)

Walls, S. C. (1991). Ontogenetic shifts in the recognition of siblings and neighbours by juvenile salamanders *Anim. Behav.* (In Press.)

Walls, S. C. & Roudebush, R. E. (1991). Reduced aggression toward siblings as evidence of his recognition in cannabalistic salamanders. *Am. Nat.* (In Press.)

Walters, B. (1975). Studies of interspecific predation within an amphibian community. *J. Herpetol.*, **9**, 267–79.

Waser, N. M. (1991). Population structure, optimal outbreeding, and assortative mating in angiosperms. In *The Natural History of Inbreeding and Outbreeding: Theoretical and Empirical Perspectives*, ed. N. W. Thornhill. Chicago: University of Chicago Press. (In Press.)

Wassersug, R. J. (1973). Aspects of social behavior in anuran larvae. In *Evolutionary Biology of the Anurans*, ed. J. L. Vial, pp. 273–97. Columbia: University of Missouri Press.

Wassersug, R. & Hessler, C. M. (1971). Tadpole behaviour: aggregation in larval *Xenopus laevis*. *Anim. Behav.*, **19**, 386–9.

Wassersug, R. J., Lum, A. M. & Potel, M. J. (1981). An analysis of school structure for tadpoles (Anura: Amphibia). *Behav. Ecol. Sociobiol.*, **9**, 15–22.

Wells, K. D. (1978). Courtship and parental behavior in a Panamanian poison-arrow frog (*Dendrobates auratus*). *Herpetologica*, **34**, 148–55.

Wells, K. D. (1981). Parental behavior of male and female frogs. In *Natural Selection and Social Behavior. Recent Research and New Theory*, ed. R. D. Alexander & D. W. Tinkle, pp. 184–97. New York: Chiron Press.

Wells, K. D. & Bard, K. M. (1988). Parental behavior of an aquatic-breeding tropical frog, *Leptodactylus bolivianus.*, *J. Herpetol.*, **22**, 361–4.

Werner, D. I., Baker, E. M., Gonzalez, E. del C. & Sosa, I. R. (1987). Kinship recognition and grouping in hatchling green iguanas. *Behav. Ecol. Sociobiol.*, **21**, 83–9.

West, L. B. (1960). The nature of growth inhibitory material from crowded *Rana pipiens* tadpoles. *Physiol. Zool.*, **33**, 232–9.

Weygoldt, P. (1980). Complex brood care and reproductive behavior in captive poison-arrow frogs, *Dendrobates pumilio* O. Schmidt. *Behav. Ecol. Sociobiol.*, **7**, 329–32.

Wiens, J. A. (1972). Anuran habitat selection: early experience and substrate selection in *Rana cascadae* tadpoles. *Anim. Behav.*, **20**, 218–20.

Wilbur, H. M. (1976). Density-dependent aspects of growth and metamorphosis in *Bufo americanus*. *Ecology*, **58**, 196–200.

Wilbur, H. M. (1977). Interactions of food level and population density in *Rana sylvatica*. *Ecology*, **58**, 206–9.

Wilbur, H. M. (1984). Complex life cycles and community organization in amphibians. In *A New Ecology. Novel Approaches to Interactive Systems*, ed. P. W. Price, C. N. Slobodchikoff & W. S. Gaud, pp. 195–224. New York: John Wiley & Sons.

Wilbur, H. M. & Collins, J. P. (1973). Ecological aspects of amphibian metamorphosis. *Science*, **182**, 1305–14.

Williams, C. G., Bridgewater, F. E. & Lambeth, C. C. (1983). Performance of single family versus mixed family plantation blocks of loblolly pine. *Proc. 17th Southern Forest Tree Improvement Conference, University of Georgia, Athens.*

Williams, G. C. (1964). Measurement of consociation among fishes and comments on the evolution of schooling. *Mich. St. Univ. Mus. Publ., Biol. Ser.*, **2**, 349–84.

Williams, M. (1976). Rearing environments and their effects on schooling of fishes. *Publ. Staz. Zool. Napoli*, **40**, 283–54.

Williamson, I. & Bull, C. M. (1989). Life history variation in a population of the Australian frog *Ranidella signifera*: egg size and early development. *Copeia*, **1989**, 349–56.

Willson, M. F., Thomas, P. A., Hoppes, W. G., Katusic-Malmborg, P. L., Goldman, D. A. & Bothwell, J. L. (1987). Sibling competition in plants: an experimental study. *Am. Nat.*, **129**, 304–11.

Wollmuth, L. P., Crawshaw, L. I., Forbes, R. B. & Grahn, D. A. (1987). Temperature selection during development in a montane anuran species, *Rana cascadae*. *Physiol. Zool.*, **60**, 472–80.

Woodward, B. D. (1987). Interactions between Woodhouse's toad tadpoles (*Bufo woodhousii*) of mixed sizes. *Copeia*, **1987**, 380–6.

Yamazaki, K., Beauchamp, G. K., Kupniewski, D., Bard, J., Thomas, L. & Boyse, E. A. (1988). Familial imprinting determines H-2 selective mating preferences. *Science*, **240**, 1331–2.

8

Kin recognition cues of vertebrates

Zuleyma Tang Halpin

Introduction

The mechanisms of kin recognition have received much attention in recent years (e.g. reviews by Holmes & Sherman, 1983; Sherman & Holmes, 1985; Hepper 1986a; Waldman, 1987). However, most treatments of this topic have concentrated almost exclusively on the mechanisms by which animals are able to 'recognize' or classify conspecifics as either kin or non-kin. The signals or cues produced by animals and utilized by conspecifics to identify kin have, on the other hand, received only scant attention (but see Beecher, 1982; Hepper, 1986a; Waldman, 1987). Beecher (1982) stresses that 'identification' (the production of a signal that indicates the identity of the sender) is an important component of kin recognition which may have a significant impact on the fitness of the recipients of altruistic or nepotistic acts. Thus, discussions that focus only on the 'recognition' component of kin recognition address only one half of the question. An understanding of the 'identification' component and of the cues used for identification is essential if we are to develop a complete understanding of the mechanisms of kin recognition.

Halpin & Hoffman (1987) suggest that, in studying the cues used in kin recognition, two related but separate sets of questions need to be addressed: (1) whether the cues used for recognition are phenotypic labels shared by all genetic relatives (e.g. a family-specific label), or whether such cues are individually distinctive and, as such, provide no direct information on the genetic relatedness of conspecifics; and (2) whether the sensory cues used for kin recognition have a genetic basis or are environmentally acquired. In this chapter both of these questions will be considered and, where available, relevant empirical data provided. Additionally, the sensory modalities used for parent–offspring and sibling recognition among different vertebrate groups and species will be reviewed.

220

Cues used in different recognition contexts

Spatially-based recognition

For the purposes of this chapter, kin recognition is defined as the ability of an animal to use *conspecific* cues to distinguish between its kin and non-kin. This definition is in contrast to those that stress only the 'differential responsiveness' of individuals towards kin versus non-kin. Using this latter definition, 'recognition' based on spatial proximity is commonly cited as one of the mechanisms of kin recognition (e.g. Holmes & Sherman, 1983; Sherman & Holmes, 1985; Hepper, 1986a). In 'recognition based on spatial proximity' there is a high probability that animals within a given location (e.g. a nest, a certain home range) will be genetically related to one another; consequently, each animal responds preferentially (i.e. behaves nepotistically) towards all others found within that given locality. Thus, whenever an individual encounters conspecifics within the correct location, it treats them as if they were kin. The problem is that these conspecifics are treated as kin regardless of whether they are true genetic relatives or not. In other words, the fact that one animal behaves nepotistically towards another is accepted as *ipso facto* evidence that kin recognition has occurred. Not only can such an argument lead to circular reasoning but it also reflects confusion regarding the mechanisms of recognition. I would argue that animals that rely only on 'spatially-based recognition' are actually incapable of recognizing kin from non-kin, and it is precisely because of this that they are forced to rely on spatial cues to determine who should be treated *as if* they were kin. Dispersal patterns and other life-history characteristics may result in the nepotistic behaviour being directed primarily at kin, and there may even have been selection for this type of behaviour, but this does not mean that the animals engaging in the behaviours can actually recognize kin from non-kin. The behaviour of animals in such a situation can best be understood as an example of spatially-mediated nepotistic behaviour, *not* as kin recognition. The failure to distinguish between spatially-mediated nepotism and true kin recognition has resulted in semantic and theoretical confusion which has been detrimental to a clear understanding of the dynamics of kin recognition. Problems with the concept of spatially-based recognition and the potential for errors have also been discussed previously by Blaustein (1983) and Waldman (1987).

Since 'recognition' based on spatial proximity utilizes only spatial cues (i.e. the presence of a conspecific in a given locality) rather than conspecific cues, I do not consider it as an example of true kin recognition and will not discuss it further in this chapter. Hepper (1986a), however, presents an excellent discussion of the non-conspecific cues that may mediate the expression of spatially-based nepotistic behaviours in situations usually classified as spatially-based recognition.

Recognition by association

This kin recognition mechanism depends on the ability of animals to treat as kin those conspecifics with whom they have associated during certain periods of their lives. The correct identification of kin occurs because, under natural conditions, there is normally a reliable correlation between genetic relatedness and the spatial/temporal component of association. For example, for most species, under most normal conditions, all conspecifics of a similar age sharing a nest or a natal burrow can be expected to be siblings. In the sense that recognition by association depends on a spatial/temporal component, it is similar to spatially-based 'recognition'. The crucial difference is that, in recognition by association, during the period of association the animal learns the cues or labels that identify its putative kin and then uses those cues to recognize those 'kin' outside of the association context. Thus, conspecific cues are learned and used for recognition. In contrast, in spatially based 'recognition', the 'recognition' relies entirely on the presence of a conspecific in the right locality or 'area of recognition'; true kin removed from this locality cannot be recognized and, therefore, are treated as non-kin.

Although under my definition recognition by association is considered as a true form of kin recognition, certain problems are still apparent. Most importantly, the potential for error is still present. Non-kin who mistakenly end up in the 'correct' locality during the period of association will be regarded as kin; true kin who are absent during the period of association will be regarded as non-kin (see also Blaustein, 1983). Despite these problems, it is likely that recognition by association is one of the most common mechanisms of kin recognition among vertebrates (see also Beecher, 1982; Waldman, 1987). The reason may be that, while it is easy for human experimenters to manipulate the periods of association so that non-kin are made to associate together, or kin are prevented from associating, under natural conditions this would happen so rarely that errors would almost never occur and the problem would become evolutionarily trivial. This, of course, would be true only for those species in which there is normally a truly reliable correlation between genetic relatedness and association.

Species that recognize kin by association probably do so primarily by learning the individually-distinctive cues of the conspecifics with which they associate. Those individual cues are then recalled and used to recognize those same conspecifics when they are encountered at a later time and/or in a different context. Beecher (1982) coined the term 'signature system' to describe cues that can be used for individual identification because they have greater inter-individual than intra-individual variation. To be effective in identifying a large number of individuals, signature cues are likely to be made up of several different trait components which can vary from individual to individual in the pattern or combination profile of the different components

(Beecher, 1982; Waldman, 1987). It is precisely this individually unique pattern or trait combination profile that confers the 'signature' characteristic to the cue. For example, Buckley & Buckley (1970) suggested that six different components (leg colour, leg blotching, bill colour, bill-tip colour, down colour and amount of down blotching) may make up the individually-distinctive visual cue used by royal tern (*Sterna maxima*) parents for recognizing their chicks. Likewise, Beecher (1982) found that call duration, figure duration, frequency difference, figure frequency, call slope and figure shape and all important components of the 'signature calls' by which adult bank swallows (*Riparia riparia*) recognize their chicks. Odours used for individual identification among mammals also appear to consist of a large variety of chemicals which differ in their frequency and combinations from one individual to another (see Halpin, 1986 for review).

Individual cues used in kin recognition systems that rely on previous association may be of either genetic or environmental origin. In either case, the cue is perceived as an individually-distinctive combination of 'signature' traits; family similarities or shared traits based on genetic relatedness are not necessary for determining putative kinship.

Recognition by association may also, in some cases, rely on 'group' cues. Such cues may come about because all members of the group acquire a common group label, or because cues produced by the different members of the group combine to produce a group cue. Some mammals, for example, scent mark their young, providing them with a common label which actually consists of the parent's *individual* label (e.g. Halpin, 1986; also below). Domestic kittens (*Felis domesticus*) can distinguish between the odour of their own nest and those of other kittens of the same age (Freeman & Rosenblatt, 1978). Although the nature of the cues involved has not been examined, the odour cue used by the kittens may be produced by their mother or by a combination of odours from both kittens and mother.

Group cues used for kin recognition based on previous association have several interesting characteristics. Firstly, as suggested above, they are most likely to be either individually-distinctive cues which are used to 'label' group members, or a combination of cues, each of which is, in itself, an individually-distinctive cue. Secondly, group labels utilized for kin recognition by association will frequently be environmentally acquired rather than genetically determined. All individuals present during the period of association obtain the label. Thus, non-kin 'in the right place at the right time' will be labelled, while kin that are not present will not be labelled. The group label is reliably correlated only with the animal's presence in the group during the period of association, but not with its genetic similarity with other group members. Therefore, the family-specific group label indicates only previous association but conveys no direct information on genetic relatedness. Thirdly, because, under certain circumstances, environmentally acquired group cues may also allow for recognition by phenotype matching (see discussion below), the introduction of group cues into kin recognition systems, results in a blurring of clear boundaries between recognition by association and recognition by phenotype matching.

Recognition by phenotype-matching

Kin recognition by phenotype matching relies on group or family cues which are reliably correlated with genetic relatedness, although not necessarily with previous association. Animals that use phenotype matching learn the family cue from themselves or from other family members that are present during certain relevant periods of association. This learned cue is generalized so that all other individuals carrying the same cue are recognized as kin, even if they have never been previously encountered. Thus, the crucial component needed for recognition is the shared phenotypic trait which is recognized as the family cue. This family cue may consist of a single trait which reliably varies from one family to another (e.g. the presence of a visual marker which is completely family specific). More often, however, it will be composed of a series of traits or labels which are shared to a greater or lesser extent by all family members. Animals that use phenotype matching relying on such a spectrum of different traits may, therefore, also be able to assess different levels of kinship based on the degree of similarity between the labels possessed by a newly encountered individual and those that have been previously learned as the family cue.

Phenotype matching will be most effective and reliable in recognizing kin when the labels used are a direct expression of the genome. In such a case, genomic similarity can be directly assessed by comparing the degree of similarity between the labels of self or known kin and those of a stranger. Animals that utilize phenotype matching may also possess genetically-determined signature traits. However, the ability to recognize kin does not depend on learning the characteristics of individual signature profiles, as in the case of recognition by association, but rather on being able to assess the degree of overlap between the labels that make up the signatures of known kin and those of others (Beecher, 1982).

Although most labels used for kin recognition by phenotype matching are likely to have a genetic origin, it is also possible for them to be environmentally acquired. *Bufo americanus* tadpoles, for example, are apparently labelled with a maternal factor (presumably of genetic origin) which may be found in the jelly enveloping the eggs before hatching (Waldman, 1981). This maternal factor, which is shared by all sibling tadpoles, regardless of previous association, appears to facilitate sibling recognition even when tadpoles have been reared in isolation. It is also possible to conceive of a hypothetical situation where siblings recognize each other by using a 'family-specific' behaviour pattern they learned from their parents. Sibling birds, for example, could learn to imitate their father's song and use it at a later time as a common label to recognize one another. Recognition by phenotype matching would occur if these same birds are also able to recognize siblings from previous or subsequent clutches because they too share the father's common 'family-specific song'. Cases in which mothers label their offspring with scent marks may, in some cases, also result in kin recognition by phenotype matching. For example, if the characteristics of the mother's scent marks are stable and do not change over time, older offspring may be able to recognize their unfamiliar sibs from a different litter because they also carry the familiar maternal label. Such recognition, however, would be extremely labile because the label would most likely be lost after contact with the mother ceases. Thus, older siblings no longer carrying the label would not be able to recognize each other, nor would younger siblings that currently have the label be able to recognize their unfamiliar older siblings.

The use of environmentally acquired labels for kin recognition by phenotype matching is probably relatively rare. While labels whose expression is under direct genetic control will almost always indicate true genetic relationship, the same is not true of labels that are environmentally acquired. As discussed previously, under certain conditions non-kin might acquire the label, while kin might not. Thus, correlation with true genetic relatedness is comparatively lower and the potential for errors of recognition is higher. Under these conditions it would be expected that the use of genetically determined labels would be the rule whenever recognition is by phenotype matching (see also Beecher, 1982; Waldman, 1987).

Recognition by recognition alleles

The cues used for recognition by recognition alleles (Hamilton, 1964) would be quite similar to those used for phenotype matching, except that they would in all cases have a genetic origin. All individuals carrying the same recognition alleles would produce the same phenotypic trait or labels used for recognition. If the recognition alleles are reliably different from one family to another, the result would be the production, within any one family, of a 'family cue' which is perfectly correlated with genetic relatedness. However, the existence of recognition alleles has not yet been demonstrated and may be unlikely (see Bekoff, 1981; Holmes & Sherman, 1983; Blaustein, 1983; Waldman, 1987).

Template models of kin recognition

The ability to recognize kin by either association or phenotype matching depends on previous exposure to, and learning of, the cues that are used for recognition. Typically the animals learn the relevant cues after a period of association with kin, although in phenotype matching the cues may also be learned from self. The exact mechanisms by which animals recognize their putative kin after such a period of association are not completely understood. It is assumed, however, that the animal stores in its memory a 'template' or 'trait profile' of the previously encountered individuals and then compares it to the trait profile of individuals it meets at a later time.

Waldman (1987) presents four different models for kin recognition based on previous experience with recognition cues. The models differ depending on whether animals store specific trait *combinations* for each individual they encounter (e.g. ABCD versus ABXY), or whether they store only the traits, but not their specific combinations (e.g. A, B, C, D, X, Y). In the first case an individual will be recognized as kin only when its trait profile *exactly* matches the individual trait profile 'signature' of a previously encountered conspecific. This, then, is equivalent to recognition by association using only individually distinctive cues. The second case, on the other hand, allows for the use of 'group' or 'family' cues because individuals that share all or some of the traits stored in the template, regardless of their exact combination, will be recognized as members of the 'group'. Thus, using the example provided above, an individual with a trait profile of ABCX or CDXY will be recognized as kin because its traits are shared with those of self or of known relatives even if the exact signature profiles differ. Such a mechanism would allow for recognition by phenotype matching. In neither case, however, will an animal recognize as kin an individual bearing a label which is not present in the memory template. An individual with a signature profile of ABCG for example, would not be recognized because the G label is absent from the template.

It is interesting that all of these models, which describe quite well the actual kin recognition behaviour of animals, depend on the type of multi-dimensional recognition cues (i.e. cues composed of several different traits or labels) that was first proposed by Beecher (1982). While such multidimensional cues clearly provide for the individual distinctiveness required for recognition by association, Waldman's (1987) models suggest that they may also play a crucial role in recognition by phenotype matching.

Genetic versus environmental components

Cues used for kin recognition may be either genetically controlled or environmentally acquired. Some species may use only environmental or only genetic cues, but in many cases individually-distinctive or family specific cues may result from an interaction of both genetic and environmental components.

Relatively little information currently exists on the genetic or environmental bases of the kin recognition cues used by most vertebrate species. While it is often assumed that most of these cues are of genetic origin, the evidence to support this claim has come from only a small number of species. Environmentally acquired cues have also been studied in only a few species, but the available data indicate that they may be important and widespread at least among certain groups of vertebrates.

The strongest evidence that the 'signature' characteristics of recognition cues may be of environmental origin comes from studies of individual odours used for individual and kin recognition among many species of rodents (Halpin, 1986). Most important among the environmental factors that may affect such odours are bacterial flora and diet. For example, rat pups (*Rattus norvegicus*) are able to discriminate between the odours of their mothers and those of other lactating females. The pheromone responsible for this individually distinctive maternal odour is found in caecotrophe, a substance which forms as the result of bacterial action on food which accumulates in the caecum of the mother (Leon, 1975). The importance of differences in the diet is demonstrated by the fact that pups are able to discriminate between their own mother and another lactating female only when the two females are maintained on different diets; that is, pups are not able to distinguish between the caecotrophe of two different lactating females which are eating the same diet (Leon, 1975). It is not known if different rats also have different bacterial flora. However, it is certainly possible that the individually distinctive odour of caecotrophe is due partly to individual differences in diet and partly to individual differences in the bacterial flora found in each female's caecum. Bacteria are known to be the source of individual odours in other mammalian species (e.g. Gorman, 1976) and the same could be true in the rat.

It is also possible that a pregnant female's diet can affect the post-natal odour preferences of her young. Hepper (1988) has found that the offspring of a female rat fed garlic while pregnant, showed a preference for the odour of garlic post-natally, despite the fact that they were not exposed to garlic after their birth. These results suggest that fetal exposure to diet odours could have important effects on the ability of rat pups to recognize their mother and/or siblings.

Individual differences in diet also play a role in parent–offspring recognition in the Mongolian gerbil (*Meriones unguiculatus*). The ability of gerbil pups to recognize their mothers appears to depend on individually different maternal odours that are at least partly influenced by the mother's diet (Skeen & Thiessen, 1977).

In the spiny mouse (*Acomys cahirinus*), environmental and genetic factors may interact in complex ways to produce the individually unique odour 'signatures' used for kin recognition (Porter, 1988). Pups respond preferentially to the odours of their own mothers as compared to those of another lactating female only when the two females are maintained on different diets (Porter & Doane, 1977). Furthermore, mothers probably label their own pups and use their own diet-specific individually distinctive odours to distinguish their pups from strangers. Mothers more readily retrieve the pups of other females fed the same diet as their own than they do pups born to females fed a different diet (Doane & Porter, 1978).

Sibling recognition in spiny mice likewise may be influenced by the labels pups receive from their mothers. Sibling pups that suckle from the same mother appear to recognize each other, while siblings that suckle from different mothers do not (Porter & Moore, 1981; Porter *et al.*, 1984). This is true even when the amount of previous contact between the pups is kept constant for the two conditions. Additionally, unrelated, unfamiliar pups nursed by the same female preferentially associate with each other, rather than with other, unrelated and unfamiliar pups nursed by a different female. The maternal labels used for sibling recognition in these studies may be diet-specific odours and/or other odours contained in the mother's milk, saliva, urine or glandular secretions (Porter, 1988).

The pup's *own* diet has also been shown to affect sibling recognition in this species. After weaning, spiny mouse pups are preferentially attracted to the odours of other pups that have been fed the same diet as themselves (Porter & Doane, 1979). More specifically, Porter *et al.* (1989) have shown that pups prefer to associate with unrelated, unfamiliar pups fed their same diet as compared to unrelated, unfamiliar pups fed a different diet, and with their familiar siblings fed their same diet as compared to familiar siblings fed a different diet. Furthermore, pups do not show a significant preference for their familiar siblings when they are eating a different diet, compared to unfamiliar, non-siblings eating the same diet. Clearly then, diet is an extremely important factor in the ability to recognize siblings (measured by preferential attractiveness) among spiny mice.

However, diet does not act alone, and genetic factors have also been shown to be important. For example, weaned pups housed with an unrelated, unfamiliar pup from a different litter later prefer to associate with their cagemate's sibling over another pup of the same age even though both stimulus pups were equally unfamiliar and were eating the same diet as the subject (Porter, 1988). Additionally, pups prefer a familiar sibling fed its same diet as compared to an unfamiliar non-sibling also fed the same diet, and a familiar sibling fed a different diet as compared to an unfamiliar non-sibling fed a different diet (Porter *et al.*, 1989). Clearly, then, spiny mouse pups can recognize their siblings (and the siblings of a familiar conspecific) even when diet is held constant; these results make sense only if there are genetic similarities among siblings which are capable of overriding the effects of diets. Thus, in this species, genetic and environmental factors appear to interact in a complex fashion to mediate sibling recognition.

It is also interesting that while spiny mice pups are obviously able to detect genetic similarities among their siblings, they are also able to detect individual differences among them. For example, pups prefer to associate with their familiar siblings as compared to unfamiliar siblings (Porter *et al.*,

1986). This suggests that the genetic similarities used to distinguish kin from non-kin probably result from an overlap in the characters or traits which make up each individual's distinctive signature. The unique individual 'signature' is used for individual recognition while the similarities among the signatures convey information on genetic relatedness.

The spiny mouse is the only species for which the interaction between environmentally acquired and genetically-influenced cues has been examined in detail. It is likely, however, that kin recognition in many other species will depend on similar interactions. Among mammals, in particular among those species that depend heavily on olfactory cues, diet-dependent odours may commonly interact with genetic factors to produce distinctive individual signatures which can be used for kin recognition. Even among humans, for example, the odours of identical twins can be distinguished more easily when the twins have eaten different diets than when they have eaten the same diet (Wallace, 1977). Thus, the ability of human parents to distinguish between the odour of their own child and that of a stranger, as well as between the odours of two of their own children (Porter & Moore, 1981), may well result from an interaction of genetic and environmentally acquired components of the individually distinctive odour signature.

In some species, cues of either genetic or environmental origin may be used to label kin who then express it as an acquired cue. The best examples of this are the many cases of maternal labelling of young. Many mammals, for example, are known to mark their young (see above). In some cases these labels may be of purely genetic origin, in others they may be environmentally-influenced (as in the case of the spiny mouse already discussed) and even in others it may result from an interaction of both. Unfortunately, there is insufficient evidence at the present time to determine which of these possibilities is likely to be the most common. In the toad *Bufo americanus*, the mother appears to label her young with a cue presumed to be of genetic origin and found either in the cytoplasm of the eggs or in the jelly surrounding the eggs. Thus, all tadpoles from a given clutch receive the same label and later appear to use it for sibling recognition. Tadpoles reared in isolation can distinguish (as measured by preferential association) between paternal and full siblings but not between maternal and full siblings (Waldman, 1981). It appears, therefore, that full siblings share an identifying label with maternal siblings, but not with paternal siblings. Thus, tadpoles may match the phenotype of other unfamiliar conspecifics to this label in determining whether or not to accept them as kin.

Not all anurans, however, use maternal labels for kin recognition. Blaustein & O'Hara (1982b) found that *Rana cascadae* tadpoles are able to discriminate full siblings from *both* maternal and paternal half-siblings. Moreover, they also discriminate between maternal or paternal half-siblings and non-siblings. Even when they were reared without any egg jelly, or with jelly from an egg clutch produced by a female other than their mother, they were still able to distinguish between their paternal half-siblings and non-siblings. These results strongly suggest that, in this species, a genetic factor, independent of any cues that might be contained in the egg jelly, is responsible for the ability of tadpoles to discriminate between siblings and non-siblings, as well as between siblings of different degrees of relatedness.

The results of kin recognition studies in the Norway rat (*Rattus norvegicus*) also support the existence of genetic factors. Hepper (1987) found that rats are able to distinguish among kin of different degrees of relatedness. Specifically, he reports that, when tested with unfamiliar but genetically related conspecifics, rats spend greater amounts of time investigating less closely related conspecifics in the following order: cousins are investigated more than half-siblings, which, in turn, are investigated more than full siblings; rats which are both unfamiliar and unrelated are investigated even more than any of the above kin categories. Given that all the stimulus conspecifics were unfamiliar and that it is unlikely that a common, environmentally acquired label could have been shared in different degrees by these different categories of kin, the most parsimonious explanation is that the rats possess a genetically-influenced family cue which varies in relation to the degree of genetic relatedness.

Likewise, birds also appear able to distinguish kin based on genetically mediated cues that provide information on degrees of relatedness. Bateson (1982), for example, has found that Japanese quail (*Coturnix coturnix japonica*) can also discriminate among siblings, first cousins, third cousins and unrelated conspecifics.

Probably the clearest examples of recognition cues known to be under genetic control are the individually distinctive odours produced by the H-2 major histocompatibility complex (MHC) of the house mouse, *Mus musculus* (Yamazaki et al., 1979, 1980, 1982, 1983; Yamaguchi et al., 1981). The H-2 complex, located on chromosome 17 of the mouse, is remarkable for its polymorphism and rates of mutation of its loci, particularly H-2K and H-2D (Klein, 1979; Beauchamp et al., 1985). Klein (1979) estimates that there are a minimum of 10 different loci in the H-2 complex and that there are at least 56 alleles at H-2K and 45 at H-2D. The alleles at these two loci alone can occur in 2500 combinations. This amount of variability, combined with that at other H-2 loci, is certainly sufficient to account for the individually distinctive odour signatures produced by the mouse.

The H-2 complex, then, is certainly sufficient to account for the individual odours necessary for kin recognition by association. Additionally, the number of loci and the amount of variability at the H-2 complex suggest that an overlap in the odours produced by genetic relatives could also be used for phenotype matching. This possibility, however, has not been investigated directly; most studies of the mouse MHC have utilized laboratory-produced congenic strains in which all individuals are genetically identical, except for alleles at the H-2 complex. Thus, it is not known whether the variability at the H-2 can also be used to determine the degree of overall genetic relatedness or to recognize unfamiliar kin in the absence of previous association. In this regard, it is interesting that, based on previous experience, mice are able to distinguish between two unfamiliar conspecifics that differ only in their H-2 types. Specifically, Beauchamp *et al.* (1988) have shown that cross-fostered male mice preferentially mate with females whose H-2 type differs from that of the male's foster parents. This indicates that, at the very least, male mice are able to match their parents' H-2 type with that of unfamiliar females and then avoid mating with females of identical H-2 type. A similar mechanism could hypothetically function in phenotype matching of closely related kin.

Although the H-2 region is likely to be the most important genetic complex coding for individuality of odours in mice, it is certainly not the only one. Mice from unrelated inbred strains that share the same H-2 type appear to produce individually distinctive odours (Beauchamp *et al.*, 1985), suggesting that other genes may also be involved in odour production. Different alleles at the T locus of chromosome 17 are likewise known to code for differences in odours (Lenington & Egid, 1985) and, in some cases, the expression of these odours may be further influenced by correlations with the H-2 locus (Egid & Lenington, 1985; Lenington *et al.*, 1988). Additionally, the Qa-Tla region of chromosome 17 may also play a role in individual odour production (Yamazaki *et al.*, 1982).

The major histocompatibility complex may code for individually distinctive odours in other species of rodents as well. Recently, Brown *et al.* (1987) have reported that the H-2 locus is also responsible for odour individuality in the rat. However, studies on the rat are still in their infancy, and it is not yet known how comparable the rat system is to that of the mouse.

Sensory modalities of recognition cues

Cues used for kin recognition among vertebrates may be mediated by any of the usual sensory modalities used in communication (i.e. vision, hearing, olfaction, taste and touch). In this section the kin recognition cues of vertebrates specifically with regard to their sensory modalities will be examined. The emphasis is on the sensory cues used rather than on recognition mechanisms (e.g. phenotype matching, recognition by association). Therefore, I review exclusively those studies that directly examine the sensory modality of the recognition cues. Aspects of recognition cues (such as their genetic or environmental origin, or the various trait components of cues) that have been discussed previously, will not be discussed again in this section.

Fishes

Among fishes, very few studies have directly addressed questions relevant to the mechanisms or cues of kin recognition. Less than a handful of studies have looked specifically at the ability of fishes to distinguish between siblings and non-siblings, or at the ability of parents to recognize their own young as compared to other unrelated young of their own species. Likewise, studies on the ability of young to recognize parents have not addressed the crucial question of whether they can distinguish between their own parents and other, unrelated, conspecific adults.

The complex parental behaviours found among fishes of the family Cichlidae make this an ideal group in which to study parent–offspring interactions and recognition. Myrberg (1966, 1975) reviewed the evidence on parental recognition of young among cichlids, and reported that several species can discriminate chemically between own young and those of another species. Furthermore, he found that in the convict cichlid, *Cichlasoma nigrofasciatum*, both visual and chemical cues appear to be important in the discrimination. However, in all of these studies, cichlid parents were required to choose only between their own young and those of another species; tests to determine whether parents could also discriminate between own young and unrelated *conspecific* young were not done. The results, therefore, can be accepted only as a demonstration that these cues are used in *species* recognition of young, but not necessarily in kin recognition.

The importance of chemical cues in parent–offspring kin recognition has, nonetheless, been demonstrated in another cichlid species, *C. citrinellum*. McKaye & Barlow (1976) reported that both male and female parents prefer the water-borne odours of their own young, as compared to those from alien conspecific young when the two are paired in a water olfactometer. Parents approach, spend significantly more time, and perform more parental-like behaviours on the side of the olfactometer containing the water coming from their own young. Thus, although the significance of this discriminatory ability under natural conditions is unclear (*C. citrinellum* is known to accept and adopt alien young), McKaye & Barlow (1976) have demonstrated conclusively that, in the laboratory, this species can discriminate between own and unrelated conspecific young based solely on chemical cues.

Little is known about the ability of cichlid young to recognize their parents. Barnett (1977a) showed that *C. citrinellum* young prefer water containing chemical cues from their mothers, as compared to clean water. Hay (1977) reported that in *C. nigrofasciatum*, fry prefer some parental models to others and that they may learn the individual characteristics of models and show an increased attraction to them over time. These two studies, therefore, suggest that young cichlids may use both chemical and visual stimuli to identify their parents. However, no firm conclusion regarding kin recognition can be drawn because neither study looked at whether young can also discriminate between their own parents and other adults of their own species.

Among non-cichlid fishes, an early study (MacGintie, 1939) found that blind goby parents (*Typhlogobius californiensis*) will not eat their own young, but do eat alien conspecific young. Barnett (1977b) proposes that, since these gobies are blind, chemical cues are probably involved in their ability to discriminate between own and alien young. While this suggestion may well prove to be correct, at present the involvement of other sensory modalities (e.g. tactile or acoustical) cannot be excluded.

Sibling recognition in fish has been examined only among salmon. Quinn & Busack (1985) reported that coho salmon juveniles (*Oncorhynchus kisutch*) recognize and prefer sibling water over that of non-siblings, regardless of whether the siblings are familiar or unfamiliar. Chemical cues contained in the water are sufficient for discrimination to occur. In another experiment, Selset & Døving (1980) found that char (*Salmo alpinus*) are attracted to odorants produced by smolts from their own population as compared to those of smolts from other populations. Although this discrimination cannot be classified as sibling recognition, it may reflect the ability to distinguish between chemical cues from closely related kin, as a result of inbreeding in the same population, versus those from unrelated individuals from different populations.

In summary, research on kin recognition in fishes is still in its infancy. Additional studies on parent–offspring and sibling recognition (not to mention on the cues used for discrimination) are badly needed. While the studies reviewed above suggest that chemical cues may be the most important in kin recognition among the fishes, this conclusion may change as more information becomes available.

Amphibians

Although the mechanisms of kin recognition have been studied more extensively among amphibians than among many other groups (e.g. see Waldman, this volume), only a few studies have looked at the cues used for recognition. Studies of sibling recognition have been limited to anuran amphibians, while parental recognition of young has been examined only in salamanders. In both cases, however, olfaction appears to be the most important sensory modality involved.

Among both frogs and toads, sibling recognition has been shown to depend on water-borne chemical cues. Blaustein & O'Hara (1982a) found that water-borne cues were sufficient for cascade frog (*Rana cascadae*) tadpoles to recognize their siblings; visual cues were not involved and auditory cues were unimportant. Likewise, Waldman (1985) demonstrated that when given a two-way choice in a Y-maze, toad tadpoles (*Bufo americanus*) preferentially orient toward water containing chemical cues from their siblings, as compared to water containing cues from non-siblings. Furthermore, blocking a tadpole's external nares destroyed its ability to distinguish between siblings and non-siblings, while removal of the blockage restored the ability to recognize siblings.

In the latter experiment (Waldman, 1985), it was also interesting that the preferential response towards siblings occurred only when the choice was between water containing cues from siblings and water containing cues from non-siblings. When given a choice between sibling water and distilled water, tadpoles did not show a significant preference for the sibling water. On the other hand, when tested with non-sibling versus distilled water, tadpoles showed a significant preference for the distilled water. Thus, these results strongly suggest that the apparent attraction shown towards siblings as compared to non-siblings may actually result from an avoidance response to chemical cues produced by non-siblings rather than to a true attraction to cues produced by siblings.

Parental recognition of young (actually egg clutches) has been reported in at least one salamander species. Forester (1979) reported that female mountain dusky salamanders, *Desmognathus ochrophaeus*, show a preference for their own egg clutches over those of other females. Olfactory cues appear to be most important, although it is possible that visual or tactile cues could also be involved. In a second experiment, Forester *et al.* (1983) examined whether the chemical cues used for clutch recognition are produced by the eggs, or are produced by the female and only secondarily deposited on the eggs as olfactory markers. The evidence strongly suggests that the cues are produced by the eggs themselves. Females do not discriminate between two conspecific clutches when one has been 'marked' by the test female and the other has not. They do, however, show a significant preference for their own clutch as compared to an alien conspecific clutch which has also been 'marked' by the female. In this experiment 'marking' by the test female was simulated by placing the egg clutch on a filter paper substrate containing the female's odours. While the results of this experiment are highly suggestive, it is still possible that female odours are normally deposited *on* a clutch and that they produce a more concentrated stimulus capable of overriding the cues presented on the filter paper. It is also possible that non-chemical cues (e.g. tactile or visual cues) also affected the results of this study.

Recently, research in several other North American laboratories, has begun to examine sibling recognition among salamander larvae. Comparisons of the results of these studies with those obtained with anuran amphibians should prove to be extremely interesting. Furthermore, additional studies designed to elucidate the exact nature of the cues involved in a broader sampling of frog and toad species, as well as in salamanders, would undoubtedly result in a more sophisticated understanding of kin recognition cues in this taxon.

Birds

With only a few exceptions, most studies of kin recognition among birds have dealt with parent–offspring recognition, examining either the ability of parents to recognize their own young, or the ability of young to recognize their parents. Some avian groups have been studied extensively, but others have received only cursory attention. Additionally, despite the relatively large body of literature on parent–offspring interactions, most bird species have not been studied at all, and data are completely lacking for many families and genera.

Parent–offspring recognition has been studied more extensively in gulls, terns, and swallows than in any other groups. Beer (1969, 1970) reported that laughing gull chicks (*Larus atricilla*) are able to recognize their parents' calls and respond to them preferentially as compared to their responses to the calls of other adult conspecifics. Parents, on the other hand, do not appear to distinguish between the calls of their own chicks and those of alien chicks (Beer, 1979); parents tested with recorded playbacks of own and alien chick calls respond to both in the same manner. However, when an experiment was performed using live chicks hidden behind a barrier, Beer (1979) found that chicks respond only to the calls of their own parents and the parents then orient their parental behaviours towards those chicks that responded. In summary, in *L. atricilla* young appear to recognize the calls of their parents but parents do not recognize the calls of their young. Preferential behaviour of parents towards own young occurs because parents respond only to chicks that answer parental calls and the chicks only answer when their own parents call.

A similar situation appears to exist in herring gulls (*L. argentata*). Knudsen & Evans (1986) found that herring gull chicks recognize their parents by the adult's 'mew call' while parents cannot recognize the calls of their own chicks from those of alien chicks. Instead, parents discriminate between own and alien chicks based only on the behaviours and positive responses of own chicks to the parental calls.

In ring-billed gulls, *L. delawarensis*, the dynamics of parent–young recognition are similar but may also be more complex. Young ring-billed gulls distinguish the mew calls of their parents from those of other adults by the time the chicks have reached four days of age (Evans, 1970a). However, Evans (1970a) also found that chicks tested with playback recordings of parent and alien calls, are better at recognizing their parents when the two calls are presented simultaneously as compared to when they are presented sequentially. These results suggest that individual differences in parental calls may be sufficiently subtle that chicks do better when they can simultaneously compare calls rather than when they hear them one at a time.

As in the other species discussed above, parent ring-billed gulls cannot distinguish between the calls of their own and alien chicks, and the young's calls appear to be unimportant in parental recognition of young. Miller & Emlen (1975) surgically muted young ring-billed gulls and found that they were still accepted and cared for by their parents. On the other hand, when the appearance of the young was altered by changing their colour patterns with a felt tip pen, the responses of the parents were quite different. In some cases the parents initially strongly rejected and attacked their own young, in others the parents responded with ambivalent behaviours; in only a few cases did the adults fully accept their altered chicks (Miller & Emlen, 1975). Yet even in those cases in which the chicks were initially rejected, the rejection lasted only for a few hours, after which the chicks were once again fully accepted. Throughout the period of negative responses by the adults, the chicks continued to respond positively to their parents, and it appears that the chicks' behaviour ultimately overrides the effects of their altered appearance. Thus, in this species, and possibly in other *Larus* species, even though parents cannot recognize the calls of their young, they can discriminate between own and alien young based on visual morphological characters. Ultimately, however, the behaviour of the chicks appears to be the most important determinant for the preferential behaviour of adults towards their own chicks.

At least in one species, the black-billed gull (*L. butleri*) the responses of chicks towards their parents' calls appear to depend partially on learning. Adult *L. butleri* begin making mew calls as soon as the chicks hatch. Evans (1970b) played recordings of parent calls and alien adult calls to chicks of different ages and found that only 40 per cent of one-day-old chicks respond preferentially to their parents' calls. By the time they reach four days of age, 80 per cent of the chicks prefer the parental calls. Since chicks are capable of responding on day 1, it is likely that the improvement in responsiveness is due to learning rather than to a maturational effect. Sound spectrographic analysis of adult mew calls in this species confirmed that calls of different birds are individually different (Evans, 1970b).

Among terns, also members of the family Laridae, young recognize their parents by their 'fish calls'. Adults typically give the 'fish call' when they return to the nest with fish to feed the young. It is less clear whether adult terns are also able to recognize calls made by their own chicks, or whether, in a manner similar to that found in the gulls, they rely on the chicks' responses to distinguish between own and alien chicks.

Using recorded calls, Stevenson *et al.* (1970) found that common tern chicks (*Sterna hirundo*) recognize and respond to their parents' calls in preference to those of strange adults. Additionally, sound spectrographic analysis revealed individual differences in the calls of different adults (Stevenson *et al.*, 1970). Busse & Busse (1977, cited in Colgan, 1983) reported similar results in the arctic tern (*S. paradisaea*). Chicks respond at a significantly higher frequency to their parents' calls as compared to the calls of other adults.

In the sandwich tern (*S. sandvicensis*), behavioural studies have not been performed, but spectrographic analysis of adult fish calls revealed sufficient inter-individual variation to allow for parental recognition by young (Hutchinson *et al.*, 1968). Since young in this species appear to recognize their parents and respond to parental calls in the same fashion as do common and arctic tern chicks, it is reasonable to assume that parental recognition in sandwich terns also relies on individual differences in the adults' calls.

Although based on small sample sizes, the results of experiments exchanging royal tern chicks (*S. maxima maxima*) between nests, suggest that chicks respond to their parents' calls by calling in turn, and that parents use the chicks' calls to recognize their own chicks (Buckley & Buckley, 1972). The fact that adults do not appear able to recognize their own chicks when the chicks are silent, suggests that parents actually recognize their young by the chicks' calls. However, as discussed below, other interpretations are also possible.

In a second study examining recognition of young by their parents, Davies & Carrick (1962) reported that crested tern (*S. bergii*) adults recognize their own young after the latter have reached two days of age. Although spectrographic analysis showed inter-individual variation in the chicks' calls, Davies & Carrick (1962) concluded that the considerable amount of intra-individual variation in the calls preclude a definitive conclusion regarding the role of the chicks' calls in parental recognition of their young.

Given the results reported in gulls, caution should be exercised in interpreting the results of the preceding two experiments. It is possible that parents cannot distinguish between the calls of their own versus alien chicks, but that they respond parentally based on the responses of the chicks to the parent's calls. Since neither Buckley & Buckley (1972) nor Davies & Carrick (1962) used recorded calls, the behaviour of the chicks in both experiments could have affected the parents' responses. If adult terns recognize their own chicks based only on the responses given by chicks to their own parents, the situation in *Sterna* would be directly comparable to that found among the gulls. Furthermore, it is also of interest in this regard that Buckley & Buckley (1970) suggest that parent royal terns may recognize their own chicks based on individual differences in morphological characteristics such as coloration of legs, bills and down (see discussion in preceding section). It is not known whether similar traits are used for recognition in other species of terns.

Among swallows, parent–young recognition is based on acoustical cues in virtually all the species that have been studied. However, strong species differences exist in the ability of parents to recognize offspring and in the ability of offspring to recognize parents. These differences appear to have evolved as a result of different selective pressures in colonial species which show mixing of young, as compared to less colonial species in which young are more likely to remain in separate nests (see also Beecher, 1988, and this volume).

Among colonial species such as the bank (*Riparia riparia*) and cliff (*Hirundo pyrrhonota*) swallows, parents are clearly able to distinguish between the calls of own and alien young by the time the young become mobile. For example, Beecher *et al.* (1981) showed that bank swallow parents recognize their own chicks by the time the chicks are 16–17 days old. When presented simultaneously with recordings of their chicks' calls and those of alien chicks, parents preferentially.approached only the calls of their own chicks. Furthermore, acoustical analysis of chick calls revealed greater inter-individual than intra-individual differences, thereby confirming that calls show sufficient individual distinctiveness to serve as 'signature calls' mediating parental recognition of young (see also discussion in preceding section). Stoddard & Beecher (1983) report similar results in the cliff swallow. They also found that plumage patterns on the faces of cliff swallow fledglings are individually different, and suggest that parents may also be able to recognize their own chicks by these patterns. However, experiments aimed at testing this hypothesis were not performed.

Young bank and cliff swallows are also able to recognize the calls of their parents. Sieber (1985) found that the calls of adult bank swallows contain signature characters which vary more between individuals than within individuals. Parents use these calls when leading their young to and from the nest, and the young respond preferentially to the calls of their own parents. Similar results, showing that young recognize their own parents' calls from those of unfamiliar adults, have been found in the cliff swallow (Beecher *et al.*, 1985).

Unlike the situation found in colonial swallows, solitary or facultatively colonial species, such as the rough-winged swallow (*Stelgidopteryx serripennis*) and barn swallow (*Hirundo rustica*), do not appear to have evolved parent–offspring recognition (Beecher, 1982, 1988). Burtt (1977) did report that tree (*Iridoprocne bicolor*) and barn swallow parents recognize the calls of their own young over those of alien young. However, Medvin & Beecher (1986) seriously challenged his conclusions in the case of the barn swallow. They used both cross-fostering and playback experiments and found no evidence that parents can recognize the voice of their offspring. Chicks, however, did show evidence that they could recognize the voice of their parents, but their responses were less robust than those seen in bank and cliff swallow chicks. Medvin & Beecher (1986) conclude that, in Burtt's (1977) experiments, parents preferentially cared for their own chicks, not because they recognized their chicks' calls, but rather because they were responding to the behaviours of their chicks. Medvin & Beecher (1986) suggest that chicks confronted by alien adults show fear behaviours which elicit attacks from the adults; chicks confronted by their own parents show approach and non-fear responses which result in the parents approaching and feeding the young.

Despite the relatively broad body of literature on parent–offspring recognition in swallows, sibling recognition has been examined in only one species, the bank swallow. Beecher & Beecher (1983) used playback experiments to show that young bank swallows respond preferentially to the calls of their siblings as compared to those of unrelated swallows of a similar age. Furthermore, the swallows appear to learn the calls of siblings; swallows artificially exposed to the calls of non-siblings during early development, respond to these familiar non-sibling calls in the same manner as they do to sibling calls. Thus, they respond preferentially to calls that are familiar regardless of whether these come from genetically related or unrelated individuals.

Among the other passerines, only the blackbirds and jays have been examined. Redwing blackbird (*Agelaius phoeniceus*) females do not recognize their young before seven days post hatching. Evidence of recognition begins to appear by the time chicks are seven days old, and it is well developed by 10 days post hatching. At this time females approach their own displaced young but reject alien young (Peek *et al.*, 1972). Furthermore, Peek *et al.* (1972) suggest that the 'location calls' given by the young serve as the basis for recognition. These calls exhibit signature characteristics with clear inter-individual variation. In the European blackbird (*Turdus merula merula*), Messmer & Messmer (1956, cited in Colgan, 1983) found that mothers produce individually distinctive calls that are recognized by their young.

Pinyon jays (*Gymnorhinus cyanocephalus*) parents and young recognize each others' calls using individually distinctive call characteristics (McArthur, 1982). When they reach 14 days of age, nestlings begin to produce 'begging calls', while parents produce 'approach calls'. Sonographic analysis shows that both types of calls are highly individualistically different, and can, therefore, serve as signature calls. McArthur (1982) also found that the youngs' calls gradually change as they get older, suggesting that parents are able to track the calls of their young over time.

Among the anatids, Cowan (1973) showed that Canada geese (*Branta canadensis*) goslings can learn to respond to the calls of particular female geese. Early exposure to an individual female's call, accompanied by a moving pendulum, results in positive responses of goslings to that call, as compared to the calls of other females. This suggests that, under natural conditions, goslings may be able to distinguish their mother's calls from those of other adults. The apparent importance of including the swinging pendulum in the training process may mean that visual cues associated with movement are necessary for learning to occur. Canada geese goslings have also been shown to recognize siblings (Radesater, 1976). Although the cues responsible for recognition were not examined in this study, based on previously unpublished studies Radesater (1976) concludes that both vocal and visual cues are used but that visual cues may be the more important. Common eiders (*Somateria mollissima*) are another anatid species in which the young are reported to recognize their mothers by their calls (Colgan, 1983).

As in the case of Canada geese, domestic chicks (*Gallus gallus*) can learn to discriminate between the calls of two broody hens and show a preference for calls that are paired with a moving stimulus (Evans & Mattson, 1972; Falt, 1981). Cowan & Evans (1974) also found that chicks develop selective responses, (such as approaching and feeding) to individual maternal calls. They suggest that, under normal conditions, chicks learn to respond preferentially to their own mother's calls and that this then helps to maintain the integrity of the family unit.

The importance of calls in parent–offspring recognition has also been demonstrated in several other avian species. Adelie penguin (*Pygoscelis adeliae*) young are maintained in communal creches while their parents go out to hunt for food. On their parents' return, young Adelies respond only to their own parents' calls. Experiments using recorded calls confirmed that young recognize their parents' calls; chicks responded preferentially by approaching only the loudspeakers playing own parent calls, while not approaching loudspeakers playing the calls of other adults (Penney, 1962). Observational field studies suggest that young King penguins (*Aptenodytes patagonica*) also recognize the calls of their parents (Stonehouse, 1960, cited in Hutchinson et al., 1968). Tschanz (1964, cited in Hutchinson et al., 1968) found evidence of individual differences in the calls of common murres (*Uria aalge*). Additionally, using playbacks of recorded adult calls, he was able to show that young murres preferentially respond only to the calls of their parents. Young of the galah cockatoo (*Cacatua roseicapilla*) have also been reported to discriminate between the calls of their parents and those of other adults (Rowley, 1980).

Mammals

Studies of kin recognition among mammals have concentrated on sibling recognition in some groups and parent–offspring recognition in others. Ungulates, rodents and primates have been most extensively studied, and the cues used for recognition in each of these groups appear to be different.

One of the earliest studies examining the cues involved in maternal recognition of young in an ungulate was the classical study by Klopfer *et al.* (1964) on maternal imprinting in goats. At the time, Klopfer and his co-workers suggested that goat mothers imprint on the odours of their newborn kid and are, thereafter, able to distinguish between their own and alien kids. The work of Klopfer & Gamble, (1966) showing that anosmic females are unable to discriminate between own and alien young, provided additional support for the hypothesis that females recognize their young by olfactory cues. More recently, however, Gubernick (1981) challenged important aspects of this conclusion. He found that mother goats will accept alien kids that have had no contact with their mothers, but reject alien kids that have been licked by another female. Unlicked alien kids are accepted even after the female has already licked and accepted her own young. Thus, the critical variable in acceptance or rejection of alien kids appears to be whether the kid has had previous contact with, or been licked by, another female. These findings led Gubernick to propose that mother goats label their own kids and reject only kids that have been labelled by another female. He suggests that the labels may be found in the female's saliva and may be related to the rumen microfauna of each individual female. Alternatively, the label may be found in mother's milk and may be transferred to the young as they nurse. Gubernick (1981) proposes that, under normal conditions, all mothers label their own kids shortly after birth and then use their individually distinctive olfactory labels to discriminate between their own and alien kids. Thus, at present, there appears to be little doubt that goat mothers recognize their offspring by odours; the only question which remains under discussion is whether the odours are produced by the kids themselves or by their mothers. If Gubernick's conclusion is correct, this would constitute another clear example of an environmentally acquired label which may be influenced by the action of bacterial flora (see discussion in preceding section).

The sensory cues that mediate parent–offspring recognition in sheep (*Ovis aries*) have received much attention. The complexity of the results, and the fact that several different modalities appear to be involved, make the results somewhat more difficult to interpret.

Lindsay & Fletcher (1968) first demonstrated that ewes recognize their own lambs. They suggested that both hearing and vision are important, particularly in recognition over long distances, and that olfaction may also play a role at the time of suckling. Bouissou (1968) confirmed the importance of olfactory cues by demonstrating that mothers that had undergone olfactory bulbectomy fail to recognize their own young and accept all lambs that attempt to suckle from them. The importance of hearing in long distance recognition has also been confirmed. Poindron & Carrick (1976) showed that ewes respond preferentially to the recorded bleats of their own lambs. Other studies, employing different methodological techniques (e.g. Morgan *et al.*, 1975; Alexander, 1977, 1978; Alexander & Stevens, 1981) have also supported the importance of auditory and olfactory cues in the recognition of lambs by their mothers.

Visual cues have also been shown to be important. Alexander (1977) changed the appearance of lambs by either dyeing them black or shearing them and found that this affected the ability of ewes to recognize their own young. Furthermore, lambs also appear to rely, at least partly, on vision to recognize their mothers. Alexander (1977) and Alexander & Shillito Walser (1978) demonstrated that lambs that are less than one week old use both auditory and visual cues to recognize their mothers, but that vision becomes more important as the lambs get older. By the time they are three weeks old, lambs appear to have learned their mother's appearance and approach only those females that look like their mothers.

Only one study has examined sibling recognition in sheep. Using a T-maze, Shillito Walser et al. (1983) demonstrated that lambs can discriminate between siblings and non-siblings, and show a preference for siblings. Although the sensory cues mediating sibling recognition were not tested, the results suggested that vocal, visual and olfactory cues may all play a role.

Parent–offspring recognition has been looked at in only a few other ungulate species. In reindeer (*Rangifer tarandus*), Espmark (1971) showed that females respond to the tape recorded 'distress calls' of their own offspring, but not to those of alien young. Likewise, reindeer young respond to their own mother's calls but not to those of other females. Among pigs (*Sus domesticus*), mothers appear to discriminate between own and alien young by using olfactory cues; anosmic females are unable to make this discrimination (Meese & Baldwin, 1975, cited in Leon, 1983).

Mares and foals of the domestic horse (*Equus caballus*) have been shown to use both visual and olfactory cues for parent–offspring recognition. It is not yet clear, however, if auditory cues are also used. Based on a series of experiments in which they attempted to alter the appearance and/or the odours of mares and their foals, Wolski et al. (1980) concluded that long range parent–offspring recognition is based on postures and behaviours, while short range recognition is mediated by odours. Surprisingly, playback experiments using recorded 'neighs' of females or foals failed to provide any evidence of recognition; both mares and foals responded to all neighs indiscriminately. It is possible, however, that a different vocalization, the 'nicker', a low frequency, low amplitude, pulsated sound used when two horses are in close proximity, may play a role in parent–offspring recognition. Wolski et al., (1980) did not test 'nickers' but both they and Tyler (1972), based on field observations, suggest that it may play a role in short-range recognition.

Although the ability of mothers to recognize their own offspring has been reported in a variety of bat species and it has been suggested that vocal communication is involved (see Balcombe, 1990), the sensory cues used in recognition have been examined experimentally in only a few species. Female Mexican free-tailed bats (*Tadarida brasiliensis mexicana*) find their offspring in creches which may consist of millions of individuals. Playback experiments have demonstrated that Mexican free-tailed bat mothers are able to discriminate the isolation calls of their own pups from those of unrelated young; pups are unable to recognize the echolocation calls of their mothers, but may respond to other calls made by the mothers while they search (Balcombe, 1990). This report is consistent with Gelfand & McCracken's (1986) earlier finding that the isolation calls of Mexican free-tailed bat pups are individually different. Olfaction may also play a role in mother–offspring recognition in bats. Gustin & McCracken (1987), for example, reported that Mexican free-tailed bat mothers may mark their own offspring with secretions from a muzzle gland, and then use the scent from these secretions to recognize their own young; pups, however, do not appear able to recognize their mothers by their odours. The importance of auditory and olfactory cues in kin recognition in bats merits further study using experimental techniques that can distinguish among the different sensory modalities involved (Balcombe, 1990).

Although sibling recognition among rodents has received considerable attention, only a relatively small number of studies have looked specifically at the cues that are involved. Porter *et al.* (1978), showed that spiny mice with zinc sulfate induced anosmia lose the ability to discriminate between siblings and non-siblings; thus, they were the first to demonstrate the importance of olfactory cues in sibling recognition among rodents. In a more recent experiment, Porter *et al.* (1986) found that spiny mice exposed to only one of their siblings preferentially associate with that one familiar sibling as compared to unfamiliar siblings. The importance of chemical cues was confirmed by showing that spiny mice rendered anosmic by treatment with zinc sulfate lost the ability to discriminate between familiar and unfamiliar siblings. As discussed in the previous section, both diet and genotypic differences appear to contribute to the individual and family-specific odours that mediate sibling recognition in this species.

In the Norway rat (*Rattus norvegicus*), results from several studies indicate that sibling recognition is mediated by olfactory cues. Hepper (1983) found that rats can recognize siblings even when they have had no contact post-natally with any of their siblings. In this experiment rats had to choose between siblings or non-siblings in an olfactometer. Since visual contact was prevented and a white noise generator was used to dampen auditory cues, it is likely that the preference shown for siblings was based primarily on chemical cues. A second experiment (Wills *et al.*, 1983) confirmed these results. Rat pups were given a choice of substrates that had been soiled or marked by stimulus pups. Since the stimulus pups were not present in the test apparatus, only olfactory information contained in the substrate was available to the test pups. Under these conditions, pups were able to discriminate between siblings and non-siblings, regardless of their degree of familiarity, as well as between familiar and unfamiliar siblings, and between familiar and unfamiliar non-siblings.

Olfactory cues have also been demonstrated to be important in other rodent species. Zinc sulfate induced anosmia destroys the ability of thirteen-lined ground squirrels (*Spermophilus tridecemlineatus*) to discriminate between siblings and non-siblings (Holmes, 1984). Mongolian gerbil (*Meriones unguiculatus*) juveniles are attracted to the saliva of their siblings as compared to that of non-siblings. Since the pups respond to saliva alone, in the absence of the actual stimulus animals, it is clear that sibling recognition can be based solely on chemical cues (Block *et al.*, 1981).

Parent–offspring recognition has been studied in a variety of rodents and olfactory cues again appear to be of primary importance. Beach & Jaynes (1956) demonstrated that Norway rat females retrieve their own pups over strange pups and that lesioning of the olfactory bulbs destroys their ability to make this discrimination. Rat pups are also able to distinguish between their own mother and other lactating females, but as previously discussed, their ability to do so is at least partly dependent on differences in the diets of the females (Leon, 1975). The ability of Mongolian gerbil pups to discriminate between their mothers and other females is also mediated by odours, and diet again appears to be an important component (Skeen & Thiessen, 1977). In spiny mice, chemical cues dependent on the diets of the mothers likewise appear to be important in the ability of young to recognize their mothers, as well as in the ability of mothers to recognize their young (Doane & Porter, 1978).

Porter *et al.* (1973) found that guinea pig (*Cavia porcellus*) pups do not discriminate between their own mothers and unfamiliar lactating females. Females, on the other hand, when tested at 48 hours post-partum, show a preference for their own litter as compared to an unrelated litter of the same age. The probable importance of chemical cues is indicated by the fact that females also prefer an odour that has been applied to their own litter as compared to an unfamiliar odour.

Among other mammals, Petrinovich (1974) showed that northern elephant seal (*Mirounga angustirostris*) mothers recognize the vocalizations of their own pups. Given a choice of a recording of their own pup calls and those of an alien pup, the female responds preferentially to the calls of her own pup. Domestic kittens orient to the odours of their own nest as compared to those from the nest of another litter of the same age (Freeman & Rosenblatt, 1978). It is unclear, however, whether the olfactory cues being used are maternal in origin or are produced by the kittens in the litter. It is possible that both cues are used and that the ability of kittens to find their nests by odours represents both sibling and offspring–mother recognition and that the odours involved are group odours. Domestic puppies have also been shown to recognize their siblings. Based on the behaviours of 4.5 to 5.5-week-old puppies of several different breeds, Hepper (1986b) suggests that visual cues are most important at a distance but that chemical cues are used at closer range.

Kin recognition among non-human primates relies on a variety of cues and sensory modalities. Vocal and chemical cues appear to be particularly important in parent–offspring recognition.

Among macaques, both mothers and young may use vocalizations to recognize one another. Thus, Hansen (1976) found that juvenile rhesus monkeys (*Macaca mulatta*) that have been separated from their mothers respond preferentially to the 'coo calls' of their mothers, but not to those of other females. Upon hearing their mothers, juveniles immediately vocalized back and approached the source of the sound. Since the mothers were not visible to the young, and the young responded only after hearing their mothers' calls, it is highly unlikely that visual or olfactory cues influenced the results. In Japanese macaques (*M. fuscata*), mothers can distinguish between the vocalizations of their own young and those of alien young. Mothers respond more strongly to the playback recordings of the 'coo calls' of their own offspring than they do to the calls of other juveniles (Pereira, 1986).

It is interesting, however, that, in some species of macaques, mothers and offspring do not appear able to recognize each other by calls. For example, Simons *et al.* (1968) and Simons & Bielert (1973) found no evidence that female pigtailed macaques (*M. nemestrina*) respond preferentially to the calls of their own offspring, nor that young respond preferentially to their mothers' calls. Furthermore, Hansen (1976) also failed to find preferential responsiveness of female rhesus monkeys to the calls of their young. In this latter experiment the calls made by females in response to the vocalizations of their own versus alien young were the only behaviours monitored. Therefore, it is possible that recognition did occur but that it was not expressed by calling. The importance of other cues (e.g. visual and olfactory) in mother–offspring recognition in *M. fuscata* and in the ability of mothers to recognize their young in *M. mulatta* has not been examined.

Playback experiments have demonstrated that female vervet monkeys (*Cercopithecus aethiops*) also recognize the calls of their own young (Cheney & Seyfarth, 1980). Mothers responded faster and approached the speakers more often when they heard the recorded calls of their own young as compared to the calls of other young. Additionally, females appeared to recognize the calls of individual young, even when they were not their own young; when a juvenile's call was played, the other females in the group would often turn and look at the mother of the juvenile making the vocalizations.

In squirrel monkeys (*Saimiri sciurus*), mothers can recognize their own young by their calls. Kaplan *et al.* (1978) found that mothers who could hear but not see their infants, vocalized more and approached and spent more time near the source of the call in response to their own infant's calls as compared to the calls of unrelated infants. Symmes & Biben (1985) confirmed these results by using playback recordings of own versus alien young. It is unclear whether squirrel monkey infants can also recognize the calls of their mothers, but Kaplan *et al.* (1977) found that they are able to discriminate between their own mothers and other lactating females based solely on olfactory cues.

Sibling recognition in monkeys has been examined under controlled experimental conditions only in the pigtailed macaque. The results of Frederickson & Sackett (1984) suggest, but do not conclusively demonstrate, that visual cues are sufficient for sibling recognition in this species.

Humans are clearly able to recognize kin based on a variety of cues and combinations of cues. Personal experience suggests that visual cues (e.g. facial features, hair length and colour, skin colour, and posture) are most important, followed by auditory cues (e.g. voice pitch, tone or texture). Although sometimes one may be able to recognize unfamiliar kin because of a 'family resemblance', in most cases recognition appears to depend on individually distinctive cues, of either genetic or environmental origin, which are learned as a result of social experience. Thus, the cues that are used to recognize kin from non-kin are generally no different from the cues used to distinguish among two or more individuals.

Most studies on the cues used for kin recognition in humans have concentrated on the use of olfactory cues in parent–offspring recognition. The positive results obtained in these studies are particularly interesting because most people are relatively unaware of the importance of odours in human communication.

The ability of women to recognize the odours of their newborn children is well established. Schaal *et al.* (1980), Porter *et al.* (1983), and Kaitz *et al.* (1987) have all demonstrated that mothers can identify the odours of garments worn by their own newborn infants as compared to those worn by unfamiliar infants. In all of these experiments the garments worn by own and alien infants were identical and/or they were presented to the mothers in such a way that they could smell but not see them. Thus, other sensory modalities were controlled for and it can be concluded that the discrimination was based solely on olfactory cues. The ability of mothers to recognize the odours of their infants was already present at 24 to 48 hours post-partum, and after only one hour, or less, of exposure to the baby.

Using a different experimental procedure, Russell (1983) and Russell *et al.* (1983) showed that shortly after delivery women can also identify the odours of their babies merely by sniffing the tops of their heads. Interestingly, fathers could not identify their babies' odours even 24 to 48 hours after birth.

Newborn babies have also been shown to recognize the odours of their own mothers as compared to those of other newly parous females. Babies exposed to breast pads previously worn by their own mothers make suckling movements and/or quickly quiet down; on the other hand, when exposed to the breast pads of other lactating women they either turn away from them or ignore them (McFarlane, 1975; Russell, 1976; Schaal *et al.*, 1980).

Visual characteristics are also utilized by mothers to recognize their newborn babies. Porter *et al.* (1984) found that, at 33 hours post-partum, mothers could pick out photos of their own children from those of three other neonates. In all cases the babies presented together in a session had been matched according to age, sex, amount and colour of hair, and colour of the swaddling blanket. It was not determined in this experiment whether mothers had learned the visual characteristics of their babies, or whether they were relying on family similarities to make the identification.

Among older children and their parents, olfactory cues may also provide sufficient information to allow for parental recognition of young. Using only olfactory cues, parents can distinguish the odours of their own children over those of unfamiliar children of the same age (Porter & Moore, 1981). The same study showed that parents can also distinguish between the odours of two of their children, suggesting that individually-distinctive odour differences are important in this discrimination.

The importance of olfactory cues in human kin recognition is also supported by the fact that children can distinguish the odours of their siblings from those of non-siblings (Porter & Moore, 1981). Additionally, Porter *et al.* (1985) found that unrelated individuals can match human mothers with their children based solely on olfactory cues. This clearly suggests the existence of a shared family odour among humans.

Before ending this section, it should be mentioned that, in a number of mammalian species, females may mark their young with secretions from specialized scent glands. For example, female marking of young has been reported in the rabbit, *Oryctolagus cuniculus* (Mykytowycz & Dudzinsky, 1972), the Mongolian gerbil, *Meriones unguiculatus* (Wallace *et al.*, 1973), the dwarf mongoose, *Helogale undulata rufula* (Rasa, 1973), and the flying phalanger, *Petaurus breviceps papuanus* (Schultze-Westrum 1969). The ability of females to use their own scent marks to distinguish between own and alien young has not been examined. However female rabbits will attack their own young if they are smeared with secretions from a strange female (Mykytowcyz, 1968). The responses of females to alien young smeared with the females' own odours were not tested. In the Mongolian gerbil, females prefer to retrieve pups marked with the female's odour as compared to unmarked pups (Wallace *et al.*, 1973), but it is not known whether they also show a preference for pups marked with own secretions compared to pups marked with secretions from another lactating female.

Discussion and conclusions

Like all communication systems, kin recognition involves three separate components: a sender, a signal and a receiver. Until recently, most studies of kin recognition have emphasized exclusively the receiver or recognition component of the system. Under the influence of kin selection theory (Hamilton, 1964), the focus has been on how the receiver in a kin recognition system is able to recognize its kin. In other words, the question that was most frequently asked addressed the neural mechanisms by which the receiver analyzes information on genetic relatedness. Specifically, most early studies focused on whether the recognition mechanism is a fixed, genetically-determined, neural template that is independent of learning (e.g. Hamilton's 1964 'recognition alleles'; Dawkin's 1976 'green beard effect'), or whether it is a mechanism that relies on learning. Even after it became generally acknowledged that recognition by association and recognition by phenotype matching are probably the two most common mechanisms of kin recognition (e.g. Holmes & Sherman, 1983), the emphasis remained on the receiver. The question then became whether the receiver has the ability to match the phenotype of an unfamiliar kin to the phenotype of self and of familiar kin, or whether the receiver needs to have direct previous experience with each particular kin before it is able to recognize them.

Beecher (1982, 1988) has expanded this analysis by examining the selective pressure operating on the sender in different ecological or social environments. He stresses the need to go beyond hypotheses that address only the receiver or recognition component of communication systems. He proposes (Beecher, 1988) that kin recognition should evolve only when the benefits to *both* sender and receiver are greater than the costs. In situations in which it is to the sender's benefit to accurately indicate its identity and genetic relatedness to the receiver (e.g. in species with colonial nesting where parents could mistakenly feed non-related young), natural selection should act on the sender (i.e. the young in the preceding example) to favour the evolution of distinct, clear-cut, and easily discriminable 'signature' cues. When the costs exceed the benefits because it is too risky for the sender to identify itself (e.g. in species where young could be misplaced to the wrong nest and be attacked by adults that recognize it as non-kin), natural selection should favour cues that help to hide the true identity and genetic relatedness of the sender from the receiver. Thus, in our hypothetical example, the young would not provide information on genetic relatedness, thereby preventing the alien adults from recognizing it as unrelated. Because in such a case parents cannot distinguish between own and alien young, they are more likely to treat all young found at the nest as if they were kin. It is also noteworthy that, in such a system of parent–offspring recognition, the ability to recognize kin may be assymetrical. That is, in the above examples, parents may not be able to recognize their own young, but the young might very well be able to recognize their parents.

This, of course, is precisely the situation found in parent–offspring recognition among the gulls discussed in the preceding section (see also Beecher, 1988, and this volume).

With only a few exceptions to date, only a small number of reviews have examined the actual signals used in kin recognition. In discussing the selection pressures that operate on the sender, Beecher (1982, 1988) simultaneously addressed some of the selection pressures that operate on the signals themselves. Hepper (1986a) also discussed some of the characteristics of the signals used as cues in kin recognition systems. In this chapter I have focused specifically on the signals used in kin recognition by examining their information content (i.e. individually distinctive versus family specific cues), their origins (i.e. environmentally-acquired versus genetically influenced) and their sensory modalities.

In terms of information content, the available data suggest that many (perhaps even most) of the cues used for kin recognition are individually distinctive. Each animal's individual 'signature' probably results from the variation in the characteristics or combinations of the individual component traits that make up each cue. This variation from one individual to another confers sufficient uniqueness to the cue to make it individually distinctive and, therefore, useful as a 'signature' for kin recognition.

The variation in the component traits of cues may also explain, in some cases, the ability to phenotype match. Although each individual may have its own unique signature, the signatures of more closely related individuals may be more similar than the signatures of genetically unrelated individuals. Furthermore, the degree of similarity may, in some cases, be positively and directly correlated to the degree of kinship. Once an animal has learned the individual signatures of its known kin (and perhaps the characteristics that these individual signatures have in common), it may also be able to assess the likelihood that a newly-encountered, unfamiliar animal is also its kin by comparing the degree of overlap between the component traits of the signature cues of the unknown individual and those of its known kin.

Cues that are not individually distinctive may, nonetheless, still convey information on genetic relatedness or family specificity. For example, all members of a clutch could carry a label, whether genetic or environmentally acquired, that identifies them as belonging to a common clutch (e.g. Waldman, 1981). In such a situation, the animals may not be able to distinguish between their siblings as individuals, but they can distinguish between their siblings and non-siblings. Furthermore, cues that convey information on family relatedness need not be identical from one individual to another. Cues that show greater intra-individual than inter-individual variation would be useless as individual identifiers. Yet these same cues could provide information on family relatedness if there is less intra-family than inter-family variation, and/or if the cues produced by all the members of a given family share a common family-specific marker.

Although only a small number of species has been investigated with regards to the origins of kin recognition cues, it is already clear from the data reviewed in previous sections that both genetic and environmental factors may be involved. Animals may be able to recognize their kin using only environmentally acquired or only genetically influenced cues. However, many animals may be able to use both types of cues, perhaps simultaneously; moreover, in some species, environmental and genetic factors may interact to produce the unique individual or family-specific cues needed for kin recognition. Such interactions can be quite complex, with each factor providing additive information which is then interpreted and used by the receiver to assess genetic relatedness. A different but also interesting type of interaction is exemplified by those species in which mothers mark or label their offspring, usually with a common olfactory cue. The label produced by the mother may well have a genetic basis, but it is acquired environmentally by the young. Such cues can also be acquired environmentally by the mother (e.g. diet dependent odours) who then passes them to her offspring.

Additional and more thorough studies on the origins of kin recognition cues are necessary. It is much too easy to examine a species and determine that it uses genetically influenced or environmental cues but then neglect to search for the alternative factor or for interactions between factors. Indeed, it might prove to be quite worthwhile to re-examine in this regard even those species that have already been studied. Not only may there be complex interactions between genotypic and environmental factors, but very subtle environmental influences may affect kin recognition in ways that are neither expected nor well-understood (see also Porter, 1988).

While recognition by association and recognition by phenotype matching are often presented as distinct alternatives with little relation to one another, these two mechanisms are, in fact, quite similar. Since both mechanisms depend on the learning of cues that identify kin, both are dependent on association. The only meaningful difference between them is the type of cue that is learned during the period of association. In recognition by association the animal learns the unique, individually-distinctive phenotypes (signatures) of those kin with which it associates; in future encounters it recognizes as kin only those specific individuals. By contrast, in phenotype matching the animal learns the family-specific characteristics of its own phenotype or of the phenotypes of those kin with which it associates; it then assesses genetic relatedness by comparing the phenotypes of strangers to those of self or known kin. Porter (1988) suggests that recognition by association is equivalent to recognition by 'direct familiarization', while recognition by phenotype matching represents recognition by 'indirect familiarization'. Thus, it could be argued that recognition by association and recognition by phenotype matching are nothing more than different forms of the same 'association' or 'familiarization' mechanism (see also Halpin & Hoffman, 1987;

Porter, 1988). Furthermore, recognition by association and by phenotype matching do not differ in the origin of the cues upon which they rely. In both cases the cues may be of either genotypic or environmental origin.

In reviewing studies on the sensory modalities used for kin recognition among vertebrates, an interesting pattern became apparent. The majority of studies of parent–offspring recognition (with the notable exception of the work by Beecher and his colleagues) does not appear to have been greatly influenced by the interest in kin selection and kin recognition which followed Hamilton's (1964) classic paper. In fact, most of these studies appear to take for granted that parents and offspring should recognize each other and the emphasis is primarily on determining the cues and sensory modalities involved in the recognition. By contrast, studies of sibling recognition began in earnest only after the appearance of Hamilton's paper and the focus has been almost exclusively on the perceptual and decision making mechanisms by which animals recognize their siblings. Indeed, despite the very large number of studies of sibling recognition, the actual cues that are used for recognition have been examined in only a limited number of species. As a result, while there is an extensive literature on the cues used for parent–offspring recognition, there is considerably less for sibling recognition.

A problem which many of the studies of the sensory modalities used for kin recognition have in common is that they examine only one modality. Yet it is not unreasonable to believe that many species may produce redundant information on genetic relatedness and that more than one sensory modality may be employed. For example, while most studies suggest that calls play a major role in avian kin recognition, other studies indicate that, in at least some species, visual cues may also be important. Olfactory cues appear to be most important for rodents, yet it is certainly possible that other cues, such as perhaps ultrasound, could provide information on kin relatedness. In fact, Beauchamp *et al.* (1985) suggest that the mouse H-2 complex which produces individually distinctive odours, might also have pleiotropic effects resulting in individual differences in ultrasound production. Furthermore, as evidenced by studies that report assymetries in kin recognition, the cues used may be very subtle and 'recognition' may occur only secondarily as a result of the overall behaviour or 'disposition' of one of the animals. Thus, in most species of gulls, parents appear unable to recognize their chicks' calls, yet they respond preferentially to their own chicks as a result of the chicks selective responses to their parents' calls. Even when chicks are muted and altered in appearance, the parents continue to 'recognize' them (i.e. behave preferentially towards them) because the chicks respond in a positive and confident manner to their own parents as compared to other adults. Evidence that animals may use several different sensory channels to communicate information on genetic relatedness comes from the few species (e.g. sheep, horses and dogs) in which more than one sensory modality has been examined; kinship relatedness in all of these species was found to be communicated by more than one sensory modality. Thus, more studies that focus on the complexity and potential redundancy of kin-recognition cues are necessary.

In summary, although identification signals are crucial components of kin-recognition systems, they have received relatively little attention until recent years. Specifically, proximate questions on the nature of the cues used for kin identification, and ultimate questions on their evolution, have only recently begun to be addressed (Beecher, 1982, 1988; Hepper, 1986a). Yet the information that is already available on the diversity and complexity of kin-recognition cues has made it clear that the study of this signal component should be considered a major challenge for investigations of kin-recognition systems in the future.

Acknowledgements

I am extremely grateful to Connie Quinlan and Michael Arduser for their invaluable assistance in finding and xeroxing references. I also thank Cheryl Wilke, Michael Phillips, Chris Barfield, and George Taylor for bringing relevant references to my attention.

References

Alexander, G. (1977). Role of auditory and visual cues in mutual recognition between ewes and lambs in Merino sheep. *Appl. Anim. Ethol.*, **3**, 65–81.

Alexander, G. (1978). Odour and the recognition of lambs by Merino ewes. *Appl. Anim. Ethol.*, **4**, 153–8.

Alexander, G. & Shillito Walser, E. E. (1978). Visual discrimination between ewes by lambs. *Appl. Anim. Ethol.*, **4**, 81–5.

Alexander, G. & Stevens, D. (1981). Recognition of washed lambs by merino ewes. *Appl. Anim. Ethol.*, **7**, 77–86.

Balcombe, J. P. (1990). Vocal recognition of pups by mother Mexican free-tailed bats, *Tadarida brasiliensis mexicana. Anim. Behav.*, **39**, 960–6.

Barnett, C. (1977a). Chemical recognition of the mother by the young of the cichlid fish, *Cichlasoma citrinellum. J. Chem. Ecol.*, **3**, 463–8.

Barnett, C. (1977b). Aspects of chemical communication with special reference to fish. *Biosci. Commun.*, **3**, 331–92.

Bateson, P. P. G. (1982). Preference for cousins in Japanese quail. *Nature*, **295**, 236–7.

Beach, F. A. & Jaynes, J. (1956). Studies in maternal retrieving in rats: recognition of young. *J. Mammal.*, **37**, 177–80.

Beauchamp, G. K., Yamazaki, K. & Boyse, E. A. (1985). The chemosensory recognition of genetic individuality. *Sci. Am.*, **253**, 86–92.

Beauchamp, G. K., Yamazaki, K., Bard, J. & Boyse, E. A. (1988). Preweaning experience in the control of mating preferences by genes in the major histocompatibility complex of the mouse. *Behav. Genet.*, **18**, 537–47.

Beecher, J. M. & Beecher, M. D. (1983). Sibling recognition in bank swallows

(*Riparia riparia*). *Z. Tierpsychol.*, **62**, 145–50.

Beecher, M. D. (1982). Signature systems and kin recognition. *Am. Zool.*, **22**, 477–90.

Beecher, M. D. (1988). Kin recognition in birds. *Behav. Genet.*, **18**, 465–82.

Beecher, M. D., Beecher, I. M. & Hahn, S. (1981). Parent–offspring recognition in bank swallows (*Riparia riparia*): II. Development and acoustic basis. *Anim. Behav.*, **29**, 95–101.

Beecher, M. D., Stoddard, P. K. & Loesche, P. (1985). Recognition of parents' voices by young cliff swallows. *Auk*, **102**, 600–5.

Beer, C. G. (1969). Laughing gull chicks: recognition of their parents' voices. *Science*, **166**, 1030–2.

Beer, C. G. (1970). On the responses of laughing gull chicks (*Larus atricilla*) to the calls of adults. I. Recognition of the voices of the parents. *Anim. Behav.*, **18**, 652–60.

Beer, C. (1979). Vocal communication between laughing gull parents and chicks. *Behaviour*, **70**, 118–46.

Bekoff, M. (1981). Mammalian sibling interactions: genes, facilitative environments, and the coefficient of familiarity. In *Parental Care in Mammals*, ed. D. J. Gubernick & P. H. Klopfer, pp. 307–46. London: Plenum Press.

Blaustein, A. R. (1983). Kin recognition mechanisms: phenotypic matching or recognition alleles? *Am. Nat.*, **121**, 749–54.

Blaustein, A. R. & O'Hara, R. K. (1982a). Kin recognition cues in *Rana cascadae* tadpoles. *Behav. Neural. Biol.*, **36**, 77–82.

Blaustein, A. R. & O'Hara, R. K. (1982b). Kin recognition in *Rana cascadae* tadpoles: maternal and paternal effects. *Anim. Behav.*, **30**, 1151–7.

Block, M. L., Volpe, L. C. & Hayes, M. J. (1981). Saliva as a chemical cue in the development of social behaviour. *Science*, **211**, 1062–1064.

Bouissou, M. F. (1968). Effet de l'ablation des bulbes olfactifs sur la reconnaissance du jeune par sa mere chez les Ovins. *Rev. Comp. Anim.*, **32**, 77–83.

Brown, R. E., Singh, P. B. & Roser, B. (1987). The major histocompatibility complex and the chemosensory recognition of individuality in rats. *Physiol. Behav.*, **40**, 65–73.

Buckley, P. A. & Buckley, F. G. (1970). Color variation in the soft parts and down of royal tern chicks. *Auk*, **87**, 1–13.

Buckley, P. A. & Buckley, F. G. (1972). Individual egg and chick recognition by adult royal terns (*Sterna maxima maxima*). *Anim. Behav.*, **20**, 457–62.

Burtt, E. H., Jr. (1977). Some factors in the timing of parent-chick recognition in swallows. *Anim. Behav.*, **25**, 231–9.

Cheney, D. L. & Seyfarth, R. M. (1980). Vocal recognition in free ranging vervet monkeys. *Anim. Behav.*, **28**, 362–7.

Colgan, P. (1983). *Comparative Social Recognition*. New York: John Wiley & Sons.

Cowan, P. J. (1973). Parental calls and the approach behavior of young Canada geese: a laboratory study. *Can. J. Zool.*, **51**, 647–50.

Cowan, P. J. & Evans, R. M. (1974). Calls of different individual hens and the parental control of feeding behaviour in young *Gallus gallus*. *J. Exp. Zool.*, **188**, 353–60.

Davies, S. & Carrick, R. (1962). On the ability of crested terns, *Sterna bergii*, to recognize their own chicks. *Aust. J. Zool.*, **10**, 171–7.

Dawkins, R. (1976). *The Selfish Gene*. Oxford: Oxford University Press.

Doane, H. M. & Porter, R. H. (1978). The role of diet in mother-induced reciprocity in the spiny mouse. *Devel. Psychobiol.*, **11**, 271–7.

Egid, K. & Lenington, S. (1985). Responses of male mice to odors of females: effects of T- and H-2 locus genotype. *Behav. Genet.*, **15**, 287–95.

Epsmark, Y. (1971). Individual recognition by voice in reindeer mother–young relationship. Field observations and playback experiments. *Behaviour*, **40**, 295–301.

Evans, R. M. (1970a). Imprinting and mobility in young ring-billed gulls, *Larus delawarensis*. *Anim. Behav. Monog.*, **2**, 193–248.

Evans, R. M. (1970b). Parental recognition and the 'mew call' in black-billed gulls (*Larus butleri*). *Auk*, **87**, 503–13.

Evans, R. M. & Mattson, M. E. (1972). Development of selective responses to individual maternal vocalizations in young *Gallus gallus*. *Can. J. Zool.*, **50**, 777–80.

Falt, B. (1981). Development of responsiveness to individual maternal 'clucking' by domestic chicks (*Gallus gallus domesticus*). *Behav. Proc.*, **6**, 303–17.

Forester, D. C. (1979). Homing to the nest by female mountain dusky salamanders (*Desmognathus ochrophaeus*) with comments on the sensory modalities essential to clutch recognition. *Herpetologica*, **35**, 330–5.

Forester, D. C., Harrison, K. & McCall, L. (1983). The effects of isolation, the duration of brooding, and non-egg olfactory cues on clutch recognition by the salamander, *Desmognathus ochrophaeus*. *J. Herpetol.*, **17**, 308–14.

Frederickson, W. T. & Sackett, G. P. (1984). Kin preferences in primates (*Macaca nemestrina*): relatedness or familiarity? *J. Comp. Psychol.*, **98**, 29–34.

Freeman, N. C. G. & Rosenblatt, J. S. (1978). Specificity of litter odors in the control of home orientation among kittens. *Devel. Psychobiol.*, **11**, 459–68.

Gelfand, D. L. & McCracken, G. F. (1986). Individual recognition in the isolation calls of Mexican free-tailed bat pups (*Tadarida brasiliensis mexicana*). *Anim. Behav.*, **34**, 1078–86.

Gorman, M. L. (1976). A mechanism for individual recognition by odour in *Herpestes auropunctatus*. *Anim. Behav.*, **24**, 141–5.

Gubernick, D. J. (1981). Mechanisms of maternal 'labelling' in goats. *Anim. Behav.*, **29**, 305–6.

Gustin, M. K. & McCraken, G. F. (1987). Scent recognition between females and pups in the bat *Tadarida brasiliensis mexicana*. *Anim. Behav.*, **35**, 13–19.

Halpin, Z. T. (1986). Individual odors among mammals: origins and functions. *Adv. Study Behav.*, **16**, 40–70.

Halpin, Z. T. & Hoffman, M. D. (1987). Sibling recognition in the white-footed mouse, *Peromyscus leucopus*: association or phenotype matching? *Anim. Behav.*, **35**, 563–70.

Hamilton, W. D. (1964). The genetical evolution of social behaviour, I and II. *J. Theor. Biol.*, **7**, 1–52.

Hansen, E. W. (1976). Selective responding by recently separated juvenile rhesus monkeys to the call of their mother. *Devel. Psychobiol.*, **9**, 83–8.

Hay, T. F. (1977). Filial imprinting in the convict cichlid *Cichlasoma nigrofasciatum*. *Behaviour*, **65**, 138–60.

Hepper, P. G. (1983). Sibling recognition in the rat. *Anim. Behav.*, **31**, 1177–91.

Hepper, P. G. (1986a). Kin recognition: functions and mechanisms. A review. *Biol. Rev.*, **61**, 63–93.

Hepper, P. G. (1986b). Sibling recognition in the domestic dog. *Anim. Behav.*, **34**, 288–9.

Hepper, P. G. (1987). The discrimination of different degrees of relatedness in the rat: evidence for a genetic identifier? *Anim. Behav.*, **35**, 549–54.

Hepper, P. G. (1988). Adaptive fetal learning: prenatal exposure to garlic affects postnatal preferences. *Anim. Behav.*, **36**, 935–6.

Holmes, W. G. (1984). Sibling recognition in thirteen-lined ground squirrels: effects of relatedness, rearing association, and olfaction. *Behav. Ecol. Sociobiol.*, **14**, 225–33.

Holmes, W. G. & Sherman, P. W. (1983). Kin recognition in animals. *Am. Sci.*, **71**, 46–55.

Hutchinson, R. E., Stevenson, J. G. & Thorpe, W. H. (1968). The basis for individual recognition by voice in the sandwich tern (*Sterna sandvicensis*). *Behaviour*, **32**, 150–7.

Kaitz, M., Good, A., Rokem, A. M. & Eidelman, A. I. (1987). Mothers' recognition of their newborns by olfactory cues. *Devel. Psychobiol.*, **20**, 587–91.

Kaplan, J. N., Cubicciotti, D. & Redican, W. K. (1977). Olfactory discrimination of squirrel monkey mothers by their infants. *Devel. Psychobiol.*, **10**, 447–53.

Kaplan, J. N., Winship-Ball, A. & Sim, L. (1978). Maternal discrimination of infant vocalizations in squirrel monkeys. *Primates*, **19**, 187–93.

Klein, J. (1979). The major histocompatibility complex of the mouse. *Science*, **203**, 516–21.

Klopfer, P. H., Adams, D. K. & Klopfer, M. S. (1964). Maternal imprinting in goats. *Proc. Natn. Acad. Sci., USA*, **52**, 911–14.

Klopfer, P. H. & Gamble, J. (1966). Maternal 'imprinting' in goats: the role of chemical senses. *Z. Tierpsychol.*, **23**, 588–92.

Knudsen, B. & Evans, R. M. (1986). Parent-young recognition in herring gulls (*Larus argentatus*). *Anim. Behav.*, **34**, 77–80.

Lenington, S. & Egid, K. (1985). Female discrimination of male odors correlated with male genotype at the T-locus: a response to T-locus or H-2 locus variability? *Behav. Genet.*, **15**, 53–67.

Lenington, S., Egid, K. & Williams, J. (1988). Analysis of a genetic recognition system in wild house mice. *Behav. Genet.*, **18**, 549–64.

Leon, M. (1975). Dietary control of the maternal pheromone in the lactating rat. *Physiol. Behav.*, **14**, 311–19.

Leon, M. (1983). Chemical communication in mother young interactions. In *Pheromones and Reproduction in Mammals*, ed. J. G. Vandenbergh, pp. 39–77. New York: Academic Press.

Lindsay, D. R. & Fletcher, I. C. (1968). Sensory involvement in the recognition of lambs by their dam. *Anim. Behav.*, **16**, 415–17.

MacGintie, G. E. (1939). The natural history of the blind goby, *Typhlogobius californiensis* Steindachner. *Am. Midl. Nat.*, **21**, 489–505.

McArthur, P. D. (1982). Mechanisms and development of parent-young vocal recognition in the pinon jay (*Gymnorhinus cyanocephalus*). *Anim. Behav.*, **30**, 62–74.

McFarlane, A. (1975). Olfaction in the development of social preferences in the neonate. In *Parent–Infant Interaction. CIBA Found. Symp.*, **33**, 103–113.

McKaye, K. R. & Barlow, G. W. (1976). Chemical recognition of young by the Midas cichlid, *Cichlasoma citrinellum*. *Copeia*, **1976**, 276–82.

Medvin, M. B. & Beecher, M. D. (1986). Parent–offspring recognition in the barn swallow (*Hirundo rustica*). *Anim. Behav.*, **34**, 1627–39.

Miller, D. E. & Emlen, J. T., Jr. (1975). Individual chick recognition and family integrity in the ring-billed gull. *Behaviour*, **52**, 124–43.

Morgan, P. D., Boundy, C. A. P., Arnold, G. W. & Lindsay, D. R. (1975). The roles played by the senses of the ewe in the location and recognition of lambs. *Appl. Anim. Ethol.*, **1**, 139–50.

Mykytowycz, R. (1968). Territorial marking by rabbits. *Sci. Am.*, **218**, 116–26.

Mykytowycz, R. & Dudzinsky, M. L. (1972). Aggressive and protective behaviour of adult rabbits *Oryctolagus cuniculus* (L.) towards juveniles. *Behaviour*, **43**, 97–120.

Myrberg, A. A., Jr. (1966). Parental recognition of young in cichlid fishes. *Anim. Behav.*, **14**, 565–71.

Myrberg, A. A., Jr. (1975). The role of chemical and visual stimuli in the preferential discrimination of young by the cichlid fish, *Cichlasoma nigrofasciatum* (Gunther). *Z. Tierpsychol.*, **37**, 274–97.

Peek, F. W., Franks, E. & Case, D. (1972). Recognition of nest, eggs, nest site, and young in female red-winged blackbirds. *Wilson Bull.*, **84**, 243–9.

Poindron, P. & Carrick, M. J. (1976). Hearing recognition of the lamb by its mother. *Anim. Behav.*, **24**, 600–2.

Penney, R. L. (1962). Voices of the Adelie. *Nat. Hist.*, **71**, 16–24.

Pereira, M. E. (1986). Maternal recognition of juvenile offspring coo vocalizations in Japanese macaques. *Anim. Behav.*, **34**, 935–7.

Petrinovich, L. (1974). Individual recognition of pup vocalizations by northern elephant seal mothers. *Z. Tierpsychol.*, **34**, 308–12.

Porter, R. H. (1988). The ontogeny of sibling recognition in rodents: the superfamily Muroidea. *Behav. Genet.*, **18**, 483–94.

Porter, R. H. & Doane H. M. (1977). Dietary-dependent cross-species similarities in maternal chemical cues. *Physiol. Behav.*, **19**, 129–31.

Porter, R. H. & Doane H. M. (1979). Responses of spiny mouse weanlings to conspecific chemical cues. *Physiol. Behav.*, **23**, 75–8.

Porter, R. H., & Moore, D. (1981). Human kin recognition by olfectory cues. *Physiol. Behav.*, **27**, 493–5.

Porter, R. H., Fullerton, C. & Berryman, J. C. (1973). Guinea pig maternal-young attachment behaviour. *Z. Tierpsychol.*, **32**, 489–95.

Porter, R. H., Wyrick, M. & Pankey, J. (1978). Sibling recognition in spiny mice (*Acomys cahirinus*). *Behav. Ecol. Sociobiol.*, **3**, 61–8.

Porter, R. H., Cernoch, J. M. & McLaughlin, F. J. (1983). Maternal recognition of neonates through olfactory cues. *Physiol. Behav.*, **30**, 151–4.

Porter, R. H., Cernoch, J. M. & Balogh, R. D. (1984). Recognition of neonates by facial-visual characteristics. *Pediatrics*, **74**, 501–4.

Porter, R. H., Cernoch, J. M. & Balogh, R. D. (1985). Odor signatures and kin recognition. *Physiol. Behav.*, **34**, 445–8.

Porter, R. H., Matochik, J. A. & Makin, J. W. (1986). Discrimination between full-sibling spiny mice (*Acomys cahirinus*) by olfactory signatures. *Anim. Behav.*, **34**, 1182–8.

Porter, R. H., McFayden-Ketchum, S. A. & King, G. A. (1989). Underlying bases of recognition signatures in spiny mice (*Acomys cahirinus*). *Anim. Behav.*, **37**, 638–44.

Quinn, T. P. & Busack, C. A. (1985). Chemosensory recognition of siblings in juvenile coho salmon, *Oncorhynchus kisutch*. *Anim. Behav.*, **33**, 51–6.

Radesater, T. (1976). Individual sibling recognition in juvenile Canada geese. *Can. J. Zool.*, **54**, 1069–72.

Rasa, A. E. (1973). Marking behaviour and its social significance in the African dwarf mongoose *Helogale undulata rufula*. *Z. Tierpsychol.*, **32**, 293–318.

Rowley, I. (1980). Parent offspring recognition in a cockatoo, the galah, *Cacatua roseicapilla*. *Aust. J. Zool.*, **28**, 445–56.

Russell, M. J. (1976). Human olfactory communication. *Nature*, **260**, 520–2.

Russell, M. J. (1983). Human olfactory communications. In *Chemical Signals in Vertebrates*, 3, ed. Müller-Schwarze, D. & Silverstein, R. M., pp. 259–73. New York: Plenum Press.

Russell, M. J., Mendelson, T. & Peeke, H. V. S. (1983). Mothers' identification of their infants' odor. *Ethol. Sociobiol.*, **4**, 29–31.

Schaal, B., Montagner, H., Hertling, E., Bolzoni, S., Moyse, A. & Quichon, R. (1980). Les stimulations olfactives dans les relations entre l'enfant et la mere. *Reprod. Nutr. Dev.*, **20**, 843.

Schultze-Westrum, T. G. (1969). Social communication by chemical signals in flying phalangers *Petaurus breviceps papuanus*. In *Olfaction and Taste III*, ed. C. Pfaffman, pp. 268–77. New York: Rockefeller University Press.

Selset, R. & Døving, K. B. (1980). Behaviour of mature anadromous char (*Salmo alpinus* L.) towards odorants produced by smolts of their own population. *Acta. Physiol. Scand.*, **108**, 113–22.

Sherman, P. W. & Holmes, W. G. (1985). Kin recognition: issues and evidence. *Fortschr. Zool.*, **31**, 437–60.

Shillito Walser, E., Hague, P. & Yeomans, M. (1983). Preference for siblings or mother in Dalesbred and Jacob twin lambs. *Appl. Anim. Ethol.*, **9**, 289–97.

Sieber, O. J. (1985). Individual recognition of parental calls by bank swallow chicks (*Riparia riparia*). *Anim. Behav.*, **33**, 107–16.

Simons, R. C. & Bielert, C. F. (1973). An experimental study of vocal communication between mother and infant monkeys (*Macaca nemestrina*). *Am. J. Phys. Anthropol.*, **38**, 455–62.

Simons, R. C., Bobbit, R. A. & Jensen, G. D. (1968). Mother monkeys (*Macaca nemestrina*) responses to infant vocalizations. *Percept. Mot. Skills*, **27**, 3–10.

Skeen, J. T. & Thiessen, D. D. (1977). Scent of gerbil cuisine. *Physiol. Behav.*, **10**, 463–6.

Stevenson, J. G., Hutchinson, R. E., Hutchinson, J., Bertram, B. C. R. & Thorpe, W. H. (1970). Individual recognition by auditory cues in the common tern (*Sterna hirundo*). *Nature*, **226**, 562–3.

Stoddard, P. K. & Beecher, M. D. (1983). Parental recognition of offspring in the cliff swallow. *Auk*, **100**, 795–9.

Symmes, D. & Biben, M. (1985). Maternal recognition of individual infant squirrel monkeys from isolation call playbacks. *Am. J. Primatol.*, **9**, 39–46.

Tyler, S. J. (1972). The behaviour and social organization of the New Forest ponies. *Anim. Behav. Monogr.*, **5**, 85–196.

Waldman, B. (1981). Sibling recognition in toad tadpoles: the role of experience. *Z. Tierpsychol.*, **56**, 341–58.

Waldman, B. (1985). Olfactory basis of kin recognition in toad tadpoles. *J. Comp. Physiol. A*, **156**, 565–77.

Waldman, B. (1987). Mechanisms of kin recognition. *J. Theor. Biol.*, **128**, 159–85.

Wallace, P. (1977). Individual discrimination of humans by odors. *Physiol. Behav.*, **19**, 577–9.

Wallace, P., Owen, K., & Thiessen, D. D. (1973). The control and function of maternal scent marking in the Mongolian gerbil. *Physiol. Behav.*, **10**, 463–6.

Wills, G. D., Wesley, A. L., Sisemore, D. A., Anderson, H. N. & Banks, L. M. (1983). Discrimination by olfactory cues in albino rats reflecting familiarity and relatedness among conspecifics. *Behav. Neural Biol.*, **38**, 139–43.

Wolski, T. R., Houpt, K. A. & Aronson, R. (1980). The role of the senses in mare–foal recognition. *Appl. Anim. Ethol.*, **6**, 121–38.

Yamaguchi, M., Yamazaki, K., Beauchamp, G. K., Bard, J., Thomas, L. & Boyse, E. A. (1981). Distinctive urinary odors governed by the major histocompatibility locus of the mouse. *Proc. Natn. Acad. Sci. USA*, **78**, 5817–20.

Yamazaki, K., Yamaguchi, M., Baranoski, L., Bard, J., Boyse, E. A. & Thomas, L. (1979). Recognition among mice: evidence for the use of a Y maze differentially scented by congenic mice of different major histocompatibility types. *J. Exp. Med.*, **150**, 755–60.

Yamazaki, K., Yamaguchi, M., Boyse, E. A. & Thomas, L. (1980). The major histocompatibility complex as a source of odors imparting individuality among mice. In *Chemical Signals*, ed. D. Müller-Schwarze, & R. M. Silverstein, pp. 267–73. New York: Plenum Press.

Yamazaki, K., Beauchamp, G. K., Bard, J., Thomas, L. & Boyse, E. A. (1982). Chemosensory recognition of the phenotypes determined by the T1a and H-2K regions of chromosome 17 of the mouse. *Proc. Natn. Acad. Sci., USA*, **79**, 7828–31.

Yamazaki, K., Beauchamp, G. K., Wysocki, G. H., Bard, J., Thomas, L. & Boyse, E. A. (1983). Recognition of H-2 types in relation to the blocking of pregnancy in mice. *Science*, **221**, 186–8.

9

Recognizing kin: ontogeny and classification

Peter G. Hepper

Introduction

Much evidence has now accumulated to demonstrate that individuals respond differentially to conspecifics according to their genetic relatedness (e.g. Hepper, 1986a; Fletcher & Michener, 1987; see also this book). Furthermore, this differential responding is not confined to one particular behaviour but is found in a diverse variety of situations and behaviours (see this book). This strongly suggests that individuals have some means which enable them to identify genetic relatedness. It is the aim of this chapter to explore how individuals recognize their kin. Previously (e.g. Hepper, 1986a; Porter, 1987; Waldman, 1988) this ability has been considered under the general rubric of the 'mechanisms' of kin recognition. This chapter will discuss factors which contribute to the individual recognizing its kin and will, by addressing these factors, enable the underlying basis of kin recognition to be elucidated. Although the chapter will concentrate on mammalian kin recognition it is hoped that the considerations presented will be applicable to other animal groups. I shall first discuss present approaches to mechanisms of kin recognition.

In Hamilton's seminal papers of 1964 (a,b), as well as demonstrating the fitness benefits to an individual of responding differentially to kin and non-kin, a number of ways were proposed by which individuals would be able to discriminate between kin and non-kin in social situations. From this discussion (Hamilton, 1964b) and that of others (e.g. Alexander, 1979; Bekoff, 1981; Hölldobler & Michener, 1980) four basic 'mechanisms' have been proposed to explain how individuals recognize their kin (Holmes & Sherman, 1983). These have become firmly rooted in the literature and I shall outline each briefly.

1 Spatial location. The individual recognizes kin not on the basis of any cues presented by its conspecifics but rather on the basis of

cues in its environment. Thus individuals learn about cues present in their environment, e.g. nest site, and respond differentially on the basis of these cues. Individuals encountered either physically in, or within a certain distance of, the particular environmental cue are responded to as kin whereas those encountered outside are responded to as non-kin. For example, adult kittiwakes will, until the time of fledging, accept any young placed in their nest as their own, responding purely on the basis of presence in the nest (Cullen, 1957, see also Beecher this volume). Cues such as home territory or particular landmarks within the home territory could also be used to recognize kin.

2 Association or familiarity. This is often proposed to be the most common mechanism used in the recognition of kin (e.g. Holmes & Sherman, 1983). Individuals learn during their development cues from the most familiar, or most commonly encountered, conspecific(s) in their environment. Recognition by this means presumably relies on the individual responding on the basis of individual cues presented by their conspecifics. Individuals thus respond to familiar individuals as kin and unfamiliar individuals as non-kin. Evidence for the existence of this mechanism is taken from studies which have reared genetically unrelated individuals together from birth. For example, Hepper (1983a) separated rat pups at birth and formed new litters of unrelated pups. When these were later given a choice between unrelated but familiar conspecifics and unrelated and unfamiliar conspecifics, individuals preferred their familiar littermates. The preference for familiar individuals is taken as evidence for recognition by association or familiarity.

3 Phenotype matching. Recognition of kin is again based upon cues presented by conspecifics. In this case individuals learn the cues of conspecifics and assimilate these to form a single template. Upon encountering another conspecific, individuals match the cues presented by that individual to that of their template, the degree of matching determining the coefficient of relatedness. This mechanism enables the identification of kin whom the individual has not previously encountered and makes the assumption that related individuals possess a shared cue. Evidence for the existence of this mechanism is obtained from studies which have examined the response of related individuals who have never previously experienced one another. This has been achieved by using littermates separated at birth (Hepper, 1983a, 1987a), paternal half-siblings (Kareem & Barnard, 1982) or individuals from successive litters (Grau, 1982). Individuals are given a choice between a related but unfamiliar individual and an unrelated and

unfamiliar conspecific. Observation of discriminative responding is taken as evidence for recognition by phenotype matching.

4 Recognition genes. In this case the recognition of kin is encoded directly by an individual's genes and requires no experience. Two possibilities have been considered under this general heading. *First*, based on an idea proposed by Hamilton (1964b), individuals may possess a supergene which would affect '(a) some perceptible feature of the organism, (b) the perception of that feature, and (c) the social response consequent upon what was perceived' (p. 25). More recently this has been termed the 'green beard' effect (e.g. Dawkins, 1982). A *second* possibility is that the recognition gene determines only the cue of the individual and the classification of this cue as a particular degree of relatedness, the response consequent upon what was perceived being determined by a separate gene (Hepper, 1986b). Whilst there have been many theoretical arguments considering the possibility of a recognition gene (e.g. Crozier, 1987; Hepper, 1986a,b; Ridley & Grafen, 1981; Rothstein & Barash, 1983), the fact remains that conclusive experimental demonstrations of the existence of a recognition gene will prove extremely difficult to achieve (Hepper, 1986b).

The above mechanisms have been widely accepted as being *the* mechanisms involved in the recognition of kin and have tended to exclude other possibilities. Although authors stress that these mechanisms are not mutually exclusive and may operate in conjunction with one another (e.g. Bekoff, 1981; Blaustein & O'Hara, 1986) other potential mechanisms are neglected. Furthermore, with the growing acceptance of these mechanisms, experiments are conducted with the aim of demonstrating the existence of a particular mechanism, or of making comparisons between mechanisms rather than attempting to determine how recognition is achieved. Recently these mechanisms of kin recognition have been the subject of increased criticism (e.g. Hepper, 1986a; Porter, 1988; Waldman, 1987; Waldman *et al.*, 1988). In particular attention has been drawn to the fact that the mechanisms provide little ability to determine the underlying basis of kin recognition (Hepper, 1986a). Furthermore, the two most commonly proposed mechanisms of kin recognition, those of association/familiarity and phenotype matching pose another problem; should they be considered as two separate mechanisms or as different aspects of the same recognition process? Association/familiarity refers to the developmental process by which individuals gain information about their kin, i.e. who they learn from (the most familiar individual) but provides very little insight as to how this information is used. Phenotype matching on the other hand refers to how individuals classify particular cues as those of kinship, by matching to a template, but says very little on how this template develops. Thus whether the two constitute separate recognition mechanisms is questionable; it may be more correct to consider both as components of a single recognition process.

Previous studies of kin recognition may generally be criticized because they have paid little attention to the role of developmental experiences in the ability to recognize kin and in particular how these experiences affect later recognition (although see Hepper, 1986a; Porter, 1988; Waldman, 1987). Overlooking these developmental aspects ignores the fact that single ontogenetic experiences may influence recognition in different ways. One experiment which illustrates this is an unpublished study performed at this laboratory.

Rat pups were reared from birth in pairs unrelated to one another and to the lactating female, 'the mother'. Individuals at 12 days of age were given a two-choice preference test for their littermate (unrelated but familiar) and an unrelated but unfamiliar conspecific of the same age, the two test individuals being siblings. The pups showed a preference for their familiar littermate (cf. Hepper, 1983a). Since both test individuals were siblings it is unlikely that a genetic cue was involved, and recognition is most probably based on individual cues. This result supports the association/familiarity mechanism and is in line with previous experiments adopting similar methodology (e.g. Hepper, 1983a; Kareem & Barnard, 1982). In a subsequent test pups were given a choice between a sibling of their littermate (a pup which the test animal had never experienced, i.e. unrelated and unfamiliar) and another pup both unfamiliar and unrelated to the other pups. Pups here showed a preference for the sibling of their littermate. This result supports the phenotype matching hypothesis and corresponds with other experiments using similar methodology in the rat (e.g. Hepper, 1983a). Thus the same pup, with this single rearing experience demonstrates recognition by both association/familiarity and phenotype matching. These results demonstrate that developmental experiences may influence recognition in a number of ways and illustrate the problems of treating phenotype matching and familiarity as wholly separate mechanisms.

Given the problems with the above approaches certain authors have adopted new categorizations of possible mechanisms in order to better understand how individuals recognize their kin. Porter (1988) has proposed two categories. 'Direct familiarity', where individuals learn about kin by direct association with them (cf. association/familiarity) and 'indirect familiarity' where individuals may recognize kin with whom they are unfamiliar by the fact they have associated with kin who bear similar cues to these unfamiliar kin (cf. phenotype matching). This approach acknowledges the fact that single ontogenetic experiences may lead to recognition by different means. On the other hand, Waldman (1988) has offered an approach based upon the type of cues that are learned categorizing recognition as 'direct' or 'indirect'. For direct recognition, individuals use the cues presented by conspecifics whereas for indirect recognition cues not directly presented by conspecifics, e.g. environmental cues, are used. Although both approaches are considerable improvements over previous categorizations, they still do little to elucidate the underlying mechanisms involved in the recognition of kin.

Kin recognition: underlying mechanisms

The remainder of this chapter will introduce a new approach which I believe can increase our understanding of kin recognition by enabling its underlying basis to be determined. Kin recognition is presented as a single process comprising a number of components all of which must be considered in attempts to determine how recognition is accomplished. There is no suggestion that individuals will use all of these processes at any one time or in all situations.

Before presenting this approach it is worth considering what the goal of kin recognition must be. Irrespective of how classification is achieved, when learning takes place or what cues are involved, the end result must be the ability to identify conspecifics as kin or non-kin, or as being of a certain degree of relatedness to the individual. The process of kin recognition must provide the information which enables individuals to respond differentially to kin and non-kin. For it is only by responding differentially on the basis of genetic relatedness that individuals can obtain the benefits in fitness as espoused by Hamilton's kinship theory (1964a,b). Any means an individual uses to achieve this can be considered a mechanism of kin recognition (I shall return to this point at the end of the chapter).

Consideration of how individuals recognize their kin must examine both the role of the sender of the cues and that of the receiver of these cues (Beecher, 1982). Here I shall concentrate upon the receiver, i.e. the individual decoding the cue and making a decision on kinship on the basis of this. Previous chapters in this volume by Beecher and Halpin have dealt with aspects relating to the sender of kinship information.

I shall be somewhat restrictive in my usage of the term kin recognition and refer solely to the perception of the cue and its classification in terms of a degree of relatedness. Figure 9.1 presents a model of the kin recognition process indicating the components which may contribute to individuals recognizing their kin. Very briefly this process involves the individual perceiving cues which may arise from its environment, its conspecifics, or, its own internal state. The cues are processed by the classification system which produces an outcome in terms of a measure of relatedness. This, in combination with a variety of other factors, is used to determine the individual's behaviour. The classification process, which consists of an

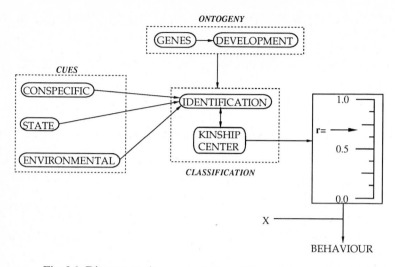

Fig. 9.1 Diagrammatic representation of the components involved in recognising kin. The outcome of this process is an assessment of kinship, which along with other factors (X) is used to determine the individual's behaviour.

identification system(s) and an integrative kinship center, will be influenced by an individual's genes and will also be shaped by the developmental history of the individual. This model presents the various components that should be considered when determining how individuals recognize their kin. Each aspect of this model will be discussed in detail.

Cues

Perhaps the essential feature of any kin recognition cue is that it accurately reflects relatedness, irrespective of the nature of the cue (see also Halpin this volume). Thus, as long as the cue is associated with and accurately determines relatedness it may be used in the recognition of kin. Individuals may use three types, or classes, of such cues. *Environmental cues* include those from the nest site, markers of the home range or territory or salient cues within the individual's home range. *Conspecific cues*, produced directly by the individual's conspecifics may be mediated by genetic factors (Hepper, 1987a), by environmental factors such as diet (Porter *et al.*, 1989), or possibly by micro-organisms (Albone, 1984; Hepper, 1988). Finally, a class of cue which is often overlooked is that of *state of the animal*. Parental state markedly affects the behaviour of the individual and cues of state should be considered in any system of kin recognition (see Elwood this volume).

Discussion of the role of kinship cues is often restricted to the question of how they reflect relatedness. However, there are a number of other features which may influence the recognition of kin. Given the fact that a number of different cues (e.g. environmental, conspecific) may provide information regarding kinship the relative importance of each in the recognition process should be determined. At different times during the individual's life, or in particular situations, certain cues may assume greater importance. When two or more cues are available, as is probably the case most of the time, the relative importance of each may change depending upon the animal's situation or condition. Thus, for a pregnant female soon after giving birth, state, i.e. recently parturient, may be the most important cue. Whilst in this post-partum state the female will respond to any newborn individual as her own, overriding any other cues from the environment or conspecifics which may be present. Similarly the individual may respond on the basis of environmental cues alone, such as presence in the nest, irrespective of other cues. The crucial point is that at particular times certain cues may assume greater importance because of the individual's state or situation, whilst at other times these same cues may not be so important. As long as these cues accurately reflect relatedness, kin recognition is possible. Little attention has been paid to the importance of particular cues during the various stages of life in the developing individual.

Similarly relationships between cues should be considered as this could have a profound effect on the recognition process. Using a human illustration, seeing your next door neighbour in your own street is not unusual and would not stimulate much reaction or close inspection, i.e. conspecific and environmental cues may be said to be in 'harmony'. However, seeing your neighbour when you are on holiday on the other side of the world would elicit a very different reaction and much more careful scrutiny, since conspecific and environmental cues would be mismatched, or could be considered to be in 'disharmony'. Such differences in the relationship between cues may greatly influence the recognition process. For example in the rat, an unfamiliar animal introduced into an individual's home cage (or territory) elicits a much more immediate and different reaction from the host than when the same animal is introduced in an unfamiliar environment. Thus, individuals in addition to using the cue itself may also be assessing the relationship between cues in their recognition of kin. Many studies of kin recognition test animals in unfamiliar environments, i.e. different from those in which they are currently living and in which they learned about kin (e.g. Hepper, 1983a) even though some habituation to the apparatus has usually taken place. This may influence the recognition process and affect the outcome (cf. Dewsbury, 1988). Very little research has attempted to explore these relationships and their implications for recognition or the animal's response.

Ontogeny

If one assumes that individuals require some knowledge regarding their kin before being able to recognize them (this need not necessarily be the case, as will be discussed later) then the question of how they obtain this information is vital and must be addressed. Yet examination of the ontogeny of kin recognition has received very little attention. I shall consider two main influences, that of genes and that of development, through which individuals may obtain information regarding kinship.

Genes

An individual's genes may predispose the individual to recognize its kin, i.e. individuals may possess a genetically determined recognition system. However, what would be the characteristics of a genetic recognition system? Many authors have proposed a recognition gene which encodes the cue, the classification of the cue, and the response (e.g. Hamilton, 1964b). More plausible perhaps is a gene which simply encodes the cue and the identification of the cue (e.g. Hepper, 1986b). Indeed there is good reason to believe that the response will be determined separately and will simply use the information provided by a genetically determined identification system (Hepper, 1986a,b).

There is much evidence to suggest the existence of cues which are genetically determined. Using female hybrids of white-handed and pileated gibbons Brockelman & Schilling (1984) found that the 'loud and stereotyped great call' of these hybrids exhibited strong genetic determination. Female hybrids when raised in groups with genetically dissimilar parents, developed a call distinct from that of their parents. Experiential influences gained through interacting with parents are important in establishing the timing, maturation or context of the call, but exert little effect on the final call pattern. Studies which have reared individuals apart and later found that such individuals recognize related but unfamiliar conspecifics (e.g. Grau, 1982; Hepper, 1983a; Kareem & Barnard, 1982) may indicate that related individuals possess some genetically determined cue(s). Hepper (1987a) found an inverse linear relationship between investigation time (time spent sniffing a conspecific) and relatedness in the rat which may indicate the presence of a genetic cue which is distributed in accordance with relatedness. Further evidence for genetically determined cues is provided by studies of the major histocompatibility complex (MHC), where rats and mice differing only in single loci of the MHC, produce discriminably different odours (e.g. Brown *et al.*, 1987; Singh *et al.*, 1987; Yamazaki *et al.*, 1979; see also chapter by Boyse *et al.* this volume). Evidence for a genetically determined identification system is much harder to obtain. Indeed, definitive proof could only be obtained if the individual were separated from all cues of relatedness, including self, throughout its development. This may prove impossible to achieve.

Although evidence for a genetic determination of kin recognition is lacking, there is little doubt that inherent recognition processes for other behaviours exist. For example, individuals recognize a natural predator without prior experience of that predator (Cattarelli, 1982; Courtney *et al.*, 1968; Mollenauer *et al.* 1974). Studies of the cricket (*Teleogryllus* sp.) indicate that there may be 'genetic coupling' between the production of the call by the male and the preference for a particular type of call by the female (Doherty & Hoy, 1985; Hoy, 1974; Hoy, Hahn & Paul, 1977). The most persuasive evidence for 'recognition' determined by genes is that exemplified by self-recognition in invertebrates (e.g. Ertman & Davenport, 1981; Grosberg & Quinn, 1986; Keough, 1984; McClay, 1971; Scofield *et al.*, 1982). For these organisms the ability to recognize self may also serve the function of recognizing kin (Hepper, 1986a). Furthermore the loci of the genes responsible have been identified in a few cases, e.g. Grosberg & Quinn (1986) found that larvae of the sessile colonial ascidian, *Botryllus schlosseri*, are differentiated on the basis of shared alleles at the polymorphic histocompatibility loci, a site involved in historecognition and colony fusion. There is however little evidence in mammals for the existence of a system of genetic recognition of conspecifics. It should be noted that recognition between related individuals reared apart, more usually attributed to phenotype matching, could be interpreted as evidence for the existence of a recognition gene (Blaustein, 1983).

Whilst many writers have expressed opinions on the likelihood of recognizing kin via a genetic mechanism and on how such a mechanism could operate if it existed (e.g. Blaustein, 1983; Crozier, 1987; Dawkins, 1982; Hepper, 1986b; Ridley & Grafen, 1981; Rothstein & Barash, 1983), little attention has been paid to the consequences of possessing such a mechanism. The major problem for a genetically determined system is its permanence. Any system which is incapable of being changed by the experiences of the individual has considerable drawbacks.

Although siblings are on average related to one another by $r = 0.5$, this can theoretically range from 0 to 1. It is possible that within sibling groups certain individuals will be genetically very similar whilst others may be very dissimilar. Such differences, reflected in the formation of the genetic representation of kin, would mean that individuals would respond to some siblings more favourably, since these would be closer in genetic constitution, but less favourably to others who would be less similar in their genetic constitution; all individuals would nevertheless be siblings. Whether individuals do respond differently to individual siblings has yet to be determined. An individual's representation of kin may consequently be dissimilar to the majority of its kin and individuals may incorrectly identify kin as non-kin.

This does not present a problem if the individual has large numbers of relations since at least some may be similar but if individuals have only a few relations, individuals may find themselves with a recognition system which is not representative of their kin. Problems are also encountered by any genetic system if the cue denoting kinship is susceptible to alteration. For example, whilst many odours have a strong genetic component (Hepper, 1988) they can also be influenced by dietary factors (e.g. Porter *et al.*, 1989) or micro-organisms (Albone, 1984). The influences of such environmental factors may alter the odour such that it is substantially different from that to which the individual's genetically determined recognition system is predisposed. Furthermore, if the cue changes as the individual develops then the represen-tation of kin provided by an individual's genes must also change. Overall a genetic system of recognition could be expected to function more effectively where individuals have large numbers of kin present, as in the case of insects, rather than in mammals where numbers of relations may be limited, and where the cue identifying kinship is stable and independent of environmental or developmental influences. Whilst these factors would tend to argue against a genetic system in mammals, obtaining proof of any genetic mechanism poses considerable difficulties (Hepper, 1986a,b), and the exis-tence of a genetically determined kin recognition system may remain a possibility for some time to come.

Development
If the information regarding kinship is not encoded by an indivi-dual's genes then the individual may have to learn from its kin in order to obtain this information. This raises a number of questions. One assumption which can be made is that individuals at the start of the learning process are ignorant as to kinship relationships. Individuals may thus be considered as a *tabula rasa* with respect to kinship and will therefore learn from the most salient object(s) or feature(s) in their environment. This raises the question as to how individuals ensure that they learn about their genetically related conspecifics? Contingencies must be arranged such that any learning is derived from kin and there exist a number of possible means whereby this could be achieved.

Learning could be initiated in the womb. There is much evidence that individuals are not passive in the womb but that they can and do respond to changes in their environment (e.g. Hepper, 1989; Smotherman & Robinson, 1989). Research indicates that prenatal learning may play an important role in the development of kin recognition, not only in mammals but in other major groups of animals. For example, newly emerged callow workers of the ant *Cataglyphis cursor* exhibit a preference for nestmates which is acquired during larval life (Isingrini *et al.*, 1985; Isingrini & Lenoir, 1986). Tadpoles appear to learn from cues derived from the mother in the egg cell mass prior to hatching; this influences preferences after hatching (Waldman, 1981). Birds within the egg have been shown to respond to the calls of their parents and respond to these calls after hatching (e.g. Shindler, 1984; Tschanz, 1968). There is much evidence in mammals for responsiveness in the womb (see Hepper, 1989, 1990). Newborn rat pups exhibit a preference for their own compared to other amniotic fluid, a preference which may be important in later recognition (Hepper, 1987b), especially since altering the composition of this fluid in the womb greatly changes the preferences of the individual after birth (Hepper, 1990). Similar findings have been reported in sheep where the amniotic fluid appears to play a role in the mother's recognition of her offspring (Lévy & Poindron, 1984). Furthermore evidence from humans indicates prenatal experiences may be important in the formation of social preferences after birth (e.g. Hepper, 1989) and in the recognition of the mother (e.g. DeCasper & Fifer, 1980; DeCasper & Spence, 1986).

The mammalian uterus provides an excellent environment in which to start learning about kin. There is no likelihood of encountering non-kin, at worst mammalian individuals may encounter maternal half-siblings $r = 0.25$ and thus any learning from cues within the womb will be restricted to those of kin. The learning of auditory cues, which can originate externally and penetrate to the womb may be restricted to cues provided by kin, especially the mother. For although human newborns prefer the voice of their mother they show no preference for that of their father (DeCasper & Prescott, 1984). Implantation of microphones into the fetal environment of sheep indicates that it is the mother's 'voice' which is the loudest sound heard by the fetus in its uterine environment. Newborn lambs may, therefore, be expected to be familiar with the voice of their mother (Vince & Billing, 1986). Although the duration of prenatal learning has yet to be established (Hepper, 1989; although see Smotherman, 1982) uterine learning may act as an important priming process, ensuring that any future learning after the prenatal period is restricted to that from genetically related individuals.

Postnatal learning presents the individual with greater problems in ensuring that information is acquired from genetically related conspecifics since the chances of encountering unrelated individuals are greatly increased. Individuals must, therefore, be careful to ensure that any learning about kinship is from related individuals. Again environmental factors can be expected to contribute greatly to achieving this end. As mentioned above if individuals have been able to learn about their kin prenatally this may have given them a rudimentary idea of who their kin are. Thus, individuals may have a preference for related individuals before birth and this may direct future learning from related individuals after birth.

Individuals may also learn from their caretaker. There is much evidence that parents recognize their offspring (e.g. Colgan, 1983; Spencer-Booth, 1970) and will care preferentially for them (Hepper, 1986a). Thus newborns, by learning from their parents can ensure they learn from genetically related individuals. Individuals other than the parents may also play a role in looking after the infant (e.g. see Hrdy, 1976; Macdonald & Moehlman, 1982; Nishida, 1983). Such individuals are often related, e.g. older siblings, aunts, so that once more learning from caretakers can ensure that the young learn from kin (Hepper, 1986a). The actions of the individual's caretakers may also result in newborn young being kept together and unrelated individuals kept away. For example, recently parturient rats become extremely aggressive and keep all other conspecifics away from their litter; furthermore they selectively retrieve their own young (Hepper, 1983b). Both actions serve to maintain the genetic integrity of the litter and ensure that newborns are surrounded by genetically related littermates, thus promoting learning from kin. Learning from self may also provide information about kin and such learning provides possibly the best and most reliable opportunity for learning about kin. Although such learning has received little attention in the literature (but see Aldhous, 1989), studies which have demonstrated recognition between genetically related individuals may be the result of learning from self.

Given the fact that the individual has a range of learning sources consideration should be given as to which source should be used. The most important factor is that the individual learns from genetically related individuals. Only by learning prenatally or from self can the individual be certain of this. In all other cases there is a possibility of learning about non-kin, although environmental contingencies can normally be expected to greatly reduce the chances of this occuring. The aim of any learning is to produce a representation of kin which can be used to discriminate kin from non-kin or different classes of kin. If the individual learns a general representation assimiliated from the features of a number of individuals or environmental cues, care must be taken to ensure the end product does in fact represent kin. The individual could learn from self but a representation based solely on self-features could be too restrictive and might lead to the rejection of related individuals as non-kin. Alternatively if the individual learns from a wide variety of kin, the resulting representation might be too broad and would lead to the acceptance of non-kin as kin or the classification of distant kin as more closely related. Thus individuals who assimilate information about kin could be expected to strike a balance between learning an overly restrictive or overly inclusive representation of kinship. Individuals who maintain individual representations of kin face a separate problem in that such representations must be associated with some other factor, familiarity has often been proposed (e.g. Bekoff, 1981), to enable relatedness to be determined (see later).

The timing of such learning may exert an important influence ensuring that the individual does in fact learn about genetically related individuals.

The essential criterion for individuals as to when learning takes place, is that they should be able to recognize kin by the time non-kin are first encountered. Learning need not necessarily be complete by then, but such a system should exist to enable non-kin to be discriminated, allowing any further learning to be from related individuals. A number of studies demonstrate that recognition (as determined by discriminative responding) appears just prior to the time of encountering non-kin (e.g. Elsacker *et al.*, 1988). In Belding's ground squirrels the dam appears to recognize her own young at the age at which she would first encounter non-kin in the natural environment (Holmes, 1988). Prior to this time the dam does not discriminate between her own and alien young.

The development of cues to be used in the recognition of kin will have important ramifications for any learning process. For example, if the cue is complete and impervious to change soon after birth individuals could learn this immediately. If, however, the cue undergoes some development then the recognition process should not end but be continually updated, assimiliating any changes in the cue. Human appearance changes over time yet we have little difficulty coping with this and we do not retain our initial picture of the person but continually update it to incorporate our most recent encounters. Generally there is little reason to assume that recognition is a 'static' process such that once learned it remains unchanged for the duration of the individual's life. It may be that as the individual's capabilities improve, its ability to learn and use this information improves. Individuals may progress from the use of group cues, to utilizing genetic cues possessed by all kin, and finally discriminating on the basis of individual cues. For example young of the phalanger (*Petaurus breviceps*) first recognize conspecifics (other than the mother) on the basis of group membership but as they develop individual cues are used to recognize conspecifics (Schultze-Westrum, 1965). Little attention has been paid to these issues, with researchers concentrating upon the initial development of this ability and not studying changes over the individual's lifetime.

The speed taken by individuals to acquire their initial representation of kin can be expected to be influenced by two factors, both governed by the timing of initial non-kin encounters. The *first* is concerned with whether the individual is precocial or altricial. For precocial individuals, who develop quickly and become mobile soon after birth, the chances of encountering non-kin soon after birth are high and a rapid form of learning may be expected to occur. For altricial species however, where there is a long period of development and individuals remain with the mother for some time, non-kin may not be encountered for some time after birth; a less rapid learning process may thus be expected to take place. A *second* factor is the difference between seasonal and all year round breeders. In the former there are clearly defined times of the year when large numbers of young will be present, and it will be advantageous to learn as quickly as possible after birth since the chances of encountering unrelated young will be high. Indeed studies of seasonal breeders, (e.g. Klopfer *et al.*, 1964; Poindron *et al.*, 1979) indicate that mother–offspring recognition is formed soon after birth. For all year round breeders the requirements to learn quickly may not be as great since individuals will not be exposed to conspecifics of the same age to the same extent. Combining both these factors creates two extreme groups. There will be species which are highly precocial and breed seasonally who may be expected to learn about their kin very rapidly, whereas at the other extreme will be species which are altricial and breed all year round where learning about kin may take place over a longer period (cf. Gubernick, 1981). A direct comparison between members of such groups has yet to be undertaken. However, the harbour seal, a species which is both precocial and a seasonal breeder appears to establish mother–offspring recognition within five minutes of birth (Lawson & Renouf, 1987).

The question as to what type of learning is involved has received very little attention in the kin recognition literature (although see Hepper, 1986a). The details of the underlying mechanisms of learning will not be considered. Rather the consequences of different types of learning in influencing the development of recognition will be discussed. For present purposes learning about kin may be divided into two broad catagories according to whether such learning is dependent or independent of experience.

Learning which is *independent* of experience refers to situations where the individual learns the most salient cue in its environment irrespective of its context or how the individual experiences the cue. This may be the result of some form of 'physiological imprinting' where exposure to the stimulus causes a permanent change in the nervous system of the individual. For example Hirsch & Spinelli (1971) reared cats in the dark for three weeks and then exposed the right eye to only vertical lines and the left eye to only horizontal lines. Cells in the visual cortex were found that responded best to vertical lines shown to the right eye and horizontal lines to the left eye. It may be that experience of the cues of kinship similarly affects certain cells causing permanent increased responsiveness to cues of kinship (see also later). However, wherever the change occurs the responsiveness of the individual's neural system is permanently altered. For such learning to be effective requires that the cue learned is unchanging, since there would be little benefit in altering a perceptual system to respond to some cue if this cue then changes. Alternatively, learning may be the result of 'simple exposure', with the most salient stimuli in the individual's environment being learned. Imprinting may be an example of this (e.g. see Bateson, 1979, 1981), where exposure to a particular stimulus causes that stimulus to become more familiar and attractive whilst other stimuli, due to a narrowing of prefer- ences, become less attractive (Bateson, 1981). Leon (1980) proposed a similar model to account for early olfactory learning in the rat. Here the pup's olfactory system, because of its immaturity and inability to habituate, is continually stimulated by olfactory cues and these stimuli, because of their prolonged access, become familiar. However as the system matures and the ability to habituate develops, novel stimuli no longer have the prolonged access to the olfactory system and have less chance of becoming familiar. For both 'physiological imprinting' and 'simple exposure', learning is independent of other factors; individuals simply learn the most salient cue in their environment. Thus other contingencies, as described earlier, must operate to ensure the individual learns about kin.

The second broad class of learning mechanisms are *dependent* upon the experiences of the individual and may be considered as examples of associative learning. Here cues associated with particular events are selectively learned. Thus young infants may learn the characteristics of their mother during feeding because this suckling is a reinforcing event. Learning in this way may increase the chances of learning about kin because it requires an interaction to take place. The individual's mother selectively responds to her own offspring, feeding only her own pups, thereby ensuring that if the pups learn whilst sucking they will learn from a related individual. Aversive events may also influence recognition. For example, attempts by young to suck from an alien mother may be rebuffed, and this may influence the learning of kinship, or non-kinship cues. Learning by this means is more selective, and individuals by learning from conspecifics with whom they engage in reinforcing activities, may increase the likelihood of learning from kin. Studies of olfactory learning have indicated that, whilst simply exposing an odour to pups induces a preference for the odour, maintaining this preference requires it to be associated with conspecifics (Galef, 1982) or its exposure to be accompanied by (reinforcing?) stimulation (Sullivan & Leon, 1986).

It should be noted that none of the above learning mechanisms are mutually exclusive. Indeed it may be that all operate to a greater or lesser extent within the individual. For example, individuals who at first respond to all conspecifics in the nest as kin may be doing so as a result of exposure, whereas more detailed recognition may arise later through associative learning perhaps in connection with feeding behaviour. Little attention has been paid to how individuals learn about their kin. One study which partially examined this used mothers and their offspring which had been differentially scented with artificial odours (Hepper, 1987c). Pups were then given a choice between the odour of their mother and that of their littermates. It was hypothesized that individuals, because of the importance of the mother, would learn preferentially from her and exhibit a preference for her odour. The results, however, ran counter to this, pups preferring the odour of their siblings to that of their mother. It may be that siblings themselves are a source of reinforcement, (see Porter *et al.*, 1987) or that learning from siblings instead of the mother provides a better representation of kin (Hepper, 1987c).

In summary, if individuals have to form a representation of kin prior to being able to recognize them, there are a variety of ways this may be achieved. Individuals are completely non-selective in the formation of their representation of kin and thus contingencies must be arranged to ensure it is acquired from genetically related individuals. The preceding discussion has indicated how this might be achieved. None of the above strategies are mutually exclusive and individuals may use all, none, or a particular combination in their recognition of kin. Indeed it may be that different means are used for recognizing different kin, e.g. the individual's mother may be recognized on the basis of individual cues acquired immediately after birth via an imprinting-type process, whereas siblings may be recognized on the basis of genetically shared cues which are learned during development. Future research has to address these issues. It should be noted that once the cue(s) have been learned then any conspecific presenting these cues will be identified. Thus, individuals not previously encountered will be recognized due to the fact that they share cues with conspecifics (or the individual itself) whom that individual has previously encountered.

Whilst a direct action of genes on the recognition process may be unlikely, it is possible that genes may exert an indirect influence. Genes may affect the timing of learning, determining the onset and offset of any sensitive period involved in the learning of kinship. Indeed genes which predisposed individuals to learn during a period when only kin were present would give the individual a distinct advantage over those unable to do so. The imprinting literature suggests that there are particular times when individuals are more sensitive to learning particular cues (e.g. Bateson, 1979, 1981). These times may be genetically preprogrammed and appear to correspond to times when an individual's kin will be present (Hepper, 1986a). Individuals may also possess a genetic preference to respond to individuals of their own species. For example, chicks appear to be predisposed to orient to adult fowl (Johnson & Horn, 1987), a predisposition which may increase the chances of learning from kin. Genes may also impose some selectivity on what is learned through the formation of 'filters' which may be more sensitive to stimuli from kin than from non-kin (cf. Grobstein, 1988, neural specificity). Thus, whilst not directly influencing the recognition system by encoding kinship information, genetic influences may make it more likely that information from genetically related individuals will be preferentially acquired.

Classification

The classification process can be regarded as that neural event which converts the cues perceived by the individual into a 'value' of genetic relatedness. At present very little is known about the processes involved in the classification of conspecifics as kin, non-kin or a particular class of kin. The aim of the following discussion is to consider how a possible classification system might function, and to discuss some recent research which may throw some light on the neural events involved. Classification may be viewed as comprising two components; identification, which processes the sensory cues to determine relatedness, and a 'kinship center', which integrates this information to produce a coefficient of relatedness, and in conjunction with other information, is used to determine the individual's behaviour. I start with the assumption that the individual possesses some stored representation of its kin (this need not necessarily be the case, see later). Upon encountering a conspecific, the cues presented by that individual are compared to this stored representation. For any particular encounter with a conspecific a variety of cues, e.g. visual, auditory, environmental, etc., may be available which provide information about kinship. It may be that a single cue provides sufficient information to determine relatedness but the more cues that are used, the greater the probability of successful kin identification. Each cue will have access to its own stored representation of kinship, i.e. visual cues being compared to a stored representation of conspecific visual cues, etc., and will be classified accordingly. Support for a process of this kind is provided by observations of humans suffering brain damage and is perhaps best illustrated by the clinical condition known as prosopagnosia (e.g. Meadows, 1974; Whiteley & Warrington, 1977). This results from damage to the visual association cortex (consisting of parts of the temporal, parietal and occipital lobes). Patients with such damage may be unable to identify someone by sight but can do so readily on hearing their voice. Not only does this indicate that cues in different modalities may have access to their own classification or memory store but also that single cues can be sufficient for identification. The information provided by these cues is further processed by the 'kinship center', which has a number of roles. It integrates all the information provided by the identification processes into a single coefficient of relatedness which is then used by the individual to determine its behaviour. By assessing all the information provided by the identification processes it can detect any disparities which exist. For example, if the identification of auditory cues produced a coefficient of relatedness of 0 (unrelated) but the identification of olfactory cues produced a coefficient of relatedness of 0.5 (sibling), this disparity would be detected and further investigation undertaken to determine more fully the coefficient of relatedness of this conspecific. The 'kinship center' will receive information from other 'brain centers' which may be used to influence the assessment of relatedness. This may be achieved by differentially weighting the input from the various identification processes, giving greater importance to the information from certain cues depending on the individual's situation and circumstances.

Thus the classification system comprises many cue specific identification processes which feed into a central 'kinship center', which integrates the information, and may selectively weight information from certain cues, or instigate further investigation by detecting disparities. The outcome of these processes is a single value, a coefficient of relatedness for the conspecific encountered.

What is stored by the identification processes? Writers on kin recognition have opted for the term 'template'. There are, however, a number of alternative ways in which information may be stored (Pinker, 1985), for example as a template, as a list of features, as fourier transformations, as structural descriptions, etc. All of these, including templates, have their own problems (Smyth *et al.*, 1987) and much work has been undertaken examining these systems, e.g. Pinker (1985). No attempt will be made to rehearse these issues regarding the storage of information for recognition processes other than to urge caution in the use of the term 'template' in discussions on kin recognition. Until more is known about how such information is stored by the individual it may be better to avoid a term which implies one particular means of information storage and retrieval. To avoid any such problem I shall refer to that which has been learned regarding kinship as the stored representation (SR). Further research is required to determine the composition and structure of this.

Whatever the nature of this stored representation, it requires that the individual possesses enough 'memory' to store it. If recognition is based upon individual representations then enough memory space must be available to store these individual representations. In terms of memory storage a single SR assimilated from a number of individuals would require less space.

At the identification stage, the assessment of kinship is determined by a comparison between the information derived from the individual's perceptual system(s) and that provided by its own internal representation of kinship. There are essentially two ways such an assessment can be achieved. Individuals may possess an all or nothing system whereby if the cue has a certain number of features it passes a threshold and the source is classified as kin, if not then it is regarded as non-kin (see Gamboa, 1988, cue similarity threshold model). Mammals appear to be able to determine different degrees of relatedness, e.g. Hepper (1987a), which would indicate that matching is not all or nothing but rather a graded process. That is, if the cue overlaps with 50 per cent of the individual's stored representation then the appropriate degree of relatedness is $r = 0.5$, if it matches 25 per cent then $r = 0.25$. This ability to determine the degree of matching enables discriminations between different degrees of relatedness to be made. Since individuals use a variety of cues in their assessment of relatedness, it may be that different assessment strategies are used dependent upon the cue. For example, environmental cues, such as nest site may only allow a bimodal assessment (either in the nest or not) so that an all or none method may suffice. Conspecific odour cues on the other hand which may accurately reflect the degree of relatedness may be assessed by examining the amount of matching to determine the conspecific's relatedness. Similar processes may function within the kinship center. The final coefficient of relatedness may be determined by an all or nothing method, i.e. if enough of the identification processes classify their cues as those of kin and this passes a threshold then the individual is classified as kin, if not then the individual is recognized as non-kin. More likely, however, the fact that individuals appear to be able to discriminate different degrees of relatedness (e.g. Hepper, 1987a), suggests that the kinship center determines the degree of relatedness from the information provided by the identification processes. A single general SR of kinship has advantages over a number of individual SRs in determining the degree of kinship. With the former, the degree of matching between the perceived cue(s) and the SR will determine the degree of relatedness. This, however, will not be the case for individual SRs. Here each individual will match with its own SR equally i.e. 100 per cent. Thus, the degree of matching will not provide information of the degree of kinship and some additional factor will have to be associated with it to determine relatedness. This may be achieved by familiarity, such that each individual SR has a degree of familiarity associated with it and it is this that determines the degree of relatedness. Exactly how this could be achieved however has yet to be determined.

   The process of identification is a neural one and it may be that physiological and neurological investigations will uncover the whereabouts of the kinship center and the sites of the SR of kinship. A number of alternatives have been proposed for how information may be stored, e.g. differential neuronal activity patterns (Grobstein, 1988), alterations of synaptic efficacy (Kandel & Schwartz, 1982), biophysical properties of neurones (Alkon, 1984), hormonal milieu (Erulkar *et al.*, 1981) or receptor distribution (Freidhof & Miller, 1983). At present we are some way from determining how information regarding kin is stored physiologically. Although little research has attempted to explore the underlying neurological, physiological or pharmacological basis of kinship recognition, studies from closely related areas may be relevant to future research in this field and I shall touch upon these now. The underlying neural basis of kin recognition may be studied by adopting two broad strategies.

Neural specificity
   One may examine whether particular areas or cells are responsive to kinship stimuli. Two areas of research are relevant here. *First* consider the role of the intermediate and medial part of the hyperstriatum ventrale (IMHV) in imprinting. The chapter by Johnson in this volume discusses this in some detail. It appears this area is specifically involved in the learning of information about particular conspicuous objects such as the chick's mother

(Johnson & Horn, 1987), but is not involved in the 'innate' predisposition to orient to objects resembling adult fowl (Johnson & Horn, 1987). Furthermore, this area appears to be concerned only with learning about conspecifics. Lesions to this area destroy the ability to learn about conspecifics but do not affect performance in other associative learning tasks (Johnson & Horn, 1986; McCabe *et al.*, 1982). This brain region may be specifically involved in learning about the characteristics of individual birds. A similar neural site for learning about kin may also exist. The *second* area of interest is the study of face recognition which has demonstrated the presence of cells which respond specifically to faces. The temporal cortex of the monkey contains a region (the anterior superior temporal sulcus) in which cells respond preferentially to the faces of monkeys or humans (Bruce *et al.*, 1981; Perrett *et al.*, 1982). Furthermore, certain cells within this region respond selectively to the faces of particular individuals who were familiar to the monkeys (Perrett *et al.*, 1984). Recent studies using sheep have also found cells within the temporal cortex which respond preferentially to the faces of sheep (Kendrick & Baldwin, 1987). The pattern of cell firing appears to respond to features of the individual animal such as dominance, breed or familiarity. Further investigations may find that particular cells, or patterns of cell firing, respond specifically to kin.

Alterations in neural responsiveness

The second strategy is to document the changes that occur in the neural structure, etc., as a result of learning about kin. Research, by Leon and co-workers, examining the neurobiology of early olfactory learning in the rat, is relevant here. These studies are performed at the time the rat pup is learning about its kin by means of olfactory cues (Hepper, 1983a,b) and may elucidate the neural changes resulting from learning about kinship. Nineteen-day old rat pups which have previously been exposed to the odour of peppermint from days 1–18 accompanied by tactile stimulation (stroking with a paint brush) show a dramatic increase in 2-deoxyglucose uptake in regions of the glomeruli complex of the olfactory bulb, compared to the same regions in control pups which have not previously been exposed to this odour (Coopersmith & Leon, 1984). This response is not due to any change in respiration rate (Sullivan *et al.*, 1988), but is specific to the odour exposed (Coopersmith *et al.*, 1986) and is longlasting (Coopersmith & Leon, 1986). Interestingly the enhanced neural response is particular to certain types of learning; for example, although conditioning an aversion to peppermint results in behavioural avoidance it does not result in any change in the glomerular response (Coopersmith *et al.*, 1986). Furthermore, the response only occurs upon exposure to the odour when coupled with the tactile stimulation. If the pup is not stroked then no change in neural responding is found (Sullivan & Leon, 1986) indicating that associative processes are involved (cf. earlier where associative learning was argued to have advantages for the learning of kinship information, pp. 273). Further studies revealed that certain mitral cells (the main output neurones of the olfactory bulb) in peppermint experienced pups, exhibited a decrease in excitatory and an increase in inhibitory firing, compared to controls (Wilson, *et al.*, 1985, 1987). Similar patterns of increased glomerular activity and differential neural firing patterns may be found as a result of exposure to, or learning about, kin.

Much of the above discussion, and that in the preceeding sections on ontogeny, has argued that individuals possess or develop a stored representation of kin and that this is necessary for recognizing kin. This need not be the case. Individuals may be able to recognize conspecifics with no long-term stored representation of kinship. There are two ways which this could be achieved.

Self-matching

This was termed the armpit effect by Dawkins (1982). Upon encountering another conspecific the individual compares its own cue with that presented by its conspecific, i.e. the individual sniffs its own armpit and then that of its conspecific. Again the degree of matching between the two cues determines relatedness. No long-term stored representation of kinship is necessary. Individuals simply require the ability to store their own cue and that of their conspecific for long enough for a comparison to take place. Little research has attempted to examine this as a means of recognizing kin although it should be noted that findings of recognition between genetically related individuals may be as a result of self-matching. Such a system would be very accurate, since individuals can be assured that their own cue is representative of their genetic identity. Because no learning is involved individuals avoid the problem of learning from unrelated individuals.

Dishabituation

Recognition by this means may play an important role in the assessment of kinship (Hepper, 1986a). Individuals constantly exposed to the cue of self and other close kin may habituate their response to these cues, i.e. no longer respond to these cues. This may be as the result of either peripheral adaptation of the sensory receptors or via a more central habituation process. On perceiving a novel cue, presented by an unrelated conspecific, the individual dishabituates. Dishabituation may first enable the degree of relatedness to be determined; the greater the dishabituation the more unrelated is that conspecific. Also, it may serve an arousing function and alert the individual to the presence of unrelated conspecifics enabling the individual to undertake a full investigation of the novel conspecific. Little work has investigated the role of habituation in kin recognition (although see studies by Brown *et al.*, 1987; Singh *et al.*, 1987 which have used habituation as a technique for demonstrating MHC produced odour differences in urine). Dishabituation is a simple process and may be the initial starting point for the determination of kinship by alerting individuals to the presence of unrelated conspecifics and allow further investigation, for example by self-matching or comparison to a SR to be undertaken.

Kin recognition, kin discrimination and familiarity

One important distinction drawn by authors (e.g. Byers & Bekoff, 1986; Hepper, 1986a; Waldman, 1987) is that between kin recognition and kin discrimination. Kin recognition is the process of classifying conspecifics to produce a degree of relatedness and has been regarded as the neural process(es) which produce a coefficient of relatedness (e.g. Byers & Bekoff, 1986). Kin discrimination on the other hand refers to the observed conse-quences of kin recognition, i.e. differential responsiveness to kin and non-kin. Since, at present, the neural processes involved in recognizing kin are unknown, demonstrations of kin recognition can only be obtained by observing the individual's behaviour. Thus, kin discrimination is used to infer that individuals can recognize their kin. A distinction between the neural event of recognizing kin (kin recognition) and the exhibition of differential behaviour (kin discrimination) is necessary since evidence that individuals show no kin discrimination does not logically mean that indivi-duals are unable to recognize their kin. For example, adult male rats, given a free choice in an open field between a related and unrelated oestrous female, show little differential responding to the two, i.e. they show no kin discrimi-nation. This could be interpreted as evidence that these animals are unable to recognize their kin. However in a two-choice preference test using a T-maze the same males exhibit discriminative responding, spending more time investigating the unrelated female, thus providing evidence of kin recogni-tion (Hepper, unpubl. obs.).

Whilst the ability to recognize kin may not necessarily result in differential responding, it is assumed that kin discrimination is the result of kin recognition. This also need not necessarily be the case. At the beginning of this chapter I proposed that any means which enabled individuals to respond differentially to kin and non-kin could be regarded as a mechanism of kin recognition. However, it is possible that kin discrimination could be achieved without kin recognition, as defined as that neural process which converts the cues to produce a degree of relatedness. For example, if environmental preferences were genetically determined, then kin would be observed inter-acting more by virtue of the fact they shared similar environmental prefer-ences. If patterns of behavioural responding were similarly genetically determined, it would be expected that individuals would interact more frequently with kin than non-kin because of the shared similarities in abilities or response patterns. In both cases individuals would be responding differentially to kin, i.e. exhibiting kin discrimination, without possessing any neural processes classifying individuals as kin. If in the natural environment, however, these 'mechanisms' did allow successful differentiation of kin from non-kin, individuals would gain the benefits espoused by Hamilton's kinship theory (1964a,b), and thus they should be considered as mechanisms of kin recognition. Much more research is required to determine whether differential responding between kin can be achieved without recognizing them as such.

Recently a major question has arisen as to whether kin discrimination is a result of familiarity rather than kin recognition. For example, Welker *et al.*, (1987) conclude from their study of association in the crab-eating monkey (*Macaca fascicularis*) that 'To explain the close affiliation among relatives in primate groups, it is hence only necessary for a mechanism to have evolved for differentiation between animals of different grades of familiarity.' (p. 220). In a similar vein Porter (1987) concludes 'familiarity thus appears to be more important than genetic relatedness *per se* for the development of sibling recognition in spiny mice' (p. 192). Similar conclusions have been reached by studies comparing mechanisms of phenotype matching and association/familiarity (e.g. Grau, 1982; Kareem & Barnard, 1982).

These comments address two separate issues, (a) the evolution of kin recognition and (b) the mechanisms used to recognize kin. With regard to the evolutionary issue, to propose that close affiliation (kin discrimination) may be explained by familiarity (Welker *et al.*, 1987) overlooks the ultimate causation of such affiliation. Hamilton's kinship theory (1964a,b) gives clear reasons why responding differentially to kin and non-kin is advantageous to fitness and it would thus be expected that a means to enable the recognition of kin would evolve. However, it is difficult to see why a mechanism for familiarity should evolve. This is perhaps most clearly illustrated by studies of inbreeding avoidance, where it is argued that prepubertal association (familiarity) is the important factor in the avoidance of inbreeding (e.g. Ågren, 1984; Dewsbury, 1982; Hill, 1974). Whilst there are clear reasons for kin to avoid inbreeding (Hepper, 1986a) there is little reason for familiar individuals to do so. Furthermore, as discussed throughout this chapter, contingencies are arranged to ensure that individuals do learn from kin. Thus, individuals may learn prenatally (Hepper, 1989), immediately after birth (e.g. Lawson & Renouf, 1987) or during a sensitive period (e.g. Bateson, 1981). These, and other contingencies, all act to ensure that individuals learn about genetically related individuals and not simply familiar conspecifics. It is difficult to see how such learning strategies could have evolved if learning about familiar individuals was the ultimate goal.

As regards the second issue, that concerning the mechanism of kin recognition, it is important to clarify what mechanisms of familiarity and genetic relatedness are. Genetic relatedness presumably refers to a recognition system determined by an individual's genes whereas familiarity refers to a system which has to be learned. In this sense it is true that kin recognition is a result of familiarity rather than genetic relatedness since given the problems of a genetically determined system it is most likely that individuals do learn about their kin rather than relying on a genetically programmed system. However, other than indicating that kin recognition is learned, the term familiarity is too broad to provide any useful information regarding the underlying basis of kin recognition and has previously been criticized for leaving many important questions of how, when and why learning takes place, unanswered (Hepper, 1986a). Furthermore, familiarity ignores the ultimate causation of kin recognition and the fact that in the natural situation the familiar individuals who are learned about are kin.

In conclusion, explanations of association or differential responding by familiarity fail to explain why such behaviour should occur in the first place and provide little information as to the underlying basis of recognition.

Conclusions

The above discussion has attempted to illustrate how individuals can recognize their kin and what the underlying basis of such behaviour might be. Previously proposed mechanisms have been so all-inclusive that they are able to explain little about how recognition is achieved. A wide variety of processes have been described which may underly the ability to recognize kin. Little research has attempted to determine how many (if any) of these influence recognition and its development, and future work must concentrate on elucidating the underlying basis of kin recognition if we are to understand fully how differential responding to kin and non-kin occurs.

It is perhaps pertinent to present one final consideration which has been greatly overlooked in the kin recognition literature. At present, confirmation of kin recognition relies on observations of discriminative responding, and as mentioned at the outset of this chapter, such behaviour has been observed in a wide variety of behaviours and times throughout an individual's life. However, because the same individual exhibits differential behaviour to kin and non-kin at different ages and in different situations does not mean that similar recognition processes are being used throughout. Siblings may recognize one another by a completely different means than do parents and their offspring; newborns, juveniles and adults may all use very different strategies for recognizing kin. It is often implied in discussions of the mechanism of kin recognition that individuals have a single means for recognizing kin, which is used in all situations. However, given the wide range of possible means of recognizing kin, recognition of kin may, and probably is, achieved by different means dependent upon the individual's age and situation.

Acknowledgements

I thank Ken Brown, Fiona Hepper and Ian Sneddon for their comments on this manuscript.

References

Ågren, G. (1984). Incest avoidance and bonding between siblings in gerbils. *Behav. Ecol. Sociobiol.*, **14**, 161–9.

Albone, E. S. (1984). *Mammalian Semiochemistry*. Chichester: John Wiley & Sons.

Aldhous, P. (1989). The effects of individual cross-fostering on the development of intrasexual kin discrimination in male laboratory mice, *Mus musculus* L. *Anim. Behav.*, **37**, 741–50.

Alexander, R. D. (1979). *Darwinism and Human Affairs*. London: Pitman.

Alkon, D. L. (1984). Calcium mediated reduction of ionic currents: a biophysical memory trace. *Science*, **226**, 1037–45.

Bateson, P. P. G. (1979). How do sensitive periods arise and what are they for? *Anim. Behav.*, **27**, 470–86.

Bateson, P. P. G. (1981). Control of sensitivity to the environment during development. In *Behavioral Development. The Bielefield Interdisciplinary Project*, ed. K. Immelmann, G. W. Barlow, L. Petrinovich & M. Main, pp. 432–53. Cambridge: Cambridge University Press.

Beecher, M. D. (1982). Signature systems and kin recognition. *Am. Zool.*, **22**, 477–90.

Bekoff, M. (1981). Mammalian sibling interactions. Genes, facilitative environments, and the coefficient of familiarity. In *Parental Care in Mammals*, ed. D. J. Gubernick & P. H. Klopfer, pp. 307–46. New York: Plenum.

Blaustein, A. R. (1983). Kin recognition mechanisms: Phenotype matching or recognition alleles. *Am. Nat.*, **121**, 749–54.

Blaustein, A. R. & O'Hara, R. K. (1986). Kin recognition in tadpoles. *Sci. Am.*, Jan, 90–6.

Brockelman, W. Y. & Schilling, D. (1984). Inheritance of stereotyped gibbon calls. *Nature*, **312**, 634–6.

Brown, R. E., Singh, P. B. & Roser, B. (1987). The major histocompatibility complex and the chemosensory recognition of individuality in rats. *Physiol. Behav.*, **40**, 65–73.

Bruce, C., Desimone, R. & Gross, C. G. (1981). Visual properties of neurones in a polysensory area in superior temporal sulcus of the macaque. *J. Neurophysiol.*, **46**, 369–84.

Byers, J. A. & Bekoff, M. (1986). What does 'kin recognition' mean? *Ethology*, **72**, 342–5.

Cattarelli, M. (1982). Transmission and integration of biologically meaningful olfactory information after bilateral transection of the lateral olfactory tract in the rat. *Behav. Brain Res.*, **6**, 313–37.

Colgan, P. (1983). *Comparative Social Recognition*. New York: John Wiley & Sons.

Coopersmith, R., Henderson, S. R. & Leon, M. (1986). Odor specificity of the enhanced neural response following early odor experience in rats. *Develop. Brain Res.*, **27**, 191–7.

Coopersmith, R., Lee, S. & Leon, M. (1986). Olfactory bulb responses after odor aversion learning by young rats. *Develop. Brain Res.*, **24**, 271–7.

Coopersmith, R. & Leon, M. (1984). Enhanced neural response to familiar olfactory cues. *Science*, **225**, 849–51.

Coopersmith, R. & Leon, M. (1986). Enhanced neural response by adult rats to odors experienced early in life. *Brain Research*, **371**, 400–3.

Courtney, R. J., Reid, L. D. & Wasden, R. E. (1968). Suppression of running times by olfactory stimuli. *Psychon. Science*, **12**, 315–16.

Crozier, R. H. (1987). Genetic aspects of kin recognition: Concepts, models and synthesis. In *Kin Recognition in Animals*, ed. D. J. C. Fletcher & C. D. Michener, pp. 55–73. Chichester: John Wiley & Sons.

Cullen, E. (1957). Adaptations in the kittiwake to cliff nesting. *Ibis*, **99**, 275–305.

Dawkins, R. (1982). *The Extended Phenotype*. Oxford: Freeman.

DeCasper, A. J. & Fifer, W. P. (1980). Of human bonding; newborns prefer their mothers voices. *Science*, **208**, 1174–6.

DeCasper, A. J. & Prescott, P. A. (1984). Human newborns' perception of male voices: preference, discrimination and reinforcing value. *Develop. Psychobiol.* **17**, 481–91.

DeCasper, A. J. & Spence, M. J. (1986). Prenatal maternal speech influences newborns' perception of speech sound. *Inf. Behav. Develop.*, **9**, 133–50.

Dewsbury, D. A. (1982). Avoidance of incestuous breeding between siblings in two species of *Peromyscus* mice. *Biol. Behav.*, **7**, 157–69.

Dewsbury, D. A. (1988). Kinship, familiarity, aggression, and dominance in deer mice (*Peromyscus maniculatus*) in seminatural enclosures. *J. Comp. Psychol.*, **102**, 124–8.

Doherty, J. & Hoy, R. (1985). Communication in insects III. The auditory behavior of crickets: some views of genetic coupling, song recognition, and predator detection. *Q. Rev. Biol.*, **60**, 457–72.

Elsacker, L. V., Pinxten, R. & Verheyen, R. F. (1988). Timing of offspring recognition in adult starlings. *Behaviour*, **107**, 122–30.

Ertman, S. C. & Davenport, D. (1981). Tentacular nematocyst discharge and 'self-recognition' in *Anthopleura elegantissima* Brandt. *Biol. Bull.*, **161**, 366–70.

Erulkar, S. D., Kelly, D. B., Jurman, M. E., Zemlan, F. P., Schneider, G. T. & Krieger, N. R. (1981). Modulation of the neural control on the clasp reflex in male Xenopus by androgens: a multidisciplinary study. *Proc. Natn. Acad. Sci., USA*, **78**, 5876–80.

Fletcher, D. J. C. & Michener, C. D. (1987). *Kin Recognition in Animals*. Chichester: John Wiley & Sons.

Friedhof, A. J. & Miller, J. C. (1983). Clinical implications of receptor sensitivity modulation. *Ann. Rev. Neurosci.*, **6**, 121–48.

Galef, B. G. (1982). Acquisition and waning of exposure-induced attraction to a non-natural odor in rat pups. *Develop. Psychobiol.*, **15**, 479–90.

Gamboa, G. J. (1988). Sister, aunt–niece, and cousin recognition by social wasps. *Behav. Genet.*, **18**, 409–23.

Grau, H. J. (1982). Kin recognition in white-footed deermice (*Peromyscus leucopus*). *Anim. Behav.*, **30**, 497–505.

Grobstein, P. (1988). On beyond neuronal specificity: problems in going from cells to networks and from networks to behavior. *Adv. Neur. Behav. Develop.*, **3**, 1–58.

Grosberg, R. K. & Quinn, J. F. (1986). The genetic control and consequences of kin recognition by the larvae of a colonial marine invertebrate. *Nature*, **322**, 456–9.

Gubernick, D. J. (1981). Parent and infant attachment in mammals. In *Parental Care in Mammals*, ed. D. J. Gubernick & P. H. Klopfer, pp. 243–305. New York: Plenum.

Hamilton, W. D. (1964a). The genetical evolution of social behaviour I. *J. Theor. Biol.*, **7**, 1–16.

Hamilton, W. D. (1964b). The genetical evolution of social behaviour II. *J. Theor. Biol.*, **7**, 17–52.

Hepper, P. G. (1983a). Sibling recognition in the rat. *Anim. Behav.*, **31**, 1177–91.

Hepper, P. G. (1983b). *Kin Recognition in the Rat*. Ph. D. Thesis, University of Durham.

Hepper, P. G. (1986a). Kin recognition: functions and mechanisms. A review. *Biol. Rev.*, **61**, 63–93.

Hepper, P. G. (1986b). Can recognition genes for kin recognition exist? In *The Individual and Society*, ed. L. Passera & J-P. Lachaud, pp. 31–5. Toulouse: Privat, IEC.

Hepper, P. G. (1987a). The discrimination of different degrees of relatedness in the rat: evidence for a genetic identifier. *Anim. Behav.*, **35**, 549–54.

Hepper, P. G. (1987b). The amniotic fluid: an important priming role in kin recognition. *Anim. Behav.*, **35**, 1343–6.

Hepper, P. G. (1987c). Rat pups prefer their siblings to their mothers: possible implications for the development of kin recognition. *Q. J. Exp. Psychol.*, **39B**, 265–72.

Hepper, P. G. (1988). The discrimination of human odour by the dog. *Perception*, **17**, 549–54.

Hepper, P. G. (1989). Foetal learning: Implications for psychiatry? *Brit. J. Psychiat.*, **155**, 289–93.

Hepper, P. G. (1990). Fetal olfaction. In *Chemical Signals in Vertebrates* V, ed. D. W. Macdonald & S. Natynzcuk. Oxford: Oxford University Press. (In Press.)

Hill, J. L. (1974). *Peromyscus*: effect of early pairing on reproduction. *Science*, **186**, 1042–4.

Hirsch, H. V. B. & Spinelli, D. N. (1971). Modification of the distribution of receptive field orientation in cats by selective visual exposure during development. *Exp. Brain Res.*, **13**, 509–27.

Hölldobler, B. & Michener, C. D. (1980). Mechanisms of identification and discrimination in social Hymenoptera. In *Evolution of Social Behaviour: Hypotheses and Empirical Tests*, ed. H. Markl, pp. 35–58. Weinheim: Verlag Chemie.

Holmes, W. G. (1988). Kinship and the development of social preferences. In *Handbook of Behavioral Neurobiology. 9. Developmental Psychobiology and Behavioral Ecology*, ed. E. M. Blass, pp. 389–413. New York: Plenum.

Holmes, W. G. & Sherman, P. W. (1983). Kin recognition in animals. *Am. Sci.*, **71**, 46–55.

Hoy, R. R. (1974). Genetic control of acoustic behaviour in crickets. *Am. Zool.*, **14**, 1067–80.

Hoy, R. R., Hahn, J. & Paul, R. C. (1977). Hybrid cricket auditory behaviour: evidence for genetic coupling in animal communication. *Science*, **195**, 82–4.

Hrdy, S. B. (1976). Care and exploitation of non-human primate infants by conspecifics other than the mother. *Adv. Study Behav.*, **6**, 101–58.

Isingrini, M. & Lenoir, A. (1986). La reconnaissance coloniale chez les hyménoptères sociaux. *Année Biol.*, **25**, 219–54.

Isingrini, M., Lenoir, A. & Jaisson, P. (1985). Preimaginal learning as a basis of colony-brood recognition in the ant *Cataglyphis cursor*. *Proc. Natn. Acad. Sci., USA*, **82**, 8545–7.

Johnson, M. H. & Horn, G. (1986). Dissociation of recognition memory and associative learning by a restricted lesion of the chick forebrain. *Neuropsychologia*, **24**, 329–40.

Johnson, M. H. & Horn, G. (1987). The role of a restricted region of the chick forebrain in the recognition of individual conspecifics. *Behav. Brain Res.*, **23**, 269–75.

Kandel, E. R. & Schwartz, J. H. (1982). Molecular biology of learning; modulation of transmitter release. *Science*, **218**, 433–43.

Kareem, A. M. & Barnard, C. J. (1982). The importance of kinship and familiarity in social interactions between mice. *Anim. Behav.*, **30**, 594–601.

Kendrick, K. M. & Baldwin, B. A. (1987). Cells in the temporal cortex of conscious sheep can respond preferentially to the sight of faces. *Science*, **236**, 448–50.

Keough, M. J. (1984). Kin-recognition and the spatial distribution of larvae of the bryozoan, *Bugula neritina* (L). *Evolution*, **38**, 142–7.

Klopfer, P. H., Adams, D. K. & Klopfer, M. S. (1964). Maternal imprinting in goats. *Proc. Natn. Acad. Sci., USA*, **52**, 911–14.

Lawson, J. W. & Renouf, D. (1987). Bonding and weaning in Harbor seals, *Phoca vitulina*. *J. Mammal.*, **68**, 445–9.

Leon, M. (1980). Development of olfactory attraction by young Norway rats. In. *Chemical Signals: Vertebrates and Aquatic Invertebrates*, ed. D. M. Müller-Schwarze & R. M. Silverstein, pp. 193–209. New York: Plenum.

Lévy, F. & Poindron, P. (1984). Influence du liquide amniotique sur la manifestation du comportement maternel chez la brebis parturiente. *Biol Behav.*, **9**, 271–8.

McCabe, B. J., Cipolla-Neto, J., Horn, G. & Bateson, P. (1982). Amnesic effects of bilateral lesions placed in the hyperstriatum ventrale of the chick after imprinting. *Exp. Brian Res.*, **48**, 13–21.

McClay, D. R. (1971). An autoradiographic analysis of the species specificity during sponge cell reaggregation. *Biol. Bull.*, **141**, 319–30.

Macdonald, D. W. & Moehlman, P. D. (1982). Co-operation, altruism, and restraint in the reproduction of carnivores. In *Perspectives in Ethology 5*, ed. P. P. G. Bateson & P. H. Klopfer, pp. 433–68. New York: Plenum.

Meadows, J. C. (1974). The anatomical basis of prosopagnosia. *J. Neurol., Neurosurg. Psychiat.*, **37**, 489–501.

Mollenauer, S., Plotnik, R. & Synder, E. (1974). Effects of olfactory bulb removal on fear responses and passive avoidance in the rat. *Physiol. Behav.*, **12**, 141–4.

Nishida, T. (1983). Alloparental behaviour in wild chimpanzess of the Mahale mountains, Tanzania. *Folia Primatol.*, **41**, 1–33.

Perret, D. I., Rolls, E. T. & Caan, W. (1982). Visual neurones responsive to faces in the monkey temporal cortex. *Exp. Brain Res.*, **47**, 329–42.

Perrett, D. I., Smith, P. A. J., Potter, D. D., Mislten, A. J., Head, A. S., Milner, A. D. & Jeeves, M. A. (1984). Neurones responsive to faces in the temporal cortex: studies of functional organization, sensitivity to identity and relation to perception. *Hum. Neurobiol.*, **3**, 197–208.

Pinker, S. (1985). Visual cognition: an introduction. In *Visual Cognition*, ed. S. Pinker, pp. 1–63. Cambridge, Mass: MIT Press.

Poindron, P., Martin, G. B. & Hooley, R. D. (1979). Effects of lambing induction on the sensitive period for the establishment of maternal behaviour in sheep. *Physiol. Behav.*, **23**, 1081–7.

Porter, R. H. (1987). Kin recognition: Functions and mediating mechanisms. In *Sociobiology and Psychology: Ideas, Issues and Applications*, ed. C. Crawford, M. Smith & D. Krebs, pp. 175–203. Hillsdale: Lawrence Erlbaum Assoc.

Porter, R. H. (1988). The ontogeny of sibling recognition in rodents: superfamily Muroidea. *Behav. Genet.*, **18**, 483–94.

Porter, R. H., McFadyen-Ketchum, S. A. & King, G. A. (1989). Underlying bases of recognition signatures in spiny mice, *Acomys cahirinus. Anim. Behav.*, **37**, 638–44.

Porter, R. H., Sentell, S. W. & Makin, J. W. (1987). Effects of intranasal ZnSO₄ irrigation are mitigated by the presence of untreated littermates. *Physiol. Behav.*, **40**, 97–102.

Ridley, M. & Grafen, A. (1981). Are green beard genes outlaws? *Anim. Behav.*, **29**, 954–5.

Rothstein, S. I. & Barash, D. P. (1983). Gene conflicts and the concepts of outlaw and sheriff alleles. *J. Soc. Biol. Struct.*, **6**, 367–79.

Schultze-Westrum, T. (1965). Innerartliche verständigung durch düfte beim gleitbeutler *Petaurus breviceps papuanus* Thomas (Marsupialia, Phalangeridae). *Z. vergl. Physiol.*, **50**, 151–220.

Scofield, V. L., Schlumpberger, J. M., West, L. A. & Weissman, I. L. (1982). Protochordate allorecognition is controlled by a MHC-like gene system. *Nature*, **295**, 499–502.

Shindler, K. M. (1984). A three year study of fetal auditory imprinting. *J. Wash. Acad. Sci.*, **74**, 121–4.

Singh, P. B., Brown, R. E. & Roser, B. (1987). MHC antigens in urine as olfactory recognition cues. *Nature*, **327**, 161–4.

Smotherman, W. P. (1982). In utero chemosensory experience alters taste preferences and corticosterone responsiveness. *Behav. Neur. Biol.*, **36**, 61–8.

Smotherman, W. P. & Robinson, S. R. (1989). *Behavior of the Fetus.* New Jersey: Telford.

Smyth, M. M., Morris, P. E., Levy, P. & Ellis, A. W. (1987). *Cognition in Action.* London: Lawrence Erlbaum Assoc.

Spencer-Booth, Y. (1970). The relationships between mammalian young and conspecifics other than mothers and peers: a review. *Adv. Study Behav.*, **3**, 119–94.

Sullivan, R. M. & Leon, M. (1986). Early olfactory learning induces an enhanced olfactory bulb response in young rats. *Develop. Brain Res.*, **27**, 278–82.

Sullivan, R. M., Wilson, D. A., Kim, M. H. & Leon, M. (1988). Behavioral and neural correlates of postnatal olfactory conditioning. I. Effect of respiration on conditioned neural responses. *Physiol. Behav.*, **44**, 85–90.

Tschanz, B. (1968). Trotellummen. *Z. Tierpsychol.*, **4**, 1–103.

Vince, M. A. & Billing, A. E. (1986). Infancy in the sheep: the part played by sensory stimulation in bonding between the ewe and lamb. *Adv. Inf. Res.*, **4**, 1–37.

Waldman, B. (1981). Sibling recognition in toad tadpoles: the role of experience. *Z. Tierpsychol.*, **56**, 341–58.

Waldman, B. (1987). Mechanisms of kin recognition. *J. Theor. Biol.*, **128**, 159–85.

Waldman, B. (1988). The ecology of kin recognition. *Ann. Rev. Ecol. Syst.*, **19**, 543–71.

Waldman, B., Frumhoff, P. C. & Sherman, P. W. (1988). Problems of kin recognition. *Trends Ecol. Evol.*, **3**, 8–13.

Welker, A., Schwibbe, M. H., Schäfer-Witt, C. & Visalberghi, E. (1987). Failure of kin recognition in Macaca fascicularis. *Folia Primatol.*, **49**, 216–21.

Whiteley, A. M. & Warrington, E. K. (1977). Prosopagnosia: a clinical, psychological and anatomical study of three patients. *J. Neurol. Neurosurg. Psychiat.*, **40**, 394–430.

Wilson, D. A., Sullivan, R. M. & Leon, M. (1985). Odor familiarity alters mitral cell response in the olfactory bulb of neonatal rats. *Develop. Brain Res.*, **22**, 314–17.

Wilson, D. A., Sullivan, R. M. & Leon, M. (1987). Single unit analysis of postnatal olfactory learning: altered mitral cell responsiveness to learned attractive odors. *J. Neurosci.*, **7**, 3154–62.

Yamazaki, K., Yamaguchi, M., Baranoski, L., Bard, J., Boyse, E. A. & Thomas, L. (1979). Recognition among mice. Evidence for the use of a Y-maze differentially scented by congenic mice of different major histocompatibility types. *J. Exp. Med.*, **150**, 755–60.

10

Parental states as mechanisms for kinship recognition and deception about relatedness

Robert W. Elwood

Parental care is one form of parental investment and investing in one young detracts from the ability of a parent to invest in others (Trivers, 1974). Parents are thus expected to allocate investment to one young until the cost of giving that care exceeds the benefit in terms of survival of the infant or, more correctly, in terms of survival of the half set of genes a diploid organism passes on to its sexually produced offspring. Each parent should attempt to maximize its production of viable offspring, that is by judicious allocation of resources it should attempt to maximize its genetic contribution to the next generation. Investment in unrelated young, however, will detract from the ability of an animal to produce its own offspring and hence will lead to a reduction in fitness. Mechanisms are, therefore, expected to evolve to ensure that investment is not wasted. It is the purpose of this chapter to examine some of these mechanisms by which parents reduce the possibility of investing in non-kin.

A parent could also suffer a loss of fitness if it were to harm its own offspring. This is a possibility because males and females of many species commonly utilize conspecific infants as food. For example, this occurs in fish (reviewed by Dominey & Blumer, 1984), gulls (reviewed by Mock, 1984), and rodents (Elwood, 1977; Sherman, 1981). It is important that parents should avoid cannibalizing their own offspring except in extreme circumstances (Labov et al., 1985). Males of several species may be infanticidal in another context, as a strategy to bring the mothers of the dead infants into oestrus, thus enabling earlier mating than would otherwise be possible, e.g. in langur monkeys (Hrdy, 1974) and lions (Bertram, 1975). Even if the date of oestrus is not changed the date of birth of the next litter may be brought forward in rodents by the death of all or part of the existing litter because of normal delayed inplantation when suckling a litter, e.g. lemmings (Mallory & Brooks, 1978) and mice (Labov, 1980). It is important that this form of infanticide should not be directed towards the male's own young and again mechanisms are anticipated to have evolved to minimize this possibility.

Parents may recognize and thus direct parental care towards their own infants using one or more of several mechanisms. In particular, parents may learn the features of their offspring and use these features to discriminate between kin and non-kin (e.g. Beecher *et al.*, 1981). However, this review concentrates on those situations in which infants are encountered for the first time; situations in which familiarity cannot account for the protection of kin. The possible mechanisms by which kin may be protected are:

1 Genetic marker: the parent may recognize the infant by some feature of the infant. It may be by phenotype matching to self or to close kin, e.g. previous infants (Dilger, 1962) or by matching to the sexual partner. The possibility of recognition genes is not precluded. The essential feature is that it is the infant *per se* that is recognized not some non-specific aspect of the stimulus situation (Hepper, 1986).

2 Association: adults may avoid harming young that are associated with a previous sexual partner, i.e. it is the partner which is recognized. In species in which maternal care is the norm it is most likely to be the male recognizing the mother of the infants but in species in which males care for eggs or young it is theoretically possible for the female to recognize the male care-giver.

3 Location: animals may respond differently to infants located in their own nest or territory or in the nest or territory of a sexual partner, compared to those located elsewhere.

4 State-mediated recognition: one or both parents may be brought into a parental state at about the time of birth of their own infants. At this time they may respond parentally to infants whilst ignoring or killing infants at times when their own young are unlikely to be encountered.

The first three of these mechanisms are discussed in recent reviews (e.g. Hepper, 1986; Holmes & Sherman, 1983; Porter, 1985) but discussion of the final mechanism is conspicuously absent (except in Hepper, 1986). It is this final possibility which is the subject of this chapter.

Onset of the maternal state

The onset of maternal care has been studied extensively in mammals (reviewed by Rosenblatt & Siegel, 1981), especially in rodents (reviewed by Rosenblatt & Siegel, 1983) and the present review concentrates on this latter group (see Picman & Belles-Isles, 1987, for avian examples).

With the exception of some laboratory strains of mice (Noirot & Goyens, 1971) virgin female rodents tend not to be maternal. For example virgin rats tend either to avoid unrelated test pups or to kill them (Jakubowski & Terkel,

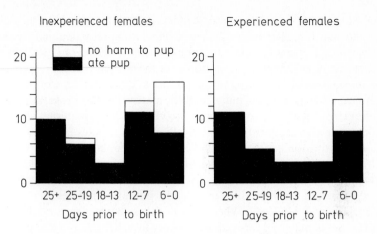

Fig. 10.1 This figure shows number of females tested at each reproductive state by the height of each column (25 + = non pregnant). (From Elwood, 1977.)

1985; Rosenblatt, 1967) and the response of virgin hamsters (Richards 1966), gerbils, (Elwood, 1977) and wild mice (McCarthy, Bare & vom Saal, 1986) is to cannibalize test pups. This avoidance or cannibalism of unrelated infants stops prior to parturition and a dramatic change in maternal responsiveness occurs in the last few days of gestation (Fig. 10.1). For example, female gerbils tested at this time commonly show a startle response on encountering the test pup but then alternate between sniffing and licking the pup and licking their own genital region in a manner similar to that seen during parturition (Elwood, 1977). This may be followed by retrieving the pup to the nest, nest-building and even assuming the lactation position. Clearly, the female enters a maternal state prior to parturition at which time she is receptive to both kin and non-kin.

Similar changes occur in the responses of pregnant rats. Whereas virgin rats require four to six days exposure to pups to elicit maternal responses those in mid-pregnancy require three to five days (Cosnier & Couturier, 1966; Fleming & Rosenblatt, 1974; Rosenblatt, 1967; Rosenblatt & Siegel, 1975). In the final day of gestation, however, rats are almost immediately maternal towards test pups (Bridges, Rosenblatt & Feder, 1978) and infanticide is inhibited in those strains in which it is exhibited by virgins (Peters & Kristal, 1983).

This onset of maternal responsiveness occurs in a wide variety of mammals and appears to be mediated by the hormonal changes seen during pregnancy (Rosenblatt & Siegel, 1981, 1983). In the rat, progesterone levels increase during pregnancy but show a sudden drop in the final four days. This contrasts with the sudden increase in levels of both oestrodiol and prolactin at the same time (Rosenblatt & Siegel, 1983). Long-term treatment with these hormones enhances maternal responsiveness (Zarrow *et al.*, 1971) but it is doubtful if they alone are responsible for the normal onset of maternal care. Some substance, occurring in the plasma at about the time of parturition, appears to be a powerful initiator of the maternal state. Cross transfusion of blood from newly parturient females to virgins, causes maternal responses in the latter after a short latency, but blood from females 24 hours before or after parturition has little effect (Terkel & Rosenblatt, 1972).

In wild mice, infanticidal females become inhibited from infanticide within one hour of being injected with prostaglandin, whereas injection with oxytocin not only inhibits infanticide but also initiates maternal care (McCarthy *et al.*, 1986). Whether or not these hormones normally mediate the onset of maternal care requires further investigation but it appears likely that they do, at least in mice. The important point to be made here is that several hormones have been implicated in the onset of the maternal state and the precise mechanism is likely to be complex. It is this physiological change that mediates the temporal change in behaviour and, for most of the time, ensures that females do not waste parental investment on non-kin. This allows some species to use cannibalism as a means of obtaining food, while ensuring that they avoid harming their own infants when they are present.

Non-pregnant female gerbils that have reared a previous litter also cannibalize test pups but again change to maternal responses in late pregnancy (Fig. 10.1). This indicates that the maternal state is temporary, its onset being due to the hormonal state but its maintenance is due to stimuli from the pups. Removing the pups at birth results in a more rapid return to cannibalism than if the litter remains with the female in both the gerbil (Elwood, 1981) and the hamster (Siegel & Greenwald, 1978). The return to infanticide is delayed, however, if hamsters interact with their pups for some days after birth (Siegel & Greenwald, 1978). In some species it is the unexpected loss of a brood or litter which seems to stimulate the parent to be cannibalistic towards unrelated infants, e.g. in gulls (Davis & Dunn, 1976) and ground squirrels (Sherman, 1981).

Onset of the paternal state

A male encountering a particular infant for the first time may be able to respond differently to kin and non-kin by the same four means as is theoretically possible for females, i.e. genetic marker, association, location and state-mediated recognition. With respect to the final hypothesis it is obvious that the same physiological state cannot pertain in both males and females because the male will not experience the same hormonal changes associated with egg production and/or pregnancy. Nevertheless, males do enter a paternal state prior to the birth of their young (Elwood, 1977) but this does not preclude the possibility that other mechanisms have an effect in kin recognition. Indeed there is evidence that they too may have a role in directing paternal responses and these will be considered briefly before examining physiological changes in males. Again the review will concentrate primarily, although not entirely, on rodents.

Genetic marker

The hypothesis that males can discriminate between related and unrelated pups at the first encounter has been tested using three laboratory strains of mice (Paul, 1986). Mated males were tested in their home cages, each with four unfamiliar one-day-old pups that were either their own offspring or those of another male. There was no difference in male responsiveness between the three strains but, overall, significantly more males killed the unrelated pups compared to their own pups (36% and 0% respectively). These data demonstrate an ability to recognize infants by some genetic marker, however, other experiments have failed to confirm these results. For example, in a recent study in this laboratory (Elwood & Kennedy, 1991) the responses of mated CS1 males to (a) unfamiliar own pups, (b) unfamiliar unrelated pups of the same strain or (c) unfamiliar pups of a different strain (CBA) were observed. There was no difference in the responses of males to these three categories of infant (3%, 4% and 3% infanticidal, respectively). Similar results were found with CBA males. In addition, Brooks & Schwarzkopf (1983) using a procedure virtually identical to that of Paul (1986) in which mated male C57B1/6J mice were offered DBA pups, reported that none of the males harmed the pups. Clearly, in this case there was no recognition of a genetic marker and similar results are reported by Parmigiani (1989). Thus, the study of Paul (1986) is the only one of several studies to report a genetic recognition of young that mediates infanticidal responses in mice. More work is required to establish the reasons for these contradictory results. Until this is done it is difficult to make firm conclusions about the ability of males to recognize their offspring *per se*, without prior experience of those young.

Association

Male collared lemmings (Mallory & Brooks, 1978) and mice (Labov, 1980, Huck *et al.*, 1982), placed into a cage with a former mate and her young, were unlikely to harm those young but they were likely to be infanticidal when placed in the cage of a strange female with her young. Furthermore, male mice were more likely to kill their own offspring if those offspring were fostered to a strange female rather than if males encountered unrelated young in the nest of a previous sexual partner (Huck *et al.*, 1982). Thus, there is the possibility that males can recognize females as familiar or not and respond to the pups accordingly. However, caution is required in evaluating these results as the response of the female towards the male might differ depending upon whether the male is the stud male or not (Ostermeyer, 1983; Svare & Gandelman, 1973). For example, stud males are attacked only very briefly whereas strange males receive a more severe attack (Mallory & Brooks, 1978) and there is the possibility that some pup-killing may be due to redirected aggression on the part of males receiving prolonged attack (although very severe attack by the female often saved the pups). A second reason for caution is that some other studies have failed to show any difference between males placed with previous sexual partners and those placed with strange females (vom Saal & Howard, 1982; Parmigiani, 1989; Elwood & Kennedy, 1991). One recent study showed the effect in wild mice that had prior infanticidal experience but not in those without prior infanticidal experience (McCarthy & vom Saal, 1986). The reasons for these inconsistencies are not yet clear.

Location

There has been one report of infanticide by male mice being more likely to occur in a novel cage rather than the home cage (Beilharz, 1975) but not in novel compared with familiar rooms (Brooks & Schwarzkopf, 1983). A recent experiment in this laboratory found no difference in infanticide by males encountering pups in their previous home cage or in a strange cage (Elwood & Kennedy, 1991). These cages were currently inhabited by the previous mate and a strange female, respectively, although the females were removed prior to the test. In this case there was no effect of familiarity of location and/or the odours therein. Similar findings are reported by Parmigiani (1989). It thus seems that location does not play a major role in determining the male's infanticidal responses, at least in rodents.

State mediated recognition

The possibility that a paternal state might exist was first investigated in the Mongolian gerbil (Elwood, 1975a, 1977), a species that shows considerable paternal care towards its own young (Elwood, 1975b; Elwood & Broom, 1978). The subjects were males that had been housed with females for some weeks but that had not yet produced their first litter. Each female was removed from the cage and a single, unrelated newborn placed in the cage with the male. Males with non-pregnant mates were likely to attempt to cannibalize the test pup whereas those with pregnant mates were more likely to be paternal. Indeed, no male housed with a female in the final six days of gestation harmed the test pup (Fig. 10.2). Since (a) none of the pups were related to the male, (b) all females were removed at the time of testing and (c) all tests were performed in the home cage, none of the three hypotheses discussed above can account for this onset of paternal care. A similar onset of

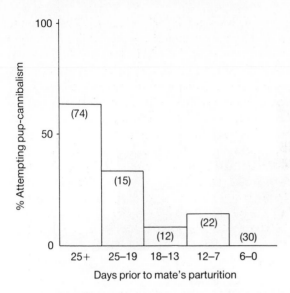

Fig. 10.2 Responses of naive males toward test pups at different stages in their mate's pregnancy (25 + = non-pregnant). Numbers in parentheses indicate sample size for each group. (From Elwood & Ostermeyer, 1984a.)

paternal care has been reported for the house mouse (Elwood, 1986; Labov, 1980; vom Saal, 1984, 1985) (Fig. 10.3) and the rat (Brown, 1986; Mennella & Moltz, 1988). It thus seems, at least in rodents, that males are brought into a parental state at a time when their own young are likely to be encountered. Note that this occurs both in species in which the main function of infanticide appears to be obtaining food, e.g. the gerbil (Elwood & Ostermeyer, 1984a,b), and in which infanticide appears to be a form of male–male competition, e.g. the house mouse (Labov, 1980).

What causes this paternal state?

Three main hypotheses have been put forward to account for the onset of a paternal state in rodents.

Firstly, it was proposed that a pheromone, released by the pregnant female, might cause the change in the male's behaviour (Elwood, 1977). Experiments on gerbils in which males were separated from their pregnant mates by wire mesh failed to provide any evidence of a pheromone effect (Elwood & Ostermeyer, 1984a). Furthermore, urine from pregnant females, dripped into the cages of males, failed to influence the infanticidal responses of those males. Males maintained on the soiled bedding of pregnant females or on clean bedding failed to show any differences in infanticide (Elwood &

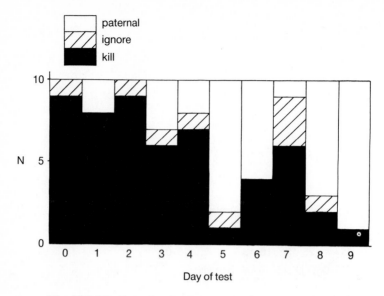

Fig. 10.3 Number of males that killed or ignored pups or acted paternally are shown for the day of mating and on subsequent days until day 9. Each male was tested once only and the female remained with the male until just prior to the test. (From Elwood, 1985.)

Ostermeyer, 1984a). Experiments on the possible role of airborne phero-mones in inhibiting infanticide in mice also failed to show any effect (Elwood, 1985). Recent work on rats, however, has demonstrated a pheromonal effect (Mennella & Moltz, 1988). Sexually naive Wistar rats provided with bedding soiled by pregnant females were less likely to be infanticidal (2/10) than were those provided with bedding soiled by non-pregnant females (7/10). Neither the source or the nature of the pheromone has been identified but the authors conclude that it is probably of high molecular weight as there was no evidence of airborne effects in their colony (Mennella & Moltz, 1988).

Secondly, the possibility that copulation may have an effect has been considered. This was regarded as unlikely in the gerbil because males that have their pregnant mates removed from the cage show a return to pup-cannibalism (Fig. 10.4) and thus copulation cannot account for the mainten-ance of the inhibition of cannibalism (Elwood, 1975a, 1980). Studies on rats have failed to show any effect of copulation on infanticide in some strains (Brown, 1986) but have in others (Mennella & Moltz, 1988) although the methods used in these studies differed greatly. In the mouse there is considerable evidence that copulation does have an effect (e.g. vom Saal & Howard, 1982; Huck et al., 1982; Elwood, 1985; Soroker & Terkel, 1988). In these studies on mice, copulation has been shown to markedly reduce the

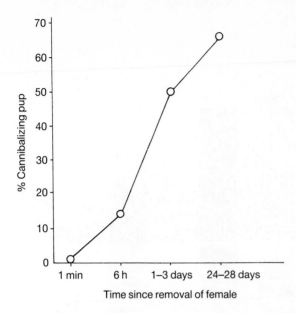

Fig. 10.4 The percentage of naive males that attempted to cannibalize the test pup is shown for four groups that differ in the time since the removal of the pregnant female. (From Elwood, 1980.)

proportion of males that are infanticidal, however, this has not been the case in every experiment (Elwood & Ostermeyer, 1984c). One problem in these studies has been the separation of effects due to copulation from those possibly due to postcopulatory interactions but this was achieved in a study by vom Saal (1985). In that study males were allowed to (a) interact with a non-oestrous female (b) mount and intromit with a receptive female, or (c) intromit and ejaculate with a female. Only this latter group showed a reduction in infanticide and an onset of paternal care when tested 20 days later, indicating that some change was brought about by ejaculation.

In another study, previously infanticidal males were (a) allowed to mate, the female being removed on the day of copulation or (b) not allowed to mate (Elwood, 1985). When tested at a later date there was a marked difference in paternal responsiveness with those that had copulated, showing a greater paternal responsiveness than those that had not (Fig. 10.5 groups 2 and 3). These data are thus in agreement with those of vom Saal (1985). However, the not mated group of Elwood (1985) were subsequently mated, the female removed, and the males tested at a later date. Again, copulation appeared to enhance paternal responsiveness but not as much as seen in the other group that had copulated (Fig. 10.5 compare groups 2 and 3a). It thus appears that differences in experimental procedure between these two groups had

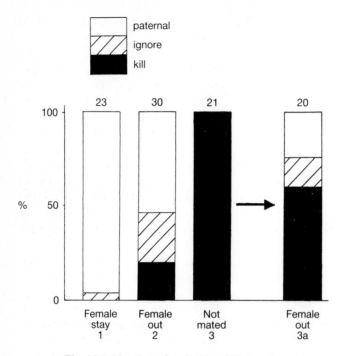

Fig. 10.5 Number of males that killed or ignored pups or acted paternally when tested 18 days after each male had mated and remained with the female (1), if the female was removed after mating (2) or if the males had not mated (3). This final group were subsequently mated and tested again (3a). (From Elwood, 1985.)

influenced their responsiveness even though both groups had copulated. The procedures differed in respect of the period of isolation prior to mating and in the number of infanticidal experiences. Subsequent experiments, however, showed that the duration of isolation did not influence the effectiveness of copulation in enhancing paternal responsiveness but additional infanticidal experiences markedly reduced the effectiveness of copulation in bringing about the onset of paternal care (Elwood, 1986). Prior infanticidal experience also appeared to reduce the effectiveness of copulation in wild mice when males were tested by placing them with a female and her litter but not when tested by placing a newborn pup into the males' cage (McCarthy & vom Saal, 1986). It may be concluded from these studies that copulation in mice has an effect in reducing infanticide in males and in initiating paternal care but that the magnitude of this effect varies with the methods used. The magnitude of the effect also varies with the genotype of the mice, as CSI males show an effect but CBA males do not, even when tested with identical methods (Kennedy & Elwood, 1988). Intra-uterine position of the developing embryo also influences the effectiveness of copulation in adulthood in eliminating infanticide (Perrigo & vom Saal, 1989).

The final hypothesis concerning the onset of the paternal state concerns effects brought about by post-mating interactions between the sexes. It has already been noted that infanticide is only inhibited in mated male gerbils if they remain in the presence of their pregnant mates (Fig. 10.4). Thus, experience after copulation is clearly important in the maintenance of the inhibition in that species. It has been suggested that social subordination of the male by the female may be a causal factor in infanticide inhibition (Elwood & Ostermeyer, 1984a). Observations on mated pairs of gerbils suggest that the female becomes dominant in late pregnancy temporarily ousting the male from the nest at that time (Elwood, 1975b). Furthermore, male gerbils that are subordinate to other males show a lower tendency to commit infanticide than do those that are dominant (Elwood & Ostermeyer, 1984b).

There is also evidence that post-copulatory interactions are important in the onset of paternal care in mice (Elwood, 1985; Elwood & Ostermeyer, 1984c; Parmigiani, 1989). In one experiment, females were either removed from males on the day of copulation or they remained after copulation (Elwood, 1985). If tested seven days after copulation there was no difference between these groups in their responsiveness to pups but when tested 13 and 18 days after copulation (Fig. 10.6), in the group where the female had remained, the males were more paternal than were those which had had the female removed.

Unlike the gerbil, a pregnant female mouse need only remain with the male for a short time after mating for him to remain paternal. Indeed, just one day of cohabitation immediately after copulation enhances paternal care when tested at a later date (Elwood, 1985). Cohabitation without copulation, however, is not effective on infanticide inhibition, as shown by pairing males with ovariectomized females (McCarthy & vom Saal, 1986). It is not clear, however, if it is the absence of copulation which accounts for this latter result or whether it is the omission of cohabitation with a pregnant female. That the important factor is cohabitation with a pregnant female has been demonstrated by studies on wild mice (Soroker & Terkel, 1988) and similar results have been obtained in studies on rats (Brown, 1986). Previously infanticidal male rats were placed into one of four groups, (a) not mated, (b) mated and left for 24 hours with an oestrus-induced ovariectomized female, (c) mated with an oestrus-induced ovariectomized female and allowed to cohabit with that female and (d) mated and allowed to cohabit with an intact female. Only this last group showed an inhibition of infanticide, indicating that in rats neither copulation nor copulation followed by cohabitation with a non-pregnant female is effective in reducing infanticide. Mennella & Moltz (1988), however, reported an effect due to copulation as well as cohabitation

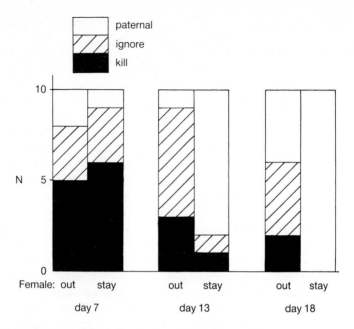

Fig. 10.6 Number of males that killed or ignored pups or acted paternally when tested with pups different times after mating. In some groups the females remained until just prior to testing and in others they were removed. (From Elwood, 1985.)

with a pregnant female. What then is the factor associated with pregnant females that brings about the onset of paternal care?

Mice are similar to gerbils in that dominant males are more likely to be infanticidal than are subordinates (Huck *et al.*, 1982; vom Saal & Howard, 1982). It is thus possible that the female subordinates the male after mating and that this is a mechanism for the onset of paternal care. This, however, requires that infanticidal males cease to be infanticidal if they are subordinated, and this has been tested (Elwood, 1986). Previously infanticidal males were paired with other males for 10 minutes per day for five days and their interactions noted. Those that were subordinate in these interactions were less likely to be infanticidal than those that were dominant. Furthermore, those subordinate males that still killed the test pups had not been subordinated to the same extent as had those that were no longer infanticidal. These data clearly demonstrate a shift in responsiveness to pups due to a shift in social status even though the dominance/subordinancy tests had occurred 14–18 days prior to testing with the pups.

The question thus arises: do female mice subordinate their mates following copulation? In particular, does this occur on the day after mating since cohabitation during this period also enhances paternal responsiveness? Observations on mated pairs have shown that in the hours following copulation the female is aggressive towards the male (Elwood, 1986) and, although the levels of female aggression towards the male decline by the following day, she appears to be dominant. Thus, marked changes in the dominance–subordinancy relationship of the pair are coincident with the time during which the female's presence is most effective in initiating the onset of paternal care. Furthermore, males with relatively large mates are more likely to be paternal 13 days after copulation than are those with relatively small mates (Elwood, 1986). Again this finding is consistent with the hypothesis that the onset of paternal care is due to males being subordinated by their mates after mating.

The physiological state

It may be concluded from the above that copulation has an inhibiting effect on infanticide in mice and possibly in rats but probably not in gerbils, but that in all three species cohabitation with the pregnant female inhibits infanticide and facilitates the onset of paternal care. There is good, albeit circumstantial, evidence that social subordination is an important factor in mice and gerbils and that there is some pheromonal effect in rats. Presumably these factors cause some relatively long lasting, but not irreversible, physiological change. The precise physiological mechanism, however, is unknown but it is probable that hormonal changes are responsible and testosterone is a likely candidate. Testosterone levels during embryonic development influence adult infanticidal tendencies (vom Saal, 1984) and males that have been gonadectomized at birth show low levels of infanticide that increase after treatment with testosterone in adulthood (Gandelman & vom Saal, 1975). It is interesting to note that in mice, ejaculation (Batty, 1978) and social subordination (Bronson & Marsden, 1973) are both followed by a decline in plasma testosterone. It is thus possible that this single physiological change mediates the inhibitory effects of these two behavioural interactions. If this is the case the male must be responsive to a temporary change in testosterone levels with the effect taking place some time later. Alternatively, or in addition, plasma prolactin may mediate these changes as male mice housed with pregnant females have been reported to have higher levels of this hormone than do isolated males (Parmigiani, 1989). The key point, however, is that males are brought into a paternal state prior to the birth of their own infants and thus avoid harming close kin.

Recent experiments have investigated whether isolated and sexually naive mice, that were either infanticidal or non-infanticidal, differed in certain ways. These indicated that (a) newly pregnant females were more likely to show pregnancy block when exposed to infanticidal males (Elwood & Kennedy, 1990), (b) newly parturient females were more likely to attack infanticidal males (Elwood et al., 1991) and (c) that infants produced more ultrasonic vocalizations and were more likely to move away from urinary odours of infanticidal compared to non-infanticidal males (Elwood et al., 1990). Parmigiani et al. (1988) have demonstrated a difference in female attack towards sexually naive males (likely to be infanticidal) and males that were rearing litters. These data are all congruent with the hypothesis that there is some physiological difference between these types of male that can be detected by other animals.

Female strategies to cuckold males

Whatever the precise physiological mechanism, it is clear that specific interactions between the male and female ensure that the male does not harm his offspring and thus both partners benefit in terms of genetic contribution to the following generation. This form of kin 'recognition', however, is not specific to the actual relatedness of the male to the infants. The possibility thus exists that a male may be equally paternal to all members of a litter of mixed paternity or may even help rear an entire litter of which he is not the father. Furthermore, in certain circumstances, females could benefit by cuckolding males into helping rear her infants and, to induce paternal responses, females might employ the normally mutually beneficial mechanisms that initiate a paternal state. This would be to the male's disadvantage if, by giving time to rear non-kin, he were to lose mating opportunities with other females (Werren et al., 1980).

One way a female may cuckold a male is to mate with that male even though her chances of conception are low (Hrdy, 1974). For example, pregnant langur monkeys show pseudo-oestrus if a new male takes over the troop and by mating with the new male may 'confuse paternity' and thus reduce the risk of her infant being killed (Hrdy, 1979). Pseudo-oestrus has also been reported in lions in similar circumstances of male take-over of the pride (Packer & Pusey, 1983). Female water voles may move location after becoming pregnant and thus move away from the fathers of their litters (Jeppsson, 1986). It has been reported that pregnant females that move are significantly more likely to show pseudo-oestrus and mate again than are those that remain in the locality of the father. It is assumed that this is an attempt to reduce the infanticidal tendencies of the male(s) in the new location (Jeppsson, 1986), that is, the female attempts to manipulate the males' kin recognition to her advantage.

An avian example of this comes from a study of the dunnock (Davies, 1985). Dunnocks are inconspicuous birds with a variable mating system and one female may commonly associate with two males. In this case the alpha male attempts to guard the female and thus obtain all the matings. The female, however, frequently attempts to escape from the alpha male to mate with the beta male. If the beta male mates with the female he helps to rear all the young in the nest. If, however, the beta male is excluded from mating with the female (i.e. cases in which mating was not observed) he does not help rear the young. In these cases, the eggs are likely to be damaged or removed or the nestlings killed. It thus appears that the beta male's responses towards the young are dependent only upon the possibility that he is related to them, rather than actual discrimination of relatedness (Davies, 1985). It is thus to the female's advantage to mate with the beta male in order to elicit his help in rearing her offspring and she appears to go to some lengths to obtain that mating. In this case the amount of paternal care given to the young is correlated with the time that the male spent in the vicinity of the female during the mating period (Burke *et al.*, 1989).

A second way by which a female may manipulate the responses of a male in such a way as to cuckold him is by domination. Maternal aggression by pregnant and newly parturient females has been recognized as a means of protecting infants from the infanticidal tendencies of conspecifics of both sexes (Ostermeyer, 1983; Paul, Gronek & Politch, 1980; Wolff 1985). These studies assumed that the young are protected by the actions of the female to keep intruders away from the young. One study on rats, however, suggests that males may be 'rendered non-infanticidal' by the post-partum aggression of the female (Jakubowski & Terkel, 1985). Similar suggestions have been made for marsh wrens (Picman & Belles-Isles, 1987). Those studies concentrated on the father of the infants, however, one recent study on fish suggests a similar means of cuckolding males (Yanagisawa & Ochi, 1986). In anemone fish both sexes help to rear the young but in cases in which the fathers were removed, females obtained help from other males. The initial response of a female towards a new male was one of attack, the male was subsequently butted whenever he approached the eggs. These males assumed submissive postures, and it was concluded that by means of butting, females were persuading or compelling their new mates to undertake egg-care duty (Yanagisawa & Ochi, 1986).

These studies thus suggest that females may manipulate males into a non-infanticidal or even paternal state even though the male is not related to her infants. This manipulation may involve the mechanism by which males normally avoid harming their own offspring but instead act paternally. That is, non-specific mechanisms for kin recognition may be manipulated to the advantage of other conspecifics or even to other species with respect to brood parasitisim (Rothstein, 1982).

References

Batty, J. (1978). Acute changes in plasma testosterone levels and their relation to measures of sexual behaviour in the male house mouse (*Mus musculus*). *Anim. Behav.*, **26**, 349–57.
Beecher, M. D., Beecher, I. M. & Hahn, S. (1981). Parent–offspring recognition in

bank swallows (*Riparia riparia*). II. Development and acoustic basis. *Anim. Behav.*, **29**, 95–101.

Beilharz, R. G. (1975). The aggressive response of male mice (*Mus musculus* L.) to a variety of stimulus animals. *Z. Tierpsychol.*, **39**, 141–9.

Bertram, B. C. R. (1975). Social factors influencing reproduction in wild lions. *J. Zool.*, **177**, 463–82.

Bridges, R. S., Rosenblatt, J. S. & Feder, H. H. (1978). Stimulation of maternal responsiveness after pregnancy termination in rats: effects of time of onset of behavioral testing. *Horm. Behav.*, **9**, 156–69.

Bronson, F. H. & Marsden, H. M. (1973). The preputial gland is an indicator of social dominance in male mice. *Behav. Biol.*, **9**, 625–8.

Brooks, R. J. & Schwarzkopf, L. (1983). Factors affecting incidence of infanticide and discrimination of related and unrelated neonates in male *Mus musculus*. *Behav. Neur. Biol.*, **37**, 149–161.

Brown, R. E. (1986). Social and hormonal factors influencing infanticide and its suppression in adult male Long-Evans rats (*Rattus norvegicus*). *J. Comp. Psychol.*, **100**, 155–61.

Burke, T., Davies, N. B., Bruford, M. W. & Hatchwell, B. J. (1989). Parental care and mating behaviour of polyandrous dunnocks *Prunella modularis* related to paternity by DNA fingerprinting. *Nature*, **338**, 249–51.

Cosnier, J. & Couturier, C. (1966). Comportement maternel provogue chez les rattes adultes castrees. *C. R. Seanc. Soc. Biol.*, **160**, 789–91.

Davies, N. B. (1985). Co-operation and conflict among dunnocks, *Prunella modularis*, in a variable mating system. *Anim. Behav.*, **33**, 628–48.

Davis, J. W. F. & Dunn, E. K. (1976). Intraspecific predation and colonial breeding in lesser black-backed gulls, *Larus fuscus*. *Ibis*, **118**, 65–77.

Dilger, W. C. (1962). The behavior of lovebirds. *Sci. Am.*, **206**, 88–98.

Dominey, W. J. & Blumer, L. S. (1984). Cannibalism of early life stages in fishes. In *Infanticide: Comparative and Evolutionary Perspectives*, ed. G. Hausfater & S. B. Hrdy, pp. 43–64. New York: Aldine.

Elwood, R. W. (1975a). *Paternal and Maternal Behaviour of the Mongolian Gerbil, and the Development of the Young*. Unpublished PhD thesis, University of Reading.

Elwood, R. W. (1975b). Paternal and maternal behaviour of the Mongolian gerbil. *Anim. Behav.*, **23**, 766–72.

Elwood, R. W. (1977). Changes in the responses of male and female gerbils (*Meriones unguiculatus*) towards test pups during the pregnancy of the female. *Anim. Behav.*, **25**, 46–51.

Elwood, R. W. (1980). The development, inhibition and disinhibition of pup-cannibalism in the Mongolian gerbil. *Anim. Behav.*, **28**, 1188–94.

Elwood, R. W. (1981). Postparturitional re-establishment of pup-cannibalism in female gerbils. *Dev. Psychobiol.*, **14**, 209–12.

Elwood, R. W. (1985). The inhibition of infanticide and the onset of paternal care in male mice, *Mus musculus*. *J. Comp. Psychol.*, **99**, 457–67.

Elwood, R. W. (1986). What makes male mice paternal? *Behav. Neur. Biol.*, **46**, 54–63.

Elwood, R. W. & Broom, D. M. (1978). The influence of litter size and parental behaviour on the development of Mongolian gerbil pups. *Anim. Behav.*, **26**, 438–54.

Elwood, R. W. & Kennedy, H. F. (1990). The relationship between infanticide and pregnancy block in mice. *Behav. Neur. Biol.*, **53**, 277–83.

Elwood, R. W. & Kennedy, H. F. (1991). Selectivity in paternal and infanticidal responses by male mice effects of relatedness, location and previous sexual partners. *Behav. Neur. Biol.* (In Press.)

Elwood, R. W. & Kennedy, H. F. & Blakely, H. M. (1990). Responses of infant mice to odors of urine from infanticidal, non-infanticidal and paternal male mice. *Dev. Psychobiol.*, **23**, 309–17.

Elwood, R. W., Nesbitt, A. A. & Kennedy H. F. (1991). Maternal aggression and the risk of infanticide in mice. *Anim Behav.* (In Press.)

Elwood, R. W. & Ostermeyer, M. C. (1984a). Infanticide by male and female Mongolian gerbils: ontogeny, causation and function. In *Infanticide: Comparative and Evolutionary Perspectives*, ed. G. Hausfater & S. B. Hrdy, pp. 367–86. New York: Aldine.

Elwood, R. W. & Ostermeyer, M. C. (1984b). The effects of food deprivation, aggression and isolation on infanticide in the male Mongolian gerbil. *Aggress. Behav.*, **10**, 293–301.

Elwood, R. W. & Ostermeyer, M. C. (1984c). Does copulation inhibit infanticide in rodents? *Anim. Behav.*, **32**, 293–305.

Fleming, A. S. & Rosenblatt, J. S. (1974). Maternal behaviour in the virgin and lactating rat. *J. Comp. Physiol. Psychol.*, **86**, 957–72.

Gandelman, R. & vom Saal, F. S. (1975). Pup-killing in mice: the effects of gonadectomy and testosterone administration. *Physiol. Behav.*, **15**, 647–51.

Hepper, P. G. (1986). Kin recognition: functions and mechanisms a review. *Biol. Rev.*, **61**, 63–93.

Holmes, W. G. & Sherman, P. W. (1983). Kin recognition in animals. *Sci. Am.*, **71**, 46–55.

Hrdy, S. B. (1974). Male–male competition and infanticide among the langurs (*Prebytis entellus*) of Abu' Rajasthan. *Folia Primatol.*, **22**, 19–58.

Hrdy, S. B. (1979). Infanticide among animals: a review, classification and examination of the implication for the reproductive strategies of females. *Ethol. Sociobiol.*, **1**, 13–40.

Huck, U. W., Soltis, R. L. & Coopersmith, C. B. (1982). Infanticide in male laboratory mice: effects of social status, prior sexual experience, and basis for discrimination between related and unrelated young. *Anim. Behav.*, **30**, 1158–65.

Jakubowski, M. & Terkel, J. (1985). Transition from pup killing to parental behavior in male and virgin female albino rats. *Physiol. Behav.*, **34**, 683–86.

Jeppsson, B. (1986). Mating by pregnant water voles (*Arvicola terrestris*): a strategy to counter infanticide by males? *Behav. Ecol. Sociobiol.*, **19**, 293–6.

Kennedy, H. F. & Elwood, R. W. (1988). Strain differences in the inhibition of infanticide in male mice (*Mus musculus*). *Behav. Neur. Biol.*, **50**, 349–53.

Labov, J. B. (1980). Factors influencing infanticidal behavior in wild male house mice (*Mus musculus*). *Behav. Ecol. Sociobiol.*, **6**, 297–303.

Labov, J. B., Huck, U. W., Elwood, R. W. & Brooks, R. J. (1985). Current problems in the study of infanticidal behaviour of rodents. *Q. Rev. Biol.*, **60**, 1–20.

Mallory, F. F. & Brooks, R. J. (1978). Infanticide and other reproductive strategies in the collared lemming, *Discrostonyx groenlandicus*. *Nature*, **273**, 144–6.

McCarthy, M. M., Bare, J. E. & vom Saal, F. (1986). Infanticide and parental behavior in wild female house mice: effects of ovariectomy, adrenalectomy and administration of oxytocin and prostaglandin F_2X. *Physiol. Behav.*, **36**, 17–23.

McCarthy, M. & vom Saal, F. (1986). Inhibition of infanticide after mating by wild male house mice. *Physiol. Behav.*, **36**, 203–9.

Mennella, J. A. & Moltz, H. (1988). Infanticide in rats: male strategy and female counter strategy. *Physiol. Behav.*, **42**, 19–28.

Mock, D. W. (1984). Infanticide, siblicide and avian nestling mortality. In *Infanticide: Comparative and Evolutionary Perspectives*, ed. G. Hausfater & S. B. Hrdy, pp. 3–30. New York: Aldine.

Noirot, E. & Goyens, J. (1971). Changes in maternal behaviour during gestation in the mouse. *Horm. Behav.*, **2**, 207–15.

Ostermeyer, M. C. (1983). Maternal aggression. In *Parental Behaviour of Rodents*, ed. R. W. Elwood, pp. 151–79. Chichester: John Wiley & Sons.

Packer, C. & Pusey, A. E. (1983). Adaptations of female lions to infanticide by incoming males. *Am. Nat.*, **121**, 716–28.

Parmigiani, S. (1989). Maternal aggression and infanticide in the house mouse: consequences on the social dynamics. In: *House Mouse Aggression: A Model for Understanding the Evolution of Social Behaviour*, ed. P. F. Brain, D. Mainardi & S. Parmigiani, pp. 161–78. London: Harwood Academic Press.

Parmigiani, S. (1989). Inhibition of infanticide in male house mice (*Mus domesticus*): is kin recognition involved? *Ethol. Ecol. Evol.*, **1**, 93–8.

Parmigiani, S., Sgoifo, A. & Mainardi, D. (1988). Parental aggression displayed by female mice in relation to the sex, reproductive status and infanticidal potential of conspecific intruders. *Monitore Zool. Ital. (NS)*, **22**, 193–201.

Paul, L. (1986). Infanticide and maternal aggression: synchrony of male and female reproductive strategies in mice. *Aggress. Behav.*, **12**, 1–11.

Paul, L., Gronek, J. & Politch, J. (1980). Maternal aggression in mice: Protection of young as a by-product of attacks at the home site. *Aggress. Behav.*, **6**, 19–30.

Perrigo, G. H. & vom Saal, F. S. (1989). Mating-induced regulation of infanticide in male mice: fetal programming of a unique stimulus-response. In *Ethoexperimental Approaches to the Study of Behaviour*, ed. R. J. Blanchard, P. F. Brain, D. C. Blanchard & S. Parmigiani, pp. 320–36. Dordrecht: Kluwer.

Peters, L. C. & Kristal, M. B. (1983). Suppression of infanticide in mother rats. *J. Comp. Psychol.*, **97**, 167–77.

Picman, J. & Belles-Isles, J. C. (1987). Intraspecific egg destruction in marsh wrens: a study of mechanisms preventing filial ovicide. *Anim. Behav.*, **35**, 236–46.

Porter, R. H. (1985). Kin recognition: a selective overview. *Proceedings of the 19th International Ethological Conference*, Vol. 3, 59–69.

Richards, M. P. M. (1966). Maternal behaviour in the golden hamster: responsiveness to young in virgin, pregnant, and lactating females. *Anim. Behav.*, **14**, 310–13.

Rosenblatt, J. S. (1967). Non-hormonal basis of maternal behaviour in the rat. *Science*, **156**, 1512–14.

Rosenblatt, J. S. & Siegel, H. I. (1975). Hysterectomy-induced maternal behaviour during pregnancy in the rat. *J. Comp. Physiol. Psychol.*, **89**, 685–700.

Rosenblatt, J. S. & Siegel, H. I. (1981). Factors governing the onset and maintenance of maternal behavior among nonprimate mammals. The role of hormonal and nonhormonal factors. In *Parental Care in Mammals*, ed. D. J. Gubernick & P. A. Klopfer, pp. 13–76. New York: Plenum Press.

Rosenblatt, J. S. & Siegel, H. I. (1983). Physiological and behavioural changes during pregnancy and parturition underlying the onset of maternal behaviour in rodents. In *Parental Behaviour of Rodents*, ed. R. W. Elwood. Chichester: John Wiley & Sons.

Rothstein, S. I. (1982). Mechanisms of avian egg recognition: which egg parameters elicit responses by rejector species? *Behav. Ecol. Sociobiol.*, **11**, 229–39.

Sherman, P. W. (1981). Reproductive competition and infanticide in Belding's ground squirrels and other animals. In *Natural Selection and Social Behaviour*, ed. R. D. Alexander & D. W. Tickle, pp. 311–31. New York: Chiron.

Siegel, H. I. & Greenwald, G. S. (1978). Effects of mother-litter separation on later maternal responsiveness in the hamster. *Physiol. Behav.*, **21**, 147–9.

Soroker, V. & Terkel, J. (1988). Changes in incidence of infanticidal and parental responses during the reproductive cycle of male and female wild mice *Mus musculus*. *Anim. Behav.*, **36**, 1275–81.

Svare, B. & Gandelman, R. (1973). Postpartum aggression in mice: experiential and environmental factors. *Horm. Behav.*, **4**, 323–34.

Terkel, J. & Rosenblatt, J. S. (1972). Humoral factors underlying maternal behaviour at parturition: cross transfusion between freely moving rats. *J. Comp. Physiol. Psychol.*, **80**, 365–71.

Trivers, R. L. (1974). Parent–offspring conflict. *Am. Zool.*, **14**, 249–64.

vom Saal, F. S. (1984). Proximate and ultimate causes of infanticide and parental behavior in male house mice. In *Infanticide: Comparative and Evolutionary Perspectives*, ed. G. Hausfater & S. B. Hrdy, pp. 401–24. New York: Aldine.

vom Saal, F. S. (1985). Time-contingent change in infanticide and parental behavior induced by ejaculation in male mice. *Physiol. Behav.*, **34**, 7–15.

vom Saal, F. S. & Howard, L. S. (1982). The regulation of infanticide and parental behaviour: implications for reproductive success in male mice. *Science*, **215**, 1270–2.

Werren, J. H., Gross, M. R. & Shine, R. (1980). Paternity and the evolution of male parental care. *J. Theor. Biol.*, **82**, 619–31.

Wolff, J. O. (1985). Maternal aggression as a deterrent to infanticide in *Peromyscus leucopus* and *P. maniculatus*. *Anim. Behav.*, **33**, 117–23.

Yanagisawa, Y. & Ochi, H. (1986). Step-fathering in the anemonefish *Amphiprion clarkii*: a removal study. *Anim. Behav.*, **34**, 1769–80.

Zarrow, M. X., Gandelman, R. & Denenberg, V. H. (1971). Prolactin: is it an essential hormone for maternal behaviour in mammals? *Horm. Behav.*, **2**, 343–54.

11

Fetal learning: implications for the development of kin recognition

Scott R. Robinson and William P. Smotherman

Since Hamilton's oft-cited prediction regarding the unequal distribution of behaviour among genetically related (kin) and unrelated conspecifics (non-kin) (Hamilton, 1964), the phenomenon of kin recognition has been repeatedly documented among a diverse assortment of animal species (Colgan, 1983; Fletcher & Michener, 1987; this volume). As in other maturing fields of inquiry, attention has shifted from description and documentation of recognition systems to a search for the 'mechanisms' supporting kin recognition. This search has produced a canonical list that typically comprises four classes of mechanisms: (1) spatial distribution, (2) familiarity, (3) phenotype matching, and (4) recognition alleles. These hypotheses have been offered as explanations of the dynamic control of recognition, but generally contain reference to the developmental means by which recognition cues come to exert their effects (Hepper, 1986; Fletcher, 1987). Thus, the ontogeny of kin recognition has, explicitly or implicitly, become one of the central foci of debate in this relatively young field.

In nature, among species bearing multiple offspring in a single clutch or litter (including most mammals), a necessary concomitant to reproduction is the association of mother with offspring and of littermate siblings with one another. Therefore, some degree of prior experience with related conspecifics is virtually inevitable. A common experimental design intended to isolate effects of prior experience involves cross-fostering of some offspring to unrelated litters shortly after birth. However, even if performed within moments of birth, cross-fostering cannot separate genetic factors from experiences of siblings that share the same prenatal environment (cf. Ressler, 1962).

As a rule, kin recognition in mammals is thought to be strongly influenced by postnatal familiarity, specifically by olfactory experience between siblings that share a common nest environment (Hepper, 1986; Blaustein *et al.*, 1987).

However, a few studies that have relied on data from cross-fostering designs have documented that siblings that share the same intra-uterine environment may recognize one another even if separated and reared apart soon after birth. House mice (*Mus musculus*) (Kareem, 1983) and Norway rats (*Rattus norvegicus*) (Hepper, 1983) behave differently when exposed to full siblings reared by different mothers than to unfamiliar non-siblings. In both Arctic (*Spermophilus parryii*) and Belding's ground squirrels (*S. beldingi*), sisters reared by different mothers are less aggressive toward one another than toward unfamiliar non-siblings (Holmes & Sherman, 1982). Young Richardson's ground squirrels (*S. richardsonii*) cross-fostered within 24 hours of birth behave similarly toward siblings reared together and siblings reared apart (Davis, 1982). While the results of studies such as these do not exclude mechanisms such as phenotype-matching or recognition alleles (cf. Blaustein, 1982), they are consistent with an interpretation of familiarity brought about through intra-uterine association. Only in the few studies reporting differential responsiveness to paternal half-siblings (e.g. Wu *et al.*, 1980; Kareem & Barnard, 1982, 1986) or to non-siblings reared in different postnatal environments (e.g. Holmes, 1986) can prenatal events logically be excluded as a factor contributing to discrimination.

But just how plausible is the hypothesis of prenatal experience in the development of kin recognition? Few data are available to directly address this empirical question. Examination of the available literature suggests that prenatal influences have received little more than occasional terse mention within broader discussions of the 'mechanisms' of kin recognition. Instead, this chapter is intended to elaborate the hypothesis that events that transpire before birth have the potential for influencing and even directing postnatal social preferences. Because our research has focused on learning in rodent fetuses, discussion will be directed in the main to issues of prenatal behavioural development in mammals. We begin with an admitted bias that the fetus can be considered as a viable, independent organism that resides within and interacts with a unique environment (Smotherman & Robinson, 1988a, 1988d). It exhibits behavioural adaptations that not only promote the smooth transition to postnatal life, but also ensure the survival and well-being of the fetus during its residence *in utero*. From a perspective of the behaviourally competent fetus, the hypothesis of prenatal experience is much more than speculation and demands further empirical attention (Smotherman & Robinson, 1987a).

In simplest form, the hypothesis of prenatal experience states that events that occur prior to birth influence or direct the differential distribution of postnatal behaviour among kin and non-kin. Three preconditions must exist for prenatal experience to remain a viable hypothesis as a contributing factor in the development of kin recognition: (1) sensory cues, some of which are associated with kin, must be available to the fetus *in utero*; (2) fetal sensory function must be sufficiently well developed to detect such cues when they are present; and (3) prenatal sensory events must have effects that last longer than the events themselves, that is, the effects of sensory experience *in utero* must be retained by the fetus and expressed at a later time. The following sections will discuss evidence supporting each of these preconditions and attempt to draw inferences of relevance to the study of kin recognition.

Information available to the fetus *in utero*

Although often characterized as a protected environment buffered from changing circumstances in the external world, the intra-uterine environment is a dynamic milieu that provides a rich source of stimuli to the developing fetus. It is now well documented that sensory information derived from the external environment and the mother is directly or indirectly transmitted to the fetus.

Tactile/somatosensory stimuli

The mammalian uterus is organized into one (unicornuate) or two (bicornuate or duplex) functional horns, each of which, depending upon species, may contain from one to a dozen or more conceptuses. Each conceptus consists of the fetus and other embryonic organs and tissues derived from the same zygote, including the placenta, umbilical cord, extra-embryonic membranes (amnion and chorion), and amniotic fluid. Although the fetus is surrounded by concentrically arranged envelopes consisting of amniotic fluid, amnion, and chorion, these structures, which lack sensory innervation, incompletely buffer the fetus from external tactile stimulation.

Potential sources of mechanical stimuli that may impinge upon the fetus include pressure or vibration applied to the maternal abdomen or produced by maternal movement, contractions of the uterus, and movements of siblings located elsewhere in the uterus. Because both the volume of amniotic fluid around the fetus and fetal body mass vary during gestation, the degree of physical isolation from external tactile stimuli also must vary. In the rat, the combined effects of diminishing amniotic fluid volume and increasing fetal body mass result in a marked reduction in free space within the uterus immediately before birth. At this time adjacent fetuses, even in uncrowded uterine horns, are pressed very close to one another, often with overlapping extremities, and physical buffering from intrinsic or extrinsic tactile stimulation is minimal (Smotherman & Robinson, 1988a).

Yet in spite of the abundance of mechanical stimulation that the fetus must experience during sibling, uterine and maternal activity, it seems unlikely that non-specific tactile stimuli would convey information of utility in labelling or discriminating kin. In particular, maternal and sibling characteristics derived from external surface features, such as visual appearance or surface texture, are obscured by intervening tissues and hence are unavailable to the fetus.

Visual stimuli

Visual stimuli are unlikely to convey useful information to fetuses *in utero* (Bradley & Mistretta, 1975). Only a fraction of incident light in visual range (about 2% in rats) penetrates the maternal abdomen and uterus to reach the fetus (Jacques *et al.*, 1987). At most, variation in light intensity may provide information about photoperiod or circadian photophase, but transmission of image information is precluded as a possibility. Information about environmental light also may be transduced by the mother and transmitted indirectly to the fetus by means of chemical communication (Reppert & Weaver, 1988; see discussion below). Nevertheless, fetal vision appears to have little potential for providing information of relevance for kin recognition.

Acoustic stimuli

Recent improvements in recording technology have promoted reassessment of the acoustic environment within the uterus. Early studies concluded that noise levels, produced by sound distortion and borborygmi (sounds produced by maternal physiological processes), were sufficiently high (as much as 95 dB) to effectively mask the attenuated acoustic signals originating outside of the mother (Walker *et al.*, 1971; Henschall, 1972). It is true that maternal pulse and digestive sounds contribute to a rich acoustic environment *in utero* and that maternal and extraembryonic tissues selectively attenuate specific signals, especially at frequencies above 1000 Hz (Fifer & Moon, 1988). However, new measurements have reversed the conclusion of relatively little transmission of recognizable signals from external sources. Human maternal vocalizations are recognizable in intrauterine recordings. Most background noise occurs under 300 Hz and amounts to about 50–60 dB in intensity (Fifer, 1987). Maternal vocalizations are transmitted with less attenuation and distortion than non-maternal vocalizations, probably owing to internal as well as external transmission. In fact, recordings collected from a hydrophone placed inside the amniotic sac of a sheep fetus (*Ovis aries*), suggest that maternal voice may actually be enhanced at low frequencies (Vince *et al.*, 1985). The overall effect of distortion and selective attenuation is to render specific vocal elements less recognizable, but to preserve prosodic characteristics such as intonation and temporal patterning of vocalizations (Fifer & Moon, 1988), providing adequate information to support individual vocal recognition.

Chemical stimuli

Of all sensory modalities, chemoreception holds the greatest potential for providing usable information about the characteristics of mother or intra-uterine siblings (wombmates). Fetuses are continually exposed to a diverse assortment of chemical cues derived from fetal and maternal physiology, maternal diet and chemicals in the environment. To gain access to the fetus, chemicals in solution must diffuse or be actively transported across the placenta, which separates maternal and fetal blood circulation. Because the rate of transplacental diffusion is inversely proportional to molecular weight, smaller molecules have a greater probability of reaching the fetus in high concentrations. However, chemicals with molecular weight less than 1000 readily cross the placenta (Beaconsfield *et al.*, 1980). Thus, a broad range of compounds, including simple sugars, amino acids, small peptides, hormones and other chemical messengers, products of cellular metabolism, antibodies and drugs are known to be present in the intra-uterine environment.

Maternal diet is obviously a rich source of chemical cues that may have relevance for kin recognition. In addition to simple nutrients, many foods contain distinctive compounds that occur infrequently in other food types. Onion and garlic, for instance, contain highly-diffusible sulphur compounds that contribute to their characteristic odours. Uncommon foods ingested by the pregnant mother thus could expose all wombmates to the same unusual chemical cues. If a kin group comprising multiple pregnant females were to forage in the same food patches, unique combinations of diet-derived cues could effectively label all members of a fetal cohort. Many foods also contain teratogenic compounds that produce developmental effects ranging in severity from catastrophic physical deformity to subtle behavioural modification. Prenatal exposure to mild behavioural teratogens may similarly label siblings or members of a wider kin group as different from other conspecifics lacking such prenatal exposure.

Maternal–fetal chemical communication transmits considerable information about the internal states of the mother and fetuses during gestation. A striking example of such communication involves maternal mediation of the fetal 'biological clock'. Regulation of normal light–dark rhythms has been traced to the suprachiasmatic nucleus (SCN) of the anterior hypothalamus (Moore, 1983). The maternal SCN exhibits endogenous circadian rhythmicity but is entrained by environmental light cycles. Autoradiographic neural mapping techniques have revealed that late in gestation the metabolic activity of the SCN of fetal rats and squirrel monkeys (*Saimiri sciureus*) remains in phase with the maternal SCN (Reppert & Schwartz, 1984a,b). In rats, shifting the phase of maternal SCN activity (such as by blinding the mothers) results in a corresponding shift of phase in the fetal SCN, regardless of the phase of environmental light (Reppert & Schwartz, 1983). Circadian fluctuation in levels of melatonin and other chemical messengers in maternal circulation have been implicated as a mechanism for communicating both photophase and photoperiod information to the fetus (Reppert & Weaver, 1988). By these means, individualized patterns of activity and or interaction with the external environment may be indirectly transmitted to the fetus. These patterns of circadian rhythmicity, established before birth, are maintained and reinforced through continued maternal co-ordination in the postnatal period (Duncan *et al.*, 1986).

Chemical exchange also occurs between siblings within the uterus. The strongest evidence for inter-fetal exchange of chemical stimuli is provided by the intra-uterine position phenomenon (vom Saal, 1984). When fetuses of both sexes are present within the uterus, steroid hormones produced by male fetuses have a tendency to masculinize the development of female womb-mates. In domestic cattle, which typically bear single offspring, the develop-ment of opposite sex twins often results in the production of a freemartin: a sterile or sexually deficient female offspring. Among polytocous species, such as Murid rodents, less catastrophic masculinization of female fetuses *in utero* is a well-documented effect (Meisel & Ward, 1981; vom Saal, 1981; Babine & Smotherman, 1984; Richmond & Sachs, 1984).

Two general hypotheses have been advanced to explain the fetal exchange of hormones *in utero*. An obvious path of connection between fetuses lies through the intermediary of the maternal bloodstream. Testosterone (which is first produced by male rat fetuses about 3–4 days before birth) may diffuse across the placenta, transport in maternal circulation, and cross a second placenta to enter the fetal circulation of a female sibling. This transplacental model predicts that female fetuses that reside downstream (in terms of circulation within the uterine blood vessels) from male fetuses will be most affected during development. A second hypothesis posits a direct diffusion of testosterone across the amniotic and chorionic membranes that separate adjacent fetuses. By this model, female fetuses that are adjacent but upstream are just as much at risk of masculinization as females that are adjacent and downstream. To date, experimental findings have failed to unambiguously support one mechanism over the other; it appears that hormones may be transmitted either directly or indirectly.

If complex molecules such as steroid hormones are transmitted between fetuses, it is virtually certain that other chemical products of similar or less complexity, including unique chemical signatures that contain individual or kin-relevant information, are also exchanged among siblings within the womb. If chemical characteristics of siblings vary, then the relative position of different siblings within the uterus may influence the amount of prenatal exposure to those characteristics. To our knowledge, uterine position has never been examined as a factor contributing to sibling recognition.

Sensory abilities of the fetus

Sensory stimuli convey information only in so far as they are capable of being perceived by the fetus. Considerable research progress in the last decade has advanced knowledge about fetal sensory function *in utero*.

Anatomical, physiological and some direct behavioural evidence has generally supported the concept of a sequential progression in the development of sensory function. Taction and somatosensation are the first to become functional in all species studied to date, followed in turn by chemoreception, audition and vision (Gottlieb, 1971a; Geubelle, 1984). From the foregoing discussion, it is apparent that acoustic and chemical cues bear the greatest potential for conveying kin-relevant information to fetuses *in utero*. Therefore, the following discussion will focus on fetal sensation in these two modalities.

Audition

Prenatal auditory sensitivity has been demonstrated in both avian and mammalian species that depend heavily on acoustic signals in maternal–infant communication. Moreover, functional audition appears to be restricted to species that bear relatively precocial offspring. Auditory sensitivity and behavioural responsiveness during the late incubation period has been firmly established in precocial birds such as ducks (*Anas platyrhynchos*) (Gottlieb, 1981), quail (*Coturnix coturnix* and *Colinus virginianus*) (Vince, 1973), chickens (*Gallus gallus*) (Vince *et al.*, 1970) and gulls (*Larus atricilla*) (Impekoven & Gold, 1973). Among mammals, prenatal responsiveness to sound has been investigated in most detail in humans (Rubel, 1985; Fifer & Moon, 1988). Human fetal cardiac and motor responses to external sources of sound occur as early as 24–25 weeks of gestation (Bernard & Sontag, 1947; Gelman *et al.*, 1982; Birnholz & Benecerraf, 1983; Jensen, 1984). However, among mammalian species that bear altricial offspring, such as rats and most other myomorph rodents, fetuses and newborns lack a functional auditory sense (Clopton, 1981).

Chemoreception

In comparison to hearing, the anatomical and neural structures that support some forms of chemoreception mature relatively early during mammalian development. Much evidence collected in recent years indicates that even altricial fetuses exhibit a functional chemical sense *in utero*. Three principal modalities may be involved in fetal chemoreception: taste, main olfaction, and vomeronasal olfaction.

Structural development of taste buds begins late in gestation in rats (day 20 of a 21-day gestation) and matures after birth. Taste buds appear much earlier in prenatal development in sheep (by week 8 of a 21-week gestation) and humans (by week 9 of a 40-week gestation). Electrophysiological recordings have revealed that some taste buds in sheep are functional during the last third of gestation, responding to a number of solutions applied directly to the tongue, including NH_4Cl, KCl, NaCl, LiCl, acetic acid, glycerol, saccharin and quinine hydrochloride. It therefore appears that taste is a functional sense *in utero* among precocial mammals, but not among species that bear altricial young (Mistretta & Bradley, 1986).

Olfactory stimuli are initially processed by chemoreceptors in the olfactory mucosa within the nares or in the isolated receptors within the vomeronasal organ. These two areas remain distinct in their neural projections to the main olfactory bulb (MOB) and accessory olfactory bulb (AOB), respectively. The anatomical segregation evident in the dual olfactory apparatus may reflect functional specialization: the main olfactory system is thought to respond differentially to volatile chemical stimuli, while the vomeronasal system may be more responsive to nonvolatile stimuli in solution (Wysocki *et al.*, 1980; Takagi, 1981). It is interesting that the vomeronasal organ and AOB mature earlier in development than the main olfactory system. Autoradiography using 2-deoxyglucose as a metabolic label has indicated that the AOB, but not the MOB, is functional during the late prenatal period in rats (Pedersen *et al.*, 1983), and may support olfactory discrimination of chemical cues present in amniotic fluid (Pedersen *et al.*, 1986). Recently, a third discrete anatomical region within the olfactory bulb, the modified glomerular complex (MGC), has been described (Teicher *et al.*, 1980). The time course of development of the MGC is intermediate between the AOB and MOB, which may indicate the functioning of a third olfactory system around the time of birth in rats (Pedersen *et al.*, 1986).

Direct observation of the behavioural responses of fetal rats to chemical exposure has confirmed the existence of a functional olfactory sense before birth (Smotherman & Robinson, 1988b). Experimentation and behavioural assessment with rodent fetuses has been made possible through recent improvements in techniques for insensitizing the pregnant rat and externalizing the uterus and subject fetuses into a saline bath maintained at physiological temperature and tonicity (Smotherman & Robinson, 1990). With these methods, fetuses remain healthy, active and responsive throughout observation sessions. Additional procedures, such as intra-oral cannulation, permit controlled delivery of precise volumes of sensory solutions into the mouth of a subject fetus. Infusion of opaque fluids indicates that small volumes (20μl) are likely to bring test fluids into contact with both gustatory and olfactory chemoreceptors.

Intra-oral infusion of chemical solutions that vary in sensory characteristics results in different patterns of behavioural response by the rat fetus. Novel solutions derived from botanical extracts (lemon, orange, mint) reliably evoke an increase in fetal activity following infusion into the mouth. The temporal pattern of change in activity is highly replicable: immediately after infusion fetuses exhibit a sharp spike in activity that may amount to 5- to 10-fold increase over pre-infusion levels of spontaneous movement (Fig. 11.1). This period of activation persists for 20–30 seconds, after which fetal movements begin to decline in frequency. Activity levels return to baseline

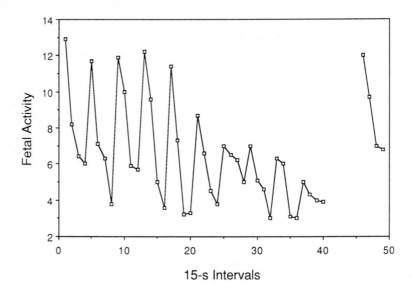

15-s Intervals

Fig. 11.1 Activity of fetuses after repeated infusions of lemon over a 10-minute observation period and a subsequent 1-minute period. Infusions were delivered at the beginning of each minute of the observation. Points represent the mean number of fetal movements per 15-seconds. Data depict the spike in fetal activity after each infusion, the waning response with successive infusions, and recovery of the response after the 5-minute delay with no infusion. (Modified from Smotherman & Robinson, 1988b.)

within 60–75 seconds after infusion. Coincident with the moment of infusion, rodent fetuses also exhibit qualitatively distinct behavioural responses. Chief among the responses to lemon infusion is facial wiping (Smotherman & Robinson, 1987b), a stereotypic action pattern isomorphic to the overhand paw-face strokes of juvenile and adult rodents that typically occur during bouts of grooming (Richmond & Sachs, 1980; Golani & Fentress, 1985) or chemical-induced aversion (Grill & Berridge, 1985).

Four lines of evidence argue that the fetal response to lemon extract and other novel solutions is mediated by olfaction. (1) Not all chemical fluids elicit a behavioural response. Isotonic saline, sucrose, and quinine hydrochloride fail to elicit either behavioural activation or facial wiping in fetal rats (Smotherman & Robinson, 1987a,b). (2) Citral, an artificial lemon scent that has been described as lacking taste components, elicits activity and wiping responses that are identical to the fetal response to lemon extract (Smotherman & Robinson, 1987a). (3) Infusion of a volatile olfactory stimulus that lacks taste qualities, cyclohexanone, in gaseous phase also elicits the characteristic fetal response (Smotherman & Robinson, 1990a). (4) Surgical transection of the fetal brain to isolate the olfactory bulbs from the rest of the telencephalon has little effect on rates of spontaneous fetal activity, yet virtually eliminates a behavioural response to lemon infusion (Smotherman & Robinson, 1990b).

For these reasons, we have argued that even altricial rodent fetuses possess a functional olfactory sense *in utero*.

There remains a question of how the fetus gains sensory access to chemical cues derived from maternal, sibling or environmental sources. Two plausible mechanisms have been proposed. Once a cue has crossed the placenta to circulate in the fetal bloodstream, it may then pass into the amniotic fluid via excretory, alimentary or respiratory tracts (Tam & Chan, 1977). Amniotic fluid undergoes continual turnover through processes of fetal micturition and ingestion (Wirtschafter & Williams, 1957; Marsh *et al.*, 1963; Lev & Orlic, 1972). In consequence, cues present in fetal circulation may be indirectly transmitted to amniotic fluid and through fetal ingestive behaviour come into contact with main olfactory or vomeronasal chemoreceptors (Pedersen *et al.*, 1986). An alternative mechanism is suggested by the discovery that chemical cues in the bloodstream can directly stimulate olfactory receptors, probably through diffusion from capillaries within the olfactory epithelium (Maruniak *et al.*, 1983b). In adult rats, this direct mechanism is sufficiently robust to support behavioural responsiveness to blood-borne odorants (Maruniak *et al.*, 1983a). The operation of either mechanism, or most likely a combination of both, ensures that many chemosensory cues are likely to be detected by the fetus *in utero*.

Fetal learning *in utero*

The third condition that must exist for fetal experience to exert an influence over postnatal recognition is for the effects of prenatal sensory events to be retained and behaviourally expressed later in life. Long-term modification of behaviour in response to sensory stimulation is the defining characteristic of learning. Therefore, fetuses must exhibit a capacity to learn *in utero*.

Exposure learning

It is logically convenient to distinguish two forms of learning that may occur before birth. The first category is exposure learning, which requires only that behavioural modification occurs following presentation of a single kind of stimulus. Exposure may involve just one sensory event (single-trial) or multiple presentations, but requires no pairing of the key stimulus with other stimuli for learning to take place. Indeed, there is no logical requirement that a behavioural response be exhibited upon the initial presentation of the stimulus, only that subsequent responsiveness to the stimulus be modified.

Perhaps the most basic evidence that fetuses are sensitive to simple stimulus exposure is habituation. Habituation is a fundamental form of learning that entails diminished responsiveness, which is not attributable to the fatigue of sensory receptors, to repeated presentations of the same stimulus (Thompson & Spencer, 1966). Ultrasonic imaging of human fetuses exposed to a repeated vibroacoustic stimulus applied to the maternal abdomen has provided some evidence that habituation may occur before birth (Leader et al., 1982). More recently, rat fetuses have been shown to exhibit a waning response to successive infusions of lemon (Fig. 11.1). Because reduced activity after a series of infusions is contingent upon olfactory context, central rather than peripheral mechanisms, and hence a habituation-like process, must be responsible for mediating the waning response (Smotherman & Robinson, 1988b).

Rat fetuses also can distinguish between novel and familiar odour cues experienced in utero. Injection of a mint solution into the amniotic fluid on day 17 of gestation results in a modified fetal response to mint infusion two days later (Smotherman & Robinson, 1988b). Altered responsiveness is evident within the first five seconds after infusion. Infusion of familiar mint (fetuses exposed to mint on day 17) and novel mint (fetuses exposed to saline on day 17) both result in an increase in overall fetal activity beginning 6–10 seconds after delivery. The initial response to novel mint is distinctly different, however, as it involves a brief suppression in fetal activity around the time of infusion. Fetuses also exhibit immediate and pronounced cardiac deceleration in response to novel stimuli, but much reduced bradycardia to familiar cues (Smotherman & Robinson, unpub. data).

The effects of prenatal exposure learning are likely to be retained until after birth. Exposure to species-typical vocalizations before and during hatching is known to contribute to the development of parental recognition in a number of precocial avian species. In a classic study of cliff-nesting guillemots (Uria alge), Tschanz (1968) demonstrated that prenatal learning occurs through exposure of embryos to parental calls as they are beginning to hatch. Pre-hatching exposure to the 'luring' call of parents promotes early postnatal discrimination between vocalizations of parents and unfamiliar adults. Elimination of pre-hatching exposure, however, prevents discrimination and sharply reduces chick responsiveness to parental calls. In similar fashion, pre-hatching auditory exposure has been found to influence postnatal responsiveness to parental vocalizations in laughing gulls (Impekoven & Gold, 1973) and mallard ducks (Gottlieb, 1971b, 1981).

Prenatal exposure to acoustic signals also appears to promote learning in some mammalian species. Guinea pigs exposed *in utero* to heterospecific vocalizations (chicken clucking) treat such sounds as familiar when tested postnatally, whereas guinea pigs that lack prenatal exposure exhibit a cardiac 'startle' response to the same stimulus (Vince, 1979). More extensive study of prenatal auditory learning has been conducted with humans. Playback of recorded intra-uterine sounds has a calming effect on human neonates (Rosner & Doherty, 1971). Newborns will alter their pattern of sucking to gain access to intra-uterine recordings of maternal heartbeat (DeCasper & Sigafoos, 1983). The robust finding that human neonates less than two days old can recognize their mother's voice, with which they have had extensive prenatal experience (Fifer & Moon, 1988), and distinguish it from other female voices (DeCasper & Fifer, 1980) strongly implies fetal learning. Further, prenatal learning is implicated by reports that newborns exhibit a preference to hear a recited children's story (De Casper & Spence, 1986) or melody (Panneton, 1985) to which they were repeatedly exposed during gestation.

Exposure learning of chemical cues has been documented through experiments conducted with fetal rats. To provide prenatal chemical experience, apple juice was introduced into the amniotic fluid on day 20 of gestation (Smotherman, 1982a). When tested as adults, prenatally exposed rats drank more apple juice than water in a two-choice preference test, but rats lacking prenatal experience showed no preference. Analogous results have been reported by Hepper (1990). Rat fetuses were exposed via intra-amniotic injection to orange essence on day 16, 18 or 20 of gestation. At 12 days of postnatal age, pups exposed to orange at the latter two prenatal ages spent more time over a dish containing orange than a dish containing either water or strawberry odour. Pups with no prenatal experience with orange failed to exhibit a preference. In both experiments, rats had no experience with apple or orange odours between the time of birth and testing. Therefore, the drinking and olfactory preferences must be attributed to prenatal experience.

More naturalistic exposure to olfactory cues apparently can support fetal learning as well. To test the effectiveness of cues derived from maternal diet, Hepper (1988) daily supplemented the food of pregnant rats with a clove of garlic from day 15–21 of gestation. Garlic contains several distinctive olfactory compounds including allyl sulphide, which is known to be incorporated into the bloodstream (Maruniak *et al.*, 1983b). When tested 12 days after birth, pups exposed prenatally to garlic spent more time over a dish containing garlic than over a dish containing onion. No preference was evident among pups that lacked prenatal garlic exposure. Preference for garlic was expressed even if pups were cross-fostered within one hour of birth to females that had never ingested garlic, indicating that the preference is almost certainly the result of prenatal experience with the olfactory characteristics of garlic.

The pattern of these findings indicates that prenatal exposure learning is a potentially important source of information in the development of postnatal auditory and olfactory preferences. Olfactory preferences may involve recognition of foods or various forms of social recognition, including infant recognition of mother, individual recognition or sibling recognition. In mammals, auditory preferences may be limited to parent–young interactions.

Associative learning

A second form of learning involves the association of a key stimulus (referred to as the conditioned stimulus or CS) with another stimulus that is known to reliably elicit a behavioural or physiological response (the unconditioned stimulus or US). Pairing of the CS and US during early experience (training trials) confers salience upon the CS, such that subsequent exposure to the CS alone is sufficient to elicit a behavioural response. As defined here, associative learning incorporates both traditional conceptions of classical (Pavlovian) conditioning and appetitive (operant) conditioning.

There is now abundant evidence that fetuses are capable of acquiring and expressing associative learning *in utero*. This was first demonstrated in our laboratory in a taste/odour aversion conditioning paradigm, which involves the pairing of a neutral CS (such as a mint odour) with a US that elicits an aversion reaction. Fetuses on day 17 of gestation that receive an intra-amniotic exposure to mint followed immediately by an intra-peritoneal (ip) injection of lithium chloride (LiCl, the US) exhibit an immediate suppression in behaviour. What little activity is apparent is dominated by lateral flexions of the body trunk ('curls'). When re-exposed to the mint CS on day 19, conditioned fetuses again suppress activity and exhibit a high incidence of curls, which ordinarily constitute a small proportion of fetal behaviour. However, fetuses in various control groups (either mint paired with ip saline or saline paired with ip LiCl on day 17) fail to show an aversive response to mint on day 19. This experiment demonstrates that rat fetuses are capable of forming learned associations between chemical stimuli as early as day 17 of gestation (only one day after they are capable of exhibiting movement), retaining the effects of this experience for two days, and expressing behavioural responsiveness to a conditioned chemical cue *in utero* (Smotherman & Robinson, 1985).

As in the case of exposure learning, prenatal manipulation of fetal sensory experience combined with postnatal behavioural assessment has firmly demonstrated that associative learning by the rat fetus can be retained to modify olfactory preferences after birth. In a series of three reports, a single pairing of intra-amniotic injection of apple juice with ip LiCl was used to bring about aversion conditioning *in utero* on day 20 of gestation. Fetuses were then allowed to complete gestation and were delivered by Caesarean section and cross-fostered. When tested before weaning, conditioned pups are less likely to attach and suckle from nipples painted with apple juice (Stickrod *et al.*, 1982a). This altered responsiveness is probably mediated by olfaction, because conditioned pups exhibit longer latencies to traverse a short runway suffused with apple odour to gain access to an anaesthetized lactating female (Smotherman, 1982b) and spend more time at the end of an experimental chamber with a lower concentration of apple odour (Stickrod *et al.*, 1982b). Pups exposed to various prenatal control treatments, such as apple paired with ip saline, saline paired with ip LiCl, or saline paired with saline, fail to show such modified behavioural responses in the presence of apple odour.

As illustrated in the preceding examples, conditioned taste/odour aversions typically are measured in terms of suppressed activity or increased latency to exhibit a particular response. However, by employing a different class of US it is possible to produce conditioned increases in activity before birth (Smotherman & Robinson, unpub. data). As described earlier, infusion of a lemon extract into the mouth of a fetal rat results in a stereotypic period of hyperactivity. By employing a dual-cannula connected to both a neutral CS (sucrose) and a solution of lemon extract, it is possible to expose the rat fetus to intra-oral pairings of the CS and an activating US. After four such pairings and a subsequent delay of five minutes, re-exposure to the sucrose alone results in a conditioned increase in overall activity (Fig. 11.2). Fetuses in various control groups fail to exhibit behavioural activation upon presentation of the sucrose CS. We have not yet examined how long such associations may be retained, but the present results are suggestive that fetal behaviour may be quite plastic and subject to long term modification through sensory association with activity-activating as well as activity-suppressing stimuli.

A model of fetal conditioning under natural conditions

The relationship of prenatal associative learning to subsequent behavioural development is of necessity more speculative than exposure learning. The principal conceptual difficulty lay in understanding how pairings of sensory events could take place within the protected environment of the uterus. Recently, we have suggested one mechanism by which conditioning could occur *in utero*.

Fetal rats exhibit a stereotypic behavioural and physiological response to hypoxia (Smotherman & Robinson, 1987c, 1988c). If blood circulation within the umbilical cord is experimentally occluded with a vascular clamp, the fetus first exhibits a sharp suppression in activity and pronounced bradycardia. Within 30 seconds, behavioural suppression gives way to

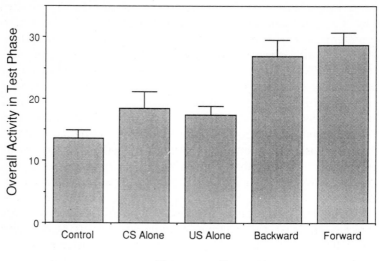

Treatment Groups

Fig. 11.2 Fetuses were exposed to one of five stimulus regimes during the first three minutes of a test session. Four trials were presented during a *training phase* where: 1. Conditioned stimulus (CS) and unconditioned stimulus (US) were paired (10 μl volume of CS infused 1–2 seconds before 10 μl of the US) – the Forward group; 2. CS and saline were paired – a CS sensitization Control group; 3. CS was presented alone – a CS Alone sensitization group; 4. US was presented alone – the US Alone sensitization group; and 5. CS was infused 30 seconds before the US – the Backward control group. After a *delay phase* of six minutes, all fetuses received one 20 μl infusion of the sucrose CS. This re-exposure was presented at the start of a one-minute *test phase*. The data in this figure represent overall levels of activity during the one-minute test phase of the experiment. Bars depict mean levels of activity while vertical lines present standard errors.

hyperactivity, which consists predominantly of lateral trunk curls and dorsal retroflexions of the head ('tosses'). This second phase of response is followed by a third phase consisting of behavioural suppression, which occurs 90–120 seconds after umbilical cord occlusion. If the clamp is removed at this time, no lasting physiological or behavioural effects are noted; fetuses apparently recover fully within a few minutes after clamp removal.

Because brief periods of cord occlusion produce dramatic behavioural and physiological effects, which are fully reversible, hypoxia may provide a transient sensory event that potentially could serve as a US to support conditioning *in utero*. We know from fetal monitoring of unremarkable pregnancies in humans that transient occlusion of the umbilical cord occurs under natural circumstances. Cord occlusion may be caused by activity of the fetus that results in twisting or wrapping of the cord, or by constriction of the cord between the fetus and a solid feature of maternal anatomy, such as the pelvis. Indeed, cord occlusion probably occurs to a greater or lesser degree in a high proportion, perhaps a majority, of otherwise normal pregnancies (Mann, 1986).

Under experimental conditions, cord occlusion is effective as a US in supporting associative learning (Smotherman & Robinson, unpub. data). To effect conditioning, fetal rats were presented with four successive infusions of a CS (sucrose). Immediately after the last infusion, the umbilical cord was clamped for two minutes, then the clamp was removed. After a brief delay, during which the fetus recovered from the period of transient hypoxia, an infusion of the sucrose CS was again presented, resulting in a conditioned suppression in fetal activity. In fact, as in previous experiments with other forms of aversion conditioning of fetuses, upon re-exposure to the CS the subjects expressed particular patterns of behaviour – lateral trunk curls – that ordinarily are elicited by the US (Fig. 11.3; bottom). Yet fetuses in control groups (exposed to brief hypoxia without prior infusion of the CS or to infusions of the CS alone) fail to exhibit characteristics of the uncondi-tioned response. We interpret these experimental results as evidence that transient hypoxic episodes can produce conditioned aversions to neutral chemical cues present *in utero*.

An additional control group in the experiment described above suggested that the same hypoxia US may support conditioned activation of fetal behaviour as well. In this group, the four infusions of the sucrose CS were delivered immediately after removal of the umbilical clamp. Thus, chemo-sensory exposure was associated with the period of recovery rather than the period of hypoxia. When re-exposed to the CS, fetuses that experienced such a backward pairing of CS and US exhibited an increase in motor activity (Fig. 11.3; top). Unlike the fetal response to hypoxia or to forward pairing, lateral trunk curls contributed little to this behavioural activation.

The potential role of transient hypoxia as an agent in fetal learning is illustrated in the following scenario. In the course of normal prenatal development, fetuses are exposed to a varying assortment of chemical cues derived from dietary, maternal and sibling sources. A subset of these cues may be present *in utero* coincident with a transient episode of hypoxia resulting from accidental umbilical cord occlusion. Following the adventi-tious association, fetuses exhibit a conditioned aversion to these chemical cues which may be retained after birth to influence postnatal preferences. The possibility that recovery from transient cord occlusion may support con-ditioned behavioural activation is particularly intriguing in this context, because recovery occurs at a time when renewed contact with maternal

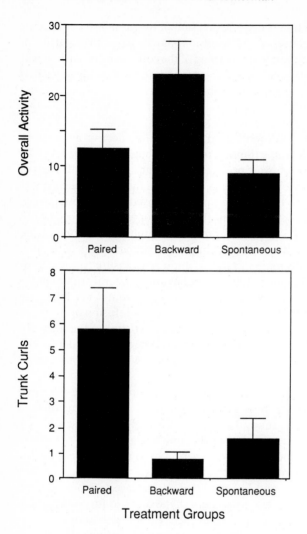

Fig. 11.3 Overall fetal activity (top) and body trunk curls (bottom) in the groups where sucrose infusions preceded application of the umbilical cord clamp (paired group) or followed removal of the umbilical cord clamp (backward group). Spontaneous levels of overall activity and body trunk curls are presented for comparison. Bars represent mean levels of activity and vertical lines depict standard errors of the mean.

circulation would provide new exposure to diet – and maternally-derived chemical cues. Elements of this scenario remain speculative, such as postnatal retention of learning resulting from hypoxia conditioning, but it remains a working hypothesis of one form of prenatal associative learning that could occur under naturalistic conditions.

Implications for kin recognition

We hope to have demonstrated in the sections above that the necessary preconditions exist for prenatal experience to play a role in the development of postnatal responses to environmental and social stimuli. (1) Fetuses are exposed *in utero* to an assortment of sensory cues, some of which potentially contain information about kin. (2) Even in altricial species of mammals, sensory development is sufficiently mature to support fetal responsiveness to tactile and chemical stimuli, and in precocial species, to acoustic stimuli *in utero*. (3) Either simple exposure to some of these stimuli before birth, or association of these stimuli with behaviour-suppressing or behaviour-activating events, may support fetal learning that can be retained after birth to influence postnatal responsiveness. What implications do these conclusions about fetal behavioural abilities have for the development of kin recognition?

Modes of prenatal influence

Two general modes may describe the potential role of prenatal experience in the development of postnatal behaviour. The first is a direct mode, in which prenatal experience determines postnatal preferences. We have already seen experimental situations, in the context of dietary and spatial preferences, that appear to follow such a direct mode. Prenatal experience with apple odour promotes consumption of fluids scented with apple in adulthood (Smotherman, 1982a). Prenatal experience with orange or garlic promotes a preferential approach toward those odours after birth (Hepper, 1988, 1990). Prenatal exposure of human infants to mother's or father's voice appears to foster recognition of their voices and preference to hear their voices over those of unfamiliar adults (Fifer & Moon, 1988).

In contrast to these direct modes of influence, an indirect mode describes prenatal experience as one link in a chain of ontogenetic events leading to the emergence of postnatal preferences. The most elegant example of such an indirect mode may be seen in the detailed studies of the development of nipple attachment in infant rats. Newborn rats must seek, find and attach to the nipple of a lactating female to survive. What cues promote recognition of the nipple by the neonatal rat? A series of experiments have demonstrated a chain of events that occur during prenatal and postnatal development that contribute to the mature expression of this adaptive behaviour.

Older pups are clearly attracted to the odour of the nipples. Anosmic pups fail to attach to nipples (Alberts, 1976). If the ventrum and nipples of the lactating female are washed, nipple attachment is sharply reduced (Hofer *et al.*, 1976; Teicher & Blass, 1976). Placement of a pup on a washed nipple or application of pup saliva reinstates the attraction of other pups to that nipple (Blass & Teicher, 1980). Analysis of components of pup saliva has identified a particular compound, dimethyl disulphide (DMDS), which appears to be responsible for attracting pups to the nipple. Application of DMDS to washed nipples is sufficient to promote nipple attachment by experienced pups (Pedersen & Blass, 1981).

Role of amniotic fluid

However, pup saliva cannot account for the first attachment of newborn rat pups to the nipple. Observation of the behaviour of mothers during parturition suggested to several investigators that the mother actively marks the nipples with attractive odour cues. Bouts of licking directed at the perineum and newborn pups is alternated with licking directed at the ventrum and nipples of the parturient female (Roth & Rosenblatt, 1967). Subsequent experiments have confirmed that washed nipples to which amniotic fluid is applied are more attractive to pups than washed, unmarked nipples (Teicher & Blass, 1977).

Prenatal manipulation of the olfactory characteristics of amniotic fluid can influence neonatal nipple attachment. Pedersen & Blass (1982) manipulated both prenatal and early postnatal exposure of rats to citral. These manipulations yielded four experimental groups: pups that were only exposed prenatally, pups that were only exposed postnatally, pups that were exposed both before and after birth, and pups that had no prior exposure to citral. Pups in all four groups were tested by painting some nipples with citral. Only pups that had experienced citral *in utero* and immediately after birth attached to citral-painted nipples. Moreover, citral-exposed pups no longer attached to normal, unwashed nipples. This experiment indicates that a combination of prenatal and postnatal exposure to an artificial odorant is necessary to direct the first attachment of pups to the nipple.

The evidence described above implies that amniotic fluid from unmanipulated fetuses contains attractive olfactory cues, perhaps a single identifiable compound such as DMDS, that promotes the first nipple attachment of newborns (cf. Hudson, 1985). However, neonatal rats can make finer discriminations on the basis of olfactory characteristics of amniotic fluid. Hepper (1987) collected amniotic fluid from different rats on day 18 of gestation and stored them on cotton swabs for later testing. After delivery, pups were presented in a two-choice test with swabs scented with amniotic fluid from intra-uterine siblings or from unfamiliar and unrelated fetuses. The direction of head turning in these altricial neonates indicated their preference for the amniotic fluid of kin over fluid from non-kin. Preference for amniotic fluid from wombmates was evident even when pups were delivered by Caesarean section and immediately washed and dried, thereby preventing postnatal exposure to amniotic fluid. Even if a single compound is responsible for the attractive properties of amniotic fluid, a more complex melange of chemical constituents renders it individually distinctive.

Suckling and postnatal recognition

Long after amniotic fluid has vanished from the neonatal environment, chemical cues associated with suckling continue to influence chemosensory preferences and, potentially, kin recognition. Young rats and spiny mice (*Acomys cahirinus*) are preferentially attracted to cues produced by a lactating female fed the same diet as their mother (Leon, 1975; Porter & Doane, 1977), and diet-specific cues are known to be transmitted to offspring through mother's milk (Galef & Sherry, 1973; Bronstein *et al.*, 1975). Moreover, spiny mouse siblings that are separated at birth and reared apart, but which suckle from the same lactating female, huddle together more often than with non-siblings or unfamiliar siblings that suckle from a different female (Porter *et al.*, 1981). Two plausible interpretations may account for recognition based upon exposure to a common lactating female: (a) cues produced by the lactating female are transferred to suckling pups, effectively marking them with the same maternal label (Gubernick, 1981), or (b) odour characteristics intrinsic to pups, such as constituents of saliva, are transferred to the nipple, permitting maternally-mediated communication between separated pups.

Fillion & Blass (1986b) have identified further effects of learning at the nipple. Male rat pups that experience odour-treated nipples during suckling exhibit altered responsiveness to estrous females scented with the same odour later in life. The effect is specific to odours associated with suckling. Citral placed on the back of the lactating mother or elsewhere within the nest environment does not affect adult sexual behaviour. But citral placed on the nipples has the effect of promoting enhanced probing behaviour in older pups (day 10) (Fillion & Blass, 1986a) and reduced latency to ejaculation in adult rats during interactions with citral-scented oestrous females (Fillion & Blass, 1986b). Early experience with citral has no facilitatory effect when males are exposed to females not in oestrous or lacking the citral scent. Similarly, males that lack early experience with citral fail to show enhanced sexual behaviour in the presence of scented-females.

An intriguing correlate of the behavioural changes reported by Fillion & Blass is found in the ontogenetic pattern of vomeronasal function. In rats, two general periods during development are characterized by high AOB activity: during the perinatal period (late in gestation and immediately after birth) and during the onset of sexual behaviour in adulthood (Shepherd *et al.*, 1987). Vomeronasal function has been implicated in the control of both maternal recognition (Teicher *et al.*, 1984) and sexual behaviour (Powers & Winans, 1975; Lumia *et al.*, 1981; Meisel *et al.*, 1982) in rats and other murid rodents. If recognition cues were similarly processed by the AOB, they may be more sensitive to early olfactory experience that occurs during the first period of AOB function. As noted earlier, the AOB is thought to be the principal means of olfaction by the rat fetus *in utero* (Pedersen *et al.*, 1986). These facts suggest that prenatal experience with odours may be especially important in the development of olfactory preferences regulated by AOB function.

Ultimately, the experimental findings about the attractiveness of amniotic fluid, the determinants of neonatal nipple attachment, and the influence of olfactory experience at the nipple for subsequent odour-directed behaviour imply continuity in behavioural development from fetus to adult. While prenatal experience with amniotic fluid may not determine postnatal social preferences, it is likely to fill a critical role in directing early olfactory learning that occurs during parturition, suckling, and interaction with littermate siblings. If so, fetal learning may be viewed as one aspect of the canalization evident in homeostatic developmental processes in general (Waddington, 1975), and in systems of kin recognition in particular. Such an epigenetic perspective of behavioural development (Kuo, 1967; Smotherman & Robinson, 1988e) seems to be lacking in the ongoing debate about phenotype-matching, recognition alleles and other putative 'mechanisms' of kin recognition, which recall earlier distinctions between nature and nurture (Oyama, 1985). At the least, we would argue that the hypothesis of prenatal experience as a contributing factor in the ontogeny of kin recognition is conceptually and empirically well-founded and should receive further attention in the future.

Acknowledgements

WPS is supported by grant HD 16102–06 and Research Career Development Award HD 00719–05 from the National Institute of Child Health and Human Development (NIH) and the NATO Collaborative Research Grants Programme 0551/88. (Correspondence regarding this chapter should be addressed to WPS, Center for Developmental Psychobiology, Department of Psychology, SUNY-Binghamton, Binghamton, NY, 13902-6000, USA.)

References

Alberts, J. R. (1976). Olfactory contributions to behavioral development in rodents. In *Mammalian Olfaction: Reproductive Processes and Behavior*, ed. R. L. Doty, pp, 67–94. New York: Academic Press.
Babine, A. M. & Smotherman, W. P. (1984). Uterine position and conditioned taste

aversion. *Behav. Neurosci.*, **98**, 461–6.

Beaconsfield, P., Birdwood, G. & Beaconsfield, R. (1980). The placenta. *Sci. Am.*, **243**, (August), 80–9.

Bernard, J. & Sontag, L. W. (1947). Fetal reactivity to tonal stimulation: a preliminary report. *J. Genet. Psychol.*, **70**, 205–10.

Birnholz, J. C. & Benecerraf, B. R. (1983). The development of human fetal hearing. *Science*, **222**, 516–18.

Blass, E. M. & Teicher, M. H. (1980). Suckling. *Science*, **210**, 15–22.

Blaustein, A. R. (1982). Kin recognition mechanisms: phenotypic matching or recognition alleles? *Am. Nat.*, **121**, 749–54.

Blaustein, A. R., Bekoff, M. & Daniels, T. J. (1987). Kin recognition in vertebrates (excluding primates): empirical evidence. In *Kin Recognition in Animals*, ed. D. J. C. Fletcher & C. D. Michener, pp. 287–331. New York: John Wiley & Sons.

Bradley, R. M. & Mistretta, C. M. (1975). Fetal sensory receptors. *Physiol. Rev.*, **55**, 352–82.

Bronstein, P. M., Levine, M. J. & Marcus, M. (1975). A rat's first bite: the nongenetic, cross-generational transfer of information. *J. Comp. Physiol. Psychol.*, **89**, 295–8.

Clopton, B. M. (1981). Neurophysiological and anatomical aspects of auditory development. In *Development of Perception: Psychobiological Perspectives*, vol. 1, ed. R. N. Aslin, J. R. Alberts & M. R. Petersen, pp. 112–37. New York: Academic Press.

Colgan, P. (1983). *Comparative Social Recognition*. New York: John Wiley & Sons.

Davis, L. S. (1982). Sibling recognition in Richardson's ground squirrels (*Spermophilus richardsonii*). *Behav. Ecol. Sociobiol.*, **11**, 65–70.

DeCasper, A. J. & Fifer, W. P. (1980). Of human bonding: newborns prefer their mothers' voices. *Science*, **208**, 1174–6.

DeCasper, A. J. & Sigafoos, A. D. (1983). The intrauterine heartbeat: a potent reinforcer for newborns. *Inf. Behav. Devel.*, **6**, 19–25.

DeCasper, A. J. & Spence, M. (1986). Newborns prefer a familiar story over an unfamiliar one. *Inf. Behav. Devel.*, **9**, 133–50.

Duncan, M. J., Banister, M. J. & Reppert, S. M. (1986). Developmental appearance of light-dark entrainment in the rat. *Brain Res.*, **369**, 326–30.

Fifer, W. P. (1987). Neonatal preference for mother's voice. In *Perinatal Development: A Psychobiological Perspective*, ed. N. A. Krasnegor, E. M. Blass, M. A. Hofer & W. P. Smotherman, pp. 111–24. Orlando: Academic Press.

Fifer, W. P. & Moon, C. (1988). Auditory experience in the fetus. In *Behavior of the Fetus*, ed. W. P. Smotherman & S. R. Robinson, pp. 175–88. Caldwell: Telford Press.

Fillion, T. J. & Blass, E. M. (1986a). Infantile behavioural reactivity to oestrous chemostimuli in Norway rats. *Anim. Behav.*, **34**, 123–33.

Fillion, T. J. & Blass, E. M. (1986b). Infantile experience with suckling odors determines adult sexual behavior in male rats. *Science*, **231**, 729–31.

Fletcher, D. J. C. (1987). The behavioral analysis of kin recognition: perspectives on methodology and interpretation. In *Kin Recognition in Animals*, ed. D. J. C. Fletcher & C. D. Michener, pp. 19–54. New York: John Wiley & Sons.

Fletcher, D. J. C., & Michener, C. D., (Eds.). (1987). *Kin Recognition in Animals*. New

York: John Wiley & Sons.

Galef, B. G. & Sherry, D. F. (1973). Mother's milk: a medium for transmission of cues reflecting the flavor of mother's diet. *J. Comp. Physiol. Psychol.*, **83**, 374–8.

Gelman, S. R., Wood, S., Spellacy, W. N. & Abrams, R. M. (1982). Fetal movements in response to sound stimulation. *Am. J. Obstet. Gynecol.*, **143**, 484–5.

Geubelle, F. (1984). Perception of environmental conditions by the fetus in utero. In *Progress in Reproductive Biology and Medicine*, vol. 11, ed. P. O. Hubinont, pp. 110–19. Basel: Karger.

Golani, I. & Fentress, J. C. (1985). Early ontogeny of face grooming in mice. *Dev. Psychobiol.*, **18**, 529–44.

Gottlieb, G. (1971a). Ontogenesis of sensory function in birds and mammals. In *The Biopsychology of Development*, ed. E. Tobach, L. Aronson & E. Shaw, pp. 67–128. New York: Academic Press.

Gottlieb, G. (1971b). *Development of Species Identification in Birds: An Inquiry into the Prenatal Determinants of Perception.* Chicago: University of Chicago Press.

Gottlieb, G. (1981). Roles of early experience in species-specific perceptual development. In *Development of Perception: Psychobiological Perspectives*, vol. 1, ed R. N. Aslin, J. R. Alberts & M. R. Petersen, pp. 5–44. New York: Academic Press.

Grill, H. J. & Berridge, K. C. (1985). Taste reactivity as a measure of the neural control of palatability. In *Progress in Psychobiology and Physiological Psychology*, vol. 11, ed. J. M. Sprague & A. N. Epstein, pp. 1–61. New York: Academic Press.

Gubernick, D. J. (1981). Mechanisms of maternal 'labelling' in goats. *Anim. Behav.*, **29**, 305–6.

Hamilton, W. D. (1964). The genetical evolution of social behavior. I. II. *J. Theor. Biol.*, **7**, 1–52.

Henschall, W. R. (1972). Intrauterine sound levels. *J. Obstet. Gynecol.*, **112**, 577–9.

Hepper, P. (1983). Sibling recognition in the rat. *Anim. Behav.*, **31**, 1177–91.

Hepper, P. (1986). Kin recognition: functions and mechanisms. A review. *Biol. Rev.*, **61**, 63–93.

Hepper, P. (1987). The amniotic fluid: an important priming role in kin recognition. *Anim. Behav.*, **35**, 1343–46.

Hepper, P. (1988). Adaptive fetal learning: prenatal exposure to garlic affects postnatal preferences. *Anim. Behav.*, **36**, 935–6.

Hepper, P. (1990). Prenatal exposure learning in the rat. *Q. J. Exp. Psychol. B.* (In Press.)

Hofer, M. A., Shair, H. & Singh, P. (1976). Evidence that maternal ventral skin substances promote suckling in infant rats. *Physiol. Behav.*, **17**, 131–6.

Holmes, W. G. (1986). Kin recognition by phenotype matching in female Belding's ground squirrels. *Anim. Behav.*, **34**, 38–47.

Holmes, W. G. & Sherman, P. W. (1982). The ontogeny of kin recognition in two species of ground squirrels. *Am. Zool.*, **22**, 491–517.

Hudson, R. (1985). Do newborn rabbits learn the odor stimuli releasing nipple-search behavior? *Dev. Psychobiol.*, **18**, 575–85.

Impekoven, M. & Gold, P. S. (1973). Prenatal origins of parent-young interactions in birds: a naturalistic approach. In *Studies on the Development of Behavior and the Nervous System*, vol. I. *Behavioral Embryology*, ed. G. Gottlieb, pp. 325–56. New York: Academic Press.

Jacques, S. L., Weaver, D. R. & Reppert, S. M. (1987). Penetration of light into the uterus of pregnant mammals. *Photochem. Photobiol.*, **45**, 637–41.

Jensen, O. H. (1984). Fetal heart rate response to controlled sound stimuli during the third trimester of normal pregnancy. *Acta Obstet. Gynecol. Scand.*, **63**, 193–7.

Kareem, A. M. (1983). Effect of increasing periods of familiarity on social interactions between male sibling mice. *Anim. Behav.*, **31**, 919–26.

Kareem, A. M. & Barnard, C. J. (1982). The importance of kinship and familiarity in social interactions between mice. *Anim. Behav.*, **30**, 549–601.

Kareem, A. M. & Barnard, C. J. (1986). Kin recognition in mice: age, sex and parental effects. *Anim. Behav.*, **34**, 1814–24.

Kuo, Z. -Y. (1967). *The Dynamics of Behavior Development: An Epigenetic View*. New York: Random House.

Leader, L. R., Baillie, P., Martin, B. & Vermuelen, E. (1982). The assessment and significance of habituation to a repeated stimulus by the human fetus. *Early Human Dev.*, **7**, 211–19.

Leon, M. (1975). Dietary control of maternal pheromone in the lactating rat. *Physiol. Behav.*, **14**, 311–19.

Lev, R. & Orlic, D. (1972). Protein absorption by the intestine of the fetal rat in utero. *Science*, **177**, 522–4.

Lumia, A. R., Meisel, R. L. & Sachs, B. D. (1981). Induction of female and male mating patterns in female rats by gonadal steroids: effects of neonatal versus adult olfactory bulbectomy. *J. Comp. Physiol. Psychol.*, **95**, 497–511.

Mann, L. I. (1986). Pregnancy events and brain damage. *Am. J. Obstet. Gynecol.*, **155**, 6–9.

Marsh, R. H., King, J. E. & Becker, R. F. (1963). Volume and viscosity of amniotic fluid in rat and guinea pig fetuses near term. *Am. J. Obstet. Gynecol.*, **85**, 487–92.

Maruniak, J. A., Mason, J. R. & Kostelc, J. G. (1983a). Conditioned aversions to an intravascular odorant. *Physiol. Behav.*, **30**, 617–20.

Maruniak, J. A., Silver, W. L. & Moulton, D. G. (1983b). Olfactory receptors respond to blood-borne odorants. *Brain Res.*, **265**, 312–16.

Meisel, R. L., Lumia, A. R. & Sachs, B. D. (1982). Disruption of copulatory behavior in male rats by olfactory bulbectomy at two, but not ten days of age. *Exp. Neurol.*, **77**, 622–4.

Meisel, R. L. & Ward, I. L. (1981). Fetal female rats are masculinized by male littermates located caudally in the uterus. *Science*, **220**, 437–8.

Mistretta, C. M. & Bradley, R. M. (1986). Development of the sense of taste. In *Handbook of Behavioral Neurobiology*, vol. 8, *Developmental Psychobiology and Developmental Neurobiology*, ed. E. M. Blass, pp. 205–36. New York: Plenum.

Moore, R. Y. (1983). Organization and function of a central nervous system circadian oscillator: the suprachiasmatic hypothalamic nucleus. *Fed. Proc.*, **42**, 2783–9.

Oyama, S. (1985). *The Ontogeny of Information: Developmental Systems and Evolution*. Cambridge: Cambridge University Press.

Panneton, R. P. (1985). *Prenatal Experience with Melodies: Effect on Postnatal Auditory Preference in Human Newborns*. Unpublished doctoral dissertation, University of North Carolina at Greensboro.

Pedersen, P. E. & Blass, E. M. (1981). Olfactory control over suckling in albino rats. In *Development of Perception: Psychobiological Perspectives*, vol. 1, ed. R. N. Aslin, J. R. Alberts & M. R. Petersen, pp. 359–81. New York: Academic Press.

Pedersen, P. E. & Blass, E. M. (1982). Prenatal and postnatal determinants of the 1st suckling episode in albino rats. *Dev. Psychobiol.*, **15**, 349–55.

Pedersen, P. E., Greer, C. A. & Shepherd, G. M. (1986). Early development of olfactory function. In *Handbook of Behavioral Neurobiology*, vol. 8, *Developmental Psychobiology and Developmental Neurobiology*, ed. E. M. Blass, pp. 163–203. New York: Plenum.

Pedersen, P. E., Stewart, W. B., Greer, C. A. & Shepherd, G. M. (1983). Evidence for olfactory function in utero. *Science*, **221**, 478–80.

Porter, R. H. & Doane, H. M. (1977). Dietary-dependent cross-species similarities in maternal chemical cues. *Physiol. Behav.*, **19**, 129–31.

Porter, R. H., Tepper, V. J. & White, D. M. (1981). Experiential influences on the development of huddling preferences and 'sibling' recognition in spiny mice. *Dev. Psychobiol.*, **14**, 375–82.

Powers, J. B. & Winans, S. S. (1975). Vomeronasal organ: critical role in mediating sexual behavior of the male hamster. *Science*, **187**, 961–3.

Reppert, S. M. & Schwartz, W. J. (1983). Maternal co-ordination of the fetal biological clock in utero. *Science*, **220**, 969–71.

Reppert, S. M. & Schwartz, W. J. (1984a). The suprachiasmatic nuclei of the fetal rat: characterization of a functional circadian clock using 14C-labeled deoxyglucose. *J. Neurosci.*, **4**, 1677–82.

Reppert, S. M. & Schwartz, W. J. (1984b). Functional activity of the suprachiasmatic nuclei in the fetal primate. *Neurosci. Letters*, **46**, 145–9.

Reppert, S. M. & Weaver, D. R. (1988). Maternal transduction of light-dark information for the fetus. In *Behavior of the Fetus*, ed. W. P. Smotherman & S. R. Robinson, pp. 119–39. Caldwell: Telford Press.

Ressler, R. H. (1962). Parental handling in two strains of mice reared by foster parents. *Science*, **137**, 129–30.

Richmond, G. & Sachs, B. D. (1980). Grooming in Norway rats: the development and adult expression of a complex motor pattern. *Behaviour*, **75**, 82–96.

Richmond, G. & Sachs, B. D. (1984). Further evidence for masculinization of female by males located caudally in utero. *Horm. Behav.*, **18**, 484–90.

Rosner, B. & Doherty, N. (1971). The response of neonates to intra-uterine sounds. *Dev. Med. Child Neurol.*, **21**, 723–9.

Roth, L. L. & Rosenblatt, J. S. (1967). Changes in self licking during pregnancy in the rat. *J. Comp. Physiol. Psychol.*, **63**, 397–400.

Rubel, E. W. (1985). Auditory system development. In *Measurement of Audition and Vision in the First Year of Postnatal Life: A Methodological Overview*, ed. G. Gottlieb & N. A. Krasnegor, pp. 53–90. Norwood: Ablex.

Shepherd, G. M., Pedersen, P. E. & Greer, C. A. (1987). Development of olfactory specificity in the albino rat: a model system. In *Perinatal Development: A Psychobiological Perspective*, ed. N. A. Krasnegor, E. M. Blass, M. A. Hofer & W. P. Smotherman, pp. 129–44. Orlando: Academic Press.

Smotherman, W. P. (1982a). In utero chemosensory experience alters taste preferences and corticosterone responsiveness. *Behav. Neural Biol.*, **36**, 61–8.

Smotherman, W. P. (1982b). Odor aversion learning by the rat fetus. *Physiol. Behav.*, **29**, 769–71.

Smotherman, W. P. & Robinson, S. R. (1985). The rat fetus in its environment:

behavioral adjustments to novel, familiar, aversive, and conditioned stimuli presented in utero. *Behav. Neurosci.*, **99**, 521–30.

Smotherman, W. P. & Robinson, S. R. (1987a). Psychobiology of fetal experience in the rat. In *Perinatal Development: A Psychobiological Perspective*, ed. N. A. Krasnegor, E. M. Blass, M. A. Hofer & W. P. Smotherman, pp. 39–60. Orlando: Academic Press.

Smotherman, W. P. & Robinson, S. R. (1987b). Prenatal expression of species-typical action patterns in the rat fetus (*Rattus norvegicus*). *J. Comp. Psychol.*, **101**, 190–6.

Smotherman, W. P. & Robinson, S. R. (1987c). Stereotypic behavioral response of rat fetuses to acute hypoxia is altered by maternal alcohol consumption. *Am. J. Obstet. Gynecol.*, **157**, 982–6.

Smotherman, W. P. & Robinson, S. R. (1988a). The uterus as environment: the ecology of fetal behavior. In *Handbook of Behavioral Neurobiology*, vol. 9, *Developmental Psychobiology and Behavioral Ecology*, ed. E. M. Blass, pp. 149–96. New York: Plenum Press.

Smotherman, W. P. & Robinson, S. R. (1988b). Behavior of rat fetuses following chemical or tactile stimulation. *Behav. Neurosci.*, **102**, 24–34.

Smotherman, W. P. & Robinson, S. R. (1988c). Response of the rat fetus to acute umbilical cord occlusion: an ontogenetic adaptation? *Physiol. Behav.*, **44**, 131–5.

Smotherman, W. P. & Robinson, S. R. (Eds.). (1988d). *Behavior of the Fetus*. Caldwell: Telford Press.

Smotherman, W. P. & Robinson, S. R. (1988e). Dimensions of fetal investigation. In *Behavior of the Fetus*, ed. W. P. Smotherman & S. R. Robinson, pp. 19–34. Caldwell: Telford Press.

Smotherman, W. P. & Robinson, S. R. (1990a). Rat fetuses respond to chemical stimuli in gas phase. *Physiol. Behav.*, **47**, 863–68.

Smotherman, W. P. & Robinson, S. R. (1990b). Olfactory bulb transection alters fetal behaviour after chemosensory but not tactile stimulation. *Dev. Brain. Res.*, (In Press).

Smotherman, W. P. & Robinson, S. R. (1991). Accessibility of the rat fetus for psychobiological investigation. In *Developmental Psychobiology: Current Methodology and Conceptual Issues*, ed. H. N. Shair, M. A. Hofer & G. Barr. Oxford: Oxford University Press. (In Press.)

Stickrod, G., Kimble, D. P. & Smotherman, W. P. (1982a). In utero taste/odor aversion conditioning in the rat. *Physiol. Behav.*, **28**, 5–7.

Stickrod, G., Kimble, D. P. & Smotherman, W. P. (1982b). Met-enkephalin effects on associations formed in utero. *Peptides*, **3**, 881–3.

Takagi, S. F. (1981). Multiple olfactory pathways in mammals: a review. *Chem. Senses*, **6**, 329–33.

Tam, P. P. L. & Chan, S. T. H. (1977). Changes in the composition of maternal plasma, fetal plasma and fetal extraembryonic fluid during gestation in the rat. *J. Reprod. Fert.*, **51**, 41–51.

Teicher, M. H. & Blass, E. M. (1976). Suckling in newborn rats: eliminated by nipple lavage, reinstated by pup saliva. *Science*, **193**, 422–5.

Teicher, M. H. & Blass, E. M. (1977). First suckling response of the newborn albino rat: the roles of olfaction and amniotic fluid. *Science*, **198**, 635–6.

Teicher, M. H., Shaywitz, B. A. & Lumia, A. R. (1984). Olfactory and vomeronasal system mediation of maternal recognition in the developing rat. *Dev. Brain Res.*, **12**, 97–110.

Teicher, M. H., Stewart, W. B., Kauer, J. S., & Shepherd, G. M. (1980). Suckling pheromone stimulation of a modified glomerular region in the developing rat olfactory bulb revealed by the 2-deoxyglucose method. *Brain Res.*, 194, 530–5.

Thompson, R. F. & Spencer, W. A. (1966). Habituation: a model phenomenon for the study of neuronal substrates of behavior. *Psychol. Rev.*, **73**, 16–43.

Tschanz, B. (1968). Trottellummen: Die Entstehung der personlichen Beziehungen zwischen Jungvogel und Eltern. *Z. Tierpsychol.*, **4**, 1–103.

Vince, M. A. (1973). Some environmental effects on the activity and development of the avian embryo. In *Studies on the Development of Behavior and the Nervous System*, Vol. I. *Behavioral Embryology*, ed. G. Gottlieb, pp. 285–323. New York: Academic Press.

Vince, M. A. (1979). Postnatal effects of prenatal sound stimulation in the guinea pig. *Anim. Behav.*, **27**, 908–18.

Vince, M. A., Billing, A. E., Baldwin, B. A., Toner, J. N. & Weller, C. (1985). Maternal vocalizations and other sounds in the fetal lamb's sound environment. *Early Human Dev.*, **11**, 179–90.

Vince, M. A., Green, J. & Chinn, S. (1970). Acceleration of hatching in the domestic fowl. *Brit. Poultry Sci.*, **11**, 483–8.

vom Saal, F. S. (1981). Variation in phenotype due to random intrauterine proximity of male and female fetuses in rodents. *J. Reprod. Fert.*, **6**, 633–50.

vom Saal, F. S. (1984). The intrauterine position phenomenon: effects on physiology, aggressive behavior and population dynamics in house mice. In *Biological Perspectives on Aggression*, ed. K. J. Flannelly, R. J. Blanchard & D. C. Blanchard, pp. 135–79. New York: Alan R. Liss.

Waddington, C. H. (1975). Canalization and the development of quantitative characters. In *The Evolution of an Evolutionist*, ed. C. H. Waddington, pp. 98–103. Ithaca: Cornell University Press.

Walker, D., Grimwade, J. & Wood, C. (1971). Intrauterine noise: a component of the fetal environment. *Am. J. Obstet. Gynecol.*, **109**, 91–5.

Wirtschafter, Z. T. & Williams, D. W. (1957). Dynamics of the amniotic fluid as measured by changes in protein patterns. *Am. J. Obstet. Gynecol.*, **74**, 309–13.

Wu, H. M. H., Holmes, W. G., Medina, S. R. & Sackett, G. P. (1980). Kin preference in infant *Macaca nemestrina*. *Nature*, **285**, 225–7.

Wysocki, C. J., Wellington, J. L. & Beauchamp, G. K. (1980). Access of urinary non-volatiles to the mammalian vomeronasal organ. *Science*, **207**, 781–3.

12

Information processing and storage during filial imprinting

Mark H. Johnson

In the absence of adult conspecifics, newly-hatched chicks will approach a wide range of bright or moving objects. After a period of exposure to such an object they form a social attachment to it. When they are close to the object the chicks emit soft calls, and if it moves away they will follow it. If subsequently exposed to a new object the chicks will run away emitting distress calls. From these observations it may be inferred that chicks learn about some of the visual characteristics of the first object to which they are exposed. This process is referred to as filial imprinting, and has been studied most intensively in precocial birds such as domestic chicks, ducklings, and goslings (Spalding, 1873; Heinroth, 1911; Lorenz, 1935, 1937; Bateson, 1966). Similar processes may also occur in mammals, but evidence for this is harder to obtain due to their comparative lack of mobility shortly after birth. This chapter will concentrate on the neural and behavioural analysis of filial imprinting in the domestic chick, a species which readily imprints on to a wide range of visual objects.

Imprinting can be reliably reproduced in the laboratory in the following way. Chicks are dark-reared until being exposed to a conspicuous object such as a rotating, illuminated red box, for a period of time, normally several hours. From between two hours to several days later, chicks are given a choice between the object to which they were exposed and a novel object. The extent to which the chick approaches each of the two objects is compared and a preference score calculated.

Many experimental investigations of imprinting have been concerned with testing the validity of the claims originally made by Konrad Lorenz. Briefly, Lorenz (1935, 1937) made three claims;

    1 *Imprinting is confined to a short well-defined period in the life-cycle, the critical time or period.*

335

Lorenz claimed that if exposure to an object did not take place within a particular critical time period a preference would not be acquired. This period was thought to be delineated by endogenous changes. More recent evidence suggests that the period is less rigid than Lorenz supposed and may only come to an end once the initial object can be discriminated from the surrounding environment and other objects. Imprinting is thus seen more as a self-terminating process (Sluckin & Salzen 1961; Bateson, 1966).

> 2 *Imprinting is irreversible.*

Jaynes (1956) pointed out that Lorenz's claim that imprinting is irreversible could be interpreted in either of two ways: once a bird has become imprinted on an object it will never again address its filial responses to dissimilar objects, or alternatively, it may come to address its filial responses to dissimilar objects but always retain information about the original object. Over the years a considerable amount of evidence has accrued to suggest that imprinting is not irreversible in the first sense (Klopfer & Hailman, 1964a,b; Klopfer, 1967; Salzen & Meyer, 1968). Later in this chapter evidence relating to the latter form of the claim will be discussed.

> 3 *Imprinting involves learning of the species rather than the characteristics of an individual.*

Originally, for Lorenz the function of imprinting was '. . to establish a sort of consciousness of species in the young bird' (Lorenz, 1937). This assertion led to the assumption that learning about an individual takes place by 'normal conditioning' (Hess, 1959). Later in this chapter evidence will be presented that suggests this view is incorrect, at least for the domestic chick.

Apart from direct tests of Lorenz's claims, research on imprinting has been motivated by two positions. One view is that imprinting is a special type of learning best characterized as 'exposure' (Sluckin, 1972) or 'template' learning (Staddon, 1983). This view is concordant with Lorenz's original notion of a 'stamping in' or 'coinage'. The opposing view is that imprinting can be accounted for in terms of associative conditioning mechanisms (e.g. Hoffman & Ratner, 1973). In the present chapter it will be argued that, although the mechanisms of learning involved in filial attachment are not the same as those involved in associative conditioning, neither are they 'special', species-specific or unique to one situation.

The neural basis of imprinting

Many neural theories of memory suppose that a particular experience or event leads to the formation or strengthening of pathways in the brain through changes in the size or number of contacts between nerve cells (e.g. Cajal, 1911; Hebb, 1949). Reasoning that such changes may be associated with protein synthesis, Bateson *et al.* (1972) studied the uptake of several precursors of proteins into different large regions of the chick brain after imprinting. The only region where there was a higher incorporation in the imprinted chicks as compared to non-imprinted controls was the forebrain roof.

This increased uptake of precursors could, of course, be due to a variety of factors other than the storage of information about the object to which the chicks were exposed. For example, while a chick is being exposed to an imprinting object it may be more active, be more 'aroused', or more attentive than chicks not exposed to such an object. In order to determine whether the biochemical changes observed could be attributed to any of these other factors a variety of control procedures were devised, including restricting learning to one side of the forebrain only (Horn *et al.*, 1973) and varying the strength of imprinting (Bateson *et al.*, 1975). The results of these studies confirmed that there is a strong relationship between the biochemical changes observed and learning (see Horn, 1985, for a more detailed account of these and subsequent neural studies). Since the biochemical changes were closely tied to the learning process of imprinting it was necessary, for more detailed analysis of these changes, to enquire whether they were localized to restricted regions within the forebrain roof.

Using an autoradiographic technique Horn *et al.* (1979) identified a restricted region immediately around the midpoint between the anterior and posterior poles of the cerebral hemispheres, which they referred to as IMHV (the intermediate and medial parts of the hyperstriatum ventrale; see Fig. 12.1). Subsequently, the same region or co-extensive areas have been localized by other groups studying visual and auditory imprinting (Kohsaka *et al.*, 1979; Maier & Scheich, 1983).

If IMHV is indeed a crucial site of plasticity for imprinting then its destruction prior to imprinting should prevent the acquisition of preferences, and its destruction after imprinting should render a chick amnesic. Experiments by McCabe *et al.* (1981, 1982) confirmed these predictions. Similar sized lesions placed elsewhere in the forebrain had no such consequences.

Neuronal changes associated with information storage

Having established that IMHV is critically involved in imprinting it is interesting to examine whether any changes take place in the structure of synapses within the region following imprinting. Horn *et al.* (1985) undertrained one group of birds on a red box (20 minutes exposure) and overtrained another group (140 minutes exposure). A third group of chicks served as dark-reared controls. Following training, samples were taken from the IMHV of each hemisphere; ultra-thin sections were cut and examined under an electron microscope. There were no differences in any measure of

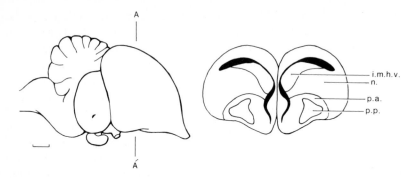

Fig. 12.1 Outline drawing of the chick brain. The vertical lines AA'
above and below the drawing of the lateral aspect (left) indicate the plane
of the coronal section outline (right) of the brain. Abbreviations: i.m.h.v.,
the intermediate and medial part of the hyperstriatum ventrale (IMHV);
n., neostriatum; p.a., paleostriatum augmentatum; p.p. paleostriatum
primitivum. Scale bar: 2 mm. (After Horn & Johnson 1989.)

synapse morphology between undertrained chicks and dark reared controls.
Chicks that had been trained for 140 minutes differed from the other two
groups in only one measure of the synapse structure: the mean length of the
postsynaptic density, the thickened part of the postsynaptic membrane,
which had increased by 17 per cent. The change occurred only on spine
synapses; there were no effects of training on synapses on dendritic shafts.
Further, the change was restricted to synapses within the left IMHV. A series
of experiments concerned with differences in function between the right and
left IMHV, and the existence of a second 'information store' are reviewed
elsewhere (Horn, 1985, 1986; Horn & Johnson, 1989).

The postsynaptic density appears to be a site of high neurotransmitter
receptor density (Fagg & Matus, 1984), so that an increase in the area of this
region of membrane specialization may be associated with an increase in the
number of ligand binding sites. In the mammalian brain, assymetric spine
synapses are considered to be excitatory and possess receptors for the
excitatory amino acid L-glutamate (Errington et al., 1987). There are several
sub-types of receptor for L-glutamate, three of which are defined by the
action of selective agonists. One of these selective agonists is N-methyl-D-
aspartate (NMDA). Following imprinting, McCabe & Horn (1988) found a
significant increase in NMDA-sensitive binding in the left IMHV of chicks
compared with dark-reared controls. Further, they found a significant
positive correlation between the degree to which a chick prefered the familiar
stimulus at testing and the degree of NMDA sensitive binding.

It would be reasonable to expect that the changes in synaptic structure and
biochemistry just described would result in changes in the spontaneous firing

of neurones within left IMHV. As a first step in analyzing the electrophysio-logical changes in left IMHV in relation to training, Payne & Horn (1984) exposed chicks to an imprinting stimulus for three hours. The amount of approach that each chick made toward the stimulus during this time was recorded. After the final hour of training, the chicks were anaesthetized, one micro-electrode penetration made through the left IMHV, and a second, simultaneous, penetration made through a visual projection area, the hyperstriatum accessorium (HA). For each region the mean firing rates of units were recorded from three or more sites. A significant negative correla-tion between total approach during exposure to the object and the mean firing rate of units within the left IMHV were found. In contrast, there was no correlation between the mean firing rates of units in HA and approach behaviour. The electrophysiological correlates of training were thus region specific, so that the changes in left IMHV neuronal activity were unlikely to have resulted from some non-specific effects of training, such as arousal, which might be expected to affect both brain regions. More recent studies have suggested that the effects of training may be even stronger on the difference in mean spontaneous firing rate between the left and right IMHV (Davey, 1988).

Stimulus-dependent neurophysiological findings

In the ablation studies mentioned earlier two different imprinting stimuli were used. In most experiments half of the chicks were trained on an illuminated rotating stuffed jungle fowl and the other half on a rotating red box illuminated from within. When presented in isolation, both stimuli elicit active approach from naive chicks and strong preferences result from a few hours of exposure to them. However, when reviewing a series of studies on the effects of IMHV lesions on the retention of filial preferences, Horn & McCabe (1984) reported that, although there were no significant differences between box-trained and fowl-trained birds within any one study, taken over several studies there was a striking stimulus-dependent effect: when IMHV was removed bilaterally chicks trained on the red box were severely impaired, whilst the preference of chicks trained on the stuffed fowl for that object were only slightly reduced.

Other manipulations have extended this initial stimulus-dependent find-ing. Bolhuis *et al.* (1986) found a strong positive correlation between plasma testosterone levels and preference for the stuffed fowl in fowl-trained birds, but no such correlation in birds trained on the red box. Furthermore, administration of testosterone prior to training increased the preference for the fowl in birds exposed to that stimulus, but was without effect on birds exposed to the box. Noradrenaline, a substance implicated in the plasticity of the brain, is present in the chick forebrain and increases with age and with experience of a visual imprinting stimulus during the first 50 hours after hatching (Davies et al., 1983). Administration of the neurotoxin DSP4 depletes forebrain noradrenalin levels by about 65 per cent in the chick. Such treatment profoundly impairs the acquisition of preference in birds exposed to the red box; birds exposed to the fowl are only slightly impaired (Davies et al., 1985). Furthermore, a significant positive correlation between preference score and noradrenaline concentration was found in a medial forebrain sample, composed mainly of IMHV, for birds exposed to the red box. No such correlation was found for birds exposed to the stuffed fowl. Taken together, the noradrenaline and testosterone evidence suggest a double dissociation of function; at least part of the neural substrate underlying preference for the stuffed fowl differs from that underlying preference for the red box.

Given these stimulus-dependent neurophysiological effects it seemed appropriate to enquire what predispositions might be influencing the behaviour of chicks in the context of filial preference. In order to investigate this issue, Johnson et al. (1985) exposed day-old domestic chicks to either the rotating red box or to the rotating stuffed fowl. Two hours or 24 hours after this period of exposure the preferences of these chicks were ascertained by simultaneously presenting both the red box and the stuffed fowl to each chick (Bateson & Wainwright, 1972). This simultaneous presentation test is thought to be more sensitive to preference than the sequential test used in the studies discussed so far (Johnson & Horn, 1986). As expected, when tested two hours after training, both the box-trained and fowl-trained chicks prefered the object to which they had previously been exposed. However, when tested 24 hours after training both box and fowl-trained chicks showed a shift in preference towards the fowl. That is, the fowl-trained birds had an even stronger preference for the fowl than their counterparts at the first testing time, while the box-trained birds tested at the later time had no preference for either object. Although these results taken in isolation could be accounted for in terms of differential forgetting, the results from a third group of chicks suggested a different interpretation. The chicks in this group were not exposed to any imprinting object prior to testing, but were merely exposed to dim overhead light at the same time as their counterparts in the other groups were being trained. Whilst being exposed to overhead light the chicks were in individual running wheels, but could not see each other. Since birds in this group were not exposed to any object it was not surprising that they had no preference for either test object two hours after being exposed to light. Contrary to expectation however, other individuals from the same group tested 24 hours after the exposure to light had a strong preference for the stuffed fowl. These results led to the initial suggestion that the filial preferences of trained chicks could be accounted for in terms of the interaction between acquired preferences on the one hand, and a developing predisposition on the other.

A specific predisposition

It has been known for some time that chicks have general predispositions to respond to particular classes of stimuli in particular ways. For example, naive chicks are predisposed to approach moving objects larger than a matchbox size, whilst they will peck at objects smaller than a matchbox (Sluckin, 1972). Although some authors have suggested that more specific predispositions may be operating in filial preference behaviour (e.g. Hinde, 1961), until recently no evidence for such a specific predisposition had been found.

The experiment of Johnson *et al.* (1985) provided evidence for the existence of a predisposition. Chicks which had not been exposed to any conspicuous object, but had merely been exposed to two hours of dim overhead light whilst in running wheels, showed a significant preference after 24 hours for the stuffed fowl over the red box. Two obvious questions arose from this observation; firstly, what are the necessary environmental conditions for the emergence of the predisposition, and second what are the characteristics of the stuffed fowl that cause it to be preferred to the red box?

The first question was addressed in a series of experiments in which it was established that even completely dark-reared birds come to prefer the stuffed fowl to the red box if allowed a period of time in running wheels (Bolhuis *et al.*, 1985). Dark-reared birds which were not allowed this period did not develop the preference. Although exposure to overhead light was not essential, exposure to patterned light accelerated the emergence of this preference. These results suggest that under conditions likely to be experienced in the natural environment, the opportunity for motor activity and exposure to patterned light, the predisposition will emerge rapidly and strongly.

In dark-reared chicks a period of time in running wheels around 24 hours after hatching appears to be essential for the fowl preference to emerge 24 hours later. In a recent study Johnson *et al.* (1989) investigated the effects on the emergence of the predisposition of allowing dark-reared chicks a period of time in the running wheels at varying ages after hatching. The results indicated that there is a critical time window between 12 hours and 36 hours after hatching within which the experience must occur. Time spent in the running wheels after this age does not result in the expression of the predisposition.

The finding that a predisposition requires experience at a particular time or stage is not without precedent. For example, Gottlieb (1981) reported that specific auditory input at a particular embryonic stage is required to ensure the maintainance of species specific call preferences in mallard ducklings. In the chick a peak of sensitivity to non-specific motor experience or arousal around 24 hours post-hatch would be consistent with reports that it is at this age that they first emerge from under the hen in the natural environment.

It has been suggested that non-specific input, such as motor experience, may elevate testosterone levels in the chick brain, which in turn may trigger the predisposition (Bolhuis *et al.*, 1985; Horn, 1985). If this is the case then this putative surge of testosterone has to occur within a particular time window for the 'validation' (Horn, 1985) of the system supporting the predisposition to occur.

The second question relating to the predisposition concerned its specificity, or the characteristics of the stuffed fowl which cause it to be preferred over the red box. One possibility is that the stuffed fowl is more attractive simply because it has greater outline or textural complexity than the box. There is some evidence that older chicks preferentially approach objects of greater visual complexity (Dutch, 1969; Berryman *et al.*, 1971), and this preference may also be important in younger chicks. In order to circumvent the problem of quantifying visual complexity (see Banks & Ginsburg, 1985), Johnson & Horn (1988) compared the preference of dark-reared chicks for an intact stuffed jungle fowl with their preference for other 'test' objects which were constructed in various ways by scrambling or degrading a similar stuffed jungle fowl. In all of the experiments in this series dark-reared chicks were placed in running wheels in darkness 24 hours after hatching. Twenty-four hours after this they were simultaneously presented with an intact stuffed fowl and a second, test object, and their preference ascertained.

The first 'test' object to which the intact stuffed jungle fowl was compared was a stuffed jungle fowl which had been partially disarticulated and reassembled in an anatomically unusual way (Fig. 12.2; top left). With this stimulus there was no significant preference for the intact fowl 24 hours after the period in running wheels suggesting that, while both objects had similarly complex outlines (the same number and shape of wings, legs, etc. protruding from the body) and were of similar 'textural' complexity (feather patterns, species markings, etc.), the fact that one object had an intact fowl outline while the other had jumbled limbs made no difference to their relative attractiveness to the chicks. The next test object investigated was similar to the previous one except that several of the elements (wings, leg, head, etc.) were separated from the trunk of the body (Fig. 12.2; top right). Again, the intact fowl was not significantly preferred over this object, reinforcing the earlier conclusion that a fowl-shaped overall outline is not a critical factor for the emerging predisposition: stuffed fowl with 'scrambled' outlines are as attractive as an intact jungle fowl. This leaves several other possibilities including: (i) the chicks could be responding to some particular features such as the eyes or beak, or for some specific arrangements (configuration) of these features; or (ii) the chicks could be responding simply to particular feather patterns or colours. The next test object was composed of the trunk pelt of a jungle fowl cut up into small squares and jumbled up with other parts of the body, before being stuck onto the cork sides of a rotating box

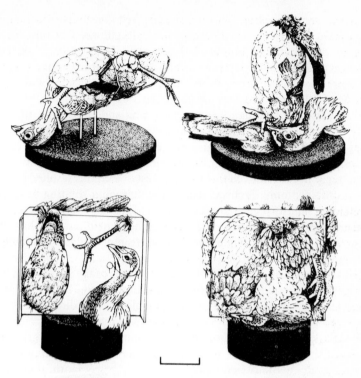

Fig. 12.2 Some of the 'test' objects used in the experiments of Johnson &
Horn (1988). Top left- the jumbled stuffed fowl; top right- the
disarticulated fowl; bottom right- the cut-up pelt; bottom left- the
'scrambled' fowl. Scale bar: 10 cm. (See text for further details. Drawings
by Priscilla Barrett.)

(Fig. 12.2; bottom right). Using this test object there was a strong preference
for the intact jungle fowl. Therefore, this test object did not contain the
essential attractive characteristics. Since the test object possessed the 'textural' complexity and colours of the intact fowl these characteristics do not
appear to be critical. Although this test object possessed individual features
of the jungle fowl, these features were not always in their correct configuration relative to each other. For example, the eye, beak and neck region were
separated. The presence of such clusters of features may be critical for the
attractiveness of the object. Another critical factor may have been the greater
complexity of the intact fowl outline compared to the simple box outline of
the test object. One way to decide which of these two factors is most
important is to create a test object which possesses the clusters of features
absent in the previous object, but retains the simple box outline. Such a test

object can be seen in Fig. 12.2 (bottom left). With this test object there was no significant preference for the intact fowl, despite the fact that the overall outline of the test object was less complex. This evidence suggests that clusters of features may be the most important characteristic.

There is some evidence from studies of adult chickens and quail that features of the head and neck are particularly important in the recognition of individuals in a dominance hierarchy (Candland, 1969) and for the elicitation of social proximity behaviour (Domjan & Nash, 1988). Could it be that features of the head are also particularly important for the chicks? To investigate this possibility all of the clusters of features except those from the head region were removed from the last test object: this resulted in an object that contained only the head region mounted on a rotating box. This test object was slightly preferred over the intact fowl. This suggests that not only is the cluster of features associated with the head an important characteristic, but that the rest of the jungle fowl body may simply act as a 'distractor' from this critical region.

Having established that the cluster of features of the jungle fowl head are critical, the next question investigated was how specific to particular species these features need to be. Although the jungle fowl is a direct ancestor of the domestic chicken (Zeuner, 1963), the Gadwall duck (*Anas strepera*) is not. When dark-reared chicks allowed time in running wheels were given a choice between these two species they had no preference, suggesting the critical facial features need not be species specific (Johnson & Horn, 1988). Furthermore, other experiments indicated that the intact fowl was not preferred to a variety of similar sized stuffed mammals such as a polecat (*Mustela putorius*). This indicates that the configuration of features associated with the head may be more important than the details of the features themselves.

In the natural environment, the specific predisposition just discussed may ensure that the chick attends toward and approaches an appropriate object, normally the chick's own mother, when it emerges from under her on the day after hatching. The young chick may then learn about the visual characteristics of the adult bird and so come to recognize her. Since features of the head and neck are particularly important for the predisposition, it may be this region about which the chick learns most. As well as this being the region of the adult chicken most likely to vary between individual birds, this would be consistent with the evidence from studies on adult chicken and quail.

The interacting systems model

The evidence on the stimulus-dependent neurophysiological effects and on the predisposition just discussed, led to the proposal (Horn, 1985; Johnson *et al.*, 1985) that there are two dissociable neural systems underlying filial preference behaviour in the domestic chick. Firstly, a predisposition for

Fig. 12.3 The two stuffed jungle fowl used as training stimuli in Johnson & Horn (1987). Half of the chicks were trained on one and half on the other. Scale bar = 10 cm. (After Johnson & Horn 1987. Drawing by Priscilla Barrett.)

the young chick to approach objects resembling adult fowl, and secondly a learning system subserved by IMHV which is engaged by a variety of conspicuous objects early in life. In the natural situation, the predisposition may serve to orient the chick preferentially toward adult fowl rather than to inanimate objects in its visual environment; the learning system is then engaged by particular objects to which the chick attends, normally the chick's own mother. Several predictions arise from these proposals.

1 If the two systems are truly dissociable then bilateral ablation of IMHV will not impair the predisposition. This prediction was confirmed by Johnson & Horn (1986): chicks with IMHV lesions respond like dark-reared birds in preference tests even if they have been exposed to an imprinting object beforehand.

2 Chicks should be capable of learning to discriminate between individual adult birds. Johnson & Horn (1987) exposed intact two-day old chicks to one of two individual stuffed fowl for four periods of 50 minutes (see Fig. 12.3). Two hours after the end of the last training session the birds were given a simultaneous choice between the two stuffed fowl. The chicks were significantly more likely to start approaching the individual fowl to which they had previously been exposed.

3 Bilateral ablation of IMHV should preclude learning about individual adult birds. In order to test this hypothesis chicks with bilateral IMHV lesions and chicks with lesions to a different part of the forebrain (HA) were trained on individual fowl and tested

as described above. Chicks with control lesions, like intact chicks, preferred the individual to which they had previously been exposed. In contrast, chicks with IMHV lesions had no preference (Johnson & Horn, 1987).

4 It should be easier to reverse preferences acquired through exposure to artificial stimuli, than preferences acquired through exposure to a fowl or hen. Since both systems are postulated to be engaged by exposure to the fowl, and only one by exposure to an artificial stimulus such as the red box, Bolhuis & Trooster (1988) predicted that chicks initially exposed to the fowl could not easily have their preference reversed by exposure to the red box. In contrast, it should be as easy to reverse the preferences of box exposed birds as it is to reverse preferences resulting from exposure to other artificial objects (e.g. Salzen & Meyer, 1968; Cherfas & Scott, 1981). In a series of experiments in which the initial attractiveness of the red box and stuffed fowl were varied, Bolhuis & Trooster (1988) established that there is a strong asymmetry in the ease of reversibility between these two stimuli in the predicted direction.

The results discussed lead to the conclusion that there is an interaction between the two systems during filial preference behaviour. There are however, a number of different forms that this interaction could take.

(a) The emergence of the predisposition could be the behavioural consequence of an increasingly specific 'filter'. This filter may increasingly restrict the range of visual stimuli about which information can be stored.

(b) The two systems may be informationally encapsulated until they 'compete' for control over motor output.

(c) When both systems are activated (during presentation of a stuffed fowl, for example) the system underlying the predisposition 'gates' or 'enables' the learning subserved by IMHV, thus ensuring that the current input is learned more rapidly or more strongly than it would be otherwise.

In the experiment by Johnson et al. (1985) chicks were exposed to the box or fowl before the predisposition had developed. However, if interaction (a) is correct then birds should not be able to learn about an artificial stimulus after the predisposition has appeared, since the specificity of the filter would not allow visual stimuli which did not meet the characteristics defined in the experiments discussed earlier by Johnson & Horn (1988). In a recent series of experiments Bolhuis, Johnson & Horn (1989) tested this prediction. The results clearly indicated that chicks are capable of learning about the characteristics of artificial stimuli after the predisposition is expressed.

Although there is no clear evidence to lead to the rejection of either interaction type (b) or (c) at present, the latter may be more consistent with the observation that the preferences of the chicks are much more easily reversible following exposure to an artificial object than following exposure to a stuffed or living hen (Boakes & Panter, 1985; Bolhuis & Trooster, 1988).

A mammalian analogy

Does the idea of two systems underlying parental recognition have any applicability to other species? Another species in which the development of parental recognition has been studied in some detail is the human infant.

Goren *et al.* (1975) demonstrated that newborn human infants will turn their head and eyes further to keep a moving face pattern in view, than to keep a variety of 'scrambled' face patterns in view or a display with no pattern. This finding challenged the established view that it takes the infant two or three months to learn about the arrangement of features that compose a face (see for example, Gibson, 1969; Maurer, 1985; Nelson & Ludemann, 1989), and accordingly came under criticism for methodological reasons. However, the result described by Goren and colleagues has recently been replicated using procedures less susceptible to criticism (Johnson *et al.*, 1991).

One way to reconcile these findings in newborns with the view that it takes several months to learn about the arrangement of features that constitutes a face, is to interpret the data in terms of two systems analogous to those just described for chicks (Johnson, 1987, 1988; Morton & Johnson, 1989; Johnson & Morton, 1991). Firstly, newborn infants are predisposed to attend to face-like patterns within their visual field, and secondly a learning system is engaged by those objects to which the infant attends – in this case faces. Elsewhere, the term Conspec has been introduced to refer the former system, i.e. that component of parental recognition which requires only non-specific input or stimulation for its expression shortly after birth or hatching (Johnson & Morton, 1991).

With regard to Conspec there are some striking similarities between chicks and infants. Both for chicks and infants there is an attraction to the correct arrangement of features associated with an adult conspecifics head. As mentioned earlier for the chick a hormonal surge may 'trigger' the expression of the predisposition (Bolhuis *et al.*, 1985; Horn, 1985). In the human infant, the surge of catecholamines associated with normal delivery may have a similar 'triggering' effect. Turning to the second, learning system, the chick is capable of discriminating between individual adult chicken in the first few days (Johnson & Horn, 1987). Although it may take the human infant three months to recognize its mother on the basis of the internal features of her face (Bushnell, 1982), the infant may be able to use more general visual cues to discriminate her from others by the second or third day after birth (Bushnell & Sai, 1987).

Associations and dissociations

In the previous section evidence was presented of a dissociation between the exposure learning subserved by IMHV and the specific predisposition. In this section evidence for a second dissociation is presented, that between exposure learning and associative learning.

The evidence is strong that the left IMHV is involved in the storage of information following imprinting. However, the question remains whether the integrity of the region is crucial in other learning situations. This question was initially investigated by McCabe *et al.* (1982). In that study chicks were exposed to a visually conspicuous object before being divided into four treatment groups: (1) received bilateral IMHV-lesions, (2) were sham-operated controls, (3) received lesions to the Wulst region (composed largely of HA, a visual projection area not previously implicated in imprinting), and (4) had lesions placed in a lateral part of the cerebral hemisphere. After recovery from surgery all chicks were given a preference test. With the exception of the IMHV-lesioned birds, all groups selectively approached the training object, and in this sense recognized it. All chicks were subsequently trained on a simultaneous visual discrimination task. This task involved the simultaneous presentation of two visual patterns. Half the chicks in each group were required to approach one pattern, the other half to approach the other pattern. Successful responses were reinforced by a blast of warm air. All groups learned this task, and there were no significant differences between groups in the speed of acquisition or in any other measure of performance. Thus, chicks with IMHV lesions are (i) severely impaired on a recognition test which is given after the chicks have been exposed to an object in the imprinting situation, and (ii) unimpaired when required to approach selectively a pattern which is associated with a reward. There are, however, at least two reasons why this experiment requires to be interpreted with caution. Firstly, since the chicks were of different ages when they were trained in the two test situations it is possible that the differences in performance were affected by maturational processes. Secondly, different stimuli were used in the two learning situations. It is not clear whether the lesion failed to affect performance in the visual discrimination task for one or both of these reasons, or because the nature of the learning tasks were different. A further experiment was, therefore, required to unravel these confounding variables.

Day-old, visually naive chicks will quickly learn to press one of two pedals in order to be presented with an attractive stimulus (Bateson & Reese, 1969). As a chick learns to associate presses of the particular pedal with the presentation of an attractive object, the chick also learns about the visual characteristics of the object. When subsequently given a choice between the reinforcing object and a novel object, the chick will preferentially approach the former. Since the two processes of recognition and association develop concurrently, this training procedure appeared to be an appropriate one to enquire whether or not the two processes could be dissociated by IMHV lesions (Johnson & Horn, 1986).

The operant training apparatus had two conspicuous chequer-board pattern pedals set into the floor. Three of the walls were painted black, and the fourth was made of wire mesh so that the chick could see the reinforcing object when it was switched on. This object, either the red box or the stuffed fowl, was illuminated and would start to rotate when a particular one of the two pedals was pressed. The chick could press a pedal merely be stepping on it. Often when a chick pressed a pedal and the stimulus appeared, the chick moved off the pedal in an attempt to approach the object. As a result the object ceased to be illuminated and stopped rotating. Pressing the other pedal had no effect. Although the position of the 'active' pedal remained fixed for each chick, the position was varied between chicks in order to counteract any right/left bias.

Chicks with lesions to IMHV did not differ significantly from sham-operated controls in various measures of performance in the operant task: (i) in the mean latency to first pedal press, (ii) in the percentage of birds in each group that reached the criterion of 9 out of 10 correct presses, (iii) in the time taken to reach criterion, or (iv) in the time spent on the active pedal. Both groups reached criterion significantly more quickly in the second of the two training sessions than in the first.

Two hours after the second training session in the operant training apparatus the chicks were given the simultaneous choice test. In this test the chicks were able to approach the object with which they had been rewarded during the operant task, or a novel object. The sham-operated chicks preferred the object which had been used as the reinforcer; chicks with bilateral lesions to IMHV had no preference. The lesioned chicks were no less active during training or testing, but unlike controls did not direct their movements selectively toward the object seen previously. That is, they behaved as if they were unable to recognize the object.

This putative dissociation between the neural representation of the visual characteristics of an object and associations involving the object is supported by certain purely behavioural studies. If the representation of an object constructed during imprinting can be utilized in the subsequent formation of associations, then we should expect that pre-exposure to the rewarding object prior to operant training would enhance the rate at which the association between the pedal press and the object is learned. However, Bateson & Reese (1968) found that birds exposed to the rewarding object prior to the operant learning task reached criterion no quicker than did birds not given prior exposure. What about the converse situation, namely does the prior formation of associations involving an object facilitate the subsequent recogition of that object? This question was investigated by Bolhuis & Johnson (1988) who studied two groups of chicks. One group, the 'response-contingent' group, were trained to press a pedal, in order to be exposed to an object. The second group of chicks did not have control over the presentation of the object, but were merely shown it whenever a partner 'response-contingent' bird pressed the pedal. Thus, although both groups were exposed to the stimulus in an identical manner, only in one group was an association involving the object being formed. If associations involving an object facilitate its subsequent recognition then the 'response-contingent' group should have a higher preference for the rewarding object after operant training. This was not the case. Taken together, the neural and behavioural dissociations provide strong evidence that exposure learning can be dissociated from the formation of associations between representations.

So far evidence for dissociations between the exposure learning subserved by IMHV and two other neural systems have been discussed. However, it would be premature to conclude from this evidence that the learning system involved in imprinting is unique. A variety of possibilities regarding the range of other tasks in which the system might be involved remains. For example, during filial preference behaviour, the integrity of IMHV is essential for the recognition of individual adult fowl (Johnson & Horn, 1987). Is the region also involved in other tasks which require the recognition of individual conspecifics? This question has recently been investigated. Animals of several species, when choosing a mate, prefer individuals which are slightly different, but not very different, from those with which they were reared, a phenomenon known as optimal outbreeding (see Barnard and Aldhous, this volume). Bolhuis et al. (1989) raised intact female chickens in small social groups with one male until they were three months old. When tested at this age the females preferred to approach and stand next to a novel male of the same strain rather than the male with which they had been reared, or a novel male of a novel strain. A second group of female chickens received bilateral IMHV lesions in their first day of life and were subsequently reared in an identical manner to the intact chickens. Although the lesioned birds appeared normal in most respects, they did not have a preference for any of the test males: they appeared to be incapable of discriminating between them (Bolhuis et al., 1989). This result is consistent with the notion of IMHV being necessary for the recognition of individual conspecifics in a variety of situations, not just in the context of filial imprinting. However, it remains possible that the impairment of IMHV-lesioned birds may be attributable to their inability to imprint during the first few days after hatching, rather than a direct consequence of IMHV damage.

Although the integrity of IMHV may be critical for tasks requiring the recognition of individual conspecifics, it also appears to be essential for the successful performance of some other tasks which do not have this requirement. One such task is known as one-trial passive avoidance learning (PAL). Newly-hatched chicks will peck at a wide variety of small objects that are presented to them, including small coloured beads. If the bead is coated with an unpleasant tasting substance, for example methyl anthranilate (MeA), approximately 75 per cent of chicks will refrain from pecking on subsequent re-presentation of the bead. In birds where the bead is initially coated with water, all but about 4 per cent will peck on subsequent re-presentation of the bead (Lee-Teng & Sherman, 1966).

A variety of biochemical and structural changes following PAL have been described within areas closely corresponding to IMHV (e.g. Kossut & Rose, 1984; Patterson *et al.*, 1986). Davies *et al.* (1988) investigated whether bilateral IMHV lesions similar to those which impair imprinting, would also impair chicks in PAL. While intact chicks, and chicks with lateral forebrain lesions, avoided the distasteful bead on subsequent re-presentation, the IMHV-lesioned chicks did not. In the PAL task the chick is required to associate some aspect of its pecking response with the unpleasant taste of the bead. Thus, the task could be described as requiring the formation of associations between representations. The finding that IMHV is essential for PAL may initially appear to conflict with the results described earlier that IMHV is not involved in other tasks that may be regarded as requiring the formation of associations. One approach to resolving this apparent paradox is to consider the integrity of IMHV as essential for the inhibiting or withholding of 'natural tendencies' or unconditioned responses.

The inhibition of spontaneous tendencies

Following an extensive review of the literature on the effects of hyperstriatal lesions on a variety of learning tasks in pigeons and other birds, Macphail (1982) proposed that the avian hyperstriatum is involved with the inhibition of previously acquired responses. Most of the evidence concerned adult birds with extensive portions of the hyperstriatal region destroyed being unable to reverse or inhibit a previously acquired response. Of course, the IMHV of chicks is only a small area within the hyperstriatum and the lesions which result in the deficits described in the present chapter are very much more restricted than most of those reviewed by Macphail. Moreover, imprinting and PAL are shown by chicks in the first few days of life when little opportunity for previous learning has occurred. However, the possibility remains that IMHV may influence behaviour by the inhibition of other neural systems in the brain albeit in a more specific manner than that resulting from the extensive lesions in adult birds reviewed by Macphail.

Turning to mammals, a more specific version of the inhibition idea has been proposed to account for the effects of limbic system damage in monkeys performing a variety of learning tasks (Gaffan *et al.*, 1984; Malamut *et al.*, 1984). Gaffan and his colleagues propose that the primate hippocampus is involved in the changing of well established 'habits' or 'innate responses' by new learning. More specifically, the hippocampus is postulated to inhibit either 'instrumental habits' acquired over many trials or 'innate responses' towards particular stimuli (Gaffan *et al.*, 1984).

Since chicks are normally dark-reared until being presented with an imprinting stimulus or being trained on an unpleasant tasting bead, it is possible that IMHV is involved when predispositions or spontaneous tendencies require to be inhibited or withheld for successful performance in a task. By this view, the deficits following bilateral IMHV lesions can be accounted for in the following way. Newly-hatched chicks will sponta-neously peck at a variety of small bright objects. For successful performance of the PAL task this spontaneous tendency requires to be supressed or inhibited. Similarly, in the absence of any naturalistic stimuli, newly-hatched chicks will approach a wide variety of bright moving objects larger than a matchbox (see Sluckin, 1972). As a chick becomes familiar with a particular object it ceases to approach others. In this sense the initial tendency to approach a range of objects is specifically diminished or inhibited. In the heat reinforcement task, chicks were rewarded with warm air for choosing the correct one of two patterns. In this case the chicks' natural tendency to approach warmth in an otherwise cold room did not require to be withheld and IMHV lesions had no effect on any measure of acquisition. Similarly, in the operant task the chicks 'worked' for the presentation of an attractive object.

Given that the selective inhibition hypothesis can account for some of the effects of bilateral IMHV lesions on early learning, it is worth enquiring if the hypothesis can also account for some other phenomena.

1 In the electrophysiological study reported by Payne & Horn (1982) there was a significant *negative* correlation between spon-taneous mean firing rate in the IMHV and approach during training for birds exposed to the red box. That is, the higher the mean firing rate in IMHV the less was the approach of the chick during training. If the frequency of firing of groups of neurones within IMHV is reflected in the impulses transmitted along its efferent pathways, and if one of its projection areas is involved in the control of approach activity, then the negative correlation implies that the efferent pathway may serve to inhibit a motor control area.

2 On the assumption that IMHV influences behaviour by the inhibition of approach, certain predictions can be made with regard to approach activity during preference tests. One of these predictions is that in a sequential preference test IMHV-lesioned

birds will approach both familiar and novel objects as much as the intact birds approached the familiar object. This indeed appears to be the case in at least one study where IMHV lesions were placed shortly after training (McCabe *et al.*, 1982). However, this pattern of testing activity does not occur in studies where the lesions were placed prior to training (McCabe *et al.*, 1981). This may be because lesions placed prior to training have non-specific effects on approach in addition to their specific effects on preference. The disinhibitory effect of lesions placed shortly after training is not a general effect of brain damage as other forebrain lesions result in equal, or even reduced, activity during testing.

Thus, at least some tentative evidence is consistent with the notion that IMHV inhibits the neural system that underlies approach behaviour. In the case of small bright objects the predisposition to peck is inhibited, and in the case of medium-sized moving objects the tendency to approach is supressed.

Concluding remarks

The neuroethological investigation of filial preference behaviour in the domestic chick has resulted in a conception of imprinting very different in some respects from that originally outlined by Konrad Lorenz. For example, Lorenz (1937) thought that the process was critical for the identification of species. The evidence presented in this chapter argues for its importance in the recognition of individuals, although some information about the characteristics of the species may be learned *en route*. In other respects, however, the neuroethological investigations discussed in this chapter suggest something of a return to views held by Lorenz. For example, the interaction between the two systems described in this chapter is somewhat foreshadowed by Lorenz's (1937; p. 267) statement,

> . . . .imprinting fills out the spaces left vacant in the picture of the proper species, outlined in the bird's perceptory world by the data given by innate perceptory patterns, very much as medieval artists in drawing astronomic maps, accommodated the pictures of the heraldic creatures of the zodiac between the predetermined points given by the position of the stars.

The other focus for research on imprinting discussed in the introduction to this chapter concerned the extent to which imprinting could be regarded as involving a 'special' or 'unique' type of learning. Research reviewed in this chapter provides evidence for the view that the neural mechanisms subserving imprinting are distinct from those critical for associative learning, although the integrity of IMHV is essential for other learning situations which require the withholding of a spontaneous tendency.

Acknowledgements

Many of the ideas developed in this chapter have originated in discussions with Gabriel Horn and Johan Bolhuis. I also wish to thank the latter and John Morton for comments on an earlier version of this chapter.

References

Banks, M. S. & Ginsburg, A. P. (1985). Infant visual preferences: a review and new theoretical treatment. *Adv. Child Dev. Behav.*, **19**, 207–46.

Bateson, P. P. G. (1966). The characteristics and context of imprinting. *Biol. Rev.*, **41**, 177–220.

Bateson, P. P. G., Horn, G. & Rose, S. P. R. (1972). Effects of early experience on regional incorporation of precursors into RNA and protein in the chick brain. *Brain Res.*, **39**, 449–65.

Bateson, P. P. G., Horn, G. & Rose, S. P. R. (1975). Imprinting: correlations between behaviour and incorporation of [14C] uracil into chick brain. *Brain Res.*, **84**, 207–20.

Bateson, P. P. G. & Reese, E. P. (1968). Reinforcing properties of conspicious objects before imprinting has occurred. *Psychon. Sci.*, **10**, 379–80.

Bateson, P. P. G. & Reese, E. P. (1969). The reinforcing properties of conspicuous stimuli in the imprinting situation. *Anim. Behav.*, **17**, 629–99.

Bateson, P. P. G. & Wainwright, A. A. P. (1972). The effects of prior exposure to light on the imprinting process in domestic chicks. *Behaviour*, **42**, 279–90.

Berryman, J., Fullerton, C. & Sluckin, W. (1971). Complexity and colour preferences of chicks of different ages. *Q. J. Exp. Psychol.*, **23**, 255–60.

Boakes, R. & Panter, D. (1985). Secondary imprinting in the domestic chick blocked by previous exposure to a live hen. *Anim. Behav.*, **33**, 353–65.

Bolhuis, J. J. & Johnson, M. H. (1988). Effects of response-contingency and stimulus presentation schedule on imprinting in the chick. *J. Comp. Psychol.*, **102**, 61–5.

Bolhuis, J. J., Johnson, M. H. & Horn, G. (1985). Effects of early experience on the development of filial preferences in the domestic chick. *Dev. Psychobiol.*, **18**, 299–308.

Bolhuis, J. J., Johnson, M. H. & Horn, G. (1989). Interacting mechanisms during the formation of filial preferences: The development of a predisposition does not prevent learning. *J. Exp. Psych.: Anim. Behav. Proc.*, **15**, 376–82.

Bolhuis, J. J., Johnson, M. H., Horn, G. & Bateson, P. (1989). Long-lasting effects of IMHV lesions on social preferences in the domestic fowl. *Behav. Neurosci.*, **103**, 438–41.

Bolhuis, J. J., McCabe, B. J. & Horn, G. (1986). Androgens and imprinting. Differential effects of testosterone on filial preferences in the domestic chick. *Behav. Neurosci.*, **100**, 51–6.

Bolhuis, J. J. & Trooster, W. J. (1988). Reversibility revisited: stimulus-dependent stability of filial preference in the chick. *Anim. Behav.*, **36**, 668–74.

Bushnell, I. W. R. (1982). Discrimination of faces by young infants. *J. Exp. Child Psychol.*, **33**, 298–308.

Bushnell, I. W. R. & Sai, F. (1987). *Neonatal Recognition of the Mothers Face.*

University of Glasgow Psychological Report No 87/1. Glasgow: Glasgow University.

Cajal, S. R. (1911). *Histologie du System Nerveux de l'Homme et des Vertebres.* Vol. 2. Paris: Maloine.

Candland, D. K. (1969). Discrimination of facial regions used by the domestic chick in maintaining the social dominance order. *J. Comp. Physiol. Psychol.,* **69,** 281–5.

Cherfas, J. & Scott, A. (1981). Impermanent reversal of filial imprinting. *Anim. Behav.,* **30,** 301.

Davey, J. E. (1988). Imprinting and spontaneous activity in the hyperstriatum ventrale of the chick: a developing asymmetry. *Neurosci. Lett.,* **32,** S48.

Davies, D. C., Horn, G. & McCabe, B. J. (1983). Changes in telencephalic catecholamine levels in the domestic chick: Effects of age and visual experience. *Dev. Brain Res.,* **10,** 251–5.

Davies, D. C., Horn, G. & McCabe, B. J. (1985). Noradrenaline and learning: the effects of the noradrenergic neurotoxin DSP4 on imprinting in the domestic chick. *Behav. Neurosci.,* **99,** 652–60.

Davies, D. C., Taylor, D. & Johnson, M. H. (1988). Restricted hyperstriatal lesions and passive avoidance learning in the chick. *J. Neurosci.,* **8,** 4662–8.

Domjan, M. & Nash, S. (1988). Stimulus control of social behaviour in male Japanese quail. *Anim. Behav.,* **36,** 1006–15.

Dutch, J. (1969). Visual complexity and stimulus pacing in chicks. *Q. J. Exp. Psychol.,* **64,** 281–5.

Errington, M. L., Lynch, M. E. & Bliss, T. V. P. (1987). Long-term potentiation in the dentate gyrus: induction and increased glutamate release are blocked by D(-)aminophosponovalerate. *Neuroscience,* **20,** 279–84.

Fagg, G. E. & Matus, A. (1984). Selective association of N-methyl aspartate and quisqualate types of L-glutamate receptor with post-synaptic densities. *Proc. Natl. Acad. Sci., USA,* **81,** 6876–80.

Gaffan, D., Saunders, R. C., Gaffan, E. A., Harrison, S., Shields, C. & Owen, M. J. (1984). Effects of fornix transection upon associative memory in monkeys: role of the hippocampus in learned action. *Q. J. Exp. Psychol.,* **36B,** 173–221.

Gibson, E. J. (1969). *Principles of Perceptual Learning and Development.* New York: Appleton-Century-Crofts.

Goren, C. C., Sarty, M. & Wu, P. Y. K. (1975). Visual following and pattern discrimination of face-like stimuli by newborn infants. *Pediatrics,* **56,** 544–9.

Gottlieb, G. (1981). Roles of early experience in species-specific perceptual development. In *Development of Perception,* vol. 1, eds. R. N. Aslin, J. R. Alberts & M. R. Peterson, pp. 5–44. London: Academic Press.

Hebb, D. O. (1949). *The Organization of Behaviour.* New York: John Wiley & Sons.

Heinroth, O. (1911). Beitrage zur Biologie nahmentlich Ethologie und Psychologie der Anatiden. Verh. 5th int. orn. kong. Berlin., 589–702.

Hess, E. H. (1959). Imprinting. *Science,* **130,** 133–41.

Hinde, R. A. (1961). The establishment of parent–offspring relations in birds, with some mammalian analogies. In *Current Problems in Animal Behaviour,* ed. W. H. Thorpe & O. L. Zangwill, pp. 175–93. Cambridge: Cambridge University Press.

Hoffman, H. S. & Ratner, A. M. (1973). A reinforcement model of imprinting. *Psych. Rev.,* **80,** 527–44.

356   M. H. Johnson

Hoffman, H. S., Searle, J. L., Toffey, S. & Kozma, F. (1966). Behavioural control by an imprinted stimulus. *J. Exp. Anal. Behav.*, **9**, 177–89.

Horn, G. (1985). *Memory, Imprinting and the Brain.* Oxford: Claredon Press.

Horn, G. (1986). Imprinting, learning and memory. *Behav. Neurosci.* **100**, 825–32.

Horn, G., Bradley, P. & McCabe, B. J. (1985). Changes in the structure of synapses associated with learning. *J. Neurosci.*, **5**, 3161–8.

Horn, G. & Johnson, M. H. (1989). Memory systems in the chick: Dissociations and neuronal analysis. *Neuropsychologia*, **27**, (Special Issue: Memory), 1–22.

Horn, G. & McCabe, B. J. (1984). Predispositions and preferences. Effects on imprinting of lesions to the chick brain. *Anim. Behav.*, **32**, 288–92.

Horn, G., McCabe, B. J. & Bateson, P. P. G. (1979). An autoradiographic study of the chick brain after imprinting. *Brain Res.*, **168**, 361–73.

Horn, G., Rose, S. P. R. & Bateson, P. P. G. (1973). Monocular imprinting and regional incorporation of tritiated uracil into the brains of intact and 'split-brain' chicks. *Brain Res.*, **56**, 227–37.

Jaynes, J. (1956). Imprinting: the interaction of learned and innate behaviour. I. Development and generalisation. *J. Comp. Physiol. Psychol.*, **49**, 200–6.

Johnson, M. H. (1987). Brain maturation and the development of face recognition in early infancy. *Behav. Brain Res.*, **26**, 224.

Johnson, M. H. (1988). Memories of mother. *New Scientist*, **1600**, 60–2.

Johnson, M. H., Bolhuis, J. J. & Horn, G. (1985). Interaction between acquired preferences and developing predispositions during imprinting. *Anim. Behav.*, **33**, 1000–6.

Johnson, M. H., Davies, D. C. & Horn, G. (1989). A sensitive period for the development of filial preferences in dark-reared chicks. *Anim. Behav.*, **37**, 1044–6.

Johnson, M. H., Driurawiec, S., Ellis, H. D. & Morton, J. (1991). Newborns preferential tracking of face-like stimuli and its subsequent decline. *Cog. Devel.* (In Press).

Johnson, M. H. & Horn, G. (1986). Dissociation of recognition memory and associative learning by a restricted lesion of the chick forebrain. *Neuropsychologia*, **24**, 329–40.

Johnson, M. H. & Horn, G. (1987). The role of a restricted lesion of the chick forebrain in the recognition of individual conspecifics. *Behav. Brain Res.*, **23**, 269–75.

Johnson, M. H. & Horn, G. (1988). Development of filial preferences in dark-reared chicks. *Anim. Behav.*, **36**, 675–83.

Johnson, M. H. & Morton, J. (1991). *Biology and Cognitive Development. The Case of Face Recognition.* Oxford: Blackwells.

Klopfer, P. H. (1967). Stimulus preferences and imprinting. *Science*, **156**, 1394–96.

Klopfer, P. H. & Hailman, J. P. (1964a). Perceptual preferences and imprinting in chicks. *Science*, **145**, 1333–4.

Klopfer, P. H. & Hailman, J. P. (1964b). Basic parameters of following and imprinting in precocial birds. *Z. Tierpsychol.*, **21**, 755–62.

Kossut, M. & Rose, S. P. R. (1984). Differential 2-deoyxglucose uptake into chick brain structures during passive avoidance training. *Neuroscience*, **12**, 971–7.

Kohaska, S., Takamatsu, K., Aoki, E. & Tsukada, Y. (1979). Metabolic mapping of the chick brain after imprinting using [14C] 2-deoxyglucose technique. *Brain*

Res., **172**, 539–44.

Lee-Teng, E. & Sherman, M. S. (1966). Memory consolidation of one-trial learning in chicks. *Proc. Natl. Acad. Sci.*, *USA*, **56**, 926–31.

Lorenz, K. (1935). Der Kumpan in der Umwelt des Vogels. *J. Ornithol.*, **83**, 137–213, 289–413.

Lorenz, K. (1937). The companion in the bird's world. *Auk*, **54**, 245–73.

Macphail, E. M. (1982). *Brain and Intelligence in Vertebrates*. Oxford: Clarendon Press.

Maier, V. & Scheich, H. (1983). Acoustic imprinting leads to differential 2-deoxy-D-glucose uptake in the chick forebrain. *Proc. Natl. Acad. Sci.*, *USA*, **80**, 3860–4.

Malamut, B. L., Saunders, R. C. & Mishkin, M. (1984). Monkeys with combined amygdalo-hippocampal lesions succeed in object discrimination learning despite 24-hour intertrial intervals. *Behav. Neurosci.*, **98**, 759–69.

Maurer, D. (1985). Infants' perception of facedness. In *Social Perception in Infants*, ed. T. N. Field & N. Fox, pp. 73–100. New Jersey: Ablex.

McCabe, B. J., Cipolla-Neto, J., Horn, G. & Bateson, P. P. G. (1982). Amnesic effects of bilateral lesions placed in the hyperstriatum ventrale of the chick after imprinting. *Exp. Brain Res.*, **48**, 13–21.

McCabe, B. J. & Horn, G. (1988). Learning and memory: Regional changes in N-methyl-D-aspartate receptors in the chick brain. *Proc. Natl. Acad. Sci.*, *USA*, **85**, 2849–53.

McCabe, B. J., Horn, G. & Bateson, P. P. G. (1981). Effects of restricted lesions of the chick forebrain on the acquisition of preferences during imprining. *Brain Res.*, **205**, 29–37.

Morton, J. & Johnson, M. H. (1989). Four ways for faces to be 'special'. In *Handbook of Research on Face Processing*, ed. A. W. Young & H. D. Ellis, pp. 47–56. Amsterdam, North Holland.

Nelson, C. A. & Ludemann, P. M. (1989). Past, current, and future trends in infant face perception research. *Can. J. Psychol.*, **43**, 183–98.

Patterson, T. A., Alvarado, M. C., Warner, I. T., Bennet, E. L. & Rosenzweig, M. R. 1986. Memory stages and brain asymmetry in chick learning. *Behav. Neurosci.* **100**, 856–65.

Payne, J. K. & Horn, G. (1982). Differential effects of exposure to an imprinting stimulus on 'spontaneous' neuronal activity in two regions of the chick brain. *Brain Res.*, **232**, 191–3.

Payne, J. K. & Horn, G. (1984). Long-term consequences of exposure to an imprinting stimulus on spontaneous impulse activity in the chick brain. *Behav. Brain. Res.*, **13**, 155–62.

Salzen, E. A. & Meyer, C. C. (1968). Reversibility of imprinting. *J. Comp. Physiol.*, **66**, 269–75.

Sluckin, W. (1972). *Imprinting and Early Learning*, 2nd edn. Methuen.

Sluckin, W. & Salzen, E. A. (1961). Imprinting and perceptual learning. *Q. J. Exp. Psychol.*, **13**, 65–77.

Spalding, D. A. (1873). Instinct, with original observations on young animals. *Macmillan's Magazine*, **27**, 282–93.

Staddon, J. E. R. (1983). *Adaptive Behaviour and Learning*. Cambridge: Cambridge University Press.

Zeuner, F. E. (1963). *A History of Domesticated Animals*. London: Hutchinson.

13

The honey bee as a model kin recognition system

Wayne M. Getz

Recognition operates at many levels in biological organisms. At the suborganismal level, immune systems manufacture antibodies that are able to recognize and bind to foreign substances (antigens), thereby initiating a process that leads to antigen destruction. At the organismal level, individuals discriminate between objects in their environments as a function of the objects', say, nutritional value. At the population level, social structures are set up by individuals who are able to classify their conspecifics in terms of belonging to a particular group or class of individuals. If group structure is based on kinship between individuals then some type of kin recognition system is usually required to maintain the integrity of kin groups.

Recognition systems at all levels involve communication of information, whether the information is stored in the stereochemistry of molecules or the morphology of body features. In the simplest recognition systems, the messenger carrying the information is the object itself (e.g. an antigen) and the entity receiving the information executes the action (e.g. a lymphocyte). In more complex recognition processes, an object encodes a message in the form of a signal that is propagated by some physical (light, sound) or chemical transport (odour) process. Communication is completed when this signal is intercepted by a sensory system, decoded, and processed by the brain (an action may be initiated or the organism may decide not to respond). This definition is not limited to biological systems: it covers machines such as bar-code readers. The most complex recognition systems we know, however, are the interactions between biological organisms. We can only fully understand the evolution of social interactions between individuals if we integrate our knowledge relating to signal production, signal perception, and the actions that are consequently initiated.

In many ways honey bees provide an ideal system for the study of kin recognition. They are easy to manage and manipulate in a laboratory or apicultural setting. We can observe all phases of their relatively short life-history. They are highly social and, despite having a brain that is several orders of magnitude smaller (numbers of neurons and synapses) than the typical mammalian brain, they demonstrate the same attributes assigned to learning in vertebrates (Bitterman, 1988). The main problem we have, which holds to various degrees in all non-human studies, is thinking about how honey bees perceive their environment.

Our ability to recognize and hence categorize objects is both facilitated and limited by perception. We can only infer what other organisms are able to perceive by studying their perceptual systems. Many of these systems, such as echolocation in bats and heat-location in snakes, are very different from anything we can experience. Our lack of certain sensory modalities may lead us to underestimate the perceptual powers associated with them because they are quite literally beyond our ken. This caveat applies even more strongly to our understanding of the perceptual world of insects. We often tend to regard insects as genetically programmed automata lacking perceptual sophistica-tion. As demonstrated by Karl von Frisch and others, even though honey bees are neurologically much simpler than humans, the olfactory acuity of these two species is remarkably similar (von Frisch, 1967, Tables 43 and 44). Obviously we do not expect honey bees to have the same cognitive abilities as humans in associating odour stimuli with stimuli from other modalities. But we should expect chemosensory perception in honey bees to function in ways that we can hardly imagine.

Evidence presented below suggests that chemosensory perception is an important modality for conveying kin recognition information in honey bees and other social insects. Further, it is probably the primary modality, since no other modalities as yet have been implicated in kin recognition. Thus, a broad understanding of kin recognition in honey bees requires that we investigate both the ethological basis of kin discrimination and the mecha-nisms for producing and perceiving chemosensory stimuli.

I begin this chapter by providing an ethological analysis of kin recognition in the honey bee. I then develop a theoretical framework for the analysis of kin recognition as a communication system in any organism.[1] The examples I draw from, however, are all centred around the honey bee. In particular, in my analysis of the genetics of cue production, I focus on the hymenopteran (ants, bees, wasps) haplodiploid genetic system; and my discussion of olfaction centres around a typical insect system. Then, before concluding, I review a selection of empirical studies that have been undertaken to explore the question of kin recognition in honey bees.

---

[1] For the sake of completeness, I have included technical details which some readers may want to omit.

Ethology

Honey bee sociobiology

A typical colony of the honey bee, *Apis mellifera*, contains several tens of thousands of female workers, one female reproductive (the queen), and – depending on the time of year – from zero to several thousand male reproductives (drones). At certain times of the year, swarming or colony fission begins with the production of queens. Female larvae destined to become queens rather than workers are reared in larger queen cells and fed with a greater quantity of 'royal jelly', a highly nutritious secretion from the mandibular glands of workers. Under normal conditions, the old queen departs a few days prior to the emergence from the pupal stage of the first new queen, leaving with roughly one half of the adult workers in the hive. In a process that is only partially understood, one of the female reproductives becomes the new queen while often one or more of the remaining female reproductives leave with a small group of workers known as an afterswarm (the number of afterswarms is typically one or two, although zero to four are possible). However, if the ability of a laying queen to lay eggs has declined for some reason (e.g. due to injury or disease), then a process of supersedure takes place in which the old queen is eliminated and replaced by a newly-reared reproductive female. This is accomplished by the workers rearing several female reproductives, the first of which to emerge from the pupal stage usually stings the other female reproductives in their cells and becomes the new queen. Before a reproductive becomes a laying queen, she embarks on several mating flights over a 2–4 day period, during which time she typically mates with 10–20 males. The sperm from these matings are stored in her spermatheca, which is only large enough to contain 4–6 million sperm (Page, 1986). Since a single male can produce up to 10 million sperm on his own, after 10 or more matings a female has stored some partial mix of sperm from the various males. Males only mate once and die shortly after mating (their endophalli separate from their bodies during mating).[2]

Males in the insect order Hymenoptera (ants, bees, wasps, sawflies) develop from unfertilized haploid eggs and females develop from fertilized diploid eggs (fusion of egg and sperm). Thus, queens are able to control the sex of their progeny by either allowing the egg passing down the oviduct to be fertilized by sperm stored in the spermatheca, or by laying an unfertilized egg.

The haplodiploid genetics of Hymenoptera and the polyandrous (i.e. multiple) mating behaviour of honey bee queens, produces a well-defined kin

---

[2] Winston (1987) provides an excellent, comprehensive presentation of honey bee reproductive biology.

structure within the hive. First, males are haploid so all their sperm are genetically identical. Females are diploid. Consequently, the random segregation of alleles during mitosis results in eggs from the same female sharing, on average, half of the same alleles (i.e. alleles that are *identical by descent – ibd*). If there is no inbreeding, then haplodiploidy implies that sisters inherit the identical genome from their father and are related to each other by $r = \frac{3}{4}$ (in sisters, 3 out of 4 genes are *ibd*: 2 out|of 2 are *ibd* from the father and 1 out of 2 are *ibd* from the mother).[3] This contrasts with a value of $r = \frac{1}{2}$ between siblings in diploid species. Because haplodiploid sisters bear this close relationship they have been termed 'super sisters' (Page & Laidlaw, 1988), a terminology we will adopt to stress this special relationship. Another unusual feature of haplodiploidy is that mothers are related to their sons by $r = \frac{1}{2}$ (only half the mother's genes are carried by their sons), but sons are related to their mothers by $r = 1$ (all of a male's genes can be found in its mother).

Second, as a result of queen polyandry, the hive contains a number of patrilines.[4] Within patrilines workers are super sisters. Between patrilines workers have the same mother and different fathers, and are thus half sisters related by $r = \frac{1}{4}$. These relationships are illustrated in Fig. 13.1.

It was a consternation to Darwin to explain how non-reproductive or sterile female worker castes in the social insects could be maintained under natural selection. Darwin, of course, knew nothing of genetics, but he was able to foresee the importance of kinship in the spread of such highly altruistic traits as sterility in female workers. Fisher, and Haldane, two of the founders of quantitative genetics, provided us with further insights into this problem, but it was Hamilton (1964) who finally articulated the principle of inclusive fitness to explain the evolution of sterility in female workers, as well as many other types of ostensibly altruistic behaviour. An individual is able to contribute copies of its genes to the next generation's gene pool either by rearing its own progeny or by assisting in rearing the progeny of its relatives. For example, an individual can expect, on average, to contribute one copy of each of its own genes to the population gene pool if it rears, say, n individuals to each of whom it is related by $r = 1/n$ (this is just a consequence of the definition of r – see footnote 3).

Thus, an individual's inclusive fitness can be defined as the number of copies of its own genes that it contributes to the gene pool in the next generation through both direct reproduction and helping to rearing nonpro-

---

[3] We define r, the relatedness of an individual A to an individual B, as the expected (not actual) proportion of genes in A that are *ibd* to genes in B. As we see below, the relatedness of A to B is not always equal to the relatedness of B to A.

[4] Unlike the employment of this term in animal breeding, it is used here to emphasize that, in the honey bee colony, paternity determines a particular worker line against the genetic background provided by a common mother. Robinson & Page (1988) prefer the more general but less ambiguous term 'subfamily'.

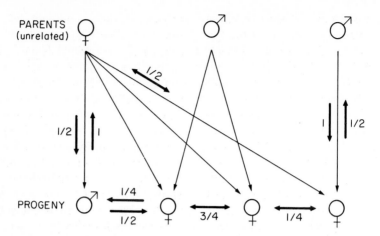

Fig. 13.1 Sex determination and relationships between individuals in haplodiploid genetic systems (see text for explanation).

geny kin.[5] Under haplodiploidy it is apparent that a female can contribute a greater proportion of her genes by rearing super sisters ($r = \frac{3}{4}$) than daughters ($r = \frac{1}{2}$) as reproductive females. Of course the situation is much more complicated than this because without some females having daughters no future generations are possible. Thus, we can only understand the elements of a particular reproductive life-history by analyzing the role of these elements in the process as a whole.

Research has shown that several patrilines of worker progeny are present in a colony at any one time (reviewed in Page, 1986). Thus the sperm from most of the males with whom a honey bee female mates must enter her spermatheca and mix to some extent. Consequently, most workers end up rearing female reproductives that are their half rather than their super sisters. All other things being equal, workers that invest in rearing super sisters will be three times more fit than workers that invest in rearing half sisters (since the genetic relationships involved are respectively $r = \frac{3}{4}$ and $r = \frac{1}{4}$). This suggests that in a population where workers are completely indifferent to whether they rear super or half sisters as queens, the introduction of a gene that biases workers to invest more effort in rearing super sisters would spread through the population, provided there are no detrimental effects from this gene to individuals or colonies. Worker nepotism with respect to rearing queens is bound to entail some cost to the reproductive efficiency of the colony as a whole. Ratnieks & Reeve (1990), however, have demonstrated

[5] There are several ways of looking at inclusive fitness, each with its own pitfalls – see Dawkins, 1982.

that if the cost is small, we can still expect nepotistic behaviour to arise and even co-exist with non-nepotistic genetic lines of workers.

Preferential rearing of super sisters, of course, requires that some mechanism exists for the workers to selectively channel their efforts. If the sperm were clumped in the spermatheca, then a temporal patriline structure would result. There is some temporal structure (Taber, 1955), but not enough to prevent the co-existence of several patrilines throughout the life of a particular queen. This leads to the question that motivates much of kin recognition research in honey bees. *Are workers in a multipatriline colony able to discriminate between super and half sisters, ultimately with regard to preferentially rearing super sisters as reproductives?* As we will discuss in more detail below, an affirmative answer requires a recognition system sophisticated enough to permit adult workers to categorize individuals – most likely at the egg or early larval stage – into super and half sisters.

The behaviour of individuals in a hive must always be evaluated within the ethos of a group of individuals co-operating to form a supra-organismic unit. This unit has evolved to solve resource and environmental problems that cannot be dealt with by individuals acting alone: that is, competition between patrilines is always within the framework of the interests of the colony as a whole. If the answer to our previous question is affirmative, then the next obvious question is: *why do queens mate with more than one male if competition between patrilines has the potential to precipitate disharmony in the functioning of the hive?*

Several hypotheses have been proposed to answer this question (Page, 1980; Page & Metcalf, 1982; Crozier & Page, 1985; Moritz, 1985; Ratnieks, 1990 a). Several of these relate to the sex determining mechanism in honey bees and the production of diploid males which, if reared as adults, are essentially infertile. In the hive, workers eat diploid drones within a day or so of their eclosion from the egg stage. Sex determination in honey bees is known to be under the control of a single locus. Heterozygosity at this locus is required for the development of a female. Lack of heterozygosity, which is homozygosity in a fertilized egg (diploid zygote) or hemizygosity in an unfertilized egg (haploid zygote) results in the development of a male. If there are n alleles associated with this sex determining locus (estimates range from 6 to 18), then a female who has mated once has a $1/n$ chance that half of her fertilized eggs are inviable diploid males. Thus, by mating many times each female reduces the risk that a substantial portion of her eggs may be inviable.

Moritz (1985) has argued that multiple mating reduces the conflict between workers and queens in the amount of effort invested in rearing male versus female reproductives. Further, recent research (Calderone & Page, 1988; Frumhoff & Baker, 1988; Robinson & Page, 1988) also indicates that across the various subfamilies that occur as a consequence of polyandry, individuals in certain genetic lines are more predisposed than others to performing one or more of various worker tasks including pollen collecting, cleaning, guarding, and undertaking (removing dead bees). Therefore, an increase in the genetic variability of a colony may correlate with an overall increase in the growth rate and probability of survival of that colony. An increase in the genetic variability that is a consequence of multiple matings may also increase a colonies resistance to disease (Hamilton, 1987; Sherman *et al.*, 1988).

The preceding discussion has focused on the most demanding recognition problem that an individual may deal with in the colony: discriminating between super and half sisters using genetic variability alone. A simpler task, because both environmental and genetic cues can be used, is the problem that guard bees face in discriminating between members and non-members of their hive. A major task of the guard bee is to keep out strange bees intent on robbing their hive. Robbing generally occurs when nectar resources are scarce. There may be many cues that allow guards to recognize a robber, including the way it approaches the hive and unfamiliar environmental odours (workers pick up the odours of the flowers they forage on and these, together with genetic components, become part of a colony gestalt odour which emerges as bees rub and groom one another). Finally, workers have to defend their hive against the attacks of other insects, including wasps and ants, and workers need to be equipped to recognize invaders.

Another important aspect of kin recognition in honey bees, discussed in some detail below, is their ability to discriminate between queens on the basis of relatedness. This ability could enable workers to select between several virgins that may emerge within a short period of each other during preparations for swarming or supersedure.

Comparative perspective

The various chapters in this book discuss kin recognition in a number of organisms, including ants and wasps. In the literature there are detailed reviews of kin recognition (Gadagkar, 1985) in primitively eusocial bees and wasps (Gamboa *et al.*, 1986a; Michener & Smith, 1987), and in highly eusocial bees and ants (Breed & Bennett, 1987; Page & Breed, 1987).[6] Here we only provide a comparative perspective on the different kind of kin recognition problems that are dealt with by ants, bees, and wasps.

Kin recognition in bees has been intensively studied in the honey bee, *Apis mellifera*, and the sweat bee, *Lasioglossum zephyrum*. In honey bees, recognition processes at the nestmate level facilitate colony defence, while recogni-

---

[6] Eusocial refers to those species in which there are at least a behavioural, if not a morphologically distinct, worker caste of individuals that help care for the young, and at least two generations of workers overlap within the same colony or nest at some point in time (Wilson, 1971). In bees, if the colonies are perennial, the species is termed highly as opposed to primitively eusocial (Michener, 1974).

tion processes at the patriline level facilitate preferential rearing of super sisters as reproductives. Recognition in the sweat bee has been observed in the contexts of mate selection by males, and nest defence by guard bees.

The work of Michener's group at the University of Kansas (reviewed in Michener & Smith, 1987) shows that female sweat bees each possess a unique genetically-correlated odour that males use to reduce inbreeding. Discrimination of kin in the context of inbreeding avoidance behaviour is known to occur in other insects including non-social insects like *Drosophila* (Spiess, 1987). More interesting in terms of kin recognition is the behaviour of the guard bees at the entrance to sweat bee nests. *L. zephyrum* nests, typically containing 2 to 20 females, are burrows usually located in earthen river banks in high density aggregations. In the larger nests, one female is queen, a few act as the principal guards, and the remaining females forage, work on nest construction or are inactive. Possibly due to the high density of nests in the nest aggregations, females often attempt to enter nests other than their own. In some species, however, females could also attempt to parasitize or usurp unguarded nests of conspecifics. Since the entrance to these nests is not much more than one bee-width in diameter, the entrance can be effectively blocked by a guard. Further, if a foreign bee does happen to enter a nest, then it is often attacked within the nest. Studies by Greenberg (1979), and Buckle & Greenberg (1981) convincingly demonstrate the role of individual learning and genetically correlated cues in discriminating foreign females.

Besides sweat bees, kin recognition has been intensively studied in only one other group of primitively eusocial insects: paper wasps of the genus *Polistes*, in temperate climates. The results discussed below are reviewed in Gamboa *et al.* (1986a). Studies have been conducted on kin recognition in workers and males, although most of the studies have centred around kin recognition in spring foundresses and gynes. Gynes are potential queens produced in colonies in late summer to found new nests in the spring, when they are referred to as spring foundresses. Gynes overwinter in small clusters on or near their natal nests. New nests are established by lone foundresses or by two or more individuals, usually sisters. Since foundresses establish new nests close to their natal nest, it was thought initially that philopatry may explain the fact that most cofoundresses are sisters. Experiments in the laboratory, however, have established that spring foundresses can discriminate between sisters and non-sisters using individually-borne recognition cues, and that foundresses are more tolerant of unfamiliar sisters joining their groups than unrelated strangers. Further experiments with spring foundresses, involving switching the brood between pairs of individuals, have established that they are able to discriminate between related and unrelated brood, although the cues themselves may have come from familiarity with the brood comb rather than the brood itself. The data from these experiments also suggests that cues for brood discrimination have a heritable component (even though the cues become embedded in the comb) that is learned by the gynes (Michener & Smith, 1987).

Studies involving gynes indicate that after exposure to their natal nest for a couple of hours after eclosion from the pupal stage, gynes of ages 1 to 3 days are able to discriminate between nestmates (sisters) and non-nestmates (Gamboa *et al.*, 1986a), and are more tolerant of their nestmates. This greater tolerance towards nestmates than non-nestmates has been identified in several species of *Polistes*, as well as the baldfaced hornet, *Dolichovespula maculata*. In a 'reciprocal nest exposure' study, Gamboa *et al.* (1986b) were able to demonstrate that the tolerance between *P. fuscatus* gynes involved discrimination of cues that have both a heritable and environmental component. Interestingly, the heritable component could be overridden by the environmental component, and the components were not additive in their effects (i.e. tolerance was no greater when both components were used for discrimination than if either component was used for discrimination).

All of the more than 10 000 species of ants (family Formicidae) are eusocial and the ability of workers to discriminate between nestmates and non-nestmates is widespread (Breed & Bennett, 1987; Carlin, 1988; Jaisson, this volume). The kin structure of nests or colonies varies considerably within Formicidae as a function of the various nest-founding and queen-mating strategies of the different species. Colonies may be monogynous (single queen) or polygynous (many queens). Queens may mate once or many times or, in some species, the colony may even reproduce asexually. Colonies may be founded by one (haplometrosis) or several queens (pleometrosis). In polygynous colonies queens are not necessarily closely related. Thus the intranidal (within nest) relatedness between workers in various species of ants can range from $r = 0.75$ in monogynous single-inseminated-queen nests (e.g. the fire ant, *Solenopsis invicta* (Ross & Fletcher, 1985)), to the observed $r \approx 0.1$ in the Australian ant, *Rhytidoponera mayri* (Crozier *et al.*, 1984). Further, ant colonies exist that contain thousands of queens, millions of workers, and are divided into discrete nests that regularly exchange workers and are connected by odour trails. Examples of such colonies, which are referred to as polydomous colonies, are the Argentine ant, *Iridomyrmex humilis*, and Pharaoh's ant, *Monomorium pharaonis*.

Kin recognition in ants has been most often studied in the context of nest defence. If internidal relatedness is high, or if the environment of conspecifics is uniform (as in Acacia ants – *Pseudomyrmex venefica*), then heritable odours or chemicals residing on the worker's cuticle may provide nestmate recognition cues. Otherwise, environmental cues may be required for nestmate recognition. In small monogynous colonies of the carpenter ant (*Camponotus* spp.), chemicals secreted by the queen appear to be used as nestmate recognition cues (Carlin & Hölldobler, 1986), while in larger colonies of this same group these cues are affected by the relatedness of the workers (Carlin & Hölldobler, 1987). Many species of ants are known to raid the colonies of their conspecifics, congenerics, as well as other species of ants. Nestmate recognition is essential to thwart a raid from, say, a neighbouring colony of conspecifics. Clashes between members of the different colonies can be extremely violent, sometimes leading to the deaths of many thousands of individuals, as in observed clashes between polydomous colonies of the European red wood ant, *Formica polyctena* (Driessen *et al.*, 1984). It appears the nestmate discrimination can be corrupted when one group of ants captures an immature group of individuals (usually eggs and larvae) and imprint their own recognition cues on these individuals. This is the case in the dulotic (slave maker) species *F. sanguinea* which enslaves its congenerics, *F. fusca* and *F. rufibarbis*. Of course, kin recognition in ants may take place in a number of contexts other than nestmate recognition. As in honey bees, this may include worker acceptance of queens and, as in honey bees and wasps, may include worker rejection of brood in cross-fostering experiments (for detailed reviews see Breed & Bennett, 1987; Carlin, 1988; Jaisson, this volume).

Evolutionary perspective

From our preceding discussion it is clear that kin recognition can enhance the fitness of individuals in a number of ways. First, reproductives can use it to promote outbreeding and avoid the loss of fitness that comes with inbreeding depression. Second, nestmate recognition is required for nest defence, without which the investment of individuals in the resources of the nest can be captured or reduced by outsiders. Third, ostensibly altruistic acts towards kin can, as we have discussed, increase the inclusive fitness of an altruist. Thus, altruism can flourish in incipient social groups built around the family, and can provide the warp for weaving the fabric of complex social systems. Through co-operation and partitioning of tasks, social species may more efficiently exploit their environment than less social species (e.g. humans). The complexity of social systems is both facilitated and limited by the ability of individuals to categorize other individuals as belonging to some group. If the groups are each of size one, we call the phenomenon individual recognition. Otherwise the categorization is at some coarser level: either siblings, say, nestmates or conspecifics.

The evolution from one level of complexity (e.g. rearing one's own young) to a greater level of complexity (e.g. the existence of a nurse caste in honey bee colonies) can only take place if the preadaptive structures are in place (e.g. communal nesting of sisters). Human society could not evolve to form tribal or larger units, without an appropriate system of communication between individuals. The communication system itself, however, evolves with the society: in humans the evolution of complex speech patterns, leading to the written word and ultimately to electronic communication, has been the key to the success of humans in developing technologies that allow them to exist at greater population densities. Insect societies also have used communication systems to expand the carrying capacity of their environments. For example, ants make extensive use of trail pheromones to guide workers to food, while honey bees, being aerial, have evolved the waggle dance for this purpose.[7]

Other forms of communication are equally important. For a honey bee colony to operate efficiently, at least some of the foragers need to know whether they should collect pollen, propolis or water, rather than nectar. There are so many types of worker interactions going on in the hive – foragers interacting with food handlers, foragers interacting with scouts, workers grooming and feeding (trophallaxis) one another, and nurse bees attending the brood and the queen – it is not inconceivable that workers are able to categorize individuals by the tasks they are currently performing. Certainly it would be inefficient for two returning foragers to try and off-load their cargo onto one another.

The most crucial communication between individuals in species of solitary Hymenoptera relates to finding a suitable mate. Olfaction in particular plays a central role in locating and accepting individuals as mates. Chemosensory systems, however, also mediate behaviours associated with the selection of food, as well as agonistic interactions between conspecifics (e.g. sweat bees determine dominance hierarchies based on a female's ability to secrete chemicals from the dufours gland – Smith & Weller, 1989). Thus chemosensory systems are well-developed in all the solitary and social species of Hymenoptera.

As a consequence of genetic variation for the production of enzymes and other biochemical products, there is a variation in the physiology and chemistry of all genetically distinct individuals. This preadaptive variation can provide a basis for kin recognition, initially in the context of mate selection. Obviously, a genetic component is not required if individuals avoid mating with nestmates. In wasps that do not construct nests or the equivalent of nests (e.g. nongregarious parasitoids) a genetic component may be necessary if mating with close kin is to be avoided. As in honey bees, it is known that sex in numerous species of hymenopterans are determined respectively by the heterozygosity and homo-/hemizygosity of a sex determining locus (Page & Metcalf, 1982). If, as in the honey bee, diploid males are relatively infertile, then one can expect mechanisms to evolve to promote outbreeding. (Note, for example, that the expected number of diploid male progeny is – depending on the number of sex determining alleles in the population – greater than 25 per cent in the case of sib-matings – see Page, 1980.) Thus selection could act to increase heritable variation in chemicals

---

[7] For a discussion of the evolution of the dance language see Gould & Gould, 1988.

that can be used to signal kinship in the mate avoidance problem. If individuals emerge in isolation, then individuals would only be able to recognize their kin if either they have an innate ability to do so or, more likely, they learn their own odour and innately avoid mating with individuals that have odours similar to their own. Of course, if individuals have nestmates, then they can learn the odour of their nestmates and avoid mating with nestmates or avoid mating with individuals that appear to have a similar odour to one or more of their nestmates. (Mechanisms for comparing the similarity of odours are discussed below.)

Once recognition of individuals, or more simply a process for classifying individuals into groups, exists in one context, it provides the preadaptive structure for the evolution of recognition in other contexts. All that is necessary is that there is a selective advantage surrounding new modes of behaviour which might arise because of a recognition system that is already in place. For example, individuals in an incipient quasisocial species (members of the same generation use the same composite nest and also co-operate in brood care – Wilson, 1971) could evolve to use a mate-avoidance recognition system as a sibling recognition system that promotes the formation of sibling-only nests. The same recognition system could then be used as a basis for the evolution of nest defence behaviour. In this way, a nestmate recognition system 'piggy backs' (to use the terminology of Crozier, 1987) on a mate-avoidance recognition system.

Using a simple diallelic one-locus genetic model, Crozier (1987) demonstrates that an agonistic kin recognition system cannot evolve if recognition entails some cost to individuals (e.g. if recognition leads to conflicts between individuals). Thus, the notion of the evolution of recognition in an agonistic context, piggy backing on other recognition systems, may be an important one. Crozier, unlike Ratnieks (1990b) however, did not consider the evolution of recognition systems under co-operative behaviour, which is another side of the kin recognition coin that itself can promote the evolution of both the labelling (Ratnieks, 1990b) and perceptual systems associated with kin recognition.

Theoretical framework

Communications system overview

Kin recognition in its broadest sense is a communication system (Fig. 13.2). This communication involves three phases: the encoding, transmission and decoding of a signal containing the kinship information. The individual encoding the information will be referred to as the *initiator*, although initiation of a signal is often passive in the sense that an individual possess a permanent olfactory or visual *signature* that can be read at will by

ENCODE   TRANSMIT & RECEIVE   DECODE   ACT

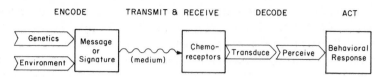

Fig. 13.2 Kin recognition in social insects viewed as a communication system.

any individual who cares to do so.[8] The individual reading or receiving the signal will be referred to as the *recipient*.

The signature itself can be regarded as a point in the space of all possible signatures. In social insects, the evidence suggests that signatures associated with kin and/or nestmate recognition (there may be more than one signature) are coded, at least in part, as a chemical signal. In a general sense, let us regard this chemical signal as a point in an *odour source space* (*OSS*). If the signal is volatile, then it will be carried in an air plume emanating from the initiator. If it is then picked up by a recipient who decodes and interprets the information, the recipient will then act in response to the 'meaning' of the information. Note that if after receiving, decoding and interpreting a signal, the recipient does nothing, then the recipient's choice of response is 'do nothing.' Of course, if a recipient 'does nothing' it is difficult for an observer to assess whether or not the signal actually 'meant' anything to the recipient.

Complex organisms receive signals using sensory devices that transduce the signal into forms that can trigger the firing of peripheral sensory neurons. Simple organisms without neural systems, however, can exhibit chemotactic or phototactic responses. Here we only consider the more complex situation of signal decoding by an organism's neural system. In arthropods and verte-brates the recipient's peripheral neurons respond to the signal by trans-forming or transducing them it into a form that can be processed by memory or other centres of their brain. Specifically, if an odour $\boldsymbol{\omega} = (\omega_1, \ldots, \omega_n)'$ is regarded as a point in an n-dimensional OSS, (i.e. ω_i is the concentration of the ith chemical in the OSS and $'$ denotes that the vector is a column rather than a row) then the transduced signal, transformed by sensory and other neurons in the olfactory lobe of the recipient, can be regarded as a point $\mathbf{r} = (r_i, \ldots, r_m)'$ in an m-dimensional *sensory response space* (*SRS*). In higher organisms, including insects, the transduced signal $\mathbf{r}$ is transmitted from the olfactory lobe to higher centres of the brain where, after further transforma-tion, it is processed in memory.

Associated with memory are neural structures performing learning tasks

---

[8] Many ethologists only designate systems as communication systems if signals are intermittent and under the active control of the initiator. See Waldman (1986) for a discussion of this point.

that facilitate the appropriate response to the particular signal on hand. Of course memory can also include species 'memory' that has evolved to become innately 'wired' into the brain of all individuals of that species. For example, honey bees innately respond to their alarm pheromone, isopentyl acetate, by extending their sting lancets. Similarly, they innately extend their proboscides when a sucrose solution is applied to their antennae.

Encoding the message

Each honey bee has a set of its own unique signatures that can be read by other honey bees (see experiments discussed in next section). This does not necessarily imply that recipients of the signals associate signatures with particular individuals. It only implies that they can classify each signature according to some criterion such as whether or not they belong to a particular group. Of course, the queen is a special individual and workers do respond differently to her than to all other individuals in the hive.

An olfactory signature can be simple or very complex, depending on the function it performs and the elements that influence its composition. Perhaps the simplest signals are sex attractants. They advertise the presence of a sexually receptive individual of a particular sex. In the silk moth, *Bombyx mori*, the species specific sex attractant is the compound bombykol produced by the female. The males are exquisitely sensitive to bombykol, probably being able to respond to stimulation from one to two hundred molecules (Boeckh, 1984). In honey bees, queens seem to produce several substances that attract males and stimulate them to mate (Winston, 1987). These and other pheromones, primarily produced by the mandibular glands, also influence the behaviour of workers, including the suppression of queen rearing, or the attraction of workers during swarming. A chemical analysis of the mandibular gland secretion reveals that each queen's profile is unique and, perhaps, provides a basis for observed worker recognition of queens and the sisters of those queens (see next section), although other glands may be involved (Moritz & Crewe, 1988a,b).

Workers, on the other hand, use the highly volatile pheromones released from their Nasanov glands to attract other workers and help orient them to the nest entrance. Although the volatile odours emanating from workers or groups of workers have sufficient heritable variability to provide a signature for kin recognition purposes (Getz et al., 1986), nonvolatile chemosensory stimuli might provide a more robust system for recognition within the hive. Volatile chemicals would distribute themselves within the hive, possibly confusing signals (Moritz, 1988) and inducing habituation of the sensory receptors to stimulation by the chemicals involved. On the other hand, nonvolatile chemicals such as waxes found on the cuticles of individuals, can be selectively sampled when individuals meet and antennate one another. Sufficient heritable variation exists in the chemicals found on the surface of workers to be used as a signature for discrimination between super and half sisters (Moritz & Southwick, 1987; Moritz, 1988), but the story is much more complicated.

First, the composition of these chemicals changes with the age of individuals (Breed *et al.*, 1985). Second, some of these chemicals are transferred between individuals during grooming, trophallaxis and other contact behaviour (Getz & Smith, 1986). Third, one should expect that the chemistry of individuals is affected by nutrition. (Not much is known about the synthesis of long chained hydrocarbons that form the basis of the cuticular chemistry, but it seems clear that the relative rates of production of various chemicals depends on the relative proportion of the constituent elements provided by different diets.) Fourth, there is bound to be a genetic component in cuticular chemical composition due to differences in the function of different isozymes in genetically distinct individuals. Finally there are effects of chemical contaminants picked up directly from the environment, such as plant waxes and floral odours permeating the cuticle (Breed *et al.*, 1988).

When it comes to colony defence, a coarse signature of both genetic and environmental factors might provide a signal for discrimination of nestmates. For discriminating between super and half sisters, however, a more refined and solely genetically determined signature is required.

The notion that a chemical signal might provide anonymity at one level, but specificity at another, has been explored by Hölldobler & Carlin (1987) in the context of social insects. For example, a volatile signal used to recognize nestmates, may contain no sibship information. The classification of signals as specific or anonymous, however, may be due as much to the sensitivity and characteristics of the decoding mechanism as to the uniformity of the signal that is encoded by individuals belonging to a particular group. We will make these notions precise by returning to our concept of an odour source space (*OSS*).

Suppose there are n chemicals, labelled $1,. . .,n$, that could potentially contribute towards a particular type of recognition signature. Clearly, an upper bound to n is the maximum number of chemicals to which at least one chemosensory receptor responds. This value of n might be extraordinarily large, but it can be reduced considerably if we focus on a particular system, such as cuticular waxes or Nasanov gland components.

To be specific, let $\omega_j = (\omega_{1j},. . .,\omega_{nj})'$ be the signature of the jth individual, where ω_{ij} is the absolute concentration of chemical i in the recognition odour phenotype of individual j. If there are k individuals in a population, then the points $\omega_1,. . .,\omega_n$ are distributed in some fashion throughout an n-dimensional odour space Ω (technically, Ω can be viewed as the positive quadrant of the n-dimensional Euclidean vector space R^n with the distance between points denoting some measure of signature similarity). The distribution of these k points could be regarded as a realization of some underlying

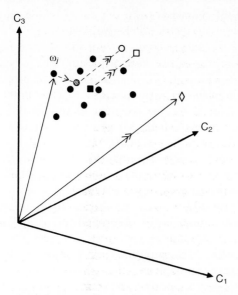

Fig. 13.3 Individual signatures in a 3-dimensional (components C_1, C_2 and C_3) kin recognition odour source space (*OSS*). The signature of each individual is denoted by a solid circle. The signature of a special individual, say the queen, is denoted by the open diamond. The mean of the solid circles is denoted by a solid square. The dotted arrow indicates how a gestalt effect (transference of chemicals between individuals) would move the *j*th individual's signature ω_j to the shaded circle, in the direction of the mean. The open circle and open square indicate where ω_j and the population mean respectively would move to if a certain amount of the queen's signature were added to the signature of each worker.

probability distribution which is determined by environmental factors, the genetics of the group, and any other factors – such as the transference of chemicals – which may affect the signature of an individual (Fig. 13.3).

In Fig. 13.3 it is apparent that the transference of labels among individuals has the effect of making each individual more like the mean, thereby reducing the amount of variation associated with the data. The addition of certain amount of the queen's signature, say, to each individual would move all the points by the same fixed amount, but keep the same absolute level of variation (as indicated in Fig. 13.3 by the movement of the mean). Since the mean is now further away from the origin (addition of two positive directional vectors), however, the relative amount of variation is reduced. Hence a gestalt effect that arises from label transference between workers is due to an absolute reduction in the variation among signatures (assuming an equal transference on average between any two workers), while a gestalt effect due to the addition of chemicals from the queen is due only to a reduction in the relative amount of variation between workers.

The question of whether a signal represents an individual or the membership of an individual within a group is more a question of the how well recipients can resolve points within an OSS than some intrinsic property of the OSS itself. Gestalt phenomena move points closer together in the OSS, thereby making it more difficult for the recipient to resolve differences between the signals from two individuals. On the other hand, repeated exposure to a number of very similar signals may eventually enable a recipient to discriminate between them. Humans, for example, often have difficulty distinguishing between identical twins whom they don't know very well; but the parents of the twins have no problem at all.

As discussed in the empirical studies below, gas chromotography is used to characterize signature distributions of both workers and queens, and provide some estimate of the variation among the signatures. The question of how well workers are able to perceive this variation, and discriminate between neighbouring points, requires a thorough investigation of olfaction in honey bees.

Heritability experiments could be used to tease apart the genetic and environmental components of a distribution of signatures in the odour space; but note that individuals from different patrilines within the same colony belong to different genetic distributions. At a broader level, individuals in different colonies belong to a combination of different genetic and environmental distributions. Whenever the distributions of two groups overlap, individuals may be categorized in the wrong group with a misclassification probability relating to the degree of overlap of the two distributions (see Getz, 1981). Further, the colony distributions may wander through the odour space Ω, as the environmental contribution changes through time; but, within that broader distribution the relative position of the patriline distributions may remain fairly constant.

The genetic components of the worker signature distributions obviously depend on the underlying distribution of alleles that effect the synthesis of each individual's signature. It is not too difficult to formulate a model for constructing the distribution of the number of alleles that are shared by individuals bearing a certain relationship to one another. This has been done for super-, full-, and half-sister relationships in both diplodiploid and haplodiploid genetic systems. The extent to which the distributions overlap can be visualized using a kingram (Getz, 1981), as illustrated in Fig. 13.4.

However, this does not provide specific information on how the distributions of signatures of various groups will overlap in the OSS unless we know how each allele contributes to the amount of each compound i, $i = 1,\ldots,n$ in the signature. Some alleles may code for enzymes involved in the synthesis of several compounds, different alleles may code for different isozymes with

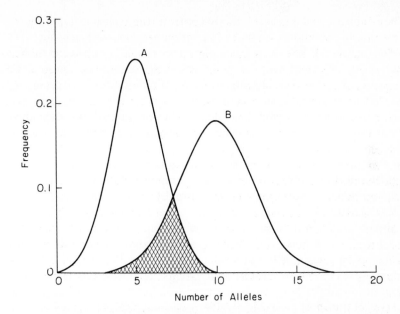

Fig. 13.4 Frequency distributions in a diplodiploid system for the number of alleles shared at 10 signature labelling loci by sisters (A) and half sisters (B) if both the parents are heterozygous and do not share any alleles at these loci. If sister and half sister encounters are equally likely, and individuals select the most probable relationship for the identified number of shared alleles, then the area of overlap of the two distributions (shaded) represents twice the probability that a misclassification will be made. (After Getz, 1981).

different activity rates, and a number of nonlinear (nonadditive) effects may occur. Of course, there is some correlation between the overlap of the distribution of alleles shared by individuals in two different kin groups and the overlap of their signature distributions in the *OSS*. Using this fact, kingram analysis of patriline overlap in diplodiploid and haplodiploid genetic systems indicates that haplodiploid systems are vastly superior to diplodiploid systems when it comes to separating out colony patrilines (Getz, 1981; cf. Fig. 13.4).

Transmitting the message
Problems associated with transmitting a message are related to the physical distance between individuals. If the message is contained in cuticular surface chemicals and is transmitted during antennal contact, then the primary problem is to obtain an adequate and representative sample of chemicals that make up the message. Workers can improve the homogeneity

of the chemicals on their cuticle through grooming, but chemicals are able to diffuse along the cuticle to make the mixture more homogeneous over different regions of the cuticle. It is known, however, that a glass rod rubbed in the region of the wax producing glands of a worker is perceived to be different from a glass rod rubbed on the thorax of that same bee (Getz *et al.*, 1988). Hence, if a chemical kin recognition signature resides on the cuticle of an individual, it may reside only on a particular part of that individual (humans recognize other humans best by looking at their faces and not at their hands).

The transmission of volatile signatures over relatively large distances poses a number of problems. According to a simple diffusion model (Bossert & Wilson, 1963), concentration falls off rapidly with distance; and, after a while concentration is too low to provide a meaningful message (the exact distance depends on the sensitivity of the recipient's sensory system). Also the signal become noisy with distance as it becomes contaminated with extraneous chemicals in the atmosphere and perhaps modified if it is a complex mixture of chemicals with components that diffuse at different rates. The diffusion model, however, is known to be inappropriate for the process of chemical transportation (Murlis, 1986). More typical is the phenomenon of 'packets' of chemicals being transported in an odour plume that travels at a rate that is relatively faster than the chemicals can diffuse. Recipients then read the message intermittently. Further, it is known that the olfactory system actually performs better under this 'flickering' process since the neurons are able to reset themselves in between stimuli (Baker, 1986).

Decoding the message

An olfactory signal essentially contains two types of information: the quality (relative proportions of different chemicals comprising the signal) and the quantity of the signal. Quantity is important in the sense that the concentration of chemicals must be sufficiently large so that neural stimulus thresholds are exceeded, but not so large that the sensory system is overloaded beyond the point where it can function properly. In some contexts, concentration or amounts may contain essential information. Such information enables female sweat bees, for example, to assess the size of an aggressor in a confrontation for possession of a nest (Smith & Weller, 1989).

In many contexts, however, concentration changes with distance or other spurious physical parameters, such as the length of time that the recipient antennae touch the cuticle of an initiator. In these circumstances it is only odour quality that carries the essential information. In our n-dimensional odour space, concentration of an odour ω_j is represented by the length of the vector connecting the point ω_j and origin (see Fig. 13.3). Quality is represented by the $n-1$ angles that define the direction of this vector in the positive quadrant of R^n. Thus, the most efficient perceptual systems are those that are able to precisely measure the angle (quality) and the length (quantity) of the vector associated with any point in the *OSS*; the power of such systems depends on their ability to resolve points that lie close together, especially those whose direction vectors are close (i.e. close in quality). The primary constraint to developing a powerful kin recognition perceptual system is that the same chemosensory system is used to perceive other types of odours including floral odours, and alarm and aggregation pheromones. Since there are many more chemical compounds in a honey bee's environment than there are sensory cells in a honey bees antennae, we can expect that only relatively few odours will be accurately perceived. Further, it would be highly unreliable for a honey bee to develop a unique sensory cell for each type of compound, because individual cells may die and single sensory cells cannot efficiently sample the environment. Also individual neurons may not fire in a strictly repeatable fashion because of sensory adaptation, and also they are susceptible to 'noise' in their environment (extraneous electrical activity from surrounding or synapsing neurons). These sampling and noise-to-signal ratio problems can only be solved by distilling out the average signal from an ensemble of neurons.

In both arthropods and vertebrates each individual olfactory receptor may be narrowly or broadly tuned. Narrowly tuned neurons typically respond to critically important chemical signals, such as sex pheromones. These receptors have given rise to the idea of 'labelled lines' or neurons that carry information regarding specific compounds from the peripheral to the central nervous system. Most olfactory receptors, however, respond to a class of compounds. A receptor may respond best to a particular compound but also less well to a related compound (response is measured in terms of the rate that voltage spikes are generated by these cells). Such receptors have been found in the American cockroach, *Periplaneta americana* (Fig. 13.5).

All of the 100 000 or so olfactory receptors (sensory cells) in the honey bee antennae project axons that converge on the order of 100 glomeruli in the olfactory lobe of the honey bee brain. From these glomeruli, a relatively small number of relay neurons emerge and carry processed olfactory information to higher centres of the brain. We assumed that the 'across fibre' firing pattern in some subset m of these relay neurons represents the quality and concentration of the odour stimulus; that is, we let the rate of firing of each of these m relay neurons define the points $\mathbf{r}$ in our *SRS*. The actual quality of the odour ω may be represented by the normalized vector $\rho = \mathbf{r}/\|\mathbf{r}\|$, and the amount (concentration) by the quantity $\|\mathbf{r}\|$ (see Getz & Chapman, 1987). Thus, the peripheral olfactory neurons, in conjunction with the glomeruli of the olfactory lobe, extract odour quality information in a process that mathematically can be represented by an odour quality mapping Q of the form (also see Fig. 13.6).

$$\rho = Q(\omega). \tag{1}$$

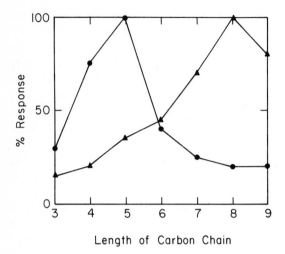

Length of Carbon Chain

Fig. 13.5 The response of two types of olfactory cells in the cockroach, *Periplaneta americana*, expressed as a percentage of the maximum excitation level, to standard concentrations of different aliphatic alcohols (labelled by the length of their carbon chains). (After Sass, 1976 and reprinted from Getz & Chapman, 1987).

The lower dimensionality of the *SRS* means that sections of subspaces of the *OSS* will be mapped onto single points in the *SRS*, leading to a loss of resolution of odour quality. However, if we select two odours at random from the *OSS*, the probability is zero that they lie in a subspace that is 'collapsed' by the map Q, although points that are comparatively distant in the *OSS* may project to be very close together in the *SRS* (e.g. points B and E in Fig. 13.6 respectively map onto points G and H). Once the *OSS* has been collapsed onto the *SRS*, we still remain with the problem that noise in the neural network system, as well the finite capacity of memory systems, limit an organism's ability to distinguish between points that lie close together in the *SRS*. Thus points that are close together might be perceived as being the same. This limitation is not always a drawback since small fluctuations in an individual's odour signature should be identified by a recipient as belonging to the same individual. Thus, a certain lack of resolution is required for robustness in identifying signatures in a noisy world.

Note that within the limits of the thresholds of the neural functions, the map $Q\omega$ defined in (Eq. 1) is actually a combination of a linear operator T ($n \times m$ matrix) and a normalization procedure (for further discussion see Getz & Chapman, 1987).

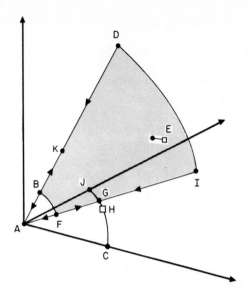

Fig. 13.6 A representation of the odor quality map Q when $n = 3$ (OSS) and $m = 2$ (SRS). All points ω lying on the sheet BDFI when normalized will lie on an arc joining K and G, while a projection of points onto the base plane will project all points on arc KG onto point G. Points lying on the sheet ABF are not sensed because they are below the concentration threshold of the system. Points lying on the extension of the sheet BDFI beyond arc DI are not mapped onto G, but are mapped elsewhere onto the arc JC because of distortions due to overloading the sensory neurons with high concentrations of odours. This may cause the quality ρ of these points to be misclassified (there may also be misclassification problems with points lying at concentrations just around the lower firing threshold of the sensory neurons). The point E, because it is close to sheet BDFI, will be projected by mapping Q onto the point H close to G. Thus the points B and E are mapped onto points that are close in the SRS, even though B and E are relatively further away in the OSS.

Perception and discrimination thresholds

The signal carried by the relay neurons is sent to other centres of the brain (e.g. the mushroom bodies in the bee brain) where memory and other perceptual processes take place. Memory processes in higher organisms are complicated and known to exhibit a number of phases, including a short labile phase during which memory is easily disrupted, and a consolidation phase where memory is more permanent (Menzel, 1979, 1983). An important aspect of memory is associative connections. Some floral odours may be associated with nectar, others with pollen, and still others with the absence of a suitable resource.

There are two complementary levels of analysis that may take place in the memory in the context of kin recognition. The first of these is signal identification or categorization. The second and more complex of these is similarity assessment. In the first case, the odour stimulus representing the signature of an initiator is either recognized as a particular individual or is categorized as a super sister or, at a more encompassing level, as a nestmate. During similarity assessment, however, the amount of difference between an odour stimulus and a 'memory template' must be evaluated. Thus, categorization is a qualitative evaluation of a stimulus while similarity assessment is a quantitative evaluation. These ideas can be more easily understood if we discuss them in the context of a simple pictorial model.

A number of different models have been proposed on how neural networks are able to recognize patterns or categorize objects. These models are typically dynamical systems describing the electrical state of a neural network and the convergence of the network to one of its several locally stable energy states. A useful intuitive notion of how this works can be provided by a plastic-sheet-and-ball model.

If a ball is placed on a perfectly horizontal plastic sheet, then it will stay exactly where it is placed (Fig. 13.7a). If the sheet is pulled down at two different points, then two basins are formed. A ball placed on the sheet no longer remains where it is placed: it rolls down towards, and finally comes to rest at, the bottom of one of the two basins formed by the distortion in the plastic sheet. If the position where the ball is placed is represented by ρ_j, (the quality component of the signal in the relay cable corresponding to odour stimulus ω), then the final position at the bottom of the basin can be regarded as a categorization of ρ_j. In a two basin sheet the categories may represent nest and non-nestmates. However, the nestmate basin may be further indented with smaller local basins that represent super sisters (i.e. 'self' basin), and various half-sister patrilines (i.e. 'non-self' nestmate basins).

The plastic properties of various parts of the sheet (some parts may stretch more than others) are determined genetically, while the initial adult configuration of the sheet (position and depths of basins) is determined ontogenetically. The basin structure of the sheet is then modified by experience. In particular, the weight of the balls placed on the sheet may stretch it, thereby either temporarily or permanently altering the shape, depth, and even number of basins. A ball placed on a flat sheet can form its own initial indentation. A ball then placed close to the position of this indentation may roll towards it, continuing with the formation of a category memory basin. Categorization then corresponds to identifying into which basin a ball finally settles.

There are many ways that a brain may make a similarity assessment. In our

(a)

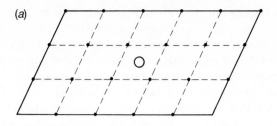

(b)

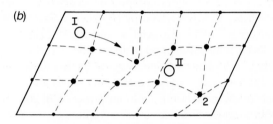

Fig. 13.7 A plastic sheet analogy of memory processes. (*a*) is a flat sheet, while (*b*) is a distorted sheet. Points on the sheet correspond to inputs into memory. Balls represent these inputs. Bottoms of basins, like points 1 and 2, represent memory templates. Ball I runs down to point 1 indicating that an input at position I will be identified with template 1. However, it is not clear whether ball II will run down to point 1 or 2. Because the sheet is plastic the weight of the balls may themselves effect the shape of the sheet (this is analogous with the fact that in many neural systems the strength of a synapse increases with stimulation). Thus stimulus I preceding stimulus II may actually alter the landscape causing stimulus II to be identified with template 1. On the other hand if stimulus II precedes stimulus I, then stimulus II may be identified with template 2. Thus a näive memory may be like a flat sheet that is molded by the stimuli that it receives. Genetic factors can influence the process by determining the degree of plasticity at different points in the sheet.

plastic sheet paradigm, the brain may be able to measure how long it takes for the ball to roll down to the bottom of the basin. The shorter the period, the more similar the initial stimulus is to the 'template' at the bottom of the basin. If the templates themselves are represented by the points at the bottom of basins, then neural networks may be able to perform the equivalent of the mathematical operation of taking inner products between two vectors; the vector representing the current stimulus and the vector representing a particular template (Getz & Chapman, 1987; Getz, 1991). The degree of similarity is then represented by the output from a network performing the inner product operation: maximum output means the incoming signal and template provide a perfect match, zero output means that the vectors are orthogonal (having no components in common). Note that two odours in the OSS giving rise to orthogonal vectors in memory does not imply that the two odours in the OSS have no compounds in common, because the mapping Q may rotate the space (i.e. the eigenvectors of the linear operator part of Q do not, in general, align themselves along the axis of the OSS).

The above view of how the olfactory system processes and perceives incoming stimuli is not consistent with a number of notions that currently exist in kin recognition. First, it seems that the olfactory system is not capable of reading each chemical component of a complex olfactory signature. In the first genetic models relating to kin recognition (Crozier & Dix, 1979; Getz, 1981, 1982; Lacy & Sherman, 1983; also see Crozier, 1987) there is a notion that kin recognition labels may correspond to alleles and that the number of labels which two individuals share may be assessed. If this were true, then discrimination mechanisms could be based on the number of alleles that two individuals share (genotype matching), and the presence of alleles in an initiator that are familiar (habituated label acceptance) or unfamiliar (foreign-label rejection) to the recipient. These types of mechanisms are only possible if labelled lines are available for the reading of individual components or if the dimension of the SRS and OSS are equal (i.e. $m = n$). The notion that genes contribute towards the ontogeny of a particular phenotype – that is, phenotypes have a heritable component (see Lacy & Sherman, 1983) – is more compatible with the idea that the dimensionality of an odour signal is reduced during processing.

Learning has been shown to be an important part of the kin recognition process in social insects (Breed et al., 1985; Getz & Smith, 1986; Moritz & Crewe, 1988a). Buckle & Greenberg (1981), for example, showed that cross-fostered guard bees learn the odour of their nestmates, and are more likely to admit the unfamiliar sisters of their nestmates than their own unfamiliar sisters to their nests. As discussed in Getz (1982), this does not necessarily mean that they are unaware of their own signature (in our plastic sheet paradigm, the memory basin representing the signatures of the cross-fostered guard bees nestmates may be larger than the memory basin representing that guard bee's own signature). Experiments described below on honey bees (Getz & Smith, 1986) indicate that an individual isolated with one other individual appears to place the same weight on knowledge of its own signature as it places on knowing the signature of a nestmate, when it comes to accepting unfamiliar individuals.

If our plastic sheet paradigm is a suitable analogy of how kin discrimination takes place, then whether a quality vector ρ_j is categorized as representing a particular relationship depends on where ρ_j lies with respect to the basins associated with the different possible relationships. Purely by chance we can expect the signatures of some individuals to be close together, even though the individuals may be unrelated (Fig. 13.4; see also Getz, 1981). Thus we can expect some misclassification to occur. To be more precise, we will discuss the problem of misclassification in the context of a similarity assessment paradigm, drawing upon some ideas proposed by Reeve (1989).

Let $\delta(\rho_1,\rho_2)$ denote a similarity assessment function implemented by the honey bee brain, where ρ_1 is an input corresponding to some odour ω and ρ_2 is a memory template. If $\delta(\rho_1,\rho_2)$ is the inner product function then, as we have mentioned, it ranges in value from 0 when ρ_1 and ρ_2 are totally dissimilar (orthogonal) to 1 when ρ_1 and ρ_2 are identical. Other measures of similarity are possible, but we can always normalize the similarity function so that it ranges over [0,1] and $\delta(\rho_1,\rho_2)=1$ implies that $\rho_1=\rho_2$ (note that Reeve developed his arguments using a dissimilarity function; i.e. $\delta(\rho_1,\rho_2)=0$ implies that $\rho_1=\rho_2$).

Now, exactly as we constructed the kingram in Fig. 13.4 using the number of labels in common as the similarity index, we can construct a kingram using the similarity index $\delta(\rho_1,\rho_2)$ as the independent variable of the probability density functions; or, equivalently, for the cumulative density functions. In Fig. 13.8 we illustrate the cumulative density functions for three different relationships for the process of matching the signature of an initiator with a template of 'self'.

The similarity levels δ_1 and δ_2 may provide suitable thresholds for decisions, respectively, relating to nestmate defence and selection of eggs or larvae for queen rearing. If it is relatively more costly to inadvertently exclude nestmates than to admit the occasional non-nestmate, then either δ_1 must evolve to a lower value δ_1' or the signature production and/or perception mechanisms must evolve to provide a greater separation between the cumulative non-nestmate and half-sister similarity curves. (For the curves in Fig. 13.8, virtually no super sisters will be rejected if the threshold is δ_1.) Similarly, if workers experience a great gain in inclusive fitness by rearing only super sisters as reproductives and, as in the honey bee, they have a relatively large number of eggs or larvae among which to choose, then the threshold δ_2 will evolve to the value δ_2', even though at δ_2' they will reject more than 75 per cent of their super sisters, they will almost never rear a half sister as a reproductive.

Reeve (1989) has suggested that many of the ideas on the evolution of optimal similarity decision thresholds are parallel to ideas in optimal foraging theory; and that an *optimal discrimination theory* can be developed along the same lines as optimal foraging theory. It is difficult, if not impossible, to assess the full ecological ramifications and hence inclusive fitness to many types of behaviour like guarding or investing effort in rearing super sisters as workers. The first of these tasks may be impossible to evaluate in a eusocial insect like the honey bee, although one may be able to get a handle on the problem in primitively social insects using some type of

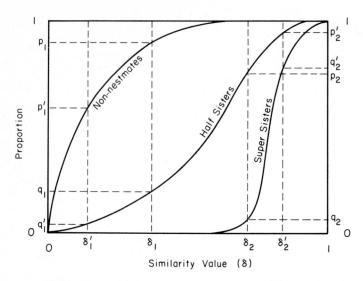

Fig. 13.8 Cumulative density functions relating to the probability that two individuals bearing the particular relationships (as labelled) have recognition signatures with a similarity value δ. If the threshold of guard bees for admitting individuals to the hive is δ_1 (assuming all guard bees process signals in the identical manner), then a proportion p_1 of non-nestmates will be refused entry while a proportion q_1 of half sisters will be refused entry (the super sisters cumulative curve indicates that all super sisters will be permitted to enter). If there is a much greater cost to excluding half sisters than there is to allowing entry to nonnestmates then δ_1 will evolve to a lower value δ_1'. Similarly, suppose the threshold δ_2 at which workers accept a reproductive as their queen results in a proportion $1-p_2$ of half sisters accepted as queens, and that this is too costly. Then δ_2 will evolve towards a value δ_2' where the benefit from making fewer errors in choosing half sisters balances with the costs of rejecting a greater proportion (q_2' v. q_2) of super sisters.

optimal discrimination analysis. The preferential queen rearing problem may be easier to analyze because the inclusive fitness of workers can be more directly tied to queen rearing than it can to guarding behaviour.

Empirical studies

Introduction

There is a binding interplay between empirical observations, theory, experiments and comparative analyses, so that studies exhibit genealogies that often become intertwined with respect to lines of investigation and experimental animal models. For example, early investigations of nestmate recognition in honey bees focused on the role of environmental chemical cues (Kalmus & Ribbands, 1952), while the first investigations of worker recognition of honey bee queens revealed that workers recognize queens based on familiarity (Boch & Morse, 1974, 1979). The first kin recognition studies in bees examined nestmate defence in sweat bees (Greenberg, 1979). Greenberg, drawing upon the mate recognition and nestmate defence work of Charles Michener and his students (see Michener & Smith, 1987) at the University of Kansas, Lawrence, USA demonstrated a genetic component to nestmate recognition in sweat bees. Breed (1981) then looked for evidence of kin related phenomena in worker choice of queens and subsequently broadened his investigations to include the question of kin discrimination in agonistic interactions between worker honey bees (Breed, 1983). Hamilton's (1964) notion of inclusive fitness and its application to the evolution of sociality in insects, provided the impetus for my own work on kin recognition (Getz, 1981; 1982) and the segregation of workers during swarming (Getz *et al.*, 1982). Page (1980) was initially interested in reproductive strategies and their relationship to the evolution of social behaviour (Page & Metcalf, 1982). All these various lines of research provided the stimulus for a concerted effort in the mid-1980s to crack the kin recognition problem in honey bees. The solution has turned out to be more elusive than originally anticipated.

For clarity, I have divided honey bee kin discrimination empirical studies into five categories: worker–worker interactions, queen–worker interactions, queen rearing, olfactory perception and chemosensory cues, and sociogenics. However, the reader should keep in mind that the results of an experiment discussed in one category might well have been the motivating factor for an experiment discussed in another category.

A number of experiments discussed below involve the use of genetic markers (primarily cuticle colour) to identify individuals from different patrilines within the same colony. Particular lines are easily bred using instrumental insemination techniques (see Mackensen & Roberts, 1948). In some experiments, the recessive colour mutant allele for cordovan has been used so that a cordovan queen (CC) inseminated with sperm from a cordovan (C) and wild type (W) drone results in one patriline of cordovan phenotypes (CC) and one patriline of wild-type phenotypes (CW). In other experiments, differences in colour between honey bees is exploited to obtain markers that do not depend on mutant alleles. Cuticle colour is a characteristic of different races of honey bees, although in the USA the races have been interbred. For example, Italian honey bees (*Apis mellifera ligustica*) have yellow stripes on the abdomen while Carniolan (*A. m. carnica*) and German honey bees (*A. m. mellifera*) are dark. Since the degree of darkness can be determined by alleles at several loci, to ensure that colour phenotypes in different patrilines do not overlap, it is important to use lines that have been selected over several generations for yellowness or darkness traits.

There are several caveats that the reader should bear in mind when interpreting the results of the experiments discussed below. First, colour markers enable the human observer to distinguish between patrilines in a colony, but such colour marker manipulations may artificially increase the average genetic distance between patrilines than is found in wild colonies. Second, colour markers may be correlated with odour recognition cues (Frumhoff, 1988 but see Carlin, et al., 1990) since both are dependent on the chemical composition of the cuticle. Accentuated differences in odour signatures could invoke kin discrimination behaviour in a spurious context. Third, it is not clear how kin discrimination levels and ensuing interactions are affected by the fact the many empirical studies use colonies with two patrilines, while the colonies of naturally mated queens consist of several or even 10 to 20 patrilines. Fourth, controlled experiments often involve creating situations that do not occur under normal circumstances (e.g. forcing colonies to rear many more queens than they normally do). Fifth, colour markers may make it difficult for the experiment to be done blind. In these cases, we have to be cautious in extrapolating the results to deduce what is happening under more normal colony conditions. Despite these shortcomings, however, the empirical results enable us to deduce what is or is not possible, and to build up a picture that is sharpened as additional results are obtained. Thus no experiment is conclusive in of itself, only results that are mutually reinforcing can be accepted with some confidence.

Worker–worker interactions

Breed (1983) was interested in whether there was a genetic component to nestmate recognition in colony defence. To address this question he collected newly emerged worker bees (those that had emerged from the pupal stage in the previous 24 hours), isolated them in groups of 10 in $\frac{1}{2}$-litre cardboard containers, and maintained them in an environmentally regulated incubator on a controlled diet for five days. In each run of the experiment he worked with two unrelated colonies, (colonies A and B), both with naturally mated queens. He then collected a set of 10 data points (assuming that all individuals survive the five days in the incubator) by selecting 11 containers, six with 10 individuals each from colony A (labelled containers A1–A6) and five with 10 individuals each from colony B (labelled containers B1–B5). He then selected an individual at random from container A6, marked it, and introduced it into one of the containers A1–A5. He now had one 'introduced' and 10 'recipient' bees, say, in container A1. For 5 minutes he observed the interaction of these 11 bees in the container. If the introduced bee was stung or grasped by the mandibles of at least one recipient bee, he scored the interaction as 'aggressive.' Breed continued to introduce individuals one at a time from A6 to the remaining containers A2–A5 and B1–B5. Thus he had five cases in which the introduced bees were sisters of the recipients (half sisters in some cases, super sisters in others), and five cases in which the introduced bees were unrelated to the recipients (the recipients all being sisters to each other).

In this experiment, Breed found aggression in 31 per cent of introductions when individuals came from the same colony, but a level of 69 per cent when individuals came from a different colony to that of the recipients. Statistically this result was significant with $p < 0.001$. In a repeat of this experiment, but eliminating most of the environmental cues by allowing queens to lay eggs in clean combs and then transferring these combs to be reared in a third (neutral) hive, Breed (1983) still obtained a significant level of discrimination (13% aggression for within colony interactions, and 56% for between colony interactions). This latter result indicates the existence of a genetic component to recognition, although it is conceivable that some environmental component could still have been present from the initial few days that the different experimental combs spent in each hive.

Following Breed's approach, Getz & Smith (1983) added the refinement of working with super sister patrilines using both cordovan and other cuticle colour markers. This eliminated the problem of workers being reared in a different comb, since both patrilines were progeny of the same queen. They also eliminated the possible complication of familiarity developing during the several hours post-eclosion period that the introduced and recipient workers spent together by removing callow adults from their cells just prior (within 24 hours) to their eclosion. This experimental design can be used to compare levels of aggression[9] among unfamiliar super sisters, half-sisters, and unrelated individuals. Results obtained by Getz & Smith (1983) indicate that discrimination exists at the super/half-sister level. Further, because all environmental differences between the two treatment groups have been eliminated, these data provide incontrovertible evidence that workers have a genetic component to their signatures that is used for discrimination purposes in the context of the arena experiments described above. These data, however, do not address the relative importance of genetic versus environmental cues in nestmate discrimination by guard bees, nor do they imply that workers can discriminate between super and half-sister within a natural hive situation. In Getz & Smith (1983), workers were only exposed to their super sisters while, in the hive, workers are exposed to both super and half-sisters and, therefore, do not get an opportunity to learn signatures of only their super sisters. As already pointed out, for workers to be able to discriminate between super and half-sisters within the same colony requires that they innately 'know' or learn their own signatures.

Using modifications of Breed's (1983) and Getz & Smith's (1983) experimental designs, the question of signature learning was addressed in two

---

[9] Many ethologists choose to use the word agonism to describe combative behaviour among workers.

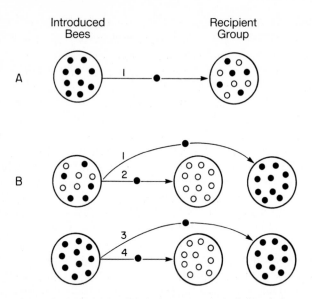

Fig. 13.9 Open and solid circles represent individuals from either two different patrilines within the same colony (Getz & Smith, 1986) or two different colonies (Breed *et al.*, 1985). In the former case (which we focus on here), an introduction of type B2 represents the introduction of a half sister from a heterogeneous group (see text for more details). (A) *Self awareness experiment*. Here we assess whether there are differences in the way an introduced bee's recipient subgroups of five super and five half sisters react towards it. Since there are behavioural differences between genetic lines, introductions B1 and B3 can be used to control for these differences by comparing how randomly designated recipient subgroups of size 5 (there is no longer a natural patriline subgrouping) respond to introduced bees. (B) *Label transference experiment*. By comparing responses to super-sister introductions 1 and 3 and/or responses to half-sister introductions in 2 and 4, it is possible to assess whether individuals from heterogeneous (i.e., mixed) recipient groups have had their signatures altered by associating with both super and half sisters.

studies in which both homogeneous groups of 10 bees and mixed groups of 10 bees (five from each of two patrilines) were set up in containers as described above (Fig. 13.9).

In one of these studies, Breed *et al.* (1985) conducted a set of self-reference[10] experiments, using two lines of bees from different colonies, following the approach illustrated in Fig. 13.9A. They did not find significant differences in levels of aggression (biting, stinging) between sisters and non-

[10] This terminology is meant to imply that individuals are able to evaluate a signature against a template of self, but does not necessarily imply that individuals have some 'higher' cognizance or awareness of self.

sisters, but they did find significantly more feeding interactions between sisters than between non-sisters. In the second of these studies, Getz & Smith (1986) used two patrilines from the same colony to conduct a set of self-reference experiments. They also did not find significant differences in the levels of aggression. These results point out how careful one has to be in interpreting data: it seems that in the experimental arena setting of Fig. 13.9A, self-reference in individuals is not manifested in an aggressive, but only in a co-operative context. This fits in with the idea that kin recognition within the hive should not strongly disrupt the underlying unity of the hive. Self-reference in the context of workers interacting within an observation hive has since been recorded by Frumhoff & Schneider (1987). Further, Evers & Seeley (1986) have identified self-reference in the context of aggressive interactions in queenless colonies where workers compete to lay unfertilized (i.e. drone) eggs.

In addition to introductions of the type depicted in Fig. 13.9A, Getz & Smith (1986) conducted a comprehensive set of introductions of the type depicted in Fig. 13.9B. If label transference occurs, whereby the signatures of individuals associating with one another become more similar, then we would expect that aggression towards super sisters is increased (comparing B1 with B3) while aggression in half sisters is decreased (comparing B2 with B4). Getz & Smith's results are consistent with this in seven of the eight comparisons illustrated in Fig. 13.10 (i.e. comparing the first and second bars and the third and fourth bars for the four sets of data). However, only one of the consistent seven comparisons is statistically significant on its own (the number of data points in each comparison is around 100 to 150) while the half-sister comparison (B3 vs. B1) in line 2 of hive 2 is quite strongly in the wrong direction. The results depicted in Fig. 13.10 are thus not convincing on their own. However, they add to a picture where it is known that the cuticular wax composition of nestmates becomes more similar as a result of close association between individuals (R. Page, pers. comm., unpub. data). Also, there is evidence for the transference of recognition labels between individuals in several ant species (Crosland, in 1989a,b – see discussions in Breed & Bennett, 1987; Hölldobler & Carlin, 1987). Further, Breed *et al.* (1985) performed one comparison of the B3 versus B1 type (see Fig. 13.9) in the context of sister and non-sister lines and found that aggression was higher towards sisters mixing with non-sisters (28%) than towards sisters coming from a homogeneous sister group (20%). Again, the number of data points (175) precluded this result from being significant, but the trend is in the right direction.[11]

---

[11] Breed *et al.* (1985) also obtained data of type B2 in Fig. 13.9 and compared the results with results that they had obtained in earlier studies on different hives. They concluded that a label transference effect was not evident. However, it is not permissible to make this type of comparison because different lines of bees exhibit different underlying levels of aggression.

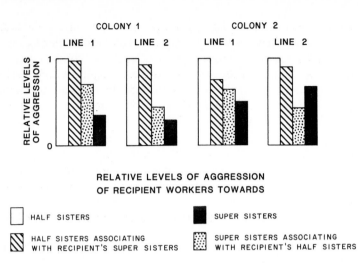

COLONY 1   COLONY 2
LINE 1   LINE 2   LINE 1   LINE 2

RELATIVE LEVELS OF AGGRESSION

RELATIVE LEVELS OF AGGRESSION
OF RECIPIENT WORKERS TOWARDS

☐ HALF SISTERS   ■ SUPER SISTERS

▨ HALF SISTERS ASSOCIATING   ▦ SUPER SISTERS ASSOCIATING
  WITH RECIPIENT'S SUPER SISTERS   WITH RECIPIENT'S HALF SISTERS

Fig. 13.10 Relative levels of aggression towards unfamiliar super and half sisters originating in both homogeneous and mixed patriline groups – see fig. 13.9.

Finally, if we compare data obtained from introductions of type A versus B3 or A versus B4 we can examine questions relating to the structure of kin and nestmate recognition templates in the honey bee brain. Recipients in A have four super sisters, five half-sisters, and themselves, while recipients in B have nine super sisters and themselves, to contribute to the formation of kin and nestmate recognition templates in their brains. If nestmate recognition is dominated by a 'template of self' (using genetic and/or environmental components of the signature), then the nestmates of an individual will have little or no influence on the process of template formation. In this case the super-sister or half-sister results from introductions of type A versus B should be similar. On the other hand, if all nestmates influence the process of a nestmate recognition template in an individual worker's brain, then the results should be different. For example, since recipient individuals in A, but not B, type introductions have half-sister nestmates to contribute towards the formation of nestmate recognition templates, the recipients in A should be more tolerant of unfamiliar half sisters than the recipients in B. On the other hand, recipient individuals in A type introductions, having fewer super-sister nestmates to contribute towards the formation of nestmate recognition templates than recipient individuals in B type introductions, might be more aggressive towards unfamiliar super sisters. The results obtained in Getz and Smith (1986) are given in Table 13.1. Since individuals in mixed recipient groups (case A) are respectively more aggressive towards unfamiliar super sisters and less aggressive towards unfamiliar half sisters

Table 13.1. *The percentage of aggressive interactions between introduced workers obtained from homogeneous groups of individuals and workers in homogeneous and mixed recipient groups (see Fig. 13.10).*

Homogeneous super sisters[a]	Mixed super sisters[b]	half sisters[b]	Homogeneous half sisters[a]
10	12	16	28
10	18	9	31
19	27	26	36
17	26	25	29

Notes:
[a] Response from random partitioning of subgroup of 5 individuals in a recipient group of 10 individuals.
[b] Response of appropriate subgroup of 5 individuals in a two-patriline recipient group of 10 individuals.

than individuals in homogeneous recipient groups (case B) (i.e. the mixed group results are intermediate between the super- and half-sister homogeneous group results in Table 13.1), we can conclude that nestmates contribute significantly towards nestmate recognition template formation.

Getz & Smith (1986) also examined the notion of a concept of self by maintaining individuals in isolation rather than in groups of 10, and then introducing one of these individuals into the container of another individual. Their results indicate that isolated worker bees are less likely to attack an unfamiliar bee if it is a super rather than half sister. Their result also suggest that a worker has a sense of self in terms of being able to assess which of its sisters are also members of its own patriline.

Getz & Smith (1986) took this question one step further by trying to assess the relative importance of self and nonself images in mediating interactions between individuals in the context of several interactions, including feeding, antennal contact and agonistic behaviour. Instead of maintaining individuals in isolation or in groups of 10, they maintained pairs of individuals, one from each patriline of a two-patriline colony. After five days (post-eclosion) they placed two of these pairs in a neutral observation arena where all interactions were recorded on video tape for subsequent analysis. Thus each individual was maintained for five days with a half sister, and then interacted for five minutes with its familiar half sister, as well as an unfamiliar super sister and an unfamiliar half sister. If, in comparing interactions between unfamiliar individuals, there are more agonistic interactions between unfamiliar super than half sisters and/or more feeding interactions between unfamiliar half than super sisters, then the data suggest that nestmate recognition templates are based more on experience with nestmates than a concept of self. If the data are the other way around, then a template or concept of self is the likely explanation. There were significant differences between the way individuals interacted with their familiar and unfamiliar half sisters, but there appeared to be no differences between the way individuals interacted with their unfamiliar super and half sisters. A consistent interpretation of these results is that both a sense of self and knowledge of a nestmate signatures contribute towards the formation of nestmate discrimination templates in the worker honey bee brain. Note that we should only refer to discrimination as kin recognition if the templates have a component of self, otherwise the process is purely a nestmate discrimination which is often correlated with interactions between kin.

The question still remains whether self and nestmate signatures are pooled in some way for recognition purposes, or whether an individual maintains a sense of self in the presence of other nestmates. Getz & Smith (1986) approached this question in the following experiment. They introduced an individual from a group of workers maintained in an incubator for five days post-eclosion into an arena containing a worker selected at random from the brood comb area of the same hive. In one hive, agonistic interactions between super and half sisters were respectively 53 and 68 per cent while in a second hive the comparable results of 80 and 89 per cent were obtained. Because sample sizes were large (400 and 275, respectively) both sets of results represent a significant level of discrimination. These results suggest that individuals within a colony retain a sense of self, despite the fact that they also learn and use the signatures of their nestmates for kin discrimination purposes. Note that the overall high levels of aggression in this experiment (especially hive 2) could be due, in part, to differences in environmental components of the signature of the recipient and introduced bees.

Evidence that honey bee workers actually exhibit some level of self-reference within a more natural hive situation has been obtained by Evers & Seeley (1986) and Frumhoff & Schneider (1987). Evers & Seeley worked with hopelessly queenless colonies: that is, colonies without a queen or brood young enough to rear into queens. In such colonies, aggression between workers can be observed; and some workers compete to lay unfertilized eggs (i.e. eggs that develop into drones). Using cordovan as a marker, Evers & Seeley were able to show greater levels of aggression between half than super sisters in all four of their observation hives (the results were highly statistically significant in three of these colonies). On the other hand, Frumhoff & Schneider gathered data on the frequencies of grooming and trophallactic (feeding) interactions between super and half sisters in five queenright observation hives, also containing two distinguishable patrilines (using racial colour markers in two colonies and the cordovan marker in the basis of

Table 13.2. *Summary of recognition findings in kin interaction studies*

A: Worker-worker interactions

Study	Interaction	Inferred recognition mechanism
Breed (1983)	Sister vs. non-sister	Genetic component
Getz & Smith (1983)	Super vs. half sister	Genetic component
Breed *et al.* (1985)	Sister vs. non-sister	Nestmates contribute towards recognition template formation
Getz & Smith (1986)	Super vs. half sister	Sense of self component comparable with nestmate component
Evers & Seeley (1986)	Super vs. half sister	Sense of self within hopelessly queenless colonies
Frumhoff & Schneider (1987)	Super vs. half sister	Sense of self within queenright colonies

B: Queen-worker interactions

Study	Interaction	Inferred recognition mechanism
Boch & Morse (1974, 1979)	Own vs. foreign queen	Familiarity component
Breed (1981)	Unfamiliar queens	Genetic component
Boch & Morse (1982)	Related vs. unrelated queen	Genetic component
Page & Erickson (1986a)	Unfamiliar queens	Regression relationship on genetic component; early emergence takes precedence over kinship

colour). They showed, in all five colonies, that there were statistically significant biases towards super rather than half sisters in at least one of the two types of interactions observed. Where the bias was not statistically significant, it was always in the direction consistent with a self-reference hypothesis.

The results of this and the next section are summarized in Table 13.2.

Queen–worker interactions

It was established in the 1970s (Boch & Morse, 1974, 1979) that honey bee workers are able to recognize their own queen when given a choice of several queens. Further, a group of workers often attack an unfamiliar queen, even though she may be closely related to them. Breed (1981) was the first to investigate whether queen recognition cues have a genetic and/or environmental component to them. He set up groups of 10 workers, much as described in the previous section, except that in each container there was a queen as well. After five or six days, two containers were paired and their queens were exchanged. The behaviour of the workers towards the introduced unfamiliar queen was observed for 10 minutes. In none of 39 cases did a worker group accept an unrelated unfamiliar queen, in 3 of 26 cases, workers accepted an unfamiliar queen that was their sister and, in one inbred worker line, workers accepted 8 of 23 sister queens to whom they were approximately related by $r = 0.95$. Although there is natural variation of levels of aggression across genetic lines, these results strongly suggest that there is a genetic component to recognition of queens.

Boch & Morse (1982) corroborated Breed's result using a different experimental situation. They set up artificial swarms with the queen confined in a small cage. They then removed the queen from the swarm, but at separate sites at a distance of two metres from the swarm they set up two other queens. After five hours they evaluated which queen the majority of workers from the initial swarm had chosen. Boch & Morse found that in 25 of 33 cases, a majority of workers selected a sister of their original queen over an unrelated queen, while in 12 out of 16 cases a majority of workers selected their original queen over a sister queen.

Following on from Breed's work, Page & Erickson (1986a) also introduced virgin queens into groups of 10 workers. The virgins had been reared in special queen rearing colonies from larvae grafted from a source colony, while the workers had been removed in the previous hour or so from areas surrounding the brood comb of the same or a different source colony. Page & Erickson found that aggression was correlated with the degree of relatedness between the workers and the unfamiliar queen.

Page & Erickson (1986a) used the virgins and workers from the above experiment in a second experiment. Two containers, each with both the workers and the introduced queen, were selected and the queen from one of these containers was then introduced into the other container. Thus, in this experiment, we have a container of 10 workers from the same patriline and a resident queen from the same or different patriline that had been placed within the worker group from 0.5 to 6 hours earlier and then a second queen was introduced. The motivation for this experiment relates to the fact the virgins often emerge within hours of one another in a natural hive situation, and the question is whether workers are more likely to accept the first or second virgin to emerge. If workers preferentially accept the first virgin to emerge, then this could be a factor in explaining why queens, which are substantially larger then workers, have a development rate that is one third faster than the development rate of a worker (egg to adult). The results of this experiment are quite complicated to interpret, because the workers may respond to the second introduction by attacking one or both queens, and the queens themselves may attack the other queens or attack workers. Also, we need to consider the kin relationship between the workers and both queens. Page & Erickson (1986a) confined their investigation to the case where the first queen was either a super or half sister to the worker group, and similarly for the second queen. Because they were working with two patriline colonies, then it follows that if the first queen was a super sister of the workers and the queens themselves were half sisters, then the second queen was a half sister to the workers as well. The essence of Page and Erickson's findings were that resident queens had a significant advantage that was greatest when they were both half sister to the workers and to the introduced queen. Since the introduced queen in this case is a super sister to the workers, it appears that time of eclosion takes precedence over kinship. The reason for this could be that virgins emerge into a multipatriline situation (not a single patriline of workers, as set up in this experiment) and it would be too divisive to the colony as a whole for the workers to squabble over virgins and risk injury to all virgins at this stage in the queen rearing game. As discussed below, there is evidence that bias during queen production takes place at the egg/larval rearing phase. Also, Page & Erickson found that an introduced queen did best, compared with other introduced queens, when the workers were her half sisters and the resident queen was her super sister. Could this be ascribed to super-sister virgins being less likely to attack one another than half-sister virgins? This is a possible explanation, although the role of the confounding effects of the relationship of the virgins to the workers is difficult to assess.

Queen rearing

The first group of experiments described here employs techniques developed for rearing large numbers of queens in special queen-rearing colonies. This technique relies on the fact that under the right set of circumstances workers will rear queens from one-day-old larvae when they are grafted from worker cells in other colonies into special queen cups into a properly stimulated colony (Laidlaw & Eckert, 1962). The basic experimental design involves setting up two queen-rearing hives, one consisting solely of individuals from colony A and the other of individuals from colony B. Then equal numbers of larvae from colonies A and B are grafted into these two queen-rearing hives and the data are analyzed to assess whether workers preferentially rear the larvae from their own colony. The experiment can also be undertaken using patrilines A and B from a single colony, in which case the test involves a comparison of rearing super versus half sisters, rather than sisters versus non-sisters.

Breed *et al.* (1984), carried out seven replications of the two colony design. In three of these replications, the two colonies were different races of bees. The queens in these colonies had mated naturally so that within colonies workers were predominantly half sisters (i.e. the average r was not much greater than 0.25), while between colonies the workers, as far as is known, were unrelated (i.e. $r = 0$). They found no evidence (Table 13.3) for the preferential rearing of sisters over non-sisters as queens. Page & Erickson (1984) carried out 13 replications of this same two-colony experiment, but they carefully controlled the genetic structure of the colonies: all workers in the same colony were super sisters ($r = 0.75$) and between colonies workers were cousins related by either $r = 0.25$ or $r = 0.31$[12] Of the larvae reared to the sealed brood stage in these colonies, 62 per cent were super sisters (i.e. 38% of reared larvae were cousins – Table 13.3). Based purely on relatedness considerations, the two sets of results can be reconciled if, in fact, workers discriminate between super and half sisters (the natural situation), but do not discriminate in the unnatural situation of sister (primarily half sister) versus non-sister. However, the contrast in results might be due to much more subtle effects, like whether the larvae were grafted with or without some of the royal jelly which normally occurs with these larvae (Visscher, 1986).

Visscher (1986), conducted a set of experiments in which he grafted both larvae and eggs, and carefully controlled for environmental recognition cues. As in Breed et al. (1984), Visscher evaluated whether workers discriminate between sisters and non-sisters. When he grafted larvae, there was no evidence of discrimination. But when he grafted eggs, 60 per cent of those reared to the cell capping stage were sisters (i.e. 10% bias – Table 13.3). To exclude the colony as a source of cues, Visscher (1986) grafted cousin eggs ($r = 0.22$ in one case and $r = 0.25$ in another) and unrelated eggs. In this case 63 per cent of the eggs reared were cousins, while 37 per cent were unrelated, to the workers that reared them to the brood capping stage. Finally, in an experiment using workers that had been conditioned for the six previous weeks to rearing only unrelated larvae as workers, Visscher (1986) then evaluated discrimination in rearing sister – but now novel – eggs over unrelated – but now familiar – eggs. In this case workers still preferred to rear sisters over non-sisters by a margin of almost 2 to 1 (64% versus 36%). These results, however, are confounded by the fact that the eggs remained for up to three days with their sisters before being transferred, thus providing an opportunity for the workers to learn the odour of these eggs.

Visscher's results suggest that worker honey bees discriminate between eggs or very young larvae, preferentially rearing kin over non-kin. Page & Erickson's (1984) and Visscher's (1986) results demonstrate that all the ingredients for preferential super-sister rearing are in place but both studies fall short of demonstrating that nepotism in queen rearing takes place in hives under natural conditions. Results obtained by Noonan (1986) took this line of study one step closer to natural conditions by demonstrating

---

[12] They actually employed a 4-colony design to control for brood viability.

Table 13.3. *Evidence for kin bias in queen rearing*[a]

Source	Breed et al. (1984)		Page & Erickson (1984)	Visscher (1986)				Noonan (1986)
Description	Queen breeder		Queen breeder	Queen breeder				Obs. hive
	Larval graft		Larval graft	Larval graft	Egg graft	Egg graft	Egg graft	Larval graft
Comparison	Sister vs. unrelated	Sister vs. unrelated	Super sister vs. cousin	Sister vs. unrelated	Sister vs. unrelated	Cousin vs. unrelated	Novel sister vs. familiar unrelated	Super vs. half sister
Measure	Larvae accepted	Queens reared	Capped larvae	Capped larvae	Capped larvae	Capped larvae	Capped larvae	Larval care[b]
Result	51:49	50:50	62:38	50:50	60:40	63:37	64:36	52:48
Bias	+1%	0	+12%	0	+10%	+13%	+14%	+5%[c]
Significance	No	No	$p < 0.05$	No	$p < 0.02$	$p < 0.02$	$p < 0.005$	$p < 0.001$

Notes:

[a] The most recent and perhaps most convincing results are those obtained by Page et al. (1989), as discussed more fully in the text.

[b] Noonan's 'feeding' and 'cell maintenance with larval care' results are combined here for those larval cells from which queens were ultimately reared.

[c] For this experiment, the expected outcome is not 50:50 but 47:53 because of differences in the larval care behaviour of workers in the different lines.

preferential super-sister rearing in two-patriline (cordovan and wild type) observation hives in especially designed plexiglass structures. After grafting one day old larvae from worker cells into queen cups, Noonan then observed the patriline membership of worker bees attending the larvae during the five day larval-rearing period. Several days after the cell had been capped, Noonan removed the capped cell and placed it in an incubator where the patriline membership of the occupant could be recorded once it had emerged. In total, Noonan (1986) obtained data on the rearing of 14 cordovan (C) and 21 wild-type (W) queens in five different colonies. Noonan used a chi-squared contingency table analysis of the four data categories of (C-workers, C-queens), (C-workers, W-queens), (W-workers, C-queens), and (W-workers, W-queens) to show a significant bias toward workers rearing super rather than half sisters. She observed 52 per cent of workers interacting with super sisters, while the expected level (from the chi-squared analysis) was only 47 per cent. Because of the large number of interactions this result was highly significant (see Table 13.3).

The previous experiments all rely on the artificial procedure of grafting eggs and/or larvae in queen cups. In colonies where the queen dies, however, workers will immediately construct emergency queen cells and rear some of the eggs or youngest available larvae into queens. Page & Erickson (1986b), exploiting this aspect of honey bee biology, set up a number of colonies each containing workers and brood from two different source colonies. They designed two sets of experiments to evaluate whether worker honey bees can discriminate between larvae during emergency queen rearing; and, if so, whether this discrimination leads to an increase in the reproductive output for the numerically dominant patriline.

In the first experiment they instrumentally inseminated related queens with sperm from a single drone – where the drones themselves were related – to obtain queenless colonies containing two patrilines of relatively closely related cousins ($r = 0.31$ between individuals in the two different patrilines). They skewed the adult worker ratio so that one patriline had an approximately two-to-one numerical advantage, but brood ratios were nearly equal. They monitored numbers of queen cells constructed on brood frames associated with the different patrilines. Their results indicated a significant level of preferential rearing, but this was not correlated with numerical dominance of the worker patriline in a particular hive; rather, certain genetic lines seemed to be preferred over others. The difficulty with interpreting this is that both a kin and a non-kin discrimination hypothesis could equally well explain the results. Specifically, either one line of larvae is inherently more desirable for all workers to rear into queens, or kin discrimination is taking place among adult workers; but the patriline that 'wins' is superior in some way other than sheer numerical dominance. For example, individuals from the winning patriline may be superior nurse bees (see section sociogenetics below where evidence is presented for a genetic basis to such behavioural tasks).

In the second set of experiments, Page & Erickson (1986b) used the cordovan cuticle colour gene to obtain the usual, visually distinctive, two half-sister patriline colonies. From these they set up queenless colonies containing the same worker and brood patriline ratios as the mother colonies. In some of these colonies, by chance, the cordovan patriline was numerically dominant while in others the wildtype patriline was numerically dominant (this relates to the way the sperm from the two drones is stored in the queen's spermatheca). The results obtained here supported the conclusions of their first experiment, although the results from this second experiment did not provide enough data to be conclusive on their own.

The empirical results listed in Tables 13.2 and 13.3 support the hypothesis that worker nepotism exists in honey bee colonies although only Noonan's (1986) study directly addresses the question of preferential queen brood care in the context of super versus half sisters. Recently, however, Page, *et al.* (1989) undertook the first study that examines super versus half sister biases over the complete brood rearing cycle (i.e. worker biases during brood care and biases in the particular queens that are raised). In this study, Page *et al.* (1989) used a malate dehydrogenase allozyme marker system to determine electrophoretically to which patrilines the queens and samples of workers and brood belonged. They found a queen rearing bias of around 10 per cent (i.e. 60:40, cf. Table 13.3), which is somewhat stronger than the bias of 5 per cent (Table 13.3) found by Noonan (1986). Their results provide the strongest evidence to date that nepotism exists in honey bee colonies since preferential queen rearing took place in colonies where the workers were free to interact as they chose (but see Carlin *et al.*, 1990). However, even this result requires a caveat: Hogendoorn & Velthuis (1988), observed both co-operative and aggressive interactions in queenright and queenless two-patriline and eight-patriline hives, and argued that nepotism only occurs in hives with relatively few patrilines. They even hypothesize that '. . . multiple mating is an evolutionary way of [honey bee queens] manipulating offspring.' Thus Hogendoorn and Velthuis's work adds credence to Ratniek's (1990b) contention that '. . . weak super versus half sister discrimination shown by honey bees [could possibly be] . . . due to low genetic recognition cue diversity, and may facilitate the evolution of highly co-operative worker behaviour' (also see Ratnieks & Reeve, 1990).

Besides a bias in queen rearing, there exist other opportunities for workers to manipulate the gene flow between generations. For example, Page & Erickson (1988) have found evidence that some workers lay drone eggs in queenright colonies, and it is possible that these eggs could also be reared preferentially by super sisters. As already mentioned, Evers & Seeley (1986) have observed greater levels of aggression among half sisters than among super sisters in hopelessly queenless colonies, although no definitive study has been made on whether workers prefer to rear their own and/or the sons of their super sisters rather than the sons of their half sisters. R. Page (pers. comm.), however, has obtained data which indicate that this type of nepotism as well takes place in hopelessly queenless colonies.

Observed biases in queen rearing must be facilitated by some type of discrimination process that is based, in part, on a concept of self and involves recognition signatures that have a genetic component. The picture is not completely clear, however, since we do not know the details of how workers are able to discriminate between super and half sisters in the egg and/or larval stage.

Olfactory perception and chemosensory cues

Foraging and scout honey bees are able to communicate the location of various resources (nectar, pollen, sites for the establishment of new hives) through variations in a dance performed on the comb of a hive or on the surface of a swarm temporarily bivouacked on the branch of a tree. This extraordinary behaviour has stimulated and assisted scientists in studying aspects of learning and perception in free flying honey bees, especially with respect to the sensory modalities of vision and olfaction. Von Frisch (1967) presented most of what was known by the mid-1960s on communication systems in honey bees. Since then several experimental methodologies have been developed for investigating learning in the context of olfactory, visual, and even magnetic field stimuli (Gould & Gould, 1988). In terms of kin recognition, we are primarily interested in chemosensation. It is impossible to summarize all the work that has been undertaken in olfaction in honey bees in the last two decades, so I will focus only on work that has a direct bearing on kin recognition.

A complete picture of olfaction in honey bees requires investigation at both the neurological and ethological levels. Ethological data are best obtained using free-flying individuals, but often more precise experiments can be undertaken using restrained individuals. Honey bees reflexively extend their proboscides if their antennae are stimulated with a drop of sugar-water. Further, a honey bee can be trained to extend its proboscis when stimulated only with an odour, using conditioning techniques involving sucrose as a reward (classical, Pavlovian, instrumental, alpha conditioning, etc.). Vareschi (1971) used conditioning techniques to explore the ability of honey bees to discriminate between a number of different organic compounds, while Bitterman et al. (1983) used conditioning techniques to study associative learning and memory in honey bees.

Getz et al. (1986) used discriminative conditioning (Bitterman et al., 1983) to demonstrate that honey bees can discriminate between super and half sisters based on differences in the genetic component of the volatile odour signature (which is not to say that volatile odours are the only components of a chemosensory signature or even that volatile odours are used for kin discrimination purposes in any particular sociobiological context).

Getz *et al.* (1988) then went on to investigate whether a discriminable chemosensory signature resides on the cuticle of workers. They demonstrated that honey bees could discriminate between the chemosensory thorax signature of different foragers from the same hive and even different newly eclosed workers from the same hive. They also tested the ability of workers to discriminate between eggs laid by the same, as well as by different, queens: using albumen to fix an egg to the tip of a glass rod. Their results were highly variable. In some cases a strong level of discrimination was apparent, while in other cases only a weak level of discrimination was obtained. Their data provide no evidence for a stronger level of discrimination if the two eggs being compared came from different rather than the same queen.

Getz *et al.* (1989) also used discriminative conditioning of the proboscis extension reflex to investigate the ontogeny of thoracic chemosensory cues. Their results indicate that the signature on the surface of adult worker's thoraces is dynamic: changes are evident on a daily scale during the first week after adult eclosion, and changes continue to occur up to the foraging stage when individuals come into contact with environmental odours external to the hive.

Breed *et al.* (1988a) used the arena type experiments (see page 386) to investigate the ontogeny of chemosensory cues in honey bees. By introducing bees within hours, rather than days, to five-day-old recipient groups of 10 individuals, they were able to establish that cues indicating relatedness develop within 12 hours of eclosion. Also, bees maintained in the hive rather than the incubator for the first five days were no longer acceptable to recipient groups maintained in the laboratory, even if the introduced bee is a sister of the recipients (Breed *et al.*, 1988a), used naturally mated queens; thus they could not distinguish between super and half sisters). Breed *et al.*, interpret this loss of acceptability as a hive effect (chemicals picked up from the hive that have an endogenous – worker wax production – and/or an environmental source). They showed that the hive effect is able to overwhelm the genetic component, if newly emerged workers spend as little as 0.5 hours in the hive before being introduced to the recipient group.

Breed *et al.* (1988b) followed the above study with a set of experiments in which workers maintained in the incubator for five days were exposed in some treatments to wax comb substrate from various hives, and in other experiments were exposed to paraffin wax with or without anise oil. From their data they conclude that naturally produced wax alters the recognition signature of each individual in the hive, while paraffin wax (which on its own has no effect) is capable of carrying odorants that can alter the recognition signature of the workers. They were also able to show that anise oil embedded in sugar candy used to feed the bees does not appear to modify the recognition signature. Thus it appears that food odours may only contribute significantly to the recognition signatures of individuals once it becomes incorporated into the honey comb wax in the hive. Such environmentally modified signatures can of course be used by guards to discriminate between nestmates, but for kin recognition within the hive, the genetic component must be distilled out. In the context of Fig. 13.3, this may not be a problem if the variation between super and half sister signatures is maintained once an additional environmental component has been added to each signature.

Gas chromatography has been used to analyze the composition of the cuticular wax signature in honey bees. Blomquist *et al.* (1980) found that the cuticular wax signature changes with age, while Carlson & Bolten (1984) and McDaniel *et al.* (1984, 1987) have demonstrated that racially correlated differences exist in cuticular wax composition. Page *et al.* (1991) have identified at least 20 compounds that could potentially contribute to a cuticular wax signature. A preliminary analysis that I have undertaken of this data indicates that greatest variation among individuals occurs in the ratio of short chained (23–27 carbon atoms) to long chained (28–33 carbon atoms) alkanes and alkenes, while a substantial level of variation is present in the ratio of the alkanes to alkenes.

Getz & Smith (1987) used discriminative conditioning to assess how well workers can discriminate between mixtures of various proportions of closely related n-alkanes, specifically tricosane (C_{23}) and pentacosane (C_{25}). Their results indicate that workers can only discriminate between the two compounds, but also between a 90:10 and 10:90 mixture of the two compounds, and between the pure compounds and a 50:50 mixture. Even stronger levels of discrimination were obtained for tests involving mixtures of undecanoic (C_{11}) and dodecanoic (C_{12}) fatty acids. Further experiments by Getz & Smith (1987, 1990) indicate that workers perceive a similarity between mixtures and their components; but, in general, workers are unable to perceive the individual components in a mixture. Thus, the results obtained by Getz & Smith (1987, 1990) indicate that workers have the chemosensory acuity to discriminate between very closely related mixtures of cuticular hydrocarbons, but in most cases as theoretical and mixture perception experiments suggest, workers are unlikely to be able to identify individual components that make up a cuticular hydrocarbon signature. Moritz & Southwick (1987) and Moritz & Crewe (1988a,b) have used a metabolic assay to assess the response of workers to volatile queen odours. Their results suggest that both mandibular and tergal glands, among others, are associated with recognition of queens.

Sociogenetics

One of the primary purposes of studying kin recognition is to understand the sociological context in which discrimination is expressed, and to see how this contributes to the adaptive fitness of individuals. If kin recognition does enable a particular patriline to assert itself by preferentially rearing its super sisters as reproductives, then this behaviour may be manifested at several points in the life cycle of individuals and the phenology of the colony, including: workers eating eggs laid by the queen in queen cells; workers preferentially taking care of super-sister larvae; workers accepting emerging virgins or influencing the outcome of conflicts between virgins; workers segregating during swarming, either remaining behind in the hive with a super-sister reproductive (primary swarming event) or leaving to accompany a super-sister reproductive (afterswarming event); and workers eating the eggs of half sisters in queenless colonies where several workers are laying unfertilized (drone) eggs.

Page & Metcalf (1982), as well as Getz *et al.* (1982), suggested that kin recognition, if it exists, might mediate competition between individuals in different patrilines in honey bee colonies. Getz *et al.* (1982) found that cordovan and wildtype patrilines segregated non-randomly during swarming, and they could not account for this in terms of a differential age distribution among the patrilines (very young workers are much less likely to swarm). They pointed out, however, that their results could be explained by a greater genetic predisposition of individuals to swarm in one patriline than in the other.

Further evidence of sociogenetic structure in an ostensibly normal hive[13] was obtained, as already discussed, by Noonan (1986) in the context of larval rearing of queens (the fact that she grafted the larvae into queen cells may be a critical departure from the natural situation), and Frumhoff & Schneider (1987) in the context of worker feeding and grooming interactions. More recently, both Frumhoff & Baker (1988) and Robinson & Page (1988) found evidence that the various tasks undertaken by workers in the hive are partly genetically determined.

Frumhoff & Baker (1988), observing cordovan and wildtype workers in two patriline colonies, were able to show in two different colonies that a much larger than random proportion of groomers came from one of the two patrilines. In a control experiment, using a CW queen (a cordovan-wildtype cross – see page 385) crossed with a cordovan drone so that within a single patriline both cordovan and wild-type worker phenotypes were present, Frumhoff and Baker (1988) found no significant deviation from random

---

[13] Restricting experimental hives to typically two patrilines is just one problem in extrapolating results to truly natural situations.

proportions. Thus they could not link discrimination to the cordovan gene. Further, Frumhoff & Baker (1988) found that within patrilines there appear to be a number of individuals that actually specialize in grooming other individuals. Frumhoff & Baker found no strong biases for nestmate feeding. A possible explanation for this is that all workers must necessarily be involved in feeding activities, but grooming is a task that can most efficiently be undertaken by specialists with experience.

Robinson & Page (1988) used allozyme markers and electrophoresis to identify paternity of workers in colonies that had been set up using instrumental insemination of queens and drones from selected allozyme lineages. Working with five different colonies, they sampled both the genetic structure of appropriately aged workers within the hive and the genetic structure of comparably aged groups of workers performing the specialized tasks of guarding and removing dead bees from the hive. Their results indicate a strong divergence of behavioural tasks among patrilines (respectively labelled S, M, and F to designate slow, medium and fast allozyme markers). In one of their colonies, for example, undertakers predominately came from the patriline with the S allozyme marker even though their control indicated a predominance of F individuals. An analysis, four months later, of this same colony revealed that undertakers were almost exclusively from the S patriline, even though now both M and F patritype individuals were more numerous.

Even when it comes to tasks that all workers participate in at some stage of their life, there are genetic differences in the propensity of workers to perform these tasks and the age at which they begin undertaking these tasks. For example, Calderone & Page (1988) selected high and low pollen-hoarding strains of bees and added 175 tagged one-day-old workers from each strain to colonies with naturally mated queens. The behaviour of these introduced workers were observed in terms of where they were located in the hive, what tasks they engaged in, how old they were when they took their first foraging flight, and how often they returned with pollen from their foraging trips. Results from two sets of such specially constructed colonies indicate substantial differences in the overall behaviour of the two strains including their spatial distribution within the hive (the latter is probably just a reflection of the fact that individuals from the different strains are performing a different mix of tasks).

Robinson & Page (1988) speculate that '. . . increased intra-colony genetic variability may have selective value, if a collection of specialized genotypes performs more efficiently than a single, more generalized genotype.' Whatever the adaptive significance of this variability is, it helps to maintain genetic separation of the different patrilines within a colony; and, this separation in turn is conducive to promoting and maintaining kin discrimination.

Conclusion

Individual honey bees co-operate with their nestmates to form a biological unit, which we refer to as a colony. Is this colony a superorganism? Wilson (1971) reviews the history of Wheeler's idea that the ant colony is a superorganism (Wheeler, 1911). Modern biology has enabled us to resolve this issue at one level: morphologically distinct cells in an organism have a clonal genetic origin, while each individual within a honey bee colony is genetically distinct. This difference is extremely important, but too often it is pushed into the background and its relevance is lost. Certainly individual queen honey bees are unable to reproduce without workers, and workers themselves are unable to mate. Thus, the inclusive fitness of individuals is in consonance with the reproductive success of a colony. However, a gene that promotes nepotism in queen rearing could conceivably invade a population, even if this gene marginally decreases the overall reproductive success of the colony as a whole.

For example, consider the situation where sometimes more than one egg is laid in a cell. This is known to occur in natural hives, especially under crowded hive conditions which often prevail around the time of swarming. This situation also occurs in queenless colonies where several workers are laying unfertilized (i.e. drone) eggs. Since only one larva can be reared in a cell, workers need to dispose of extra eggs in the cell, the most efficient mechanism being consumption of those extra eggs. This behaviour can lead to a plausible scenario, whose components are in large part supported by current evidence, related to egg eating in hopelessly queenless colonies.

In the previous section, we presented several studies that point to the fact that workers are able to discriminate (at least partially) in the hive between super- and half-sister adult workers (Table 13.2). We know that workers are able to discriminate between pairs of eggs (Getz *et al.*, 1988). We also know that workers can preferentially rear super sisters as queens. Thus, several pieces of the puzzle exist, but the actual event of workers preferentially eating eggs that are their half rather than their super sisters has not as yet been recorded.

Preferential egg eating has been observed to occur in the case of hopelessly queenless colonies where R. Page (pers. comm.), has evidence that some workers are at least eating more unfertilized eggs laid by their half sisters than laid by their super sisters. The basis for this recognition, of course, could be a maternal odour associated with the egg. If one patriline practices preferential half-sister egg eating while another patriline does not, then the hive will produce a greater proportion, than random, of drones descended from females in the 'discriminating' patriline. If all patrilines discriminate then the reproductive success of the hive is not reduced as long as workers only continue to eat excess eggs (i.e. they will not eat a viable egg if it is the only one in the cell, irrespective of kinship considerations). In this case, the behaviour may still be to the benefit of a particular line; for example, the line that discriminates the best, or that has the greatest relative proportion of workers that are egg eaters. Note that the trait to eat the only egg in a cell would have a devastating impact on a hopelessly queenless colony if it were widespread. Thus, one would expect this trait to be limited by its effects on the reproductive fitness of the hopelessly queenless colony as a whole.

Eggs very likely pick up an odour from their mothers and it is possible that workers tending these eggs discriminate between them on this basis. For example, Ratnieks & Visscher (1989) have demonstrated that workers in manipulated colonies overwhelmingly prefer to rear haploid drone eggs laid by their sisters. They hypothesize that the functional significance of this behaviour is related to workers resolving a conflict over rearing male reproductives: workers should prefer to rear their own sons (to whom they are related by $r = \frac{1}{2}$) or the sons of their super sisters ($r = \frac{3}{8}$) to rearing their brothers ($r = \frac{1}{4}$) or the sons of their half sisters ($r = \frac{1}{8}$). Since, under multiple mating or polyandry, if workers laid eggs, then other workers would end up rearing the sons of their half sisters. Thus the inclusive fitness solution for workers in a co-operative polyandrous society is to rear brothers instead of nephews. To ensure that workers do not cheat by laying eggs, other workers would be required to 'police' the hive (Ratnieks, 1988) and eat any worker eggs that are laid (Ratnieks & Visscher, 1989).

Of course in the context of queen rearing in a hive, all eggs are laid by the same queen so some non-queen-specific odour cues would have to be used if a worker is able to discriminate between eggs that are destined to become her super and half sister. Except for Noonan (1986) and Page et al. (1989), all the queen rearing experiments described above have not excluded queen odour as a possible source for egg recognition cues, while Noonan's result relates to choices between grafted one day old larvae – not eggs.

Even if workers cannot discriminate between eggs laid by the same queen, there are enough reproductive events relating to choices between eggs laid by different individuals to make egg discrimination an important contribution towards inclusive fitness. For example, when the fecundity of a queen deteriorates, several virgins are raised to supersede the queen. K. Smith and I have observed mothers and daughters simultaneously laying eggs in the same hive. In some of these cells we have observed two eggs. If one of the eggs was laid by the old queen and the other by the new queen, then a worker can enhance its reproductive fitness by eating its mother's egg if the new queen is its super sister, or eating the new queen's egg if she is the worker's half sister. (Note that workers are respectively related by $r = 0.375$ and $r = 0.125$ to the female progeny of their super and half sisters, assuming the queens and drones are unrelated, while they are related on average to their own sisters by $0.25 > r > 0.375$, assuming less than half the eggs of their mother belong to their patriline.)

We have pointed out in the introduction, and elsewhere in this chapter, that kin recognition can enhance an individual's fitness in other, more subtle, ways including choosing which reproductives to support during swarming and in afterswarms. Theory argues for the existence of kin discrimination, the empirical evidence establishes that the behavioural and perceptual components exist. Also kin discrimination has been observed under a variety of conditions, albeit in contrived situations (in many of these situations, kin discrimination could actually be interpreted as an artefact of nestmate discrimination).

Much work remains to be done before the story is satisfactorily understood. However, our understanding of the social interactions within the honey bee colony has dramatically improved due to recent findings.

Scientific programmes for the lay public invariably portray the honey colony as one of nature's remarkable achievements: thousands of programmed workers going about their business – in an environment of utopian co-operation (Gould & Gould, 1988) – building 'mathematically perfect' combs; rearing workers, queens, and drones; and communicating the location of resources using the waggle dance. This utopian view is no longer valid. The hive is a place where there is a delicate balance between co-operation and competition. Too much competition and the forces of conflict shatter the viability of the hive and eliminate the genes responsible for the discord; too much co-operation and competitive genes invade the population rendering the individuals that have them as the fittest in a colony.

Acknowledgements

Comments from Francis Ratnieks, Robert Page, Dorothea Brückner, Peter Frumhoff, Katherine Smith, and Bruce Waldman have helped to make this a better paper. All opinions and errors remain my own. Preparation of this manuscript was supported in part by NSF Grants BNS-8518037 and BNS-8809728 to WMG.

References

Baker, T. C. (1986). Pheromone-modulated movements of flying moths. In *Mechanisms in Insect Olfaction*, ed. T. L. Payne, M. C. Birch & C. E. J. Kennedy, pp. 39–48. Oxford: Oxford University Press.

Bitterman, M. E. (1988). Vertebrate-invertebrate comparisons. In *The Evolutionary Biology of Intelligence*, NATO ASI Series G, vol. 17, ed. H. J. Jerison & I. Jerison, pp. 251–76. Heidelberg: Springer Verlag.

Bitterman, M. E., Menzel, R., Fietz, A. & Schäfer, S. (1983). Classical conditioning of proboscis extension in honeybees (*Apis mellifera*). *J. Comp. Psych.*, **97**, 107–19.

Blomquist, G. J., Chu, A. J. & Remaley, S. (1980). Biosynthesis of wax in the honeybee *Apis mellifera* L. *Insect Biochem.*, **10**, 313–21.

Boch, R. & Morse, R. A. (1974). Discrimination of familiar and foreign queens by honey bee swarms. *Ann. Ent. Soc. Am.*, **67**, 709–11.

Boch, R. & Morse, R. A. (1979). Individual recognition of queens by honey bees *Ann. Ent. Soc. Am.*, **72**, 51–3.

Boch, R. & Morse, R. A. (1981). Effects of artificial odors and pheromones on queen discrimination by honey bees. *Ann. Ent. Soc. Am.*, **74**, 66–7.

Boch, R. & Morse, R. A. (1982). Genetic factor in queen recognition odors of honey bees. *Ann. Ent. Soc. Am.*, **75**, 654–6.

Boeckh, J. (1984). Neurophysiological aspects of insects olfaction. In *Insect Communication*, ed. T. Lewis, pp. 83–104. London: Academic Press.

Bossert, W. H. & Wilson, E. O. (1963). The analysis of olfactory communication among animals. *J. Theor. Biol.*, **5**, 443–69.

Breed, M. (1981). Individual recognition and learning of queen odours by worker bees. *Proc. Natl. Acad. Sci., USA*, **78**, 2635–7.

Breed, M. (1983). Nestmate recognition in honeybees. *Anim. Behav.*, **31**, 86–91.

Breed, M. & Bennett, B. (1987). Kin recognition in primitively eusocial insects. In *Kin Recognition in Animals*, ed. D. J. C. Fletcher & C. D. Michener, pp. 243–86. New York: Wiley.

Breed, M., Butler, L. & Stiller, T. M. (1985). Kin discrimination by worker honeybees in genetically mixed groups. *Proc. Natl. Acad. Sci., USA*, **82**, 3058–61.

Breed, M., Stiller, T. M. & Moor, M. J. (1988a). The ontogeny of kin discrimination in the honey bee, *Apis mellifera. Behav. Genet.*, **18**, 439–47.

Breed, M., Velthuis, H. H. W. & Robinson, G. E. (1984). Do worker honeybees discriminate among unrelated and related larval phenotypes. *Ann. Ent. Soc. Am.*, **77**, 737–9.

Breed, M., Williams, K. R. & Fewell, J. H. (1988b). Comb wax mediates the acquisition of nestmate recognition cues in honey bees. *Proc. Natn. Acad. Sci., USA*, **85**, 8766–9.

Buckle, G. R. & Greenberg, L. (1981). Nestmate recognition in sweatbees (*Lasioglossum zephyrum*): does an individual recognize its own odour or only odours of its nestmates? *Anim. Behav.*, **29**, 802–9.

Calderone, N. W. & Page, R. E. (1988). Genotypic variability in age polyethism and task specialization in the honey bee, *Apis mellifera* (Hymenoptera: Apidae) *Behav. Ecol. Sociobiol.*, **22**, 17–25.

Carlin, N. F. (1988). Species, kin and other forms of recognition in the brood discrimination behavior of ants. In *Advances in Myrmecology*, ed. J. Trager, pp. 267–95. Leiden: E. J. Brill.

Carlin, N. F., Frumhoff, P. C., Page, R. E., Breed, M. D., Getz, W. M., Oldroyd, B. P., Rinderer, T. E. & Robinson, G. E. (1990). Nepotism in the honey bee. *Nature*, **346**, 706–7.

Carlin, N. F. & Hölldobler, B. (1986). The kin recognition system of carpenter ants (*Camponotus* Spp.). I Hierarchical cues in small colonies. *Behav. Ecol.Sociobiol.*, **19**, 123–34.

Carlin, N. F. & Hölldobler, B. (1987). The kin recognition system of carpenter ants (*Camponotus* Spp.). II Larger colonies. *Behav. Ecol. Sociobiol.*, **20**, 209–18.

Carlson, D. & Bolten, A. B. (1984). Identification of Africanized and European Honeybees, using extracted hydrocarbons. *Bull. Entomol. Soc. Am.*, **30**(2), 32–5.

Crosland, M. W. J. (1989a). Kin recognition in the ant *Rhytidoponera confusa*. I Environmental odour. *Anim. Behav.*, **37**, 912–19.

Crosland, M. W. J. (1989b). Kin recognition in the ant *Rhytidoponera confusa.* II Gestalt odour. *Anim. Behav.,* 37, 920–6.

Crozier, R. H. (1987). Genetic aspects of kin recognition: concepts, models, and synthesis. In *Kin Recognition in Animals,* ed. D. J. C. Fletcher & C. D. Michener, pp. 55–73. New York: Wiley.

Crozier, R. H. & Dix, M. W. (1979). Analysis of two genetic models for the innate components of colony odour in social Hymenoptera. *Behav. Ecol. Sociobiol.,* 4, 217–24.

Crozier, R. H., Pamilo, P. & Crozier, Y. C. (1984). Relatedness and microgeographic genetic variation in *Rhytodoponera mayri,* an Australian arid-zone ant. *Behav. Ecol. Sociobiol.,* 15, 143–50.

Crozier, R. H. & Page, R. E. (1985). On being the right size; male contributions and multiple mating in social Hymenoptera. *Behav. Ecol. Sociobiol.,* 4, 217–24.

Dawkins, R. (1982). *The Extended Phenotype.* Oxford: Oxford University Press.

Driessen, G. J. J., Raalte, A. G. van & DeBruyn, G. J. (1984). Cannabilism in the red wood ant, *Formica polyctena* (Hymenoptera: Formicidae). *Oecologia,* 63, 13–22.

Evers, C. A. & Seeley, T. D. (1986). Kin discrimination and aggression in honey bee colonies with laying workers. *Anim. Behav.,* 34, 924–5.

Frisch, K. von (1967). *The Dance Language and Orientation of Bees.* Cambridge, Mass: Harvard University Press.

Frumhoff, P. C. (1988). *The Social Consequence of Honeybee Polyandry in Honey Bees,* Apis mellifera *L.* Ph.D. Thesis, University of California, Davis.

Frumhoff, P. C. & Baker, J. (1988). A genetic component to division of labour within honey bee colonies. *Nature,* 333, 358–61.

Frumhoff, P. C. & Schneider, S. (1987). The social consequence of honeybee polyandry: kinship influences worker interactions within colonies. *Anim. Behav.,* 35, 255–62.

Gadagkar, R. (1985). Kin recognition in social insects and other animals – A review of recent findings and a consideration of their relevance for the theory of kin selection. *Proc. Indian Acad. Sci. (Anim. Sciences),* 94, 587–621.

Gamboa, G. J., Reeve, H. K. & Pfennig, D. W. (1986a). The evolution and ontogeny of nestmate recognition in social wasps. *Ann. Rev. Ent.,* 31, 431–54.

Gamboa, G. J., Reeve, H. K., Ferguson, I. & Wacker, T. L. (1986b). Nestmate recognition in social wasps: the origin and acquisition of recognition odours. *Anim. Behav.,* 34, 685–95.

Getz, W. M. (1981). Genetically based kin recognition systems. *J. Theor. Biol.,* 92, 209–26.

Getz, W. M. (1982). An analysis of learned kin recognition in Hymenoptera. *J. Theor. Biol.,* 99, 585–97.

Getz, W. M. (1991). A neural network for processing olfactory-like stimuli. *Bull. Math. Biol.* (In Press.)

Getz, W. M., Brückner, D. & Parisian, T. R. (1982). Kin structure and the swarming behavior of the honeybee *Apis mellifera. Behav. Ecol. Sociobiol.,* 10, 265–70.

Getz, W. M., Brückner, D. & Smith, K. B. (1986). Conditioning honeybees to discriminate between heritable odors from full and half sisters. *J. Comp. Physiol. A,* 159, 251–6.

Getz, W. M., Brückner, D. & Smith, K. B. (1988). Variability of chemosensory stimuli within honeybee (*Apis mellifera*) colonies: a differential conditioning assay for discrimination cues. *J. Chem. Ecol.*, **14**, 249–60.

Getz, W. M., Brückner, D. & Smith, K. B. (1989). The ontogeny of cuticular chemosensory cues in worker honey bees *Apis mellifera. Apidologie*, **20**, 105–13.

Getz, W. M. & Chapman, R. F. (1987). An odor discrimination model with application to kin recognition in social insects. *International J. Neuroscience*, **32**, 963–78.

Getz, W. M. & Smith, K. B. (1983). Genetic kin recognition: honeybees discriminate between full and half sisters. *Nature*, **302**, 147–8.

Getz, W. M. & Smith, K. B. (1986). Honeybee kin recognition: learning self and nestmate phenotypes. *Anim. Behav.*, **34**, 1617–26.

Getz, W. M. & Smith, K. B. (1987). Olfactory sensitivity and discrimination of mixtures in the honeybee *Apis mellifera. J. Comp. Physiol. A.*, **160**, 239–45.

Getz, W. M. & Smith, K. B. (1990). Odorant moiety and odor mixture perception in free flying honey bees (*Apis Mellifera*). *Chem. Senses*, **15**, 111–28.

Gould, J. L. & Gould, C. G. (1988). *The Honey Bee*. New York: W. H. Freeman.

Greenberg, L. (1979). Genetic component of bee order in kin recognition. *Science*, **206**, 1095–7.

Hamilton, W. D. (1964). The genetical evolution of social behaviour. I, II. *J. Theor. Biol.*, **7**, 1–52.

Hamilton, W. D. (1987). Kinship, recognition, disease, and intelligence constraints of social evolution. In *Animal Societies: Theories and Facts*, ed. Y. Itô, J. L. Brown & J. Kikkawa, pp. 81–102. Tokyo: Japan Science Society Press.

Hogendoorn, K. & Velthuis, H. H. W. (1988). Influence of multiple mating on kin recognition by worker honeybees. *Naturwiss.*, **75**, 412–13.

Hölldobler, B. & Carlin, N. F. (1987). Anonymity and specificity in the chemical communication signals of social insects. *J. Comp. Physiol. A.*, **161**, 567–81.

Kalmus, H. & Ribbands, C. R. (1952). The origin of odors by which honey bees distinguish their companions. *Proc. Royal Soc. (B)*, **140**, 50–9.

Lacy, R. C. & Sherman, P. W. (1983). Kin recognition by phenotype matching. *Am. Nat.*, **121**, 489–512.

Laidlaw, H. H. & Eckert, J. E. (1962). *Queen rearing*. Berkeley: University of California Press.

Mackensen, O. & Roberts, W. C. (1948). A manual for the artificial insemination of queen bees. *U.S. Dept. of Agric. Bur. Entomol. Plant Q.*, *ET*–250.

McDaniel, C. A., Howard, R. W., Blomquist, C. J. & Collins, A. M. (1984). Hydrocarbons of the cuticle, sting apparatus, and sting shaft of *Apis mellifera* L. Identification and preliminary evaluation of chemotaxonomic characters. *Sociobiology*, **8**, 287–98.

McDaniel, C. A., Howard, R. W., Collins, A. M. & Brown, W. A. (1987). Variation in the hydrocarbon composition of non-africanized *Apis mellifera* L. sting apparatus. *Sociobiology*, **13**, 133–43.

Menzel, R. (1979). Behavioral access to short-term memory in bees. *Nature*, **281**, 368–9.

Menzel, R. (1983). Neurobiology of learning and memory: the honey bee as a model system. *Naturwiss.*, **70**, 504–11.

Michener, C. D. (1974). *The Social Behavior of the Bees: A Comparative Study*. Cambridge, Mass: The Belknap Press of Harvard University Press.

Michener, C. D. & Smith, B. H. (1987). Kin recognition in primitively eusocial insects. In *Kin Recognition in Animals*, ed. D. J. C. Fletcher & C. D. Michener, pp. 209–42. New York: Wiley.

Moritz, R. F. A. (1985). The effects of multiple mating on the worker–queen conflict in Apis mellifera L. *Behav. Ecol. Sociobiol.*, **16**, 375–7.

Moritz, R. F. A. (1988). Group relatedness and kin discrimination in honey bees *Apis mellifera* L. *Anim. Behav.*, **36**, 1334–40.

Moritz, R. F. A. & Crewe, R. M. (1988a). Chemical signals of queens in kin recognition of honey bees (*Apis mellifera* L). *J. Comp. Physiol. A*, **64**, 83–9.

Moritz, R. F. A. & Crewe, R. M. (1988b). Reaction of honey bee workers (*Apis mellifera* L) to fatty acids in queen signals. *Apidologie*, **19**, 333–42.

Moritz, R. F. A. & Southwick, E. E. (1987). Metabolic test of volatile odor labels as kin recognition cues in honey bees (*Apis mellifera* L). *J. Exp Zool.*, **243**, 503–7.

Murlis, J. (1986). The structure of odor plumes. In *Mechanisms in Insect Olfaction*, ed. T. L. Payne, M. C. Birch & C. E. J. Kennedy, pp. 27–38. Oxford: Oxford University Press.

Noonan, K. C. (1986). Recognition of queen larvae by worker honey bees (*Apis mellifera* L). *Ethology*, **73**, 295–306.

Page, R. E. (1980). The evolution of multiple mating behavior by honey bee queens (*Apis mellifera* L.). *Genetics*, **96**, 263–73.

Page, R. E. (1986). Sperm utilization in social insects. *Ann. Rev. Ent.*, **31**, 297–320.

Page, R. E. & Breed, M. D. (1987). Kin recognition in social bees. *Trend Ecol. Evol.*, **2**, 272–5.

Page, R. E. & Erickson, E. H. (1984). Selective rearing of queens by worker honey bees: kin or nestmate recognition. *Ann. Ent. Soc. Am.*, **77**, 587–80.

Page, R. E. & Erickson, E. H. (1986a). Kin recognition and virgin acceptance by worker honey bees (*Apis mellifera* L.) *Anim. Behav.*, **34**, 1061–9.

Page, R. E. & Erickson, E. H. (1986b). Kin recognition during emergency queen rearing by worker honey bees (Hymenoptera: Apidae). *Ann. Ent. Soc. Am.*, **79**, 460–7.

Page, R. E. & Erickson, E. H. (1988) Reproduction by worker honey bees (*Apis mellifera* L.). *Behav. Ecol. Sociobiol.*, **23**, 117–26.

Page, R. E. & Laidlaw, H. H. (1988). Full sisters and super sisters: a terminological paradigm. *Anim. Behav.*, **36**, 944–5.

Page, R. E. & Metcalf, R. A. (1982). Multiple mating, sperm utilization, and social evolution. *Am. Nat.*, **119**, 117–26.

Page, R. E., Metcalf, R. A., Metcalf, R. L., Erickson, E. H. & Lampman, R. L. (1991). Extractable hydrocarbons and kin recognition in the honey bee (*Apis mellifera* L.) *J. Chem. Ecol.* (In Press.)

Page, R. E., Robinson, G. E. & Fondrk, M. K. (1989). Genetic specialists, kin recognition, and nepotism. *Nature*, **338**, 576–9.

Ratnieks, F. L. W. (1988). Reproductive harmony via mutual policing by workers in eusocial Hymenoptera. *Am. Nat.*, **132**, 217–36.

Ratnieks, F. L. W. (1990a). The evolution of polyandry by highly eusocial bee queens: the significance of progressive versus mass-provisioning and timing of removal of diploid drones. *Behav. Ecol. Sociobiol.*, **26**, 343–8.

Ratnieks, F. L. W. (1990b). The evolution of genetic odor cue diversity in social Hymenoptera. *Am. Nat.* (In press.)

Ratnieks, F. L. W. & Reeve, H. K. (1990). The evolution of queen-rearing discrimination in social hymenoptera: effects of discrimination costs in swarming species. *J. Evol. Biol.* (In press.).

Ratnieks, R. E. & Visscher, P. K. (1989). Worker policing in the honey bee. *Nature*, **342**, 796–7.

Reeve, H. K. (1989). The evolution of conspecific thresholds. *Am. Nat.*, **133**, 407–35.

Robinson, G. E. & Page, R. E. (1988). Genetic determination of guarding and undertaking in honey bee colonies. *Nature*, **333**, 356–8.

Ross, K. G. & Fletcher, D. J. C. (1985). Comparative study of genetic and social structure in two forms of the fire ant, *Solenopsis invicta* (Hymenoptera: Formicidae). *Behav. Ecol. Sociobiol.*, **17**, 349–56.

Sass, H. (1976). Zur nervösen von Geruchsreizen bei *Periplaneta americana*. *J. Comp. Physiol.*, **107**, 49–65.

Sherman, P. W., Seeley, T. D. & Reeve, H. K. (1988). Parasites, pathogens and polyandry in social Hymenoptera. *Am. Nat.*, **131**, 602–10.

Smith, B. H. & Weller, C. (1989). Competitive signalling among gynes in social halictine bees. The influence of bee size and pheromones on behavior. *J. Insect Behav.*, **2**, 397–411.

Spiess, E. B. (1987). Discrimination among prospective mates in *Drosophila*. In *Kin Recognition in Animals*, ed. D. J. C. Fletcher & C. D. Michener, pp. 75–120. New York: Wiley.

Taber, S. (1955). Sperm distribution in the spermathecae of multiply mated queen honeybees. *J. Econ. Entomol.*, **48**, 522–5.

Vareschi, E. (1971). Duftunterscheidung bei der Honigbiene – Einzelzell-Ableitungen und Verhaltensreaktionen. *Z. vergl Physiologie*, **75**, 143–73.

Visscher, P. K. (1986). Queen rearing by honey bees (*Apis mellifera*). *Behav. Ecol. Sociobiol.*, **18**, 453–60.

Waldman, B. 1986. Chemical ecology of kin recognition in anuran amphibians. In *Chemical Signals in Vertebrates, Vol. 4: Evolution and Comparative Biology*, ed. D. Duvall, D. Müller-Schwarze & R. M. Silverstein, pp. 225–42. New York: Plenum Press.

Wheeler, W. M. (1911). The ant colony as an organism. *J. Morphology*, **22**, 307–25.

Wilson, E. O. (1971). *The Insect Societies*. Cambridge, Mass: The Belknap Press of Harvard University Press.

Winston, M. L. (1987). *The Biology of the Honey Bee*. Cambridge, Mass: Harvard University Press.

14

Mutual mother–infant recognition in humans

Richard H. Porter

Introduction

Kin recognition, especially the failure to recognize close relatives, is a commonly occurring theme in the history and literature of western civilization. Sophocles' tragic tale of Oedipus is perhaps the best known example of kin recognition gone awry. Oedipus followed the inadvertent murder of his father (Laius) by unknowingly marrying his own mother (Jocasta). According to the Old Testament (Genesis 27:27) an error in kin discrimination enabled Jacob to cheat his older brother Esau out of his birthright. By disguising himself as Esau, Jacob deceived their father (Isaac) into granting him the final blessing and inheritance intended for the first born son. Similarly, the fate of Odysseus (Ulysses) was ultimately decided by his inability to identify one of his offspring. On his final voyage, Odysseus launched an attack against Telegonus – his son born of the goddess Circe. Only after being mortally wounded by the younger man did the hero of Homer's saga learn of their relationship.

Despite such long-held fascination with the consequences of kin recognition errors, humans are evidently quite proficient at recognizing family members (as well as other individuals). Nonetheless, there is little knowledge of the scope of human kin recognition capabilities or the underlying bases of this form of social discrimination. As recently pointed out by Wells (1987), in comparison to research with various non-human species, there have been relatively few empirical studies of kin recognition in our own species. Much of our limited current understanding of human kin recognition stems from investigations of early parent–infant interactions. Since the restricted social experiences of neonates can be monitored and documented, newborns afford unique advantages as both subjects and stimuli for research on the development of social recognition. The present review focuses primarily on the results of empirical studies directly relevant to the issue of mutual parent–

infant recognition in humans and considers the adaptive significance, sensory mediation and ontogenetic mechanisms of such recognition.

Practical importance and biological significance of parent–infant recognition

The salience of accurate kin discrimination in contemporary human society is strikingly evident in cases of contested paternity and inheritance, and the need for compatible donors for effective organ and tissue transplants. Regardless of species, however, it is unlikely that individuals derive any direct benefits simply through recognizing their kin. Rather, kin recognition facilitates discriminative treatment of kin, resulting in ultimate (genetic) or more immediate advantages to the interacting individuals (Hamilton, 1964). The potential benefits associated with parent–infant recognition may differ somewhat for the two members of such dyads. As will be seen below, this asymmetry in the parent–infant relationship is a reflection of basic differences in the control of resources and obligate reliance on others for sustaining normal development and survival.

Parental recognition of offspring

From a strictly biological perspective, the advantages that parents accrue through recognizing and responding discriminatively to their own offspring are readily apparent. Newborn mammals, including humans, are dependent upon adult conspecifics (typically the mother, or possibly both parents) for the provision of food, shelter and other essential commodities. Since the necessary parental resources are limited, those parents who discriminate between their own and other neonates, and invest preferentially in the former, would tend to produce more offspring who survive to reproductive age than would parents who respond indiscriminately to newborns. There would be heavy selection pressure against wasting resources on unrelated individuals since such behaviour has the effects of depriving and thereby endangering one's own offspring. Thus, a lactating female who allows unrelated young to suckle might not be able to provide sufficient nutrients to her own offspring. Over numerous generations, the evolutionary process should, therefore, have enhanced the ability of parents to recognize their progeny.

According to this conceptualization, parents gain in an ultimate genetic sense through being able to recognize their offspring. While adults may derive considerable personal satisfaction by caring for and becoming attached to unrelated neonates, they would not thereby accrue the same genetic benefits associated with investment in their own children. Reserving parental solicitude only for one's progeny does not necessarily increase the immediate well-being of parents, but it can have a profound effect on their

Darwinian fitness or reproductive success. It is consistently observed across cultures that adults invest primarily in their own offspring – or the offspring of other close kin, in which instance the genetic arguments developed above also apply (Hamilton, 1964). Even in modern industrialized societies, human infants benefit through preferential parental investment. Of course, in the case of our own species, parental resources may go well beyond food and shelter. Those individuals who have access to more abundant parental resources (e.g. medical care, proper diet, superior education) are nonetheless more likely to survive and succeed than are those born into disadvantaged families.

Recognition of parents by offspring
Recognition of the mother (or other caregivers) is a necessary component of the attachment process. An infant can only begin to develop a unique relationship with its mother if that woman can be distinguished from other adults; social attachment implies social discrimination. Therefore, elucidation of this precursor to infant–mother attachment may provide further insights into the development of infant socialization.

Although human infants must rely upon their adult caregiver(s) for all physical needs, they are not simply passive recipients of such parental largess. Beginning at an early age, parent–infant interactions reflect a subtle mutual interchange, with parents seemingly very sensitive to signals emanating from their neonate (e.g. Lewis & Rosenblum, 1974). One means through which infants can exert some influence over the parent is to convey the impression that the parent can be recognized as a distinct individual. Recognition becomes evident as the infant begins to respond differentially to the parent as opposed to others – especially in the form of smiling and visual orientation (Robson & Moss, 1970; Maurer & Salapatek, 1976; Masi & Scott, 1982). According to anecdotal accounts, as the infant begins to show signs of recognizing the mother, the parent responds with increasingly positive attitudes towards her offspring. In a detailed investigation of maternal attachment by Robson & Moss (1970), mothers reported that being recognized by their infant was 'both reassuring and gratifying'. Approximately one-third of the mothers interviewed in this study felt that their infant 'became a person' when it began to recognize them. By strengthening maternal attachment, an infant's discriminative social responsiveness can increase the likelihood of subsequent parental care and investment. On the other hand, parents may experience difficulties in developing positive social relationships with infants who fail to respond socially to them, and such deficits may even evoke hostile reactions from the parents (Robson & Moss, 1970).

The infant's initial discriminative relationship with its mother (or with others with whom it frequently interacts) may also serve as the foundation for developing further skills for coping with the physical and social environment. Schaffer (1971) stresses that early parent–infant attachment provides a context which allows for the establishment of additional social relationships. The infant's distinctive bond with parents may be a critical first step towards eventually becoming a competent social being (Cairns, 1979). It has likewise been suggested that effective systematic exploration of the environment cannot begin until the infant becomes socially attached to the primary caregiver(s) (Bowlby, 1969; Sluckin, 1972). As the child grows increasingly mobile, the physical presence of the mother may provide a base from which to venture forth to become familiarized with the surroundings (Rheingold, 1969; Ainsworth & Bell, 1970). Once again, social recognition is important in this context since it is a necessary component of any continuing relationship between parent and offspring.

Of course, fully adequate neonatal care can be provided by adults other than the biological parents. Therefore, it is not critical that infants recognize their own parents *per se* – rather, preferential responses should be directed to the primary caregiver(s) regardless of that individual's genetic relationship to the child. Infants may reap immediate personal benefits, in the form of enhanced care, through recognizing and interacting discriminatively with their caregiver; which would have the concomitant effect of increasing the likelihood that the infants would survive to pass their genes on to future generations. As long as the infant is receiving proper care, whether or not the source of that care is the biological mother may be of little relevance from the perspective of the child. In contrast (as discussed above), for parents to gain any *genetic* advantage, it is imperative that they allot their investment to their own young – or to other closely-related neonates.

Sensory mediation: recognition of infants by mothers (and fathers)

Because of the biological significance of accurate parent–infant recognition, one might expect that discrimination of neonatal offspring could be accomplished through several sensory modalities. Redundant recognition cues would allow for the reliance on different sensory systems according to the immediate constraints posed by the environment. For example, while visual recognition might be most reliable and appropriate when parent and infant are in close physical proximity and adequate illumination is available, auditory signals may become more salient at a distance or in the dark. Empirical evidence implicating the major sensory modalities in parental recognition of neonates is summarized below.

Vision

Vision is without doubt the pre-eminent sensory modality for human social recognition. Of the various features by which an individual may be identified visually, the face is the most distinctive and conspicuous (Bruce & Young, 1986). Despite marked changes in the visual appearance of neonates during the early postpartum period, mothers are quite accurate in recognizing their newborn infant through facial features alone. Within 33 hours after delivery, mothers were tested for their ability to identify their own infant from an array of colour photographs (full-face views) of four infants of the same sex (Porter *et al.*, 1984). For each test trial, the three comparison infants were unfamiliar to the subject mother and matched as closely as possible to her infant on body weight and amount of hair. Identical results were obtained in each of two replications of this experiment – i.e. 11 of 12 mothers correctly picked their own child from amongst the four stimulus photographs. These results are especially noteworthy since the two-dimensional still photographs used in this study were less than optimal representations of the neonates' faces. It is reasonable to assume that mothers would perform even better if tested with the infants themselves as visual stimuli.

An additional series of experiments attempted to ascertain whether particular facial regions are differentially salient for individual recognition of neonates (Porter, Boyle, Hardister, & Balogh, unpub. data). As in the previous study, mothers were tested with photographs of four infants, one of whom was their own. When the eyes, nose or mouth of each stimulus photograph was masked, subject mothers were still able to identify reliably their offspring (who were less than 36-hours-old when the photographs were taken). Likewise, mothers performed at a better-than-chance level when tested with photographs of isolated facial features of neonates – i.e. eyes, nose, mouth or outline of the head. These data indicate that the entire face is not necessary for mothers to discriminate between photographs of their own and other infants, and there is sufficient variability in the major facial features of neonates within the first one to two days after birth to mediate individual recognition.

Audition
Neonatal vocalizations, which are correlated with the state and needs of infants, compose an important channel for communicating with caregivers (Vuorenkoski 1975). Parents commonly report that they are highly attuned to sounds (especially cries) produced by their neonate and may be roused from a sound sleep by even low-volume whimpers (Illingworth, 1955). But can the cries of one's own infant be discriminated from those of others of the same age?

Researchers on both sides of the Atlantic have addressed this question by testing mothers with tape recordings of infants' cries (Formby, 1967; Valanne *et al.*, 1967; Murry *et al.*, 1975; Morsbach & Bunting, 1979). The results of these studies consistently indicate that mothers can discern their own infant's cries; with many showing evidence of such auditory recognition within the first two days after delivery (Formby, 1967; Valanne *et al.*, 1967).

Green & Gustafson (1983) have recently cautioned that the subject mothers in the previous studies were presumably present when the stimulus tapes of their infant were recorded. They might, therefore, have been able to remember the particular cries during later testing rather than recognizing their offspring by the acoustic quality of the cries. To obviate this problem, the cries of 30-day-old infants were recorded when the parents were not present to hear them. In subsequent tests, 80 per cent of the mothers and 45 per cent of the fathers recognized recordings of their own infant's cries (which they had not previously heard) when presented along with the cries of three other infants of the same age.

Olfaction

It is apparent from research conducted within the last 10 to 15 years that biologically produced olfactory cues may exert a subtle, but nonetheless meaningful, influence over human social interactions (for reviews see Doty, 1981; Filsinger & Fabes, 1985; Porter *et al.*, 1988). As is true for other mammalian species, humans are able to recognize individuals by their characteristic body odours (Hold & Schleidt, 1977; Wallace, 1977; Porter & Moore, 1981). The relevance of olfactory cues for social discrimination is evident shortly after birth. In a series of methodologically similar experiments conducted in different laboratories, recently parturient mothers were tested with an array of T-shirts that had been worn and soiled by newborn infants (Schaal *et al.*, 1980; Porter *et al.*, 1983; Schaal, 1985; Kaitz *et al.*, 1987). Consistent results were obtained in all of these independent studies; mothers reliably identified T-shirts worn by their own neonates, by olfactory cues alone, within several days postpartum. Similar findings were obtained when blindfolded mothers were instructed to sniff the heads of three infants and select the one they believed to be their own (6-hours-old) child (Russell *et al.*, 1983).

Related experiments in which *fathers* attempted to identify their own offspring by odours have yielded contradictory results. Russell *et al.* (1983) tested 10 blindfolded fathers and found no evidence that they could recognize the odours of the head of their own child at 24 to 48 hours after birth. By comparison, fathers were successful in recognizing the odours of T-shirts previously worn by their one-to two-day-old neonates (Porter *et al.*, 1986). In a series of two-choice discrimination tests, 26 of 30 fathers correctly chose their own infant's shirt over that worn by an unfamiliar infant. The discrepancy in the data from these two studies could be due to a difference in the difficulty of the discrimination tasks that were employed. Whereas the subjects tested with actual infants were given a three-choice discrimination task, only two choices were available to fathers who were presented with T-shirts as odour stimuli. Furthermore, the odours of the soiled garments may have been more intense and therefore more easily discriminated than the odours of the infants' head – especially since the infants in the latter experiment (Russell *et al.*, 1983) had been washed before the test trials. Potentially recognizable olfactory signatures could have been removed at least partially by this procedure or masked by whatever substance had been employed in the washing. That mothers could still recognize infants treated in this same manner does suggest that they are more sensitive to the olfactory characteristics of their newborn than are fathers. This difference between mothers and fathers could be a reflection of generally superior olfactory capabilities in females. At ages ranging from 2 days (Balogh & Porter, 1986) through over 90 years (Doty *et al.*, 1984), female subjects consistently perform better than males on olfactory discrimination and identification tests.

Sensory mediation: recognition of parents by infants

The questions of when, and through what sensory modalities infants begin to recognize their parents pose considerable methodological difficulties. Recognition of parents is typically inferred when neonates respond to the mother or father in a manner differing from their interactions with other people. Failure to manifest preferential responsiveness to parents does not necessarily imply lack of discrimination. Testing situations must be devised that enable infants to display overt evidence of distinguishing between social stimuli despite their limited behavioural repertoire. Also, as will be seen below, investigations into the development of parent recognition have helped to elucidate the early perceptual capabilities of infants.

Vision

The human face is one of the most common and salient visual stimuli to which infants are routinely exposed (Bushnell, 1982; Carpenter, 1974). The age at which infants first begin to recognize particular individuals by facial appearance alone, however, is still the subject of debate.

Several investigators have attempted to determine whether infants can recognize photographs of their mother's face. In an early study by Fitzgerald (1968), one-, two-, and four-month-old infants were presented with videotaped recordings of black and white photographs of their mother and a stranger (adult female), along with three black and white geometric patterns. For both the one- and four-month-olds, the photographs of the two women (mother and stranger) elicited greater pupillary dilation than did the nonsocial stimuli. Only the four-month old infants displayed differential pupillary responsiveness to the faces of the mother versus the stranger.

When infants are tested with colour slides of faces, discrimination of the mother is evident at an earlier age than found in the above study with (less realistic) black and white photographs. Three-month old infants initially looked longer at an unfamiliar slide of their mother than that of a stranger (Barrera & Maurer, 1981). With repeated exposure to the mother's facial photograph, infants' preferences switched to the novel woman; which was taken as evidence that they had habituated to the slide of their mother. Similarly, four- to seven-week-old infants that were habituated to a slide of their mother's face displayed differential fixation to that same stimulus compared to a novel slide of an unfamiliar female (Bushnell, 1982). While such habituation studies reveal that infants are capable of discriminating between *particular* photographs of their mother and unfamiliar females given sufficient exposure to the former stimuli, they do not allow one to conclude that the mother's face *per se* is recognized without such systematic training procedures.

If colour slides represent a marked improvement over black and white photographs as stimuli for assessing infants' visual recognition of their mother, the parent's live face should be even better. Two- to seven-week-old full-term infants oriented towards their own mother's face (accompanied by a voice recording) for a longer period of time than to the face and voice of a stranger (Carpenter, 1974). Likewise, pre-term infants whose chronological ages were 8 and 12 weeks looked preferentially at the face of their talking mother (Masi & Scott, 1982). Stimulus faces in the above experiments were presented through an open door positioned approximately 10 to 12 inches (20 to 30 cm) from the infants. Using a similar testing procedure, Field *et al.* (1984) found that neonates less than four days old discriminated between the *silent* faces of their mother and a stranger – thereby ruling out any possible influences of voice cues. These latter authors did point out, however, that the infants might have relied upon olfactory cues to recognize their mother; a possibility that is not unlikely in the light of the literature on odour recognition discussed below. This same caveat applies as well to all of the previous experiments in which stimulus faces were presented in close proximity to the subject infants.

Maurer & Salapatek (1976) employed a testing paradigm that apparently allowed infants access only to *visual* cues from the mother's live face. Babies viewed reflected mirror images of three stimulus adults (mother, female stranger, male stranger) who were instructed 'to remain still and expressionless'. The optical image of each adult was presented in front of the infant while the adult sat behind the baby. One-month-old babies tested in this situation responded discriminatively to their own mother. Although they spent the majority of the test session oriented away from the stimulus faces, they looked at their mother's face less than at the stranger's.

Audition

Sensitivity to sounds has been documented in human fetuses as early as the beginning of the third trimester (Sontag & Wallace, 1936; Birnholz &

Benacerraf, 1983), and audition appears to be highly developed at birth (DeCasper & Fifer, 1980; Fifer, 1987). The first systematic investigation of infants' recognition of the maternal voice was reported by Mills & Melhuish (1974) who developed a procedure enabling 20- to 30-day-old infants to control the presentation of voices by sucking on a non-nutritive nipple. Both the duration and frequency of sucking responses increased when the mother's voice was contingent upon such behaviour, but *not* when sucking was correlated with a stranger's voice. In a methodologically similar study, one-month-old infants once again discriminated between their mother's voice and that of a stranger providing that the mother spoke in her 'usual fashion' (Mehler *et al.*, 1978). When mothers spoke in a monotone, their voices did not elicit discriminative sucking.

More recently, DeCasper & Fifer (1980; see also Fifer, 1987) reported that infants less than three days old preferred the voice of their mother over that of a stranger. Subject infants learned to produce either voice, by varying the manner in which a non-nutritive nipple was sucked, and worked preferentially to hear their mother. Related experiments found no evidence that infants of this same age would perform a similar operant task to hear the voice of their *father* rather than the voice of a stranger – even though infants unfamiliar with either of the stimulus voices could learn to discriminate between them (DeCasper & Prescott, 1984).

Olfaction

Olfactory stimuli may be especially salient for the mediation of overt behavioural responses during the neonatal period. Reports from several laboratories indicate that infants (ranging in age from three days to six weeks) respond discriminatively to breast odours of their nursing mother (Macfarlane, 1975; Russell, 1976; Schaal *et al.*, 1980). In his pioneering experiments, Macfarlane (1975) presented breast-feeding infants with olfactory stimuli (soiled breast pads) from their mother and an unfamiliar nursing female. Preferential orientation to the mother's breast pad was observed by six days of age. Furthermore, as early as three days after birth, infants display reduced activity (movement of the arms and nose) when the nose comes into contact with the odour of the mother compared to a similar stimulus from a stranger – suggesting that maternal odours may have a quieting effect on the neonate (Schaal *et al.*, 1980). Recognizable odours are not restricted solely to the mother's breast region; preferential responsiveness by breast-feeding infants is also elicited by odour cues from the maternal axillary region (Cernoch & Porter, 1985) and neck (Schaal, 1985). Comparable tests revealed no evidence that two-week-old infants could distinguish between the odours of their father and an unfamiliar adult male (Cernoch & Porter, 1985).

In contrast with breast-feeding infants of the same age, 12- to 18-day-old bottle-feeding infants did not respond discriminatively to axillary odours from their own mother versus an unfamiliar (bottle-feeding) woman (Cernoch & Porter, 1985). The difference between bottle- and breast-feeding infants in maternal odour recognition tests may be a reflection on differential mother–infant interactions in these two groups. Breast-feeding infants presumably have more close contact with the mother's bare flesh than do bottle-fed infants, and are thereby more likely to become familiarized with her characteristic odours.

Developmental mechanisms of parent–infant recognition

Thus far, the discussion has focused upon the phenotypic traits or characteristics ('signatures') by which parents and infants can be discriminated. The complementary question concerns the processes that enable individuals to recognize the signature of their parent or offspring. That is, how does recognition of close kin (or of their particular signatures) develop?

Familiarization

The most obvious and ubiquitous mechanism mediating recognition of one's kin is familiarization with their characteristic phenotypic features through direct exposure. Shortly after birth (with the exact timing varying according to hospital routine, mother's medication and health status, etc.), mothers typically begin to interact with their infant in a manner that facilitates the development of mutual familiarization. During the initial encounters with their newborn infant, mothers commonly adopt an *en face* orientation allowing for close visual inspection of each member of the dyad (Trevathan, 1983). At such times, the mother may also talk or sing to her infant (Klaus *et al.*, 1975). As mentioned above, breast-feeding infants have recurring prolonged periods when their nose is in close (or direct) contact with the mother's bare flesh, and are thereby exposed to her body odours. Similarly, even if mothers do not actually sniff their neonates, they are nonetheless exposed to the infant's odours – especially when kissing or nuzzling it.

It has been suggested that familiarization with recognizable maternal attributes may actually begin prior to birth. Research with rats demonstrates that postnatal olfactory preferences are influenced by prenatal exposure to chemical substances injected through the uterine wall (reviewed by Smotherman & Robinson, 1987). In addition, pups delivered by Caesarean section preferred their mother's amniotic fluid over that of an unrelated female during tests conducted shortly after birth (Hepper, 1987). Might the human fetus similarly perceive characteristic maternal chemical cues contained in the amniotic fluid (e.g. Schaal, 1988)? Several authors have further speculated that *in utero* exposure to the mother's speech may result in familiarization with the acoustic properties of her voice (DeCasper & Fifer, 1980; Fifer, 1987; DeCasper & Prescott, 1984). At present, there are no data directly in support of the hypothesized influence of prenatal experience on postnatal recognition of (human) mothers; however, this is an intriguing possibility worthy of further investigation.

To date, there have been few empirical studies of the familiarization process involved in parent–offspring recognition – at least in humans. Schaal (1985) compared olfactory recognition of neonates by mothers who had differing amounts of early contact with their infants:

> Enriched contact – Immediately after delivery, the neonate was placed on the mother's chest for 30–40 minutes and stimulated to suckle.
>
> Routine contact – Less than five minutes of immediate postpartum contact between mother and infant.

In tests conducted over the first 10 days postpartum, mothers from both groups reliably recognized the odour of their neonate's vest when presented with comparable garments from two other infants. Nevertheless, those women who had enriched contact with their neonate performed better than those in the routine contact condition on the odour recognition tests. While these data are suggestive of the importance of early exposure in the development of mother–infant recognition, the limited sample size does not allow one to draw any firm conclusions (number of mothers in the enriched contact condition = 8; see also Cherfas, 1985).

Related studies provide additional evidence that a prolonged period of familiarization is not necessary for mothers to develop the ability to recognize their infant's odour. Women who gave birth by Caesarean section, and therefore had only limited contact with their infant (mean = 2.4 hours) prior to testing, accurately discriminated between the odours of their own and another baby (Porter *et al.*, 1983). More recently, 9 of 10 mothers with no 'obvious olfactory deficits' correctly identified their newborn's odour (three-choice soiled garment test) even though they only had 10–60 minutes postnatal contact with the infant (Kaitz *et al.*, 1987; see also Russell *et al.*, 1983).

In principle, recognition of parents and offspring through the process of familiarization is no different than learning to identify unrelated individuals as a result of direct contact. It may be easier to study the development of recognition of non-kin, however, since the amount of direct exposure to such persons can be more readily controlled than that of relatives. On the other hand, it is unlikely that carefully controlled presentations of unrelated adults or infants (or of their signatures) are at all comparable to normal interactions between parents and infants. Be that as it may, Cornell (1974) found that 23-week old (but *not* 19-week old) infants rapidly become familiarized with the faces of adult strangers. Over six consecutive trials (10-seconds each), infants displayed a decrement in visual fixation to paired black and white photographs of the same adult – even when different views of the same face were presented across trials. When paired photographs of different adults of the same sex were presented over six trials, *no* decrement in visual fixation was observed. Five- to six-month-old infants also fixate longer to a novel face than to a photograph of a face to which they were previously exposed for just one minute, providing that the two stimulus faces differ in sex or age (Fagen, 1972). Of course, such demonstrations of short-term memory do not imply that these same experimental manipulations would result in long-term face recognition.

Research into adult familiarization with unrelated neonates has been restricted primarily to the recognition of cries. After listening to a single 30-second tape recording of a crying one-month old infant, many non-parent adults were able to identify that same infant from a sample including the cries of four neonates (Green & Gustafson, 1983). Based upon tests conducted with a large number of non-parents, it appears that infants differ considerably in the extent to which they can be recognized by their cries alone following brief prior exposure (Gustafson *et al.*, 1984). Cries of individual infants were not invariant, but did tend to fall within a characteristic range of acoustic parameters (e.g. peak frequencies, breaths per unit of time, spectral band energy).

Familial resemblance

Because of their underlying genotypic similarity, biological kin often bear a discernible resemblance to one another. Familial resemblance could, therefore, potentially serve as the basis for recognizing individuals as close relatives – even if there has been no prior contact with them. This process, which is commonly referred to as phenotype- or signature-matching in the animal behaviour literature (Beecher, 1982; Holmes & Sherman, 1982; Blaustein *et al.*, 1987) is actually a variation of the development of kin recognition through direct familiarization (Waldman, 1987; Porter, 1988). Rather than becoming acquainted directly with the relatives to be recognized, however, individuals learn the signatures of close kin (or their own signature) and use those familiar phenotypic traits as standards against which to compare the signatures of others. To the extent that the phenotypic signatures of strangers match or approximate those of familiar kin, the former individuals may be suspected of being relatives as well.

It is unlikely that the detection of familial resemblance plays a major role in the development of infants' early recognition of their parents. In most instances, the first person with whom the infant becomes familiar is the mother and such familiarization would allow for recognition before there is further opportunity to learn any characteristic familial traits (through exposure to other relatives or even to the infant's own phenotype). Furthermore, since the mother and father are not likely to be biological relatives, they would not manifest closely similar genetically mediated signatures (on the other hand, there is some evidence for positive assortative mating among humans, so that mates may be more similar than predicted by chance alone on a variety of phenotypic traits – Thiessen & Gregg, 1980; Russell *et al.*, 1985). Thus, learning to recognize the mother should have little effect on the infant's ability to recognize the father. This conclusion is consistent with the above mentioned experiments (Cernoch & Porter, 1985), wherein breast-feeding infants recognized the axillary odours of their own mother but showed no indications that they could discriminate between the odours of their father and an unfamiliar adult male.

Most parents have had ample opportunity to become quite familiar with a number of individuals who are close genetic kin of their neonate; these include their spouse, their own parents, and any previous offspring (i.e. siblings of the neonate). Whether newborns evince any familial resemblance is an important question for parents (Macfarlane, 1977; Boukydis, 1981) but, once again, one that has received little research attention (Wells, 1987).

It is common for parents and other family members to comment on the resemblance between the newborn infant and particular relatives – most often the father (Robson & Moss, 1970; Daly & Wilson, 1982). Daly & Wilson suggest that such claims may function primarily to assure the alleged father of his paternity and need not necessarily reflect especially striking similarities between the infant and father. Unlike the mother, who can be certain that an infant is her own, males cannot always be absolutely assured that they have sired a particular infant. Fathers might, therefore, be more sensitive to any physical signs indicating that their reputed offspring is indeed their own, and others may attempt (intentionally or otherwise) to allay any doubts in this regard by asserting an immediate father–infant resemblance.

A recently reported experiment assessed the validity of such claims of familial resemblance in the facial–visual appearance of neonates (Porter *et al.*, 1984). Adult non-relatives were tested over two successive trials for their ability to match colour photographs of mothers and their healthy full-term infants (less than 40 hours of age). On one of the trials, each subject was given a photograph of an infant and instructed to attempt to identify the mother of that infant from an array of photographs of four recently parturient women. During the second test trial, each subject was presented with the opposite task – i.e. asked to identify which of four infants was born to a particular mother. The number of subjects who correctly paired photographs of mother and child was significantly greater than expected by chance alone; indicating that infants may bear at least some resemblance to their mother as early as the second day after birth. According to subjects' comments, the nose and mouth were the facial features that most often served as the bases for matching presumed mothers and offspring.

The possibility of perceptible familial odours in humans is supported by both anecdotal and empirical evidence. Following tests for recognition of their infant's odour (discussed above), several mothers and fathers who had correctly identified their own infant commented that the selected shirt smelled like that of their spouse or older child (Porter *et al.*, 1983; Porter *et al.*, 1986). Even unrelated adult subjects can detect odour similarities between mothers and offspring as evidenced by odour-matching tests in which T-shirts soiled by mothers and their three- to eight-year-old children were paired at a greater-than-chance level of accuracy (Porter *et al.*, 1984).

. Body odours of close kin may be similar, but they are not identical. Parents that were tested with T-shirts worn by two of their own children correctly distinguished between those garments by odour cues alone (Porter & Moore, 1981). Furthermore, adult subjects learned to discriminate the hand odours of (unfamiliar) female siblings, including identical twins (Wallace, 1977). The accuracy level on such discrimination tasks was not as great, however, as that for analogous experiments using odours from *unrelated* females. Those familiar with the Old Testament may recall that Isaac relied on olfactory cues when attempting to identify which of his two sons he was speaking with. After smelling Esau's garment (being worn by Jacob), he exclaimed 'See, the smell of my son *is* as the smell of a field which the Lord hath blessed' (Genesis 27:27).

In theory, mother–child odour resemblance could develop as a function of exposure to similar environmental variables (such as comparable diet or ambient household odours) as well as genotypic similarity (or through an interaction of these two variables). To investigate further the possible influence of environmental factors on body odours, an additional experiment was conducted in which subjects attempted to match soiled T-shirts worn by husbands and wives (Porter *et al.*, 1985). The two members of each husband–wife pair shared a large portion of their meals and had otherwise been exposed to the same household odours. Since the odours of spouses were *not* reliably matched in this experiment, routine environmental influences alone do not appear to be sufficient for the development of detectable odour similarity between individuals. Based upon these experiments, it was concluded that the odour resemblance of mothers and their three- to eight-year-old children (reported by Porter *et al.*, 1985) is most likely a reflection of underlying genetic relatedness among these classes of biological kin (for a further discussion of the likelihood of genetically determined odour signatures see Kalmus, 1955; Nicolaides, 1974). It should be pointed out, however, that dietary constituents (such as garlic or strong spices) can affect one's body odour and close kin fed distinctively different diets are more readily discriminated by their olfactory signatures alone than are those fed similar diets (Wallace, 1977).

To the extent that the odour signatures of individuals are indeed genetically mediated, these signatures should remain quite constant over the lifetime and could, therefore, play a role in recognizing newborn kin. It is also possible that the neonate could acquire the mother's scent during their initial interactions and thereby be recognized by such familiar maternal cues. Several species of non-human mammals are known to 'label' their neonates in this manner (presumably with their saliva, urine, milk or other glandular secretions) and subsequently rely upon such chemical signatures to recognize those offspring (Gubernick, 1980; Porter *et al.*, 1981). While a potential role of maternal labels for recognizing human neonates cannot be discounted, chemical cues acquired from the mother do not appear necessary for olfactory identification of newborns. Mothers were tested with shirts that infants had worn under hospital gowns (which protected the stimulus shirts against odour contamination) (Kaitz, *et al.*, 1987). All infants had been thoroughly washed with water and a lightly scented soap before being dressed. Nonetheless, mothers consistently recognized their own infant's undershirt in a series of three-choice odour tests.

The visual appearance (as well as other phenotypic characteristics) of infants may deviate dramatically from normal due to any of a variety of causal agents (e.g. surgical procedures, premature delivery, metabolic disorders, chromosomal anomalies). Thus, the somewhat distorted facial features of pre-term infants are generally considered less attractive than those of full-term infants (Boukydis, 1981; Maier *et al.*, 1984). Certain inborn errors of metabolism, such as phenylketonuria and maple syrup urine disease, result in strong characteristic body- or urine-odours (Mace *et al.*, 1976). As a final example, one of the defining characteristics of *cri du chat* syndrome (an autosomal aberration) is the production of high-frequency cat-like wails (Hay, 1985).

An unhealthy or atypical infant can be the source of considerable parental anxiety, stress or guilt (e.g. Boukydis, 1981; Murphy *et al.*, 1982; Nolan & Pless, 1986). There may also be long periods of separation from the parents while the infant remains in the hospital. When the baby is finally allowed to go home, it may be perceived as especially fragile and vulnerable and might be relatively unresponsive to parental stimulation. Infants who manifest anomalous phenotypic traits, therefore, may be at risk for disturbed early parent–infant interactions (e.g. Daly & Wilson, 1980; Drotar *et al.*, 1975). Even unusual neonatal body odours could have an adverse influence on parental attitudes and responsiveness. Accordingly, there are clinical accounts of mothers who had considerable difficulty in accepting their newborn baby because the child 'did not smell right' (Curtis-Jenkins, 1975). An additional factor that could conceivably contribute to disrupted parental interactions with such infants is their deviation from parents' expectations – including lack of anticipated familial resemblance. When interviewed several months after delivery, mothers often reported that perceived similarity between their offspring and other family members (especially the father) 'helped them feel closer' to the infant (Robson & Moss, 1970). Data from such retrospective self reports are of limited value but they do beg the question whether infants who fail to meet expectations of familial resemblance elicit less positive parental responses. In keeping with this line of reasoning, adoption agencies attempt to match children with adoptive families on phenotypic traits (van den Berghe, 1979), and the probability of successful adoptions appears to be correlated with the perceived resemblance between child and foster parents (Daly & Wilson, 1982; Jaffee & Fanshel, 1970). It nevertheless must be emphasized that noticeable birth defects, or lack of infants' resemblance to family members, typically occur in the context of other variables that could have profound effects on parent–offspring interactions. Therefore, at the present time, there is no clear evidence of a meaningful influence of perceived familial resemblance *per se* on parental behaviour.

Conclusions

The development of mutual recognition is a requisite step in the establishment of the unique social bond between mothers and infants. It is evident from the data reviewed above that reciprocal mother–infant recognition can be mediated by visual, auditory or olfactory cues. During the initial encounters with their neonate, mothers, as well as fathers, become familiarized with phenotypic traits of the infant, allowing for the rapid development of individual identification. Newborn babies may also bear some resemblance to family members, which could facilitate recognition by their parents.

The few directly pertinent studies conducted thus far suggest that familiarization through direct contact may also play a primary role in the development of neonates' recognition of parents; however, firm conclusions regarding ontogenetic mechanisms must await further research. The earliest ages at which infants have been reported to discriminate reliably between their mother and other women is two days for vocal stimuli (DeCasper & Fifer, 1980) and three days for olfactory cues (Schaal *et al.*, 1980). While the results of studies of *visual* recognition of the mother are rather ambiguous, it appears that by approximately one month of age infants begin to respond discriminatively to the visual features of their own mother's face.

Acknowledgements

Preparation of this chapter was supported in part by Grant No. HD-15051 from the National Institutes of Child Health and Human Development. Helpful discussions with Dr. Benoist Schaal were made possible through a NATO collaborative research grant (5–2–05/RG No. 366/87).

References

Ainsworth, M. D. S. & Bell, S. M. (1970). Attachment, exploration, and separation: illustrated by the behavior of 1-year-olds in a strange situation. *Child Develop.*, **41**, 49–67.

Balogh, R. D. & Porter, R. H. (1986). Olfactory preferences resulting from mere exposure in human neonates. *Infant Behav. Develop.*, **9**, 395–401.

Barrera, M. E. & Maurer, D. (1981). Recognition of mother's photographed face by the 3-month-old infant. *Child Develop.*, **52**, 714–16.

Beecher, M. D. (1982). Signature systems and kin recognition. *Am. Zool.*, **22**, 477–90.

Berghe, P. L. van den (1979). *Human Family Systems*. New York: Elsevier.

Birnholz, J. C. & Benacerraf, B. R. (1983). The development of human fetal hearing. *Science*, **222**, 516–18.

Blaustein, A. R., Bekoff, M. & Daniels, T. J. (1987). Kin recognition in vertebrates (excluding primates): mechanisms, functions, and future research. In *Kin Recognition in Animals*, ed. D. J. C. Fletcher & C. D. Michener, pp. 333–57. Chichester: John Wiley & Sons.

Boukydis, C. F. Z. (1981). Adult perception of infant appearance. A review. *Child Psych. Hum. Develop.*, **11**, 241–54.

Bowlby, J. (1969). *Attachment and Loss*, vol. I *Attachment*. New York: Basic Books.

Bruce, V. & Young, A. (1986). Understanding face recognition. *Brit. J. Psychol.*, **77**, 305–27.

Bushnell, I. W. R. (1982). Discrimination of faces by young infants. *J. Exp. Child Psychol.*, **33**, 298–308.

Cairns, R. B. (1979). *Social Development: The Origins and Plasticity of Interchanges.* San Francisco: Freeman.

Carpenter, G. (1974). Mother's face and the newborn. *New Scient.*, **61**, 742–4.

Cernoch, J. M. & Porter, R. H. (1985). Recognition of maternal axillary odors by infants. *Child Develop.*, **56**, 1593–8.

Cherfas, J. (1985). How important is the family smell? *New Scient.*, **108**, 27.

Cornell, E. H. (1974). Infants' discrimination of photographs of faces following redundant presentations. *J. Exp. Child Psychol.*, **18**, 98–106.

Curtis-Jenkins, G. H. (1975). Discussion. In *Parent–infant interaction*, p. 34. Ciba Foundation Symposium 33. New York: Elsevier.

Daly, M. & Wilson, M. (1980). Discriminative parental solicitude: a biological perspective. *J. Marriage Fam.*, **42**, 277–88.

Daly, M. & Wilson, M. I. (1982). Whom are newborn babies said to resemble? *Ethol. Sociobiol.*, **3**, 69–78.

DeCasper, A. J. & Fifer, W. P. (1980). Of human bonding: newborns prefer their mothers' voices. *Science*, **208**, 1174–6.

DeCasper, A. J. & Prescott, P. A. (1984). Human newborns' perception of male voices: preference, discrimination, and reinforcing value. *Develop. Psychobiol.*, **17**, 481–91.

Doty, R. L. (1981). Olfactory communication in humans. *Chem. Senses*, **6**, 351–76.

Doty, R. L., Shaman, P., Applebaum, S. L., Giberson, R., Siksorski, L. & Rosenberg, L. (1984). Smell identification ability: changes with age. *Science*, **226**, 1441–3.

Drotar, D., Baskiewicz, A., Irvin, N., Kennell, J. & Klaus, M. (1975). The adaptation of parents to the birth of an infant with a congenital malformation: a hypothetical model. *Pediatrics*, **56**, 710–17.

Fagen, J. F. (1972). Infants' recognition memory for faces. *J. Exp. Child Psychol.*, **14**, 453–76.

Field, T. M., Cohen, D., Garcia, R. & Greenberg, R. (1984). Mother–stranger face discrimination by the newborn. *Inf. Behav. Develop.*, **7**, 19–25.

Fifer, W. P. (1987). Neonatal preference for mother's voice. In *Perinatal Development: A Psychobiological Perspective*, ed. N. A. Krasnegor, E. M. Blass, M. A. Hofer & W. P. Smotherman, pp. 111–24. Orlando: Academic Press.

Filsinger, E. E. & Fabes, R. A. (1985). Odor communication, pheromones, and human families. *J. Marriage Fam.*, **47**, 349–60.

Fitzgerald, H. E. (1968). Autonomic pupillary reflex activity during early infancy and its relation to social and nonsocial visual stimuli. *J. Exp. Child Psychol.*, **6**, 470–82.

Formby, D. (1967). Maternal recognition of infant's cry. *Develop. Med. Child Neurol.*, **9**, 293–8.

Green, J. A. & Gustafson, G. E. (1983). Individual recognition of human infants on the basis of cries alone. *Develop. Psychobiol.*, **16**, 485–93.

Gubernick, D. J. (1980). Maternal 'imprinting' or maternal 'labelling' in goats? *Anim. Behav.*, **28**, 124–9.

Gustafson, G. E., Green, J. A. & Tomic, T. (1984). Acoustic correlates of individuality in the cries of human infants. *Develop. Psychobiol.*, **17**, 311–24.

Hamilton, W. D. (1964). The genetical evolution of social behaviour, I, II. *J. Theoret. Biol.*, **7**, 1–52.

Hay, D. A. (1985). *Essentials of Behaviour Genetics*. Oxford: Blackwell.

Hepper, P. G. (1987). The amniotic fluid: An important priming role in kin recognition. *Anim. Behav.*, **35**, 1343–1346.

Hold, B. & Schleidt, M. (1977). The importance of human odour in nonverbal communication. *Z. Tierpsychol.*, **43**, 225–38.

Holmes, W. G. & Sherman, P. W. (1982). The ontogeny of kin recognition in two species of ground squirrels. *Am. Zool.*, **22**, 491–517.

Illingworth, R. S. (1955). Crying in infants and children. *Br. Med. J.*, **1**, 75–8.

Jaffee, B. & Fanshel, D. (1970). *How They Fared in Adoption: A Follow-Up Study*. New York: Columbia University Press.

Kaitz, M., Good, A., Rokem, A. M. & Eidelman, A. I. (1987). Mothers' recognition of their newborns by olfactory cues. *Develop. Psychobiol.*, **20**, 587–91.

Kalmus, H. (1955). The discrimination by the nose of the dog of individual human odours and in particular of the odours of twins. *Br. J. Anim. Behav.*, **5**, 25–31.

Klaus, M. H., Traus, M. A. & Kennell, J. H. (1975). Does human maternal behaviour after delivery show a characteristic pattern? In *Parent–Infant Interaction*, pp. 69–78. CIBA *Found. Symp.*, **33**. New York: Elsevier.

Lewis, M. & Rosenblum, L. A. (1974). *The Effect of the Infant on its Caregiver*. New York: John Wiley & Sons.

Mace, J. W., Goodman, S. I., Centerwall, W. R. & Chinnock, R. F. (1976). The child with an unusual odor. *Clin. Pediat.*, **15**, 57–62.

Macfarlane, A. (1975). Olfaction in the development of social preferences in the human neonate. In *The Human Neonate in Parent–Infant Interaction. CIBA Found. Symp.*, **33**, 103–13.

Macfarlane, A. (1977). *The Psychology of Childbirth.* Cambridge, MA: Harvard University Press.

Maier, R. A., Holmes, D. L., Slaymaker, F. L. & Reich, J. N. (1984). The perceived attractiveness of preterm infants. *Inf. Behav. Develop.*, **7**, 403–14.

Masi, W. S. & Scott, K. G. (1982). Preterm and full-term infants' visual responses to mothers' and strangers' faces. In *Infants Born at Risk: Physiological and Perceptual Processes*, ed. T. Field & A. Sostek, pp. 173–9. New York: Grune & Stratton.

Maurer, D. & Salapatek, P. (1976). Developmental changes in the scanning of faces by young infants. *Child Develop.*, **47**, 523–7.

Mehler, J., Bertoncini, J., Barriere, M. & Jassik-Gerschenfeld, D. (1978). Infant recognition of mother's voice. *Perception*, **7**, 491–7.

Mills, M. & Melhuish, E. (1974). Recognition of mother's voice in early infancy. *Nature*, **252**, 123–4.

Morsbach, G. & Bunting, C. (1979). Maternal recognition of their neonates' cries. *Develop. Med. Child Neurol.*, **21**, 178–85.

Murphy, T. F., Nichter, C. A. & Liden, C. B. (1982). Developmental outcome of the high-risk infant: a review of methodological issues. *Sem. Perinatol.*, **6**, 353–64.

Murry, T., Holien, H. & Muller, E. (1975). Perceptual responses to infant crying: maternal recognition and sex judgments. *J. Child Lang.*, **2**, 199–204.

Nicolaides, N. (1974). Skin lipids: their biochemical uniqueness. *Science*, **186**, 19–26.

Nolan, T. & Pless, I. B. (1986). Emotional correlates and consequences of birth defects. *J. Pediat.*, **109**, (Supplement), 201–16.

Porter, R. H. (1988). The ontogeny of sibling recognition in rodents: Superfamily Muroidea. *Behav. Genet.*, **18**, 483–94.

Porter, R. H., Balogh, R. D., Cernoch, J. M. & Franchi, C. (1986). Recognition of kin through characteristic body odors. *Chem. Senses*, **11**, 389–95.

Porter, R. H., Balogh, R. D. & Makin, J. W. (1988). Olfactory influences on mother–infant interactions. *Adv. Inf. Res.*, **5**, 39–68.

Porter, R. H., Cernoch, J. M. & Balogh, R. D. (1984). Recognition of neonates by facial-visual characteristics. *Pediatrics*, **74**, 501–4.

(1985). Odor signatures and kin recognition. *Physiol. Behav.*, **34**, 445–8.

Porter, R. H., Cernoch, J. M. & McLaughlin, F. J. (1983). Maternal recognition of neonates through olfactory cues. *Physiol. Behav.*, **30**, 151–4.

Porter, R. H. & Moore, J. D. (1981). Human kin recognition by olfactory cues. *Physiol. Behav.*, **27**, 493–5.

Porter, R. H., Tepper, V. J. & White, D. M. (1981). Experiential influences on the development of huddling preferences and 'sibling' recognition in spiny mice (*Acomys cahirinus*). *Develop. Psychobiol.*, **14**, 375–82.

Rheingold, H. L. (1969). The effects of a strange environment on the behavior of infants. In *Determinants of Infant Behaviour*, IV, ed. B. M. Foss, pp. 137–66. London: Methuen.

Robson, K. S. & Moss, H. A. (1970). Patterns and determinants of maternal attachment. *J. Pediat.*, **77**, 976–85.

Russell, M. J. (1976). Human olfactory communication. *Nature*, **260**, 520–2.

432   R. H. Porter

Russell, M. J., Mendelson, T. & Peeke, H. V. S. (1983). Mothers' identification of their infant's odors. *Ethol. Sociobiol.*, **4**, 29–31.
Russell, R. J. H., Wells, P. A. & Rushton, J. P. (1985). Evidence for genetic similarity detection in human marriage. *Ethol. Sociobiol.*, **6**, 183–7.
Schaal, B. (1988). Discontinuite natale et continuite chimio-sensorielle: Modeles animaux et hypotheses pour l'homme. *Année Biol.*, **xxvii**, 1–41.
Schaal, B. (1985). Contributions olfactives a l'establissement du lien mere-enfant. In *Ethologie et Development de l'Enfant*, ed. R. E. Tremblay, M. A. Provost & F. F. Strayer, pp. 187–211. Paris: Stock.
Schaal, B., Montagner, H., Hertling, E., Bolzoni, D., Moyse, A. & Quichon, R. (1980). Les stimulations olfactives dans les relations entre l'enfant et la mere. *Repro., Nutr. Develop.*, **20**, 843–58.
Schaffer, H. R. (1971). *The Growth of Sociability*. Harmondsworth: Penguin Books.
Sluckin, W. (1972). *Imprinting and Early Learning*, 2nd Ed. London: Methuen.
Smotherman, W. P., & Robinson, S. R. (1987). Psychobiology of fetal experience in the rat. In *Perinatal Development: A Psychobiological Perspective*, ed. N. A. Krasnegor, E. M. Blass, M. A. Hofer & W. P. Smotherman, pp. 39–60. Orlando: Academic Press.
Sontag, L. W. & Wallace, R. F. (1936). Changes in the rate of fetal heart in response to vibratory stimuli. *Am. J. Disabled Child.*, **51**, 583–9.
Thiessen, D. & Gregg, B. (1980). Human assortative mating and genetic equilibrium: an evolutionary perspective. *Ethol. Sociobiol.*, **1**, 111–40.
Trevathan, W. R. (1983). Maternal 'en face' orientation during the first hour after birth. *Am. J. Orthopsychiat.*, **53**, 92–9.
Valanne, E. H., Vuorenkoski, V., Partanen, T. J., Lind, J. & Wasz-Hockert, O. (1967). The ability of human mothers to identify the hunger cry signals of their own new born infants during the lying-in period. *Experientia*, **23**, 768–9.
Vuorenkoski, V. (1975). Maternal recognition of infant sounds. *Symp. Zool. Soc. Lond.*, **37**, 331–8.
Waldman, B. (1987). Mechanisms of kin recognition. *J. Theoret. Biol.*, **128**, 159–85.
Wallace, P. (1977). Individual discrimination of humans by odor. *Physiol. Behav.*, **19**, 577–9.
Wells, P. A. (1987). Kin recognition in humans. In *Kin Recognition in Animals*, ed. D. J. C. Fletcher & C. D. Michener, pp. 395–415. Chichester: John Wiley & Sons.

Author index

Species and common name index

aardvark 36
Acomys cahirinus 227, 228, 229, 243, 327
acorn woodpecker 38, 40
Adelie penguin 240
Aeshna umbrosa 204
Agelaius phoeniceus 239
Alytes obstetricans 204
Ambystoma
    maculatum 179
    opacum 179
    tigrinum 205
American beaver 35
American cockroach 377
Anas
    platyrhynchos 314, 319
    strepera 344
ant
    acacia 366
    argentine 76, 366
    Australian 366
    fire 366
    Pharoah's 366
Anteater chat 36–7
Aphelocoma c. coerulescens 52
Apis mellifera 360, 364, 371, 377, 379, 385,
    386, 387–403, 405, 406
Aptenodytes patagonica 240

baboon
    anubis 40
    Sacred 35
bald faced hornet 366
beaver 35
bee
    honey 360, 364, 371, 377, 379, 385, 386,

        387–403, 405, 406
    sweat 364–6, 385
bell miner 33
black tailed cottontail 34
blind goby 233
Bombina variegata 202, 204
Bombyx mori 371
Botryllus schlosseri 132, 133, 267
Branta canadensis 239–40
Bufo
    americanus 172, 173, 178, 180, 181, 184–
        90, 192, 194–8, 200, 201, 203, 205, 224,
        229, 234
    boreas 173, 178, 184, 189, 190, 204
    bufo 175, 194, 196
    careus 170
    woodhousei fowleri 171, 179

Cacatua roseicapilla 240
Cacicus cela 35
Camponotus
    abdominalis 66, 80, 81
    floridanus 80, 81, 82, 83, 86
    pennyslvanicus 81
    senex 66
    vagus 80, 82
Canada goose 239–40
Canis
    familiaris 245
    lupus 30, 34
    mesomelas 53
Castor
    canadensis 35
    fiber 35
cat 223, 244

448

Subject index

452